SOIL MANAGEMENT FOR SUSTAINABLE AGRICULTURE

New Research and Strategies

SOIL MANAGEMENT FOR SUSTAINABLE AGRICULTURE

New Research and Strategies

Edited by
Nintu Mandal, PhD
Abir Dey, PhD
Rajiv Rakshit, PhD

A∧P APPLE
ACADEMIC
PRESS

First edition published 2022

Apple Academic Press Inc.
1265 Goldenrod Circle, NE,
Palm Bay, FL 32905 USA

4164 Lakeshore Road, Burlington,
ON, L7L 1A4 Canada

CRC Press
6000 Broken Sound Parkway NW,
Suite 300, Boca Raton, FL 33487-2742 USA

2 Park Square, Milton Park,
Abingdon, Oxon, OX14 4RN UK

Library and Archives Canada Cataloguing in Publication

Title: Soil management for sustainable agriculture : new research and strategies / edited by Nintu Mandal, PhD, Abir Dey, PhD, Rajiv Rakshit, PhD.
Names: Mandal, Nintu, editor. | Dey, Abir, editor. | Rakshit, Rajiv, editor.
Description: First edition. | Includes bibliographical references and index.
Identifiers: Canadiana (print) 20210376902 | Canadiana (ebook) 20210376988 | ISBN 9781774630235 (hardcover) | ISBN 9781774639139 (softcover) | ISBN 9781003184881 (ebook)
Subjects: LCSH: Soil management. | LCSH: Sustainable agriculture.
Classification: LCC S591 .S65 2022 | DDC 631.4—dc23

Library of Congress Cataloging-in-Publication Data

Names: Mandal, Nintu, 1987- editor. | Dey, Abir, 1987- editor. | Rakshit, Rajiv, 1986- editor.
Title: Soil management for sustainable agriculture : new research and strategies / edited by Nintu Mandal, Abir Dey, Rajiv Rakshit
Description: First edition. | Palm Bay, FL, USA : Apple Academic Press, 2022. | Includes bibliographical references and index. | Summary: "Soil Management for Sustainable Agriculture: New Research and Strategies explores the various soil management techniques and the latest improvements in soil management. Taking a sustainable approach, the volume begins with an overview of the elementary concepts of soil management and then delves into new research and novel soil management tools and techniques. Key features: Explains how clays are a critical component in sustainable agriculture with respect to carbon sequestration in conjunction with its interaction with soil enzymes Discusses potential utilization of microbes to mitigate crop stress Presents resource conservation technologies and prospective carbon management strategies. Covers the use of smart tools for monitoring soils Shares a number of nutrient management approaches Explores nanotechnological interventions for soil management Presents techniques for the remediation of soils contaminated by metals and pesticides The recommendations and future research directions presented in this valuable book will be helpful to students, researchers, farmers, and policymakers from the disciplines of agronomy, soil science, and natural resource management around the world"-- Provided by publisher.
Identifiers: LCCN 2021056316 (print) | LCCN 2021056317 (ebook) | ISBN 9781774630235 (hardback) | ISBN 9781774639139 (paperback) | ISBN 9781003184881 (ebook)
Subjects: LCSH: Sustainable agriculture. | Soil management. | Soil science. | Soils--Environmental aspects.
Classification: LCC S596.7 .S648 2022 (print) | LCC S596.7 (ebook) | DDC 338.1--dc23/eng/20211201
LC record available at https://lccn.loc.gov/2021056316
LC ebook record available at https://lccn.loc.gov/2021056317

ISBN: 978-1-77463-023-5 (hbk)
ISBN: 978-1-77463-913-9 (pbk)
ISBN: 978-1-00318-488-1 (ebk)

Dedication

Dedicated to our teachers and mentors at the Division of Soil Science and Agricultural Chemistry, Indian Agricultural Research Institute, Pusa, New Delhi

About the Editors

Nintu Mandal, PhD

Nintu Mandal, PhD, is Assistant Professor cum Junior Scientist in the Department of Soil Science and Agricultural Chemistry, Bihar Agricultural University, Sabour, Bhagalpur, Bihar, India. He has published 20 research/review papers in national and international journals and has authored one book, *Agricultural Nanotechnology: Basics and Practicals.* He also co-authored several book chapters. He has field one patent for "A Superabsorbent Polymer (NSP) and Process for Preparing the Same." His current research areas are nanoformulations for increasing input use efficiency, soil chemistry, and clay mineralogy.

He earned his MSc and PhD degrees from the Division of Soil Science and Agricultural Chemistry, ICAR-Indian Agricultural Research Institute, New Delhi, India. During his MSc, he received an ICAR Junior Research Fellowship, followed by ICAR Senior Research Fellowship, IARI Merit Scholarship and UGC Junior Research and DST INSPIRE Fellowship during his PhD tenure. He received an Indian Society of Soil Science (ISSS) Zonal Award (North Zone) (2011) for best presentation of an MSc dissertation. Other awards include the Indian Society of Soil Science (ISSS) Best Doctoral Thesis Award 2015 for Outstanding Doctoral Research and Dr. S. P. Raychaudhary Gold Medal for Outstanding Doctoral Research.

Abir Dey

Abir Dey, PhD, is an Agricultural Research Scientist in the Division of Soil Science and Agricultural Chemistry, ICAR-IARI, New Delhi, India. He published 19 research/review papers in national and international journals. He also co-authored one book and several book chapters. His current research areas are conservation agriculture, soil carbon dynamics, quality of soil organic matter, soil fertility, and plant nutrition.

He earned his MSc and PhD degrees from the Division of Soil Science and Agricultural Chemistry, ICAR-Indian Agricultural Research Institute (IARI), New Delhi, India.. During his MSc he received an ICAR Junior Research Fellowship, followed by an ICAR Senior Research Fellowship, IARI Merit Scholarship, and UGC Junior Research Fellowship during his PhD tenure. He received the Indian Society of Soil Science (ISSS) Zonal Award (North Zone) for best presentation of MSc dissertation. He received the Golden Jubilee Award for Outstanding Doctoral Research in Fertilizer Usage by the Fertilizer Association of India and the ISSS Commendation Certificate for Best Presentation of Doctoral Research Work done in Soil Science.

Rajiv Rakshit

Rajiv Rakshit, PhD, is an Assistant Professor cum Junior Scientist in the Department of Soil Science and Agricultural Chemistry, Bihar Agricultural University, Sabour, Bhagalpur, Bihar, India. He has published 25 research/review papers in national and international journals and has co-authored five book chapters. He has also reviewed many research papers for internationally reputed journals. His research interest is focused on soil quality, soil microbial interactions, soil carbon, and nutrient dynamics. His aim is to improve the understanding of how fertilization affects nutrient cycling and microbial interactions in soils.

Dr. Rakshit received his MSc and PhD degrees from the Division of Soil Science and Agricultural Chemistry, ICAR-Indian Agricultural Research Institute (IARI), New Delhi, India. During his MSc, he received an ICAR Junior Research Fellowship followed by IARI merit scholarship during his doctoral program.

Contents

Contents

Contributors

Ifeoluwa Adesina
Department of Natural Resources and Environmental Design, North Carolina A&T State University, Greensboro, NC 27411, USA

Samrat Adhikary
Department of Agricultural Chemistry and Soil Science, Bidhan Chandra Krishi Viswadidyalya, West Bengal, India

Koushik Banerjee
ICAR-Indian Agricultural Research Institute, New Delhi 110012

Mandira Barman
Division of Soil Science and Agricultural Chemistry, ICAR-Indian Agricultural Research Institute, New Delhi 110012, India

Kasturikasen Beura
Department of Soil Science and Agricultural Chemistry, Bihar Agricultural University, Bhagalpur, |Bihar 813210, India

Pradip Bhattacharyya
Agricultural and Ecological Research Unit, Indian Statistical Institute, Giridih 815301, Jharkhand

Ranjan Bhattacharyya
ICAR-Indian Agricultural Research institute, New Delhi 110012 India

Arnab Bhowmik
Department of Natural Resources and Environmental Design, North Carolina A&T State University, Greensboro, NC 27411, USA

D. R. Biswas
Division of Soil Science and Agricultural Chemistry, ICAR-Indian Agricultural Research Institute, New Delhi 110 012, India

R.R. Burman
Division of Agricultural Extension, IARI, New Delhi

Hillol Chakdar
ICAR-National Bureau of Agriculturally Important Microorganisms (NBAIM), Kushmaur, Mau, Uttar Pradesh 275103, India

Chandini
Department of Agronomy, Bihar Agricultural University, Sabour, Bhagalpur, Bihar, India

Nilanjan Chattopadhyay
Department of Soil Science and Agricultural Chemistry, Bihar Agricultural University, Sabour, Bhagalpur, Bihar 813210

S. K. Chaudhary
Indian Council of Agricultural Research , New Delhi 110012, India

Chirsmita
Department of Soil Science and Agricultural Chemistry, Bihar Agricultural University, Sabour 813210, Bhagalpur, Bihar, India

Suborna Roy Choudhury
Department of Agronomy, Bihar Agricultural University, Sabour, Bhagalpur 813210, Bihar, India

Anupam Das
Department of Soil Science and Agricultural Chemistry, Bihar Agricultural University, Sabour, Bhagalpur 813210, Bihar, India

Bappa Das
ICAR-Central Coastal Agricultural Research Institute, Goa 403402, India

Debarup Das
Division of Soil Science and Agricultural Chemistry, ICAR-Indian Agricultural Research Institute, New Delhi 110012, India

Malay Kr Das
ICAR-National Bureau of Agriculturally Important Microorganisms (NBAIM), Kushmaur, Mau 275 103, Uttar Pradesh, India

Ruma Das
Division of Soil Science and Agricultural Chemistry, ICAR-Indian Agricultural Research Institute, New Delhi 110012, India

Samar Chandra Datta
Division of Soil Science and Agricultural Chemistry, ICAR Indian Agricultural Research Institute, New Delhi, Delhi 110012, India

Abir Dey
Division of Soil Science and Agricultural Chemistry, Indian Agricultural Research Institute, New Delhi 110012, India

Chatterjee Dibyend
ICAR National Rice Research Institute, Cuttack, Odisha 753006, India

Akhila Nand Dubey
Department of Soil Science and Agricultural Chemistry, Institute of Agricultural Sciences, Banaras Hindu University, Varanasi, Uttar Pradesh 221005, India

Brahma S. Dwivedi
ICAR-Indian Agricultural Research Institute, New Delhi, 110012, India

Bhaskar Gaikwad
ICAR-National Institute of Abiotic Stress Management, Baramati, Pune, India

Pritam Ganguly
Department of Soil Science & Agricultural Chemistry, Bihar Agricultural University, Sabour, Bhagalpur, Bihar, India

Avijit Ghosh
ICAR-Indian Grassland and Fodder Research Institute, Jhansi 284003, Uttar Pradesh, India

Samrat Ghosh
Department of Agricultural Chemistry and Soil Science, Bidhan Chandra Kirishi Viswavidhalaya, Mohanpur, Nadia, West Bengal, India

Debasis Golui
Division of Soil Science and Agricultural Chemistry, ICAR-Indian Agricultural Research Institute, New Delhi 110012

Justin George K.
Indian Institute of Remote Sensing, ISRO, Dehradun, Uttarakhand, India

Anshuman Kohli
Department of Soil Science and Agricultural Chemistry, Bihar Agricultural University, Sabour, Bhagalpur, Bihar, India

Rahul M. Kulkarni
ICAR-Central Coastal Agricultural Research Institute, Goa 403402, India

Amarjeet Kumar
Department of Soil Science and Agricultural Chemistry, Bihar Agricultural University, Sabour, Bhagalpur, Bihar, India

Arun Kumar
Department of Seed Science and Technology, Bihar Agricultural University, Sabour, Bhagalpur, Bihar, India

Upendra Kumar
ICAR-National Rice Research Institute, Bidyadharpur, Cuttack, Odisha

Ragini Kumari
Department of Soil Science and Agricultural Chemistry, Bihar Agricultural University, Sabour, Bhagalpur, Bihar, India

Manoj Kundu
Department of Horticulture (Fruit and Fruit Technology, Bihar Agricultural University, Sabour, Bhagalpur, Bihar, India

V. Lenin
Division of Agricultural Extension, IARI, New Delhi, India

Gopal Ramdas Mahajan
ICAR-Central Coastal Agricultural Research Institute, Goa 403402, India

G. S. Mahra
Division of Agricultural Extension, IARI, New Delhi

Jajati Mandal
Department of Soil Science and Agricultural Chemistry, Bihar Agricultural University, Sabour, Bhagalpur 813210,

Nintu Mandal
Nanosynthesis and Nanoformulations Laboratory, Bihar Agricultural University, Sabour 813 210, Bihar

K. M. Manjaiah
Division of Soil Science and Agricultural Chemistry, Indian Agricultural Research Institute, New Delhi 110012, India

M. C. Manna
ICAR-Indian Institute of Soil Science, Bhopal 462038, Madhya Pradesh, India

Mahesh C. Meena
Division of Soil Science and Agricultural Chemistry, Indian Agricultural Research Institute, New Delhi 110012, India

M. D. Meena
ICAR-Directorate of Rapeseed-Mustard Research, Sewar, Bharatpur 321303, Rajasthan, India

Hidayatullah Mir
Department of Horticulture (Fruit and Fruit Technology), Bihar Agricultural University, Sabour, Bhagalpur, Bihar, India

Rahul Mishra
ICAR-Indian Institute of Soil Science, Bhopal, Madhya Pradesh, India

P.C. Moharana
ICAR-National Bureau of Soil Survey and Land Use Planning, Regional Centre, Udaipur 313001, Rajasthan, India

Prithusayak Mondal
Regional Research Station (Terai Zone), Uttar Banga Krishi Viswavidyalaya, Pundibari, Cooch Behar 736165, West Bengal, India

Dayesh Murgaokar
ICAR-Central Coastal Agricultural Research Institute, Goa 403402, India

Kumar Murugan
ICAR-National Bureau of Agriculturally Important Microorganisms (NBAIM), Kushmaur, Mau, Uttar Pradesh 275103, India

R. N. Padaria
Division of Agricultural Extension, IARI, New Delhi

Rajeev Padbhushan
Department of Soil Science and Agricultural Chemistry, Bihar Agricultural University, Sabour, Bhagalpur, Bihar 813210

Kiran Patel
ICAR-Central Coastal Agricultural Research Institute, Goa 403402, India

Amit Kumar Pradhan
Department of Soil Science and Agricultural Chemistry, Bihar Agricultural University, Bhagalpur, Bihar 813210, India

Rajiv Rakshit
Department of Soil Science and Agricultural Chemistry, Bihar Agricultural University, Sabour 813210, Bhagalpur, Bihar, India

Paul Ranjan
Division of Soil Resource Studies, ICAR-National Bureau of Soil Survey and Land Use Planning, Amravati Road, Nagpur 4440033, Maharashtra, India

I. Rashmi
ICAR-Indian Institute of Soil and Water Conservation, RC-Kota, Rajasthan

Prasenjit Ray
ICAR-National Bureau of Soil Survey and Land Use Planning, Regional Centre, Jorhat, India

Shyamashree Roy
Department of Agronomy, RRS (OAZ), Majhian, Uttar Banga Krishi Viswavidyalaya, Dakshin Dinajpur 733133, West Bengal, India

Trisha Roy
ICAR-Indian Institute of Soil and Water Conservation, Dehradun, Uttarakhand, India

Bhola Nath Saha
Department of Soil Science and Agricultural Chemistry, Dr. Kalam Agricultural College,
Affiliated to Bihar Agricultural University, Kishanganj, Bihar 855007, India

Sonalika Sahoo
Division of Soil Resource Studies, ICAR-National Bureau of Soil Survey and Land Use Planning,
Nagpur 440033, Maharashtra, India

Dhruba Jyoti Sarkar
ICAR-Central Inland Fisheries Research Institute, Kolkata, West Bengal 700120, India

Sujit Sarkar
Indian Agricultural Research Institute (IARI), Regional Station, Kalimpong, West Bengal

Abhijit Sarkar
ICAR-Indian Institute of Soil Sciences, Bhopal, Madhya Pradesh

Satdev
Department of Soil Science and Agricultural Chemistry, Bihar Agricultural University, Sabour,
Bhagalpur, Bihar 813210

Abolghasem Shahbazi
Department of Natural Resources and Environmental Design, North Carolina A&T State University,
Greensboro, NC 27411, USA

Harmandeep Sharma
Department of Natural Resources and Environmental Design, North Carolina A&T State University,
Greensboro, NC 27411, USA

Anupama Singh
ICAR-Indian Agricultural Research Institute, New Delhi, Delhi 110012, India

Mahendra Singh
Department of Soil Science and Agricultural Chemistry, Bihar Agricultural University, Bhagalpur,
Bihar 813210, India

Dikchha Singh
ICAR-National Bureau of Agriculturally Important Microorganisms (NBAIM), Kushmaur, Mau,
Uttar Pradesh 275103, India

Gyan Prakash Srivastav
ICAR-National Bureau of Agriculturally Important Microorganisms (NBAIM), Kushmaur, Mau,
Uttar Pradesh 275103, India

Vivek Trivedi
Division of Soil Science and Agricultural Chemistry, ICAR-Indian Agricultural Research Institute,
New Delhi 110012

Pravin K. Upadhyay
Division of Soil Science and Agricultural Chemistry, Indian Agricultural Research Institute,
New Delhi 110012, India

Shaloo Verma
ICAR-National Bureau of Agriculturally Important Microorganisms (NBAIM), Kushmaur, Mau,
Uttar Pradesh 275103, India

Abbreviations

a.i.	active ingredient
ACC	1-aminocyclopropane-1-carboxylate
ACCD	1-aminocyclopropane-1-carboxylate deaminase
AF	aqueous flowables
AFASs	amorphous ferri-aluminosilicates
AM	arbuscular mycorrhizae
AMF	arbuscular mycorrhizal fungi
AMSR	Advanced Microwave Scanning Radiometer
AMX	amoxicillin
APS	ammonium persulfate
BAIF	Bharatiya Agro-Industries Foundation
BD	bulk density
B-nZVI	bentonite-supported nanoscale zero-valent iron
BPP	beta propeller phytases
BTT	Block Technology Team
CA	conservation agriculture
CCI	climate change initiative
CEC	cation exchange capacity
CM	chlorophyll meters
CMC	carboxymethyl cellulose
COC	clay–organic complex
COMF	colloidal organomineral fraction
CRF	controlled-release-fertilizers
CRFF	controlled release formulation of fertilizer
CRS	community radio stations
CRs	crop residues
CSIA	compound-specific isotopic analysis
CU	common urea treatment
CZNZPBC	chitosan grafted zinc containing nanoclay polymer biocomposite
DAP	diammonium phosphate
DC	dispersion concentrate
DEM	Digital Elevation Model
DLS	dynamic light scattering

DOM	dissolved organic matter
DSS	decision support system
DSSAT	Decision Support System for Agro-technology Transfer
DTPA	dithelyne-triamine-penta-acetic acid
DWT	discrete wavelet transformation
EC	emulsifiable concentrate
EC	electrical conductivity
EDaX	energy-dispersive X-ray spectroscopy
EDTA	ethylene-diamine-tetra-acetic acid
EOS	Earth observation satellite
ESA	European Space Agency
ET	evapotranspiration
FA	fulvic acid
FAO	Food and Agriculture Organization
FCO	Fertilizer Control Order
FIAM	free ion activity model
FPO	Farmers' Producer Organization
FTIR	Fourier transform infrared spectroscopy
FVC	fractional vegetation cover
GCOM-W	Global Change Observation Mission—Water
GHG	greenhouse gas
GI	germination index
GS	GreenSeeker
HA	humic acid
HAP	histidine acid phosphatases
HDTMA	hexadecyltrimethylammonium
HQ	Hazard quotient
HR-TEM	high-resolution transmission electron microscopy
IASF	Intelligent advisory system for farmers
ICT	information and communication technology
ID	induced defoliation
IFS	Integrated Farming System
IGP	Indo-Gangetic plains
Imz	imazamox
INM	integrated nutrient management
INSEY	in-season estimate of grain yield
iPOM	intra-aggregate particulate organic matter
KSB	potassium solubilizing bacteria
KSM	K solubilizing microorganisms

LAI	leaf area index
LC	labile carbon
LCC	leaf color chart
LMWOAs	low-molecular-weight organic acids
LST	land surface temperature
MBC	microbial biomass carbon
ME	microemulsion
MIRAS	microwave imaging radiometer using aperture synthesis
MRT	mean residence time
MSW	municipal solid waste
MU	matrix-based urea treatment
MW	microwave
MWD	mean weight diameter
NCC	noncomplexed clay
NCPC	nanoclay polymer composite
NDVI	normalized difference vegetation index
NE	nutrient expert
NEK	nonexchangeable K
NF	nanofertilizers
NI	nitrification index
NIR	near-infrared
NLC	nonlabile carbon
NMSA	National Mission for Sustainable Agriculture
NNI	N nutrition Index
NOEC	no-observed effect concentration
NP	nanoparticle
NPC	National Productivity Council
NPK	nitrogen–phosphorus–potassium
NPMSF	National Project on Management of Soil Health and Fertility
NPOF	National Project on Organic Farming
NSSI	normalized sunlit and shaded index
NUE	nutrient use efficiency
nZVI	nanoscale zero-valent iron
OD	oil dispersion
ODD	optical density difference
OIPM	organic–inorganic pillared montmorillonite
OM	organic matter
OMWW	olive mill waste water

PAH	polycyclic aromatic hydrocarbon
PAM	polyacrylamide
PB	permanent bed planting
PB	permanent raised beds
PBS	polybutylene succinate
PCA	principal component analysis
PCN	polymer–clay nanocomposite
PDI	polydispersity index
PGP	plant growth promoting
PGPR	plant growth promoting rhizobacteria
PHB	poly 3-hydroxybutyrate
PI	pollution index
PLFAs	phospholipid fatty acids
PLSR	partial least squares regression
PMKSY	Pradhan Mantri Krishi Sinchayee Yojana
PNM	precision nutrient management
PSMs	phosphate solubilizing microorganisms
PTF	pedotransfer function
PTMA	phenyltrimethylammonium
PVA	poly(vinyl alcohol)
PWM	precision water management
RAD	Rainfed Area Development
RCTs	resource conservation technologies
RfD	reference dose
RMSE	root-mean-square error
RP	rock phosphate
RT	reduced tillage
RVM	relevance vector machine
RW	rice–wheat
SAP	superabsorbent polymer
SC	suspension concentrates
SE	suspoemulsion
SEE	soil evaporative efficiency
SEM	scanning electron microscopy
SHC	Soil Health Card
SHM	soil health management
SHP	soil hydraulic property
SMAP	soil moisture active passive
SMC	soil moisture content

SMOS	soil moisture and ocean salinity
SOC	soil organic carbon
SOM	soil organic matter
SPAD	Soil–Plant Analysis Development
SPTF	spectral pedotransfer function
SRF	slow-release-fertilizers
SRF	slow-release formulation
SRO	short-range order
SSA	specific surface area
SSDF	sub-surface drip fertigation
SSM	surface soil moisture
SSM	sustainable soil management
SSNM	site-specific nutrient management
SSP	single superphosphate
STF	spectral transfer function
STLs	soil testing laboratories
SVR	support vector regression
TEM	transmission electron microscope
TMA	tetramethylammonium
TRRVDI	temperature rising rate vegetation dryness index
TVDI	temperature vegetation dryness index
UAV	unmanned aerial vehicle
UNFCCC	United Nations Framework Convention on Climate Change
USDA	United States Department of Agriculture
USEPA	United States Environmental Protection Agency
VIT	vegetation index/temperature
VOC	volatile organic compound
VRNAT	Variable rate N application technology
VRT	Variable Rate Technology
WCM	water cloud model
WDP	watershed development program
WHC	water holding capacity
XRD	X-ray powder diffraction
ZNCPC	zincated nanoclay polymer composite
ZnO-NPs	zinc oxide nanoparticles
Z-nZVI	zeolite-supported nanoscale zero-valent iron
ZT	zero tillage

Foreword

The world's population was 7 billion in 2011 and is expected to reach more than 9 billion by 2050. Food security will remain one of the greatest global challenges in the current century, and it will continue to rely on Earth's diverse soil resources. Soil, the living epidermis of the planet, has the ability to support the sustainable intensification of agriculture and to play a central and critical role in delivering food security. But soil is often not valued as a critical resource and its quality or soil health is generally taken for granted. As a consequence, soils are being poorly managed, particularly in countries with high population densities, and soils there are experiencing great stress. The pressure on soils is continuously increasing but due to complex nature of the soils there exists significant challenges for developing robust science to support key decisions for their efficient management in the decades to come. Being one of the most complex of materials, soils exhibit spatial variation even within small areas and it becomes difficult to scale from pores, in which biogeochemical reactions take place, to profiles and field scales so that managing soils requires integration of skills from both fundamental and applied sciences. This book, *Soil Management for Sustainable Agriculture: New Research and Strategies,* is a significant step in this direction.

Soils, the chief enablers for feeding and nourishing everyone on earth, are forgiving but it is only after decades and in some cases centuries of mismanagement that these are now showing stress. As every block of soil is a timed memory of past and present biosphere-geosphere interactions at that location, appropriate management of soils becomes difficult on scales necessary to help increase not only agricultural productivity but also many other important functions such as storage, transmission and filtration of water; transformation of nutrients; emission of greenhouse gases including carbon dioxide and nitrogen oxides; processing of waste materials; naturally sequestering carbon to help build resilience against events caused by climate changes; and acting as the sustaining medium for all terrestrial ecosystems. Soils are finite and fragile and their health is reflected in their capacity to respond to agricultural interventions and processes that can degrade it. The main anthropogenic interventions in the functioning of soils since the advent of organized agriculture has demanded a fundamental change in soil and crop management to produce more food from land already in cultivation.

Every human intervention in the form of a soil-management practice invariably represents major and sometimes irrevocable change in the nature and
properties of the original soil. The key issue is to minimize the negative
effects of such changes. Otherwise, the history of agriculture is replete with
examples in which civilizations waned or disappeared because of failure
to minimize the impact of human interventions on the soil. To reduce the
significant and growing pressure on soils, a concerted effort is required by
researchers, policy makers, and farmers across a broad range of research
disciplines, jurisdictions, and commercial markets to ensure that soils are
managed in a way that their health is enhanced and these continue performing
their functions in a sustainable manner forever in the future. It seems policy
makers, businesses, farmers, and the public are not well informed as to what
healthy soil is or what the management factors that affect soil health are.
Also, in about 450 million small-scale farms all over the world supporting
a population of roughly 2.2 billion people, mostly in developing countries,
the most effective way to lever small farmers out of poverty is to help them
increase agricultural productivity. Such poverty-alleviation policies may also
create pressures on soil that are rarely balanced with better soil-management
practices. This book is an attempt to address complex soil-management
issues in the light of recent advancements in our understanding of different
aspects of functioning of soils. It is expected that this collection of chapters
will catalyze investment and action in support of healthy soils and contribute
to the implementation of UN Sustainable Development Goal (SDG) 2 and
SDG15.

Scientific understanding about sustainable agriculture gives equal
weightage to economic, environmental, and social aspects, and is influenced
by contemporary issues, perspectives, and values. For example, climate
change was not a critical issue 20 years ago but now it is receiving increasing
attention. Conservation of resources to enhance agricultural productivity
alludes to maintaining integrity of soil as highly structured entity composed
of mineral particles, organic matter, air, water, and living organisms.
Maintaining soil functioning often translates to maintaining or increasing
soil organic matter, which functions as a substrate for microbial activity,
crucial source and sink for nutrients, and as a buffer against fluctuations in
acidity, water content, and contaminants. Soil organic matter is important
not only in relation to soil fertility, sustainable agricultural systems, and crop
productivity but also with respect to global warming. Thus chapters on soil
organic matter build-up, resource conservation technologies, and biochar
management in this book provide the latest information available and should

prove very timely. During the past decade, there is an exponential increase in the interest in using biochar in agriculture. When applied to the soil, it has been reported to not only enhance soil carbon sequestration but also to reduce bulk density, enhance water-holding capacity and nutrient retention, stabilize soil organic matter, improve microbial activities, and sequester heavy metals. Although several bottlenecks remain to be addressed before widespread production and use of biochar becomes popular, it seems that biochar will be able to play a significant role in facing the challenges posed by climate change and threats to sustainability of agroecosystems.

Soil is a dynamic system in which close interplay among abiotic and biotic entities governs several chemical and biogeochemical processes and enzyme activities that are essential for organic matter decomposition and nutrient cycling. Also, soil minerals constantly interact with organic matter and microorganisms and even enzyme activities change due to interaction with clay and soil organo-minerals. Chapters included in this book offer up-to-date information on these aspects in terms of managing soils to achieve sustainable agricultural systems.

Sustainable soil management strives to meet the needs of the present without compromising the ability of future generations to meet their needs from that soil. Therefore, policy makers and those involved in the development of sustainable agriculture management approaches always keep an eye on the emerging possibilities from new developments in the relevant areas of science. As nutrient management in agroecosystems is a key challenge for global food production, there is an urgent need to increase nutrient availability to crops grown even by smallholder farmers in developing countries. Taking into account the local economic and social conditions, fertilizers will be essential for food security. But to achieve sustainable agricultural systems, nutrients supplied by fertilizers will have to be managed following strategies that are site-specific and able to ensure optimum flow of nutrients to the crop plants but with minimal loss of nutrients to the environment. These aspects of soil-management strategies have been adequately dealt in this book through chapters pertaining to nitrogen, phosphorus, potassium, and micronutrients. Use of smartphones for managing fertilizer nitrogen in a field specific way in smallholder farms has also been described in Chapter 11 of this book.

Two chapters in the book are devoted to nanotechnological interventions in increasing nutrient use efficiency in agroecosystems. Nanomaterials offer large specific surface area to fertilizers and enhance the productivity of crops by increasing the nutrient use efficiency by facilitating site-targeted controlled delivery of nutrients. Nanofertilizers also help in minimizing the losses of nutrients from the soil–plant system. Nanotechnology has the potential to revolutionize the agricultural systems by nanostructure formulation of fertilizers coupled with mechanisms of targeted delivery or controlled release and conditional release. Nanofertilizers can also release the active ingredients in response to environmental triggers and crop demand more precisely.

It will not be possible to develop sustainable agroecosystems in the long run until and unless the knowledge and the technical competence is successfully transferred to the farmers. As nature of agriculture is region specific, sustainability in agriculture requires diverse and adaptive knowledge based on experimental science as well as farmers' on-the ground local knowledge. Chapter 12 discusses innovative extension approaches for diffusion of nutrient management technologies that immensely enhances the value of this book. Both high agricultural productivity and long-term sustainability will be achieved by encouraging innovation and promoting farmer–researcher partnerships.

Bijay Singh
*Fellow of the Indian National Science Academy
and National Academy of Agricultural Sciences*

Preface

A soil is the base of food production, and it is quite established that sustainable soil management contributes significantly to food production. Even in the United Nations, there was an agenda for restoring and improving soil health under the sustainable management goals. This sustainable concept totally relies upon the soil services or functioning. The notion of a society toward preserving soils or managing soils requires a universal implementation.

Soil Management for Sustainable Agriculture: New Research and Strategies incorporates aspects starting from elementary concepts to research output to date, with a focus on future research. This book is oriented toward a number of sustainable approaches that will encourage the beneficiaries to achieve entirely new levels of thought on soils. This compilation identifies areas where agriculture and more specifically soils can be made more sustainable globally, concomitantly reducing pressure on the natural soil functions within the areas that are farmed.

The book comprises 22 chapters and reinforces the following principal concepts: (1) Clays: The Critical Component in Sustainable Agriculture with Respect to Carbon Sequestration *vis-a-vis* Their Interaction with Soil Enzymes. (2) Potential Utilization of Microbes for Stressed Agriculture. (3) Resource Conservation Technologies and Prospective Carbon Management Strategies. (4) Smart Tools for Monitoring Soils. (5) Nutrient Management Approaches. (6) Nanotechnological Interventions. (7) Remediation of Metals and Pesticides Contaminated Soils.

The recommendations and future research directions presented in this book will be helpful to students, researchers, and policymakers from disciplines of agronomy, soil science, and broadly natural resource management. It is expected that the book will stimulate these people at all levels for sustainable soil management.

We are highly grateful to all the contributors for accepting our invitation and sharing their views and integrating their expertise in composing these chapters while considering the suggestions from the editors time to time. We highly appreciate their dedication toward their contribution. We are also highly indebted to Professor Bijay Singh, INSA Honorary Scientist and Ex-ICAR National Professor, for writing the foreword for this compilation. We finally thank the team of Apple Academic Press for their generous cooperation at every stage of the publication.

—Editors

Introduction

In the past few decades, public interest in soil health and sustainability has been increased tremendously due to enhanced recognition of the fragility of natural resources and the necessity to preserve them for societal well-being. Continued deterioration in soil health and ever-increasing population pressure on finite land resources in most of the developing countries made it imperative to enhance crop productivity per unit area. The challenge is much bigger in India, as it supports over 16% of the global population through only 2% of the world's geographical area. The per capita land availability (land-to-man ratio) continuously decreased from 0.34 ha in 1951–952, to 0.14 ha in 2012–2013, which is likely to come down further by the year 2022.

In the mid-1960s, ushering of green revolution in India enhanced crop productivity remarkably, which helped to achieve self-sufficiency in food grain production. However, this intensive cultivation in a system mode for five decades led to several soil sustainability related problems. These problems are: (1) depletion of level and quality of groundwater due to excessive and unchecked use; (2) herbicide resistance to currently used chemicals; (3) increased subsoil compaction consequent to long-term cultivation of puddled rice; (4) atmospheric pollution due to excess diesel combustion for excessive tillage; (5) inconvenience caused by excessive amount of cereal straw generated and lack of proper residue disposal technologies, leading to residue burning causing further atmospheric pollution; (6) declining soil organic matter, both in terms of quality and quantity; (7) increasing multinutrient deficiencies due to excessive mining from soils; (8) decline in nutrient-supplying capacity of soil attributed to a deterioration of soil physical, chemical and biological health. These problems pose a serious threat to long-term sustainability of crop production system *vis-a-vis* the country's food security. It is time that we innovate and adopt novel soil and crop management strategies.

This book, *Soil Management for Sustainable Agriculture: New Research and Strategies,* covers various soil management techniques and the latest improvements in the subdisciplines of soil science, namely, chemistry, fertility, microbiology, physics, mineralogy, remote sensing, nanotechnology, etc. The book encompasses chapters dedicated to novel soil-management tools and techniques that will be of topical and practical interest to students, scientists, researchers, farmers, and policymakers associated with soil management, across the globe, and especially in Indian conditions.

PART I

Clays: The Critical Component in Sustainable Agriculture with Respect to Carbon Sequestration *vis-a-vis* Their Interaction with Soil Enzymes

CHAPTER 1

Advances in Clay Research for Sustainable Agriculture

DEBARUP DAS[1*], SONALIKA SAHOO[2], RUMA DAS[1], and SAMAR CHANDRA DATTA[1]

[1]*Division of Soil Science and Agricultural Chemistry, ICAR Indian Agricultural Research Institute, New Delhi, Delhi 110012, India*

[2]*Division of Soil Resource Studies, ICAR National Bureau of Soil Survey and Land Use Planning, Nagpur, Maharashtra 440033, India*

Corresponding author. E-mail: debarup.das@icar.gov.in; debarupds@gmail.com

ABSTRACT

Sustainability of agriculture is crucial to meet the food requirement of ever-growing population. Hence, management practices ought to have the ability to augment the use efficiency of agro-inputs, minimize environmental pollution, and maintain or improve soil quality. With the advancements made in clay research, natural and modified clay minerals along with zeolites are gaining considerable importance in agricultural sector. Owing to some unique physical and chemical characteristics (surface reactivity, surface area, etc.), low cost of production, easy availability, and many other reasons, clay minerals have great potential for use in multiple aspects of agriculture. For instance, they can be used for preparation of slow-release formulations of agrochemicals to increase the use efficiencies of the latter and lower negative impact on environment. Natural and modified clay minerals and zeolite have also showed their potential for remediation of soil and water, which are polluted with various inorganic contaminants, for example, heavy metals (e.g., lead, cadmium, chromium, etc.), metalloids (e.g., arsenic), nitrate (NO_3^-), bicarbonate (HCO_3^-), etc.; various radio nuclides; and organic contaminants such as dyes, polyaromatic hydrocarbons, etc. Clay minerals and their modified products can be employed to improve soil quality by increasing carbon sequestration potential, water and nutrient holding capacity, structure, and buffering capacity of soil. Furthermore, clay minerals can also be employed to increase shelf-life and facilitate handling of agricultural produce through improving quality of packaging materials. This chapter describes the various applications of clay minerals from the viewpoint of agricultural sustainability.

1.1 INTRODUCTION

Sustainability can be defined as to "meet the needs of the present without compromising the ability of future generations to meet their own needs" (Keeble, 1988). To make agriculture a sustainable affair, we have to minimize the use of nonrenewable inputs; pick up strategies to maintain or improve environmental quality; integrate biological and ecological processes for better crop growth; and rely more upon precise, resource-efficient (i.e., nutrients and energy), and cost-effective techniques that are in congruence with nature and accessible by farmers (Tomich et al., 2004; Manjaiah et al., 2019). Out of many ways leading toward agricultural sustainability, suitable application of clay minerals could be one. Owing to numerous advancements made in clay research during the past few decades, it has been possible to use different clay minerals (without or with modifications) and also zeolite (a tectosilicate) in a number of agricultural applications, namely, nutrient delivery, pesticide delivery, remediation of polluted soil and water, as amendments in a number of specific problems, and packaging material for agricultural produce. Natural clay minerals already present in soil have vital role in sequestering carbon (C), as some of them form strong bonds with organic matter, and reduce their mineralization and subsequent emission of CO_2 into the atmosphere (Das et al., 2019a). Soil's native clay minerals can also indicate deterioration in soil health under long-term mismanagement (Das et al., 2019b).

Controlled-release fertilizers developed by means of clay minerals can enhance nutrient use efficiency through better synchronization between nutrient release from fertilizer and nutrient uptake by plants (Shaviv, 2000). Similarly, clay-based formulations for pesticides can prevent losses and cause lesser contamination of environment (Manjaiah et al., 2019). Lesser rate of application, sustainable supply of nutrients and plant protection chemicals, and lesser environmental pollution are the advantages of such clay-based controlled-release fertilizers and pesticides (Ni et al., 2010; Bhardwaj et al., 2012). Not only prevention, but also for remediation of polluted soil and water, natural and modified clay minerals and zeolite have been found useful (Manjaiah et al., 2018, 2019; Sarkar et al., 2019; Mukhopadhyay et al., 2019a, 2019b, 2020). Apart from remediation of polluted soil and water, clay minerals can also be used as amendments to alleviate a number of soil-related problems too. For instance, some clay minerals have been found useful to enhance water and nutrient holding capacity, reduce nutrient leaching and runoff, improve soil structure, buffering capacity, etc. (Karbout et al., 2015; Ajayi and Horn, 2016; Padidar et al., 2016; Tahir and Marschner, 2017; Saurabh et al., 2019). Clay minerals along with polymers can also be used to make cost-effective packaging materials for transport and marketing of agricultural produce.

Some of the recent advancements made in clay research and their findings showing the potential and role of natural and modified clay minerals along with zeolite in attaining agricultural sustainability are discussed in the following sections.

1.2 ADVANCES IN CLAY RESEARCH FOR SUSTAINABLE MANAGEMENT OF AGROCHEMICALS

Agrochemicals (fertilizers and pesticides) have been extensively used in agriculture worldwide for decades and have contributed toward substantial increase in food production. Unfortunately, only a small part of applied agrochemicals reaches the anticipated target because of their volatilization, leaching, and quick degradation, leading to serious ecological problems (Ravier et al., 2005). Pesticide contamination of surface and groundwater is a major problem as they are potentially hazardous to human, animal, and biodiversity altogether. Likewise, fertilizer leaching and volatilization cause eutrophication of water bodies, greenhouse gas emission, and various health-related problems. The focus on slow-release formulation (SRF) or controlled-release formulation (CRF) of agrochemicals has largely increased in the past few years because of the urge to realize sustainability in on-going systems, that is, to enhance economic efficiency and better health and nutrition along with minimum socio-environmental problems caused by excessive use of agrochemicals. With SRF/CRF, the agrochemical is released gradually over a longer time period, thus limiting the amount immediately available for transport processes, which, in turn, minimize volatilization and leaching losses and reduce negative impact on environment (Fu et al., 2018).

Clay minerals perform important functions in multitude of agricultural applications due to their chemical surface functionality and surface structure. Natural and modified/functionalized clays have been gaining interest owing to their potential as well as successful applications in controlled/slow-release systems and/or as modifiers (Gerstl et al., 1998; Armstrong et al., 2000; Fernández-Pérez et al., 2004; Hocine et al., 2004; Hermosín et al., 2006; Sanchez-Martin et al., 2006). Clay-based CRFs mostly involve montmorillonites (Gerstl et al., 1998; Chevillard et al., 2012a), kaolinites (Singh et al., 2009), and bentonites (Fernández-Pérez et al., 2004; Céspedes et al., 2013; Sahoo et al., 2014, 2016).

1.2.1 SUSTAINABLE MANAGEMENT OF FERTILIZERS

1.2.1.1 CLAY MINERALS AS FILLER OR CARRIER OF NUTRIENTS

Initially, clay minerals have been used as filler materials in many fertilizers. Clays such as sodium bentonite, montmorillonite, kaolin, attapulgite, sepiolite, and tectosilicate such as zeolite are used as modifying agents to improve fertilizer handling quality. By virtue of their crystalline geometry, clay (bentonite, kaolin, attapulgite, or sepiolite) or zeolite addition to urea melt or urea synthesis liquor can impart good anticaking and nonfriability characteristics to the urea granules. Palygorskite and sepiolite, when blended with bentonite, act as excellent absorbents and used as carriers for agrochemicals in agricultural applications. Clays such as sepiolite, attapulgite, and bentonite can be used in suspension application of fertilizers due to their thickening ability. Suspension formulations are fluid mixtures of solid materials suspended in concentrated solutions of a particular fertilizer. These formulations are quite popular because they are easy to apply and do not create a dust problem. Suspension formulations are generally gel-like and remain stable over extended time periods. With mild agitation, these gels can be easily redispersed and made fluid enough for pumping and uniform application on target objects (Murray, 2000). Attapulgite and sepiolite of colloidal or gel grades usually are used in fertilizer suspensions. Owing to their large adsorptive and water holding capacity (WHC), bentonites have been frequently used as carrier/diluents/additives in chemical fertilizers to provide the optimum concentration of the needed elements or as a suspension aid or as a stabilizing agent in the liquid fertilizer (Murray, 2000).

1.2.1.2 FERTILIZER-INTERCALATED CLAYS FOR SLOW RELEASE OF NUTRIENTS

Intercalation of fertilizer into clay interlayers has shown excellent potential in increasing nutrient use efficiency. Kaolin was utilized as a nutrient carrier after undergoing milling process to lower its crystallinity and incorporation of nutrients into its structure (Solihin et al., 2011). Incorporation of KH_2PO_4 and $NH_4H_2PO_4$ was done into kaolin using a milling process, where the nutrient-intercalated kaolin showed slow release of nutrients in water (Solihin et al., 2011; Yuan et al., 2014). Similarly, talc with nitrogen (N), phosphorus (P), and potassium (K) salts was able to produce SRFs of high quality (Borges et al., 2017). Recently, phlogopite activated by solvent-free ball milling was used as slow-release K fertilizer (Said et al., 2018). Montmorillonite intercalated with complexed urea and magnesium (Mg) was found to suppress nitrification of the applied urea and ultimately improved N uptake by crops (Kim et al., 2011). Pereira et al. (2012) prepared a nanocomposite-based slow-release fertilizer by intercalation of urea into montmorillonite clay at room temperature through a fast and relatively simple extrusion process. Here, the urea acted as a dispersant for montmorillonite clay. The nanocomposite had high urea content (50%–80%) and retarded the release of N up to 120 h. At lower content of montmorillonite (20% by weight), the composite showed a slow urea release. Microstructural investigations of the nanocomposites revealed that slow release of nutrient was due to urea–clay mineral interaction, and the formation of barriers restricting diffusion of free urea out of the granules (Pereira et al., 2012). Nitrogen retention can be increased by incorporating attapulgite nanoclays into conventional fertilizers (urea/NH_4Cl) (Cai et al., 2015). Santos et al. (2015) attempted to develop microspheres of chitosan–clay (montmorillonite) composites for slow release of K. In the microsphere, montmorillonite clay provides better sorption properties due to its rough and porous surface. An SRF having double-layer microsphere was developed for KNO_3 based on chitosan and montmorillonite clay, which showed slow-release property in both soil and water, and maintained K delivery up to 60 days in soil (Messa et al., 2016). Similarly, França et al. (2018) prepared microcapsule and microsphere, and Messa et al. (2020) prepared microparticles for SRF of fertilizer using chitosan/montmorillonite. A montmorillonite-urea nanocomposite (Mt-Ur) was developed as an ecofriendly slow-release fertilizer (Golbashy et al., 2017). The intercalation of urea into montmorillonite interlayer resulted in significant expansion of d_{001} spacing (of montmorillonite) to 1.71 from

1.23 nm. The nanocomposite at 1:20 ratio (Mt:Ur) exhibited sustained slow release of N over a period of more than 150 h. The slow-release behavior of the nanocomposite occurred in different steps. During initial period, urea localized on the layer surfaces was released as the absorbed urea came into contact with water. During intermediate stage, layers that were smaller in volume were expected to release urea, and at later stages, urea occupying the interlayers of montmorillonite might have been released (Golbashy et al., 2017). Recently, clay-intercalated urea was encapsulated within biodegradable superabsorbent polymer for slow release of nutrient. Nanoencapsulation of montmorillonite-intercalated urea within polyvinyl alcohol nanofiber was found to reduce fertilizer release in water and soil (Azarian et al., 2018).

Pillaring and/or encapsulation of layered compounds with nanoclays and nutrient for their slow release have been studied by some researchers. Various pillared montmorillonites acting as additives together with P fertilizers have been found to enhance their nutrient use efficiency. Three types of pillared montmorillonite, namely, organic pillared montmorillonite, inorganic pillared montmorillonite, and organic–inorganic pillared montmorillonite (OIPM), were prepared, and their potentials in increasing use efficiency of P fertilizer were tested (Ping-Xiao and Zong, 2005). Among the pillared P-fertilizers, OIPM had high bioavailability and the highest P content and P absorption capacity and could fetch highest biomass yield (Ping-Xiao and Zong, 2005). Encapsulation of urea-modified hydroxyapatite nanoparticles into the nanolayers of montmorillonite, called as nanohybrid composite, has been reported to have slow and sustained release of N up to 140 days in a pot culture study with rice crop (Madusanka et al., 2017).

1.2.1.3   CLAY–POLYMER COMPOSITES AS ENCAPSULATION OR MATRIX IN SLOW-RELEASE FERTILIZERS

Clays are made up of a set of stacked nanometric lamellae, which undergo exfoliation in certain aqueous suspensions, but tend to reagglomerate when dried (Konta, 1995; Santos et al., 1989). Clay's ionic accessibility depends on its appropriate exfoliation. Therefore, to improve the ionic accessibility of clay and to expose the clay nanostructure, surface modification is done. Use of polymers for surface modification of clay results in clay exfoliation within the polymer matrix and develop a new material known as nanoclay polymer composite/polymer nanocomposite (NCPC). Hydrogel polymers are mostly used with clay doping to form NCPCs for agricultural use. The NCPCs can be used for encapsulation, adsorption, or as matrix for nutrient release. Hydrogel-based NCPCs have advantages as fertilizer carriers, as they contain clay minerals (bentonite, montmorillonite, kaolin, or attapulgite) in high proportions, making them cost effective (Xiang et al., 2017). Clay minerals also improve swelling properties of hydrogels adding further benefit toward their use as soil conditioners (Xiang et al., 2017). Incorporation of clay into a polymer matrix can improve water absorption rate, ion exchange capacity, and mechanical resistance of the polymer or polymer composite (Bortolin et al., 2013). The nutrient release from the NCPCs is mainly controlled by desorption of nutrients through functional materials by the diffusion process or by degradation of the matrix components (Ni et al., 2013).

Some hydrogel-based composite materials composed of methylcellulose, polyacrylamide (PAM), and calcic montmorillonite were prepared by Bortolin et al. (2013). The presence of montmorillonite resulted in very high fertilizer loading in the composite structure and also retarded the nutrient release in different pH ranges. The hydrogels containing 50% calcic montmorillonite showed excellent slow fertilizer release (almost 200 times slower than pure urea) and could be used for development of SRF/CRF (Bortolin et al., 2013). A matrix-based urea treatment (MU, 195-kg N for one hectare) was tested against common urea treatment (CU, 195-kg N for one hectare) for N use efficiency, N leaching, and NH_3 emission under maize, rice, and wheat in China (Yang et al., 2017a, 2018a, 2018b). The matrix material (mixture of organic polymer and modified bentonite) was mixed with molten urea at a proportion of 5%; then, the melt of urea and matrix underwent granulation to obtain the MU. The results showed that the MU was superior to the CU in multiple aspects. Grain yields with MU were 6.3%–14.7% higher in maize, more than 11% higher in wheat, and more than 10% higher in rice crop than those with CU. Both apparent recovery efficiency and agronomic efficiency were higher with MU as compared to that with CU. Nitrogen leaching and NH_3 emission were also found to be lower with MU (Yang et al., 2017a, 2018a, 2018b).

Several biodegradable polymers [poly(3-hydroxybutyrate) (PHB), algi-nate, starch, cellulose, polybutylene succinate (PBS), etc.] with clay doping are now being used keeping in mind environmental concerns, their easy biode-gradability, and natural abundance (Chen et al., 2018). Souza et al. (2018) prepared clay–polymer composites based on montmorillonite clay and poly-mers such as PHB, starch, and glycerol. They used the composites as carrier for KNO_3 and NPK fertilizers. Fertilizers and montmorillonite were mixed together by the mechanochemical method, followed by incorporation into the polymer matrix. In the composites, montmorillonite helped in increasing the compatibility and homogeneity of PHB/starch. The combination of montmorillonite with different fertilizer results in different matrix composi-tions and characteristics. For instance, introduction of montmorillonite-NPK particles in the polymer resulted in uniformly dispersed, homogeneous, and compatible polymeric phase, whereas montmorillonite–KNO_3 combination was set preferentially on the PHB phase (Souza et al., 2018). The formula-tion released 100% and 60% of the total amount of KNO_3 and NPK fertilizer, respectively, in aqueous medium after 60 h (Souza et al., 2018). Using PHB and montmorillonite clay, another type of composites was prepared by melt processing with KNO_3 and NPK fertilizers (Souza et al., 2019). After a comprehensive nutrient release analysis from the composites, they observed that the presence of clay (also amount and type of the fertilizer) explained the release of nutrients from the composites. Potassium (K^+) and ammonium (NH_4^+) released faster than nitrate (NO_3^-) and phosphate (PO_4^{3-}) from the composites due to interactions between cations and montmorillonite clay and their diffusion through clay phase. They concluded that the composite preparations with lesser quantity of PHB and clay (25% by weight) were sufficient to encapsulate the fertilizer and slow down the release of nutrient ions. Even without encapsulation, montmorillonite containing composites were efficient in slow release of nutrients (releasing about 30% in 9 h) (Souza et al., 2019).

Rashidzadeh and Olad (2014) used montmorillonite clay in a sodium alginate PAM-based superabsorbent nanocomposite to obtain high water absorbing capacity and slow release of NPK fertilizer in soil. The nanocom-posite released 68.3% of fertilizer as compared to 73.1% by the pure polymer at the end of 30 days. Incorporation of clay mineral in hydrogels increased the porosity of the polymer and induced controlled-release features to the nanocomposite. Similarly, Sahoo et al. (2016) prepared slow-release urea fertilizer with nanoclay polymer composites (NCPCs) based on cross-linked PAM and starch-grafted polyacrylamide with different clay doping percent-ages. The study showed that, with increment in clay concentration from 6% to 18%, the urea release from NCPCs was decreased due to the augmented cross-linking density of NCPCs and amplified tortuosity of polymer matrix. Increasing the clay doping above 12% did not decrease the nutrient release as beyond that not much increase in cross-linking density of polymer occurred. NCPCs doped with 24% clay were found to release urea at a higher rate compared to NCPCs doped with 12% and 18% clay as the porosity of the composite drastically reduced at 24% clay doping resulting in surface accumulation of urea (Sahoo et al., 2016). Gharekhani et al. (2018) developed a superabsorbent nanocomposite-based slow-release multinutrient fertilizer using maize bran-g-poly(acrylic acid-co-acrylamide) and montmorillonite, which showed elevated salt resistance, WHC, and excellent slow-release behavior due to the incorporation of montmorillonite. Zhou et al. (2018) developed a novel slow-release fertilizer with montmorillonite, leftover rice-g-poly(acrylic acid), and urea; it showed considerably lower leaching loss of N (19.7%) than pure urea (52.3%). Biodegradable carboxymethyl cellulose (CMC)-based superabsorbent clay polymer composites (CMC-g-PAM/MMT) were produced using acrylamide, sodium salt of CMC, and montmorillonite for slow release of urea (Kenawy et al., 2018). The clay, due to its high ability to adsorb and intercalate urea, resulted in reduced release of urea with increasing concentration of clay in the composite. At the end of 83 days, total N released from CMC-g-PAM with 0% clay was 96%, whereas the release from CMC-g-PAM with clay doping (2.7%, 5%, 10%, and 14.4% montmorillonite by weight) was 86.3%, 78.8%, 74.3%, and 70.2%, respectively (Kenawy et al., 2018). Similarly, PBS polymer with montmorillonite melt mixed with urea can act as a slow-release urea fertilizer, which could improve lettuce crop growth over commercial fertilizer (Baldanza et al., 2018).

Kaolin can also be added to a superabsorbent polymer matrix to improve slow release of nutrients, hydrogel strength, and swelling property and lower the cost of production (Wu et al., 2003; Liang et al., 2007). Liang et al. (2007) prepared controlled-release compound NPK fertilizer having core/shell structure capable of retaining water. The fertilizer exhibited good slow release of nutrients as well as high capacity for water retention.

Zeolites can act as physical cross-linking agents responsible for porous structure of hydrogel nanocomposites. Some zeolite-based hydrogel nanocomposites for slow release of Zinc (Zn) and NPK fertilizers have been reported to optimize nutrient release (Rashidzadeh et al., 2014; Sarkar et al., 2015). In both cases, at the end of one-month release study, zeolite hydrogel composites released significantly lower amounts of nutrients than the composites without zeolite doping. Natural zeolite such as clinoptilolite could decrease nitrification rate up to 11% by retaining NH_4^+ in sites, where nitrifying bacteria were unable to oxidize NH_4^+ and, therefore, reduced N losses from soil (Mackown and Tucker, 1985). Lower volatilization loss and improved N use efficiency were observed by uniting urea with zeolite (He et al., 2002; Campana et al., 2015).

1.2.2 SUSTAINABLE MANAGEMENT OF PESTICIDES

Clay minerals can be used in various ways for development of SRF/CRF of pesticides. Ability of clays for adsorbing and later desorbing pesticides at a relatively slower rate is attributed to their large specific surface area (SSA) and ion-exchange capacity (Park et al., 2014). Therefore, various types of clay minerals were evaluated for CRF of pesticides (Zaghouane-Boudiaf and Boutahala, 2011). Clay minerals were also found to reduce leaching of pesticides (Nennemann et al., 2001).

Approaches for preparing clay-based SRF/CRF of pesticides include adsorbing or trapping the pesticide in the interlamellar space of clay particles by coagulating smectites (Nennemann et al., 2001), modification of hydrophilic clay surface to make them hydrophobic by pre-adsorbing organic cations for enhancing the affinity towards hydrophobic herbicides (Undabeytia et al., 2000; El-Nahhal et al., 2001a), sorption on thermally treated-clay (Bojemueller et al., 2001), encapsulation into a clay polymer composite, intercalation of copolymers of the pesticide (Rehab et al., 2002) and clay–gel matrix, etc.

1.2.2.1 CLAY MINERAL AS ADSORBENT OR CARRIER OF PESTICIDE

Low cost and easy access of clay minerals have made them excellent candidates for pesticide carrier and also its adsorbent (Ali et al., 2012; Sánchez-Jiménez et al., 2012; Galán-Jiménez et al., 2013). When clay minerals are incorporated with pesticides as rheological modifiers, they provide better stability over the products' lifespan. Reversible binding or adsorption of pesticides on clay minerals can increase their efficiency and reduce their volatilization and leaching losses into air, soil, and water (Hermosin et al., 2001; Nennemann et al., 2001; Carrizosa et al., 2000). Clays can be very useful in protecting unstable pesticides from volatilization and photodegradation (Margulies et al., 1992; El-Nahhal et al., 2001b). Smectite clays have been widely studied as pesticide carriers (Cruz-Guzmán et al., 2004; Celis et al., 2007; Akelah et al., 2008).

Naturally occurring cationic clays or organomodified cationic clays, and Mg/Al-layered double hydroxides or anionic clays have been emerged as soil-compatible nanoadsorbents or nanocarriers of controlled-release pesticides (Radian and Mishael, 2008; Celis et al., 2012; Chevillard et al., 2012b; Pérez-De-Luque and Hermosín, 2013; Cabrera et al., 2016;

Nuruzzaman et al., 2016). The cationic and anionic organic pesticides can be held in the interlayer spaces of these clay minerals and can be slowly released afterward (Pérez-De-Luque and Hermosín, 2013; Nuruzzaman et al., 2016). Incorporation of pesticides in the vesicles and micelles molded by surfactant molecules adsorbed on montmorillonite clays results in a slow-release system (Mishael et al., 2003; Sánchez-Verdejo et al., 2008). These nanocarrier/nanoadsorbent-based pesticide formulations are called as nanopesticides, and when applied to soil/plant leaves/rhizosphere could reduce pesticide losses and increase pesticide efficacy (Walker et al., 2018; Kumar et al., 2019). Leachates from imazamox (Imz, a herbicide) treated soil columns had 10–30% lower concentration of the herbicide when anionic clays like synthetic-layered double hydroxides or commercial cationic organoclay (Cloisite 10A) were used as host nanocarriers for Imz than when soil was treated with a commercial Imz (Khatem et al., 2019).

1.2.2.2   CLAY MINERALS FOR ENCAPSULATION OF PESTICIDES

Encapsulation–formulation of agrochemicals involves incorporation of a certain amount of a solid, liquid, or gaseous substance into a second material to develop an SRF/CRF or a better formulation than the original product. Encapsulation results in alteration of liquids or sticky solids into flowable powders, an enhanced mixing behavior due to specific adjustment of surface reactivity (reactive or immiscible materials), and improved storage, safety, and handling of the agrochemical formulations. Encapsulation can reduce the release of active ingredient by protecting it against external environment (oxidation, light, water, evaporation, etc.). Clays, before or after modification, can be used to encapsulate agrochemicals (Takei et al., 2008). The release pattern of agrochemicals from the encapsulated fraction depends on the chemical properties of both the agrochemical and the surfactants along with the type of clay and its concentration in the encapsulation (Galán-Jiménez et al., 2013). Li et al. (2012) developed an encapsulated atrazine using a carboxymethyl chitosan/bentonite composite, where 95.1% of the atrazine was encapsulated into the formulation. Bentonite in the carboxymethyl chitosan/bentonite composite slowed down the release of atrazine by blocking the wetting pores. Incorporation of bentonite into the formulation increased the time needed for release of half of the atrazine to 572 h, which was only 312 h for the formulation without bentonite. Also, the encapsulated atrazine reduced the leaching (Li et al., 2012). Rashidzadeh et al. (2017)

devised a controlled-release system by intercalating paraquat herbicide into montmorillonite and clinoptilolite clays followed by encapsulation with Alginate polymer. The release study showed that intercalation of paraquat into montmorillonite performed better than clinoptilolite for slow release of the formulation. Similarly, isoproturon loaded into microparticles of starch/montmorillonite also performs as a CRF of the pesticide (Wilpiszewska et al., 2016).

1.2.2.3 MODIFIED CLAYS AND CLAY-POLYMER COMPOSITES FOR SLOW RELEASE OF PESTICIDES

Clay mineral surfaces are hydrophilic and mostly negatively charged (Uddin, 2017). Hence, they are generally used as adsorbents for polar/hydrophilic and cationic pesticides. Anionic pesticides can be adsorbed on positively charged clay edges prevalent at acidic pH, or through H-bonds, or van der Waals attraction (Lagaly, 2001). In order to enhance affinity toward hydrophobic and anionic pesticides, surface modification of clay minerals can be done to make them hydrophobic. Intercalation of long chains of organic cations or grafting with different functional groups changes the interlayer space of clays and imparts hydrophobic character.

Sepiolite, modified with cationic surfactant ethoxylated amine, had enhanced capacity to adsorb mesotrione, an herbicide, onto its surface (Galán-Jiménez et al., 2013). Surface modification increased mesotrione solubility by about 59% in water and also the herbicide loading of the formulation. This formulation reduced the mesotrione release by 50% over commercial mesotrione formulation during the study period (Galán-Jiménez et al., 2013). Interlayer methoxy modification of kaolinite followed by amitrole loading through intercalation reduced amitrole release due to the lamellar structure of kaolinite and strong electrostatic attraction between amitrole and kaolinite (Tan et al., 2015). In the methoxy- modified kaolinite, out of total amitrole loading, 47.6% was into the interlayer space, while the rest was onto the external surface of kaolinite (Tan et al., 2015).

Various studies have been done on SRF/CRF of agrochemicals based on sodium alginate, an anionic hydrophilic biopolymer. Pure montmorillonite and montmorillonite modified with cetyl trimethyl ammonium bromide loaded with acetamiprid were used to prepare sodium alginate-exfoliated clay composite beads for CRF of acetamiprid (Yan et al., 2016). Montmorillonite (modified and unmodified) in the composite bead considerably slowed down the release of acetamiprid, due to barrier properties and adsorption capacity of the clay. About 21% and 9% of acetamiprid were released in the initial 10 h from the unmodified and modified montmorillonite alginate composite beads, respectively (Yan et al., 2016). An organobentonite (modified with a surfactant dodecyltrimethyl ammonium chloride) was used with alginate to encapsulate imidacloprid (Jiang et al., 2015). Adsorption capacity of bentonite for organic hydrocarbon was increased by the organic surfactant; also, the organobentonite in the alginate matrix efficiently reduced the dissolution of the water-soluble pesticide. The organobentonite/alginate matrix released 50% of the loaded imidacloprid in 11.5 h, while the same was completely released within only 2 h from the technical-grade imidacloprid (Jiang et al., 2015). Wang et al. (2017a) prepared a composite film containing 2,4-D with clay, Na-montmorillonite, and an alginate ion-cross-linking structure, which showed slow and sustained release of the herbicide (2,4-D). In a calcareous soil, bentonite alginate-based SRF/CRF of chloridazon reduced the herbicide presence in the leachate by 60% (Céspedes et al., 2013). Starch-alginate-based microsphere added with rice husk powder and kaolin was developed to entrap nontoxigenic *Aspergillus flavus* spores and a fungicide simultaneously (Feng et al., 2019). It was observed that addition of rice husk and kaolin to the formulation further slowed down the release profile of the fungicide and spores (Feng et al., 2019). Recently, PAM, hydrophobic derivate of sodium alginate, and montmorillonite were used to formulate stretchable double-network nanocomposite hydrogel for sustained release of λ-cyhalothrin (Wang et al., 2019a).

Hydrogel polymers having ability to capture active ingredients in their matrices and release the same at a slower rate makes them ideal carriers in SRF/CRF (Aouada et al., 2010; Feng et al., 2012; Kulkarni et al., 2012). The rate of release of active ingredient from hydrogel matrix can be modified by optimizing its network properties and water absorption capacity using filler materials such as clays, which increase the cross-linking density of the composite matrix (Sahoo et al., 2014). Sahoo et al. (2014) used pure nanobentonite for a clay–polymer (PAM) composite loaded with metribuzin as a polymer-matrix-based SRF/CRF. The metribuzin release from the SRF (bentonite–polymer composite) was 29.9% after 28 days, whereas it was 78% after only five days in case of the commercial metribuzin. Clay doping of the polymer matrix reduced the porosity of the polymer and subsequent diffusion of metribuzin in water (Sahoo et al., 2014). Kumar et al. (2017) used natural polymer guar gum with polyacrylate and bentonite clay to prepare a hydrogel composite and nanohydrogel composite for controlled release of Imazethapyr. Those prepared formulations had slower release kinetics than that of the commercial formulation.

1.3   REMEDIATION OF SOIL AND WATER POLLUTION

Apart from efficient utilization of agrochemicals, clay minerals can also be used to remediate soil and water from a wide array of contaminants/pollutants. Such contaminants/pollutants include heavy metals such as lead (Pb), chromium (Cr), cadmium (Cd), mercury (Hg), zinc (Zn), nickel (Ni), iron (Fe), copper (Cu), manganese (Mn), etc.; metalloid like arsenic (As); radionuclides such as cesium (^{134}Cs), barium (^{133}Ba), etc.; other inorganic contaminants like nitrate (NO_3^-), carbonate (CO_3^{2-}), etc.; and organic contaminants, for example, dyes, phenols, etc. Properties such as large SSA, high adsorbing power, surface charge, presence of surface functional groups available for bonding, porosity, colloidal nature, swelling ability, etc. enable clay minerals to clean up polluted soil and water by immobilizing or removing the contaminants (Yuan and Wada, 2012; Yuan et al., 2013; Aboudi Mana et al., 2017). Moreover, clay minerals have widespread occurrence in nature, can be mined easily, are generally nontoxic, and are comparatively cheaper than other materials for remediation of contaminated/polluted soil and water (Yuan et al., 2013).

Naturally occurring clay minerals such as bentonite, vermiculite, attapulgite, sepiolite, kaolinite, Fe oxides, etc. have found their use in remediation of soil and water pollution (Sarkar et al., 2019). Besides clay minerals, tectosilicate such as zeolite can be effectively used for remediation of soil and water pollution (Manjaiah et al., 2019). With advancements made in clay mineral research, it has been possible to modify clay minerals and zeolite to improve their capacity for remediation of contaminants. Clay minerals and zeolites can be functionalized or modified by various treatments such as heating, grinding, ion exchange, pillaring reactions, acid treatment, loading with organic molecules, and formation of micro- and nanocomposites. Such modifications are made to increase their efficiency for remediation of a specific kind of contaminant. *Heating* of some clay minerals up to a particular temperature may lead to redistribution of interlayer cations, dihydroxylation of silanol and aluminol groups on clay surface and edges, change in pore-size distribution, and increased SSA ultimately improving their adsorbing power for different contaminants (Akar et al., 2009; Toor et al., 2015; España et al., 2019). *Physical grinding* of naturally occurring minerals can increase their SSA and consequently their adsorption ability. Clay minerals can also be made *homoionic* (i.e., saturated with the same cation species) by replacing the electrostatically held cations on the surfaces, edges, and interlayers with a particular cation. Such homoionic clays function to serve a desired purpose based on the surface properties and the saturating cation. Another technique called *pillaring* can be used to augment the remediation efficacy of clay minerals. Through pillaring, molecular props or bulky cations are intercalated in the interlayers of suitable clay minerals to produce thermally and chemically stable pillared clays with increased porosity (Schooneheydt et al., 1999; Sarkar et al., 2019). *Acid-activated clay* minerals generally have more porosity, SSA, and surface acidity than the original material and can be used to remove phenolic compounds, dyes, and heavy metals from wastewater (Srinivasan, 2011; Al-Essa, 2018; España et al., 2019). Organic molecules or ions such as amino acids, vitamins, quaternary ammonium cations, etc., can be incorporated into clays to make hydrophobic *organoclays* with largely increased affinity for hydrophobic organic pollutants (Sarkar et al., 2012a; Ben Moshe and Rytwo, 2018; Shokri et al., 2018). Organoclays with quaternary ammonium cations in the interlayer and surface of clay minerals possess net positive charge if the quaternary ammonium cations are present in excess of the clay's cation exchange capacity (CEC). They have high affinity for negatively charged contaminants, for example, arsenate (Sarkar et al., 2010, 2012b). In general, organoclays remove contaminants by ion exchange, electrostatic attraction, and hydrophobic partitioning (Sarkar et al., 2019). However, organoclays developed by inserting surfactants into the clay structure may adversely affect the soil microorganisms by affecting their cell membranes (Sarkar et al., 2012a). *Bioreactive clays* could be an improved alternative in this regard, which utilizes clay–bacteria interaction for remediation of inorganic as well as organic contaminants from soil and water. Apart from these, composites of clays with different other materials such as polymers, metal oxides, zero valent iron, carbon, etc., have found their application in remediation of polluted soil and water (Li et al., 2018; Mukhopadhyay et al., 2020). For more details on clay modification procedures and associated environmental applications, readers can refer to *Modified Clay Minerals for Environmental Applications* by Sarkar et al. (2019).

For convenient reading, use of clay minerals for immobilization/removal of contaminants from soil and water have been divided as per the types of contaminants and discussed in the following subsections.

1.3.1 HEAVY METALS AND METALLOID

In soil, natural or modified clays can be applied with the aim to immobilize or reduce the mobility of heavy metals and metalloids so as to restrict their leaching into ground water. It also aims to reduce their ability to make them less available for plant uptake, which subsequently lowers their transfer to human and animal food chain. Effectiveness of immobilization of heavy metals and metalloids in soil depends on the adsorbing power of the applied clay, its dosage, and the heavy metal/metalloid in question. The same also largely depends on some soil properties, for example, pH or soil reaction, texture, content of organic matter, presence of metal oxides, etc. Clay minerals can also be used to remove heavy metals or metalloids from contaminated water, the extent of which may depend on the nature and amount of the clay mineral added, the heavy metal or metalloid species and its concentration, and pH of the aqueous system.

Interactions of heavy metals and metalloids with clay minerals are complex processes. A number of mechanisms can be thought off to explain such interactions, which can be broadly categorized as "surface adsorption" and "structural incorporation" (O'Day and Vlassopoulos, 2010). Surface adsorption includes inner sphere (i.e., direct bonding between the ion and atoms present on mineral surface) and outer sphere (i.e., water or hydroxyl is present between the metal ion and mineral surface) adsorption, and cation exchange apart from surface oxidation/reduction (O'Day and Vlassopoulos, 2010). Generally, affinity of clay minerals for heavy metal ions is more than that for alkali or alkaline-earth metal ions (Tiller, 1996). Adsorption sites for heavy metals could be on edges, interlayers, and surfaces of clay minerals and may depend on the heavy metal species (Inskeep and Baham, 1983; Undabeytia et al., 1998, 2002). The extent of adsorption at a particular site may also vary with the nature of the metal ions depending on pH, ionic strength, anions present in the system, etc. (Yuan et al., 2013). Besides surface adsorption, there could be structural incorporation, which includes coprecipitation (for amorphous minerals), formation of solid solution (for crystalline minerals), and microencapsulation (O'Day and Vlassopoulos, 2010).

Natural bentonite was found to have potential to immobilize Zn and Cu in a sandy alkali soil (pH 8.95) by Kumararaja et al. (2014). They observed an increase in the adsorption capacity of the soil from 0.42 mg g^{-1} to 10.50 mg g^{-1} for Zn and 1.39 mg g^{-1} to 11.76 mg g^{-1} for Cu when bentonite at the rate of 1% of soil mass was added. Values of distribution coefficient (K_d) observed in their study clearly showed many folds increase in metal (Zn and Cu) retention capacity of soil solid phase due to bentonite addition (Kumararaja et al., 2014). Hence, due to extra sorption sites offered by bentonite, concentration of the said metals would drop in soil solution, thus reducing the chances of plant uptake in polluted soils. Concentrations of Cu, Zn, and Ni in *Amaranthus blitum* grown in a metal spiked alkaline soil were reduced by 30.5% and 29.9% (Cu), 6.5% and 21.2% (Zn), and 34.4% and 40.2% (Ni) at first and second harvest, respectively, by application of unmodified bentonite at 2.5% of soil mass (Kumararaja et al., 2016). The effectiveness of different doses (1%, 2%, and 5%) of natural sepiolite for remediation of a light-textured neutral (pH 7.3) soil polluted heavily with Pb, Cd, and Zn was studied by Abad-Valle et al. (2016). Application of sepiolite noticeably lowered (even up to 60–70% at 5% level of sepiolite application) the leachability of Pb, Cd, and Zn. The same also decreased Zn concentration in alfalfa (*Medicago sativa*) shoots up to 45% and improved microbial activity in soil (Abad-Valle et al., 2016). Zhang and Pu (2011) evaluated the heavy metal remediating ability of palygorskite in a calcareous (pH 8.22) and an acid soil (pH 4.63) having moderate to relatively high contents of Cd, Cu, Zn, and Pb. In both soils, amendment with palygorskite at 2% and 5% largely reduced the exchangeable fractions of those cations. Zeolites can induce alkaline environment and promote precipitation of heavy metal cations (Oste et al., 2002; Shi et al., 2009). Zeolites are also able to retain heavy metal cations by virtue of their CEC (Shi et al., 2009), thereby reducing their concentration in soil solution. In a pot experiment with a silty clay textured soil (pH 5.25), spiked with Cd (2 mg Cd per kg dry soil), zeolite application at 1.5% and 3% rates reduced the $CaCl_2$ extractable Cd by 39.0% and 61.8%, respectively, over control soil (Bashir et al., 2018). In that study, Cd uptake by *Ipomoea aquatica* roots was 22.9% and 39.5% lower, and by shoots was 22.9% and 39.6% lower due to zeolite application at 1.5% and 3% rates, respectively, than that from the control soil (Bashir et al., 2018).

Besides heavy metal cations, natural clay minerals also have the potential to reduce bioavailability of As in contaminated soils. In a pot experiment with *Beta vulgaris* as the test crop grown on an As-contaminated soil (pH 6.49), Mukhopadhyay et al. (2017a) observed noticeable decrease in available As in soil and As concentration in plants due to the application of unmodified bentonite at different rates (0.125%, 0.25%, and 0.5%). In the same soil, adsorption efficiency of unmodified kaolinite for As was more than that of unmodified bentonite (Mukhopadhyay et al., 2017b). Bentahar et al. (2016) observed greater As removal from contaminated water with natural clays rich in Fe oxide due to the additional SSA provided by the Fe oxide. They observed As adsorption up to 1.076 mg g^{-1}.

Apart from natural clays, modified or tailored clay minerals have been evaluated by researchers for their ability to remediate heavy metal and/or metalloid contaminated soils. Kumararaja et al. (2017) studied the metal removal efficiency (from water with initial metal concentrations of 25 or 50 mg L^{-1}) and metal immobilization efficiency (in a sandy loam soil with pH 8.23 spiked with Cu, Zn, and Ni) of bentonite clay pillared with Al. Removal of Cu, Zn, and Ni from water largely increased with increase in clay amount. Adsorption of the heavy metals generally increased with pH of the system, with maximum adsorptions occurring at pH 6, 7, and 8 for Cu^{2+}, Zn^{2+}, and Ni^{2+}, respectively. Amendment with 2.5% of pillared bentonite significantly reduced the bioavailability of heavy metals in the spiked soil (Kumararaja et al., 2017). Kumararaja et al. (2018) modified bentonite by graft polymerization with chitosan, a biopolymer, and tested its metal immobilization ability in aqueous system and a sandy clay loam soil (pH 8.65) from a crop field of Rajasthan, India, which was irrigated with heavy-metal-loaded water for more than 20 years preceding sample collection. In water, the chitosan-g-poly(acrylic acid)-bentonite composite showed maximum monolayer adsorption of 48.5, 51.5, 72.9, and 88.5 mg g^{-1} for Ni, Cd, Zn, and Cu, respectively. Amendment with the said polymer was able to increase the retention capacity of the soil by 5.6, 4.9, 3.2, and 3.4 times for Ni, Cd, Zn, and Cu, respectively (Kumararaja et al., 2018). Montmorillonite-based organoclays prepared with hexadecyltrimethylammonium (HDTMA) and tetramethylammonium (TMA) were prepared and evaluated (added at the rate of 5% of soil mass) for stabilization of Cr(VI) in Cr-spiked soil (Yang et al., 2017b). They observed that loading of montmorillonite with HDTMA and TMA mostly enhanced the Cr(VI) immobilization at pH 2–3 over unmodified montmorillonite (Yang et al., 2017b). Nanoscale zero-valent iron (nZVI) has high adsorbing ability for heavy metals (Mandal et al., 2012), but easily forms aggregates and increases in size causing lowering of adsorption capacity. Clay minerals or zeolite can stabilize nZVI and prevent its oxidation and autoagglomeration, thereby enhancing its potential for removal of contaminants (Arancibia-Miranda et al., 2016; Li et al., 2017, 2018). Soliemanzadeh and Fekri (2017) prepared bentonite-supported nanoscale zero-valent iron (B-nZVI) and evaluated for adsorption of Cr(VI) in water and a sandy loam soil (pH 7.6) spiked with Cr(VI). In water, B-nZVI showed much higher adsorption capacity than natural bentonite for Cr(VI), and maximum adsorption was observed in the pH range 2–6. Moreover, B-nZVI applied at 2% and 4% rates led to significant reductions in exchangeable fraction of Cr, while increased the oxide (Fe-Mn)-bound and residual fractions (Soliemanzadeh and Fekri, 2017). Li et al. (2018) synthesized zeolite-supported nanoscale zero-valent iron (Z-nZVI), which showed maximum adsorption capacity of 85.4 mg Pb(II) g^{-1}, 48.6 mg Cd(II) g^{-1} and 11.5 mg As(III) g^{-1} at pH 6, much higher than natural zeolite. Immobilization of the same contaminants were also studied by them in an acid soil (pH 5.13) and an alkaline soil (pH 8.04) by adding Z-nZVI at different doses (0.5–3%), and observed that most of the Pb, Cd, and As in soil got immobilized when Z-nZVI at 3% rate was added (Li et al., 2018). Wang et al. (2017b) synthesized an illite–carbon nanocomposite by hydrothermal carbonization process, which showed greater adsorption capacity than illite or the carbon material used separately for Cr(VI), with maximum adsorption capacity of ~149 mg Cr(VI) g^{-1} in contaminated water.

Mukhopadhyay et al. (2017b) evaluated two unmodified (smectite and kaolinite) and three modified (Fe-exchanged smectite, Ti-pillared smectite, and phosphate bound kaolinite) clay minerals for their As adsorption properties in water as well as an As-contaminated soil (pH 6.49). They observed enhanced As adsorption with modified clays than their unmodified counterparts. Moreover, the two modified smectites showed better As adsorbing power than the modified kaolinite both in water and soil. However, the adsorption capacity of the modified clays for As was more in water than soil, because of soil heterogeneity and the fact that As adsorption depends on more number of factors in soil than in water. While comparing smectites modified with HDTMA, citric acid or chitosan for As removal from water, Mukhopadhyay et al. (2019a) observed highest efficiency (66.9%) and maximum adsorption capacity (473 µg g^{-1}) with HDTMA modified smectite.

1.3.2 RADIONUCLIDES

Radionuclides from nuclear wastes generated through nuclear power reactors, nuclear weapons, and radioactive materials used in industry, academics, medicine, and agriculture may hamper ecosystem stability. Agricultural products coming from soils contaminated with radionuclides can act as paths for the entry of those radionuclides into the food chain and seriously threaten animal and human health. Clay minerals with or without modification can be effectively utilized to get rid of radioactive materials from wastewater. In landfills, clay minerals can also act as barriers preventing the entry of leachates containing radioactive elements into subsoil and groundwater.

Seliman et al. (2014) used Egyptian bentonite as such and also after saturating with Na (Na-bent) for adsorption of radionuclides such as [152]Eu(III), [134]Cs(I), [133]Ba(II), and [90]Sr(II) in an aqueous system. They noticed the adsorption to be largely influenced by initial concentration, pH of the medium, and contact time. Moreover, for removal of [134]Cs(I) at low initial concentrations and $^{90}Sr(II)$ irrespective of initial concentration, Na-Bent was better than unmodified bentonite (Seliman et al., 2014). Sandeep and Manjaiah (2009) reported >90% adsorption of ^{134}Cs by clays such as kaolinite, halloysite, nontronite, attapulgite, mica, illite, and vermiculite even in the presence of competing solutions such as Na oxalate. Sorption of ^{134}Cs was found to be more with mica and smectite than kaolinite (Chari and Manjaiah, 2010), as kaolinite cannot hold ^{134}Cs in the interlayer spaces. Decrease in particle size increased ^{134}Cs sorption by waste mica (Chari and Manjaiah, 2010).

1.3.3   OTHER INORGANIC CONTAMINANTS

Apart from heavy metals, metalloids, and radionuclides, other inorganic species such as nitrate (NO_3^-), bicarbonate (HCO_3^-), and sulfate (SO_4^{2-}) in excess amounts in water have the potential to create problems to aquatic lives, crops by irrigation, and ultimately humans. Clay minerals with desired modifications can be used for removal of such contaminants from wastewater.

Mukhopadhyay et al. (2019b) used Fe_3O_4 nanoparticles and Fe-exchanged nanobentonite to remove NO_3^- and HCO_3^- from contaminated wastewater. They observed adsorption potential of 49.9 mg g^{-1} for NO_3^- and 3.07 me g^{-1} for HCO_3^- with Fe_3O_4 nanoparticles, which increased to 64.8 mg g^{-1} for NO_3^- and 9.73 me g^{-1} for HCO_3^- with Fe-exchanged nanobentonite (Mukhopadhyay et al., 2019b). Homoionic clays with ferrous ion in the interlayers enhance the contaminant adsorption ability of clay minerals, while such modified minerals are nontoxic to and compatible with existing microorganisms in the system to be treated (Ugochukwu et al., 2014a; Prabhu et al., 2015). Clay–polymer composites made from zeolite and natural clay deposits of Shama Mountain, Saudi Arabia using N, N-methylene-bis-acrylamide exhibited better removal of ions like Cl$^-$, SO_4^{2-}, and Na$^+$ from tannery wastewater than other adsorbents (Sallam et al., 2017).

1.3.4   ORGANIC CONTAMINANTS

Organic pollutants such as dyes, polycyclic aromatic hydrocarbons (PAHs), pesticides, antibiotics, phenols, etc. coming from various industries can pollute water, which when used for agriculture, animal husbandry, or fisheries would pose problem to human and animal health through contamination of food chain. Many researchers have used clay minerals with appropriate modifications for removal of these organic pollutants from environment.

Potential of clays for removal or degradation of organic pollutants is discussed hereunder.

A number of researchers have used modified clay minerals to remove dyes from the aqueous system. Adsorption capacity of montmorillonite for methylene blue (MB), a dye, increased to 500 mg g^{-1} from 350 mg g^{-1} after heating the clay at 300 °C (Mouzdahir et al., 2010). Chitin–clay composite made by Xu et al. (2018) could remove a maximum of 99.99% of MB (152.2 mg g^{-1}), and the adsorption was mainly dependent on pH of the aqueous system. A nanoorganoclay named Cloisite 30B modified by CTS, a biopolymer made up of N-acetylglucosamine and glucosamine was used to remove dyes such as reactive red-141 and reactive blue-21 from water (Vanaamudan and Sudhakar, 2015). The nanoorganoclay exhibited adsorption up to 439 mg g^{-1} for reactive red-141 and 476 mg g^{-1} for reactive blue-21, and around 60%–62% of that occurred within 50 min (Vanaamudan and Sudhakar, 2015). Bée et al. (2017) used magnetic CTS/clay beads in the pH range of 3–12 in the aqueous system and observed adsorption maxima of 82 mg g^{-1} for MB, with 50% of removal occurring within only 13 min. Electrostatic attraction among the permanent negative charges of clay and dye cations was thought to be crucial for the adsorption.

PAHs introduced to soil can be detrimental to environment and also to human health (Vane et al., 2014). Clay or modified clay-modulated microbial degradation of PAHs can be an efficient way for their remediation. Biswas et al. (2017a) observed that mild acid activation of smectite and palygorskite could support biodegradation of PAHs in a PAH-contaminated soil with pH 6.4. However, extensive acid activation of clays was found to not support the same (Ugochukwu et al., 2014b). Ugochukwu et al. (2014c) compared the efficiency of montmorillonites with four different interlayer cations (K, Zn, Ca, and Fe) for removal of some crude oil PAHs during biodegradation. They observed significantly enhanced biodegradation of the PAHs by Ca- and Fe-montmorillonites, while K- and Zn-montmorillonites could not enhance the PAH biodegradation. However, K- and Zn-montmorillonites could cause 45% removal of PAHs by adsorption when clay to oil ratio was 5:1 (w/w). Clay–bacteria interactions offered by bioreactive clays have the potential to enhance degradation of organic contaminants by bacteria. *Burkholderia sartisoli* is a bacterium, which can degrade phenanthrene, one kind of PAH having potential detrimental influence on environment. Biswas et al. (2017b) observed that palygorskite heated at 400 °C could support growth and metabolic activity of *Burkholderia sartisoli* and maximize the degradation of phenanthrene by mineralization through this bacterium.

Agricultural intensification has brought large increase in use of pesticides, which many times enter into surface water bodies and also ground water. One such example is atrazine, a herbicide having high potential to worsen environmental quality. It is susceptible to leaching and runoff losses and may contaminate surface and groundwater. Dutta and Singh (2015) used bentonite modified by surfactants, namely, HDTMA, phenyltrimethylammonium (PTMA), trioctylmethylammonium, and stearylkonium, and studied their atrazine removal efficiency in aqueous solution as well as wastewater collected from an atrazine manufacturing factory. In that study, unmodified bentonite showed only 9.4% adsorption at initial atrazine concentration of 1 μg mL^{-1}, and modification by PTMA could not improve the same. On the other hand, other three modifications largely improved the adsorption capacity for atrazine, which varied from 49% to 72.4%. Also, atrazine removal by the modified clays was more from the wastewater than aqueous solution of atrazine (Dutta and Singh, 2015). Gámiz et al. (2015) found that hexadimethrine–montmorillonite, a nanocomposite, had >70% removal efficiency for anionic pesticides such as mecoprop and clopyralid, by virtue of the electrostatic attraction between these anionic pesticides and NH$_4^+$ of the nanocomposite.

Since long, antibiotics such as amoxicillin (AMX) have been used profusely in many sectors, including agriculture, animal husbandry, and fisheries, which ultimately cause environmental pollution (Pan et al., 2008; Homem and Santos, 2011). Conventional approaches have been found to be inadequate to remove these mostly nonbiodegradable materials (Pan et al., 2008; Homem and Santos, 2011; Pan et al., 2008). AMX can be harmful to aquatic ecosystem even if it is present in low concentrations. It has the potential to enter into human food chain via fishes cultivated in water bodies with AMX contamination or through animal products if the drinking water used in animal husbandry is contaminated with AMX. Weng et al. (2017) used B-nZVI to remove AMX from aqueous solution and observed the removal was by adsorption and reductive degradation, the extent of which depended on initial concentrations of AMX and B-nZVI, pH, and reaction temperature.

Phenols in water pose threat to living organisms and may contaminate food chain through irrigation or drinking water. Al-Essa (2018) modified Jordanian bentonite using HCl and used it to remove phenolic compounds from olive mill waste water (OMWW). They observed highest removal of phenolic compounds at pH 6, and for maximum adsorption of phenolic compounds, 1 g of acid-activated bentonite per 10-mL OMWW was the optimal rate. Homoionic clays (e.g., Ca-montmorillonite) can also support biodegradation of organic contaminants in soil and water. However, improvement in clay's efficiency to support biodegradation of organic compounds depends on the saturating cations (Al-Essa, 2018). Hernández-Hernández et al. (2018) prepared a clay polymer composite by incorporating HDTMA-modified clay in an alginate polymer matrix. It showed adsorption of 0.334 mg g^{-1} for 4-chlorophenol. Shabtai and Mishael (2018) designed a regenerable dual-site clay (montmorillonite)-polymer (polymerized β-cyclodextrin [pCD] modified with pCD+, a cationic functional group) composite, which showed greater removal efficiencies than activated carbon for micropollutants. The composite could target *bisphenol A*, an emerging micropollutant through its incorporation inside β-cyclodextrin cavities, and also the anionic effluent organic matter compounds by electrostatic attraction (Shabtai and Mishael, 2018).

1.4   CLAY MINERALS AND CARBON SEQUESTRATION

Global warming and its effect on soil quality and crop productivity pose threat to agricultural sustainability (Friedrich and Scanlon, 2008). For sustainable agriculture, maintenance of soil quality is essential. Soil quality and crop productivity largely depend on soil organic matter (SOM), the key component of which is soil organic carbon (SOC) (Trivedi et al., 2013). Globally, soil contains massive pool of C (approximately 2344 Gt of organic C within upper three meters of soil) (Stockmann et al., 2013). The SOC largely regulates the concentration of C in the atmosphere as it acts as a source of atmospheric CO_2 (Luo et al., 2010; Smith et al., 2008). Therefore, a small change in soil C level significantly affects the C concentration in atmosphere (Eglin et al., 2010). Hence, sequestration of C in soil is indispensable to improve soil quality and reduce CO_2 emission to atmosphere. Carbon sequestration is the locking of atmospheric C for a long time in any form that prevents it to return back to the atmosphere (Singh et al., 2018). Clay minerals play an important role in stabilizing SOM as the latter remains largely associated with organomineral clusters on rough surfaces of clays (Chen et al., 2014; Das et al., 2019a). Soil clays generally possess large SSA with high surface charge density; due to which they are more effective in increasing SOC stabilization (Regelink et al., 2013). Amorphous, short-range ordered, and poorly crystalline clays are also found to retain considerable amount of C in soil (Chatterjee et al., 2013; Datta et al., 2015). The SOC interacts with clay minerals through various bonding mechanisms and get protected from microbial decomposition. The protection of organic C by clay minerals for subsequent stabilization is through physical, chemical, and biological means (Sarkar et al., 2018; Singh et al., 2018). The organic matter after binding with silt and clay form slowly available or passive pool of C and remain protected from microbial decomposition (Kleber et al., 2015). In fine-textured soils, SOM may get protection through getting into small pores or getting occluded and become physically inaccessible to the microorganisms (Hassink, 1997; Baldock and Skjemstad, 2000; Castellano et al. 2015). Chemical protection comes from the adsorption of SOM on clay mineral surfaces through hydrogen (H)-bonding, hydrophobic interactions, van der Waals attraction, electrostatic attraction (by polyvalent cation bridging and cation exchange), and ligand exchange (Kogel-Knabber and Kleber, 2011; Schmidt et al., 2011). Such adsorption is of two types: (1) *physical adsorption* through van der Waals interaction, H-bonding and hydrophobic interactions with uncharged organic molecule, and (2) *chemical adsorption* by formation of outer or inner sphere complexes through electrostatic interaction or ligand exchange. The outer sphere complexes are formed in presence of monovalent or multivalent cations, which act as bridges between clay and SOM, the bridge with multivalent cations being stronger (Setia et al., 2014). The inner sphere complex formation involves the adsorption at specific sites of mineral surfaces through the ligand exchange mechanism. The formation of clay–organic complexes (COC) through physical and chemical adsorption gives biological protection to SOM as it is resistant to microbial degradation. Generally, the functional groups present on the mineral surfaces, types of clay minerals, and the nature of organic matter determine the stability of SOM in COC (Schmidt et al., 2011; Saidy et al., 2013; Purakayastha et al., 2019; Das et al., 2019a). Besides this, climatic factors such as temperature, rainfall, etc., and different soil management practices with organic and inorganic fertilizers also play important roles in SOM stabilization by clay minerals (Singh et al., 2017).

Soil clays accumulate more C than sand and silt (Jagadamma and Lal, 2010). The amount of clay significantly influences organic C sequestration in soils especially in areas where rainfall is comparatively higher (Xu et al., 2016). There is a contradiction between the SOC level and clay content. Some research stated that SOC content increases linearly with increasing clay content (Deng et al., 2017; Johannes et al., 2017; Zhong et al., 2018), while others did not find any definite relationship between these two (Martens et al., 2004; Ren et al., 2017). Ratio of clay to SOC can strongly influence soil physical properties such as bulk density, water retention capacity and structural quality (Dexter et al., 2008; Johannes et al., 2017). Dexter et al. (2008) introduced a concept of complexed organic C, noncomplexed organic C, complexed clay and noncomplexed clay (NCC) based on the complexation of unit mass of organic C with clay minerals. Complexed organic C showed better correlation with the soil volume (1/bulk density) compared to the total SOC when the clay: complexed organic C = 10. They also found that NCC more readily disperses in soil compared with complexed clay. The maximum complexed organic C, which is the fraction of SOC corresponding to a tenth of the clay content, controls the structural properties of soils, but is no longer related to SOC above 10% of clay content (Dexter et al., 2008). The clay-to-SOC ratio is also important for management of structural quality of soil and SOC content, and the ratio of 10 indicates the limit between good and medium structural quality of soils (Johannes et al., 2017).

Besides clay content, the extent of protection as well as stabilization of SOM depends on the SSA and charge characteristics of different types of clay minerals present in soils (Robert and Chenu, 1992). Smectite-dominating soil contains higher amount of organic C with greater stability compared to the other clay minerals, because smectite has higher internal surface area with greater number of –OH binding sites offering better retention of organic molecules (Chotzen et al., 2016; Ovesen et al., 2011). The external SSAs of montmorillonite and kaolinite are $15-160\,m^2\,g^{-1}$ and $6-40\,m^2\,g^{-1}$, respectively (Saidy et al., 2013; Singh et al., 2016). Larger SSA due to smaller particle size and also greater extent of isomorphous substitution in montmorillonite are the reasons for higher adsorption sites per unit mass in montmorillonite compared with kaolinite. On the other hand, illite has comparatively higher SSA ($55-195\,m^2\,g^{-1}$) and CEC ($10-40$ cmol (p+) kg^{-1} than kaolinite and can retain comparatively higher amount of organic C by their broken edges (Six et al., 2002; Wiseman and Püttmann, 2006). According to Saidy et al. (2013), the maximum sorption capacity for dissolved organic C by phyllosilicates decrease in the order smectite > illite > kaolinite on a mass basis, whereas, on the basis of surface area, kaolinite adsorbed the most followed by smectite and illite, respectively.

Apart from crystalline phyllosilicates, poorly crystalline amorphous minerals are also important for the stabilization of SOM owing to their smaller size and larger SSA (Chatterjee et al., 2013; Zhao et al., 2017; Das et al., 2019a). Poorly crystalline minerals such as allophanes have very high SSA ($700-1500\,m^2\,g^{-1}$), leading to higher sorption of dissolved organic matter (DOM) than that by crystalline minerals such as smectite and kaolinite (Singh et al., 2016). The amorphous Fe and Al-oxides form stable complexes with DOM and reduce its degradation by soil microbes (Schwesig et al., 2003; Schneider et al., 2010). Amorphous oxides also improve the soil structure by promoting aggregation and impart stability to SOM (Oades and Waters, 1991). Sorption capacity of Fe and Al-oxides has been found to be higher in acid soil than the neutral or alkaline soils as most of the positive sites get neutralized in latter soils, thereby decreasing their ability to stabilize SOM (Kiemand Kögel-Knabner, 2002). However, positive correlation between organic C and amorphous Fe and Al-oxides was also observed in near neutral soils (Moni et al., 2010). Generally, amorphous Fe-oxides are more important for stabilization of SOM than other oxides, hydroxides, and phyllosilicates (Chatterjee et al., 2013; Zhao et al., 2017). Das et al. (2019a) also found very high positive correlation between the content of amorphous oxides and C stability in colloidal organomineral fraction (COMF) of different soils. The presence of hydrous oxide coatings on phyllosilicates significantly influences the sorption capacity of dissolved organic C in soil, though the extent of influence varies with the type of clay minerals (Saidy et al., 2013). The presence of goethite coating significantly improves the sorption capacity of kaolinitic clays, whereas ferrihydrite coating increases the sorption capacity of the illitic clays (Saidy et al., 2013). The bonding mechanisms involved in clay–organic association control the sorption and desorption of SOM by different clay minerals. According to Mikutta et al. (2007), goethite sorbs organic matter predominantly by ligand exchange, pyrophyllite by van der Waals forces and Ca^{2+} bridging, and vermiculite by Ca^{2+} bridging. Among these, the SOM bonded by ligand exchange is more protected from microbial attack than that bonded through other mechanisms.

Type of clay minerals is one of the crucial factors determining the mean residence time (MRT) of the SOM (Wattel–Koekkoek et al., 2003). Organic matter associated with kaolinite has lower turn over time (on an average 360 years) than that associated with smectite (on an average 1100 years) (Wattel-Koekkoek et al., 2003). Organic matter associated with kaolinite has more amounts of easily decomposable organic molecule (i.e., plant remnants), which are aliphatic in nature than that associated with smectite, which are mainly dominated by more resistant aromatic compounds (Wattel-Koekkoek et al., 2003). Such variation in MRT of organic matter as a part of COC strongly depends on the effective CEC of the clay mineral involved (Wattel-Koekkoek and Buurman, 2004). Allophanes with very high SSA form inner-sphere complexes with organic matter and increase their MRT in soil (Bolan et al., 2012).

Long-term cultivation with organic and inorganic fertilizers can also influence SOM stabilization by clay minerals. Long-term application of organics alone or with inorganic fertilizers was found to alter the relative abundance of poorly crystalline minerals, especially the noncrystalline Fe oxides and hydroxides in soils (Wang et al., 2019b). These poorly crystalline Fe-oxides interact with the SOC, especially the aromatic part, and enhance stability (Das et al., 2019a; Wen et al., 2019). Long-term cultivation with organics might facilitate weathering of clay minerals, decrease their crystallite size, and increase SSA resulting better interaction with SOM (Das et al., 2019a). Das et al. (2019a) observed negative correlation ($r = -0.46$, $P < 0.01$) between crystallite size of illite and C stability in COMF.

Temperature and rainfall of an area also have impact on the nature of association and retention of C by clay minerals in soil. Decomposition of organic C in soil increases with increase in temperature and moisture availability due to higher mineralization rate (Hossain, 2017; Singh et al., 2017). With increasing temperature (from 4 to 37 °C) and moisture (from 30% to 60%), SOM mineralization rate increased, and smectite dominating soil showed higher degree of mineralization compared to the kaolinite-illite-dominating soil and allophanic soil (Singh et al., 2017). The effect of moisture content and temperature on CO_2 emission and C stabilization in soils was found to be more dependent on the amount of sesquioxides than the type and SSA of clay minerals (Singh et al., 2017).

Needless to say, the nature of organics, already present in a soil or applied externally, significantly influences the soil C sequestration. However, as the foregoing discussion is on role of clay minerals on promoting C sequestration in soil, effect of the nature of the organics on soil C sequestration is not described here.

1.5 CLAY MINERALS AS SOIL AMENDMENTS

Clay can be used as an amendment to improve moisture retention capacity of sandy soils, especially in arid and semiarid regions (Dempster et al., 2012). Application of nanoclay increases moisture retention capacity of soils and also decreases soil loss by wind erosion in arid region (Padidar et al., 2016). Use of specifically modified clays (e.g., pillared clays, organoclays, nanocomposites, etc.) can prove even better in enhancing the WHC of soils as well as fertilizer use efficiency, which are very important for agricultural sustainability over long time periods (Basak et al., 2012; Sourabh et al., 2019). Besides modified clays, NCPC can also improve the WHC of soils (Sarkar et al., 2014). Superabsorbent NCPC can be used to ameliorate moisture stress under rain-fed agriculture (Verma et al., 2017; Pandey et al., 2018). Application of sepiolite clay mixed with a peat-based growing medium in nursery improves stock quality of seedlings to survive in strong water deficit areas, particularly in dryland ecosystems (Chirino et al., 2011). Nutrient leaching is a major problem in light textured sandy soils. High nutrient level in aquatic ecosystem causes eutrophication and ground water pollution. Addition of clay in such soils decreases the nutrient leaching and increases the fertilizer retention capacity and the effect is more pronounced if the clay is finely ground because it increases the overall surface area of soil (Tahir and Marschner, 2017).

Application of clay can decrease bulk density and improve aggregate stability, soil structure, nutrient availability to plants, and the extent of C sequestration in soils; however, the particle size distribution is not affected by clay addition (Tangkoonboribun et al., 2006; Ye et al., 2019). Application of clay increases the buffering ability of soils, thereby improves nutrient retention capacity (Karbout et al., 2015). Clay particles have high SSA and charge density; hence, their application can increase the surface area of soils as well as the binding sites for SOM and ultimately improve aggregation and pore configuration. Application of bentonite clay strengthens interparticle bonds in sandy soils (Ajayi and Horn, 2016), hence, improves the soil structure and subsequently plant biomass yields under tropical environment (Croker et al., 2004). Clay addition might inhibit enzyme activities in C and N cycle, but has no influence on microbial community in soil (Ye et al., 2019). Therefore, addition of clay, with or without modification, has the potential to improve many physical and chemical properties of soils and help to attain agricultural sustainability.

1.6   CLAY MINERALS FOR AGRICULTURAL PACKAGING

Many agricultural produces and food products are perishable in nature. Proper packaging of such produces is very important for prolonged shelf-life. Lots of research studies have been done so far to develop suitable food packaging materials, which not only increase the self-life of agricultural produces, but also are easily available and cost effective. The most extensively used packaging material is polymer, in which various nanomaterials such as nanoclays, graphene, and carbon nanotubes can be incorporated to improve the inherent property of the polymer material (Ray et al., 2013; Xia et al., 2019). Addition of nanoclays to polymer matrix is useful for improving its mechanical, thermal, and degradation properties and reduce the weight of the polymer–clay nanocomposites (PCNs) (Beneyto et al., 2014; Castro-Aguirre et al., 2018). Furthermore, nanoclays are relatively inexpensive, hence, reasonable to use as packaging material or container for food and beverages. The market for nanoclay-based food packaging material was USD 343 million in 2014 and anticipated to enlarge significantly through 2022 (Bumbudsanpharoke and Ko, 2019). Nanoclays such as montmorillonite and organically modified montmorillonite have very high SSA, aspect ratio, and compatibility with polymers and can be incorporated into the polymer matrix to improve their characteristics (Farhoodi, 2016; Ganguly et al., 2011). Naturally, montmorillonite has hydrophilic surface, which can be modified by organic cations to make it organophilic to increase the compatibility with polymers (polyethylene and polypropylene), which are mostly hydrophobic in nature (Majeed et al., 2013; Xia et al., 2019). Furthermore, due to large volume of organic cations, the basal spacing of organically modified clays gets increased, which subsequently increases the penetration level of polymer (Kim et al., 2013). However, recent studies indicated some adverse effects of nanoparticles on human health due to direct intake of packaged foods contaminated by nanoparticles due to their migration (Bumbudsanpharoke and Ko, 2019; Bandyopadhyay and Ray, 2018), raising concerns regarding their use as packaging material. In this regard, understanding the migration behavior of nanoclays is very important to assess the potential health hazards of PCNs present in packaging materials.

1.7 CONCLUSIONS

Clay minerals and their modified products along with zeolite are found to be very effective for slow or controlled release of agrochemicals as they can enhance the use efficiency of applied nutrient/pesticide and check economic losses and environmental damage. Natural and modified clay minerals and zeolites have huge potential of remediation of environmental pollution from both inorganic and organic contaminants. However, more studies are still needed to find cost-effective and biodegradable clay polymer formulations for economic and sustainable agricultural production systems. Clay minerals are also important for soil C sequestration and soil quality improvement and can be employed in packaging of agricultural produces. With such a widespread applicability, clay minerals and zeolite, without or with suitable modifications, appear to be very important in achieving agricultural sustainability.

KEYWORDS

- amendments
- clay minerals
- C sequestration
- packaging
- slow-release formulations
- sustainable agriculture

REFERENCES

Abad-Valle, P.; Álvarez-Ayuso, E.; Murciego, A.; Pellitero. E. Assessment of the use of sepiolite amendment to restore heavy metal polluted mine soil. *Geoderma* 2016, *280*, 57–66.

Aboudi Mana, S. C.; Hanafiah, M. M.; Chowdhury, A. J. K. Environmental characteristics of clay and clay-based minerals. *Geol. Ecol. Landsc.* 2017, *1* (3), 155–161.

Ajayi A. E.; HornInt, R. Comparing the potentials of clay and biochar in improving water retention and mechanical resilience of sandy soil. *Agrophys.* 2016, *30*, 391–399.

Akar, S. T.; Yetimoglu, Y.; Gedikbey, T. Removal of chromium (VI) ions from aqueous solutions by using Turkish montmorillonite clay: Effect of activation and modification. *Desalination* 2009, 244, 97–108.

Akelah, A.; Rehab, A.; El-Gamal, M. M. Preparation and applications of controlled release systems based on intercalated atrazine salt and polymeric atrazine salt onto montmorillonite clay. *Mater. Sci. Eng. C* 2008, *28* (7), 1123–1131. https://doi.org/10.1016/j.msec.2007.05.005.

Al-Essa, K. Activation of Jordanian Bentonite by hydrochloric acid and its potential for olive mill wastewater enhanced treatment. *J. Chem.* 2018, *2018,* 1–10.

Ali, I.; Asim, M.; Khan, T. A. Low cost adsorbents for the removal of organic pollutants from wastewater. *J. Environ. Manage.* 2012, *30*, 170–183. https://doi.org/10.1016/j.jenvman.2012.08.028

Aouada, F. A.; De Moura, M. R.; Orts, W. J.; Mattoso, L. H. C. Polyacrylamide and methylcellulose hydrogel as delivery vehicle for the controlled release of paraquat pesticide. *J. Mater. Sci.* 2010, *45*, 4977–4985. https://doi.org/10.1007/s10853-009-4180-6

Arancibia-Miranda, N.; Baltazar, S. E.; Garcı´a, A.; Mun~oz-Lira, D.; Sepu´lveda, P.; Rubio, M. A.; Altbir, D. Nanoscale zero valent supported by zeolite and montmorillonite: template effect of the removal of lead ion from an aqueous solution. *J. Hazard. Mater.* 2016, 301, 371–380.

Armstrong, A.; Aden, K.; Amraoui, N.; Diekkruger, B.; Jarvis, N.; Mouvet, C., Nicholls, P.; Wittwer, C. Comparison of the performance of pesticide-leaching models on a cracking clay soil: Results using the Brimstone Farm dataset. *Agric. Water Manage.* 2000, 44, 85–10.

Azarian, M. H.; Kamil Mahmood, W. A.; Kwok, E.; Bt Wan Fathilah, W. F.; Binti Ibrahim, N. F. Nanoencapsulation of intercalated montmorillonite-urea within PVA nanofibers: Hydrogel fertilizer nanocomposite. *J. Appl. Polym. Sci.* 2018, *135* (10), 45957. https://doi.org/10.1002/app.45957

Baldanza, V. A. R.; Souza, F. G.; Filho, S. T.; Franco, H. A.; Oliveira, G. E.; Caetano, R. M. J.; Hernandez, J. A. R.; Ferreira Leite, S. G.; Furtado Sousa, A. M.; Nazareth Silva, A. L. Controlled-release fertilizer based on poly(butylene succinate)/urea/clay and its effect on lettuce growth. *J. Appl. Polym. Sci.* 2018, *135* (47), 46858. https://doi.org/10.1002/app.46858

Baldock, J. A.; Skjemstad, J. Role of the soil matrix and minerals in protecting natural organic materials against biological attack. *Org. Geochem.* 2000, *31*, 697–710.

Bandyopadhyay, J.; Ray, S. S. Are nanoclay-containing polymer composites safe for food packaging applications?—An overview. *J. Appl. Polym.* 2019, *136*, 47214.

Basak, B. B.; Pal, S.; Datta, S. C. Use of modified clays for retention and supply of water and nutrients. *Curr. Sci.* 2012, *102* (9), 1272–1278.

Bashir, S.; Zhu, J.; Fu, Q.; Hu, H. Cadmium mobility, uptake and anti-oxidative response of water spinach (Ipomoea Aquatic) under rice straw biochar, zeolite and rock phosphate as amendments. *Chemosphere* 2018, *194*, 579–587.

Bée, A.; Obeid, L.; Mbolantenaina, R.; Welschbillig, M.; Talbot, D. Magnetic chitosan/ clay beads: A magsorbent for the removal of cationic dye from water. *J. Magn. Magn. Mater.* 2017, *421*, 59–64.

Ben Moshe, S.; Rytwo, G. Thiamine-based organoclay for phenol removal from water. *Appl. Clay Sci.* 2018, *155*, 50–56.

Bentahar, Y.; Hurel, C.; Draoui, K.; Khairoun, S.; Marmier, N. Adsorptive properties of Moroccan clays for the removal of arsenic(V) from aqueous solution. *Appl. Clay Sci.* 2016, *119*, 385–392.

Bhardwaj, D.; Sharma, M.; Sharma, P.; Tomar, R. Synthesis and surfactant modification of clinoptilolite and montmorillonite for the removal of nitrate and preparation of slow release nitrogen fertilizer. *J. Hazard. Mater.* 2012, *227–228*, 292–300.

Biswas, B.; Sarkar, B.; Naidu, R. Bacterial mineralization of phenanthrene on thermally activated palygorskite: A 14 C radiotracer study. *Sci. Total Environ.* 2017b, *579*, 709–717.

Biswas, B.; Sarkar, B.; Rusmin, R.; Naidu, R. Mild acid and alkali treated clay minerals enhance bioremediation of polycyclic aromatic hydrocarbons in long-term contaminated soil: A 14 C-tracer study. *Environ. Pollut.* 2017a, *223*, 255–265.

Bojemueller, E.; Nennemann, A.; Lagaly, G. Enhanced pesticide adsorption by thermally modified bentonites. *Appl. Clay Sci.* 2001, *18* (5–6), 277–284. https://doi.org/10.1016/S0169-1317(01)00027-8

Bolan, N.; Kunhikrishnan, A.; Choppala, G.; Thangarajan, R.; Chung, J. Stabilization of carbon in composts and biochars in relation to carbon sequestration and soil fertility. *Sci. Total Environ.* 2012, *424*, 264–270.

Borges, R.; Prevot, V.; Forano, C.; Wypych, F. Design and kinetic study of sustainable potential slow-release fertilizer obtained by mechanochemical activation of clay minerals and potassium monohydrogen phosphate. *Ind. Eng. Chem. Res.* 2017, *56*, 708–716.

Bortolin, A.; Aouada, F. A.; Mattoso, L. H. C.; Ribeiro, C. Nanocomposite PAAm/Methyl cellulose/montmorillonite hydrogel: evidence of synergistic effects for the slow release of fertilizers. *J. Agric. Food Chem.* 2013, *61* (31), 7431–7439. https://doi.org/10.1021/jf401273n

Bumbudsanpharoke, N.; Ko, S. Nanoclays in food and beverage packaging. *J. Nanomater.* 2019, *2019*, 8927167.

Cabrera, A.; Celis, R.; Hermosín, M. C. Imazamox-clay complexes with chitosan- and iron(iii)-modified smectites and their use in nanoformulations. *Pest Manage. Sci.* 2016, *72* (7), 1285–1294. https://doi.org/10.1002/ps.4106

Cai, D.; Wu, Z.; Jiang, J.; Wu, Y.; Feng, H.; Brown, I. G.; Chu, P. K.; Yu, Z. Controlling nitrogen migration through micro-nano networks. *Sci. Rep.* 2015, *4* (1), 1–8. https://doi.org/10.1038/srep03665

Campana, M.; Alves, A. C.; Anchão de Oliveira, P. P.; de Campos Bernardi, A. C.; Santos, E. A.; Herling, V. R.; Gomes de Morais, J. P.; Barioni, W. Ammonia volatilization from exposed soil and tanzania grass pasture fertilized with urea and zeolite mixture. *Commun. Soil Sci. Plant Anal.* 2015, *46* (8), 1024–1033. https://doi.org/10.1080/00103624.2015.1019080

Carrizosa, M. J.; Caldero´n, M. J.; Hermosı´n, M. C.; Cornejo, J. Organosmectites as sorbent and carrier of the herbicide bentazone. *Sci. Total Environ.* 2000, *247*, 285–293.

Castellano, M. J.; Mueller, K. E.; Olk, D. C.; Sawyer, J. E.; Six, J. Integrating plant litter quality, soil organic matter stabilization, and the carbon saturation concept. *Glob. Chang. Biol.* 2015, *21*, 3200–3209.

Castro-Aguirre, E.; Auras, R.; Selke S.; Rubino, M.; Marsh, T. Impact of nanoclays on the biodegradation of poly(Lactic Acid) nanocomposites. *Polymers* 2018, *10* (2), 1–21.

Celis, R.; Adelino, M. A.; Hermosín, M. C.; Cornejo, J. Montmorillonite-chitosan bionanocomposites as adsorbents of the herbicide clopyralid in aqueous solution and soil/water suspensions. *J. Hazard. Mater.* 2012, *209–210*, 67–76. https://doi.org/10.1016/j.jhazmat.2011.12.074

Celis, R.; Trigo, C.; Facenda, G.; Hermosín, M. D. C.; Cornejo, J. Selective modification of clay minerals for the adsorption of herbicides widely used in olive groves. *J. Agric. Food Chem.* 2007, *55* (16), 6650–6658. https://doi.org/10.1021/jf070709q

Céspedes, F. F.; Pérez García, S.; Villafranca Sánchez, M.; Fernández Pérez, M. Bentonite and anthracite in alginate-based controlled release formulations to reduce leaching of chloridazon and metribuzin in a calcareous soil. *Chemosphere* 2013, *92* (8), 918–924. https://doi.org/10.1016/j.chemosphere.2013.03.001

Chari, M. S.; Manjaiah, K. M. Radiocesium sorption-desorption on soil clays and clay-organic complexes. *Clay Res.* 2010, *29*, 23–45.

Chatterjee, D.; Datta, S. C.; Manjaiah, K. M. Clay carbon pools and their relationship with short-range order minerals: avenues to mitigate climate change. *Curr. Sci.* 2013, *105*, 1404–1410.

Chen, J.; Lü, S.; Zhang, Z.; Zhao, X.; Li, X.; Ning, P.; Liu, M. Environmentally friendly fertilizers: A review of materials used and their effects on the environment. *Sci. Total Environ.* 2018, *1*, 829–839. https://doi.org/10.1016/j.scitotenv.2017.09.186

Chevillard, A.; Angellier-Coussy, H.; Guillard, V.; Gontard, N.; Gastaldi, E. Controlling pesticide release via structuring agropolymer and nanoclays based materials. *J. Hazard. Mater.* 2012a, *205–206*, 32–39. https://doi.org/10.1016/j.jhazmat.2011.11.093.

Chevillard, A.; Angellier-Coussy, H.; Guillard, V.; Gontard, N.; Gastaldi, E. Investigating the biodegradation pattern of an ecofriendly pesticide delivery system based on wheat gluten and organically modified montmorillonites. *Polym. Degrad. Stab.* 2012b, *97*, 2060–2068.

Chirino, E.; Vilagrosa, A.; Vallejo, V.R. Using hydrogel and clay to improve the water status of seedlings for dryland restoration. *Plant Soil* 2011, *344*, 99–110.

Chotzen, RA.; Polubesova, T.; Chefetz, B.; Mishael, Y.G. Adsorption of soil-derived humic acid by seven clay minerals: A systematic study. *Clays Clay Miner.* 2016, *64*, 628–638.

Croker, J.; Poss, R.; Hartmann, C. Effects of recycled bentonite addition on soil properties, plant growth and nutrient uptake in a tropical sandy soil. *Plant Soil* 2004, *267*, 155–163.

Cruz-Guzmán, M.; Celis, R.; Hermosín, M. C.; Cornejo, J. Adsorption of the herbicide simazine by montmorillonite modified with natural organic cations. *Environ. Sci. Technol.* 2004, *38*, 180–186.

Das, D.; Dwivedi, B. S.; Datta, S. P.; Datta, S. C.; Meena, M. C.; Agarwal, B. K.; Shahi, D. K.; Singh, M.; Chakraborty, D.; Jaggi S. Potassium supplying capacity of a red soil from eastern India after forty-two years of continuous cropping and fertilization. *Geoderma*, 2019b, *341*, 76–92.

Das, R., Purakayastha, T. J., Das, D., Ahmed, N., Kumar, R., Biswas, S., Walia, S. S., Singh, R., Shukla, V. K., Yadava, M. S., Ravisankar, N., Datta, S. C., 2019a. Long-term fertilization and manuring with different organics alter stability of carbon in colloidal organo-mineral fraction in soils of varying clay mineralogy. *Sci. Tot. Environ. 684*, 682–693. https://doi.org/10.1016/j.scitotenv.2019.05.327

Dempster D. N.; Jones D. L.; Murphy D. V. Clay and biochar amendments decreased inorganic but not dissolved organic nitrogen leaching in soil. *Soil Res.* 2012, *50* (3), 216–221.

Deng, Y. S.; Cai, C. F.; Xia, D.; Ding, S. W.; Chen, J. Z. Fractal features of soil particle size distribution under different land-use patterns in the alluvial fans of collapsing gullies in the hilly granitic region of southern China. *PLoS One* 2017, *12* (3), 1–21.

Dexter, A. R.; Richard, G.; Arrouays, D.; Czyz, E. A.; Jolivet, C.; Duval, O. Complexed organic matter controls soil physical properties. *Geoderma* 2008. *144*, 620–627.

Dutta, A.; Singh, N. Surfactant-modified bentonite clays: Preparation, characterization, and atrazine removal. *Environ Sci. Pollut. Res.* 2015, *22*, 3876–3885.

Eglin, T.; Ciais, P.; Piao, S. L.; Barre, P.; Bellassen, V.; Cadule, P.; Chenu, C.; Gasser, T.; Koven, C.; Reichstein, M.; Smith, P. Historical and future perspectives of global soil carbon response to climate and land-use changes. *Tellus B.* 2010, *62*, 700–718.

El-Nahhal, Y.; Nir, S.; Serban, C.; Rabinovitz, O.; Rubin, B. Organo-clay formulation of acetochlor for reduced movement in soil. *J. Agric. Food Chem.* 2001a, *49* (11), 5364–5371. https://doi.org/10.1021/jf010561p

El-Nahhal, Y.; Undabeytia, T.; Polubesova, T.; Mishael, Y. G.; Nir, S.; Rubin, B. Organo-clay formulations of pesticides: reduced leaching and photodegradation. *Appl. Clay Sci.* 2001b, *18*, 309–326.

España, V. A. A.; Sarkar, B.; Biswas, B.; Rusmin, R.; Naidu, R. Environmental applications of thermally modified and acid activated clay minerals: Current status of the art. *Environ. Technol. Innovat.* 2019, *13*, 383–397.

Farhoodi, M. Nanocomposite materials for food packaging applications: Characterization and safety evaluation. *Food Eng. Rev.* 2016, *8* (1), 35–51.

Feng, B. H.; Peng, L. F. Synthesis and characterization of carboxymethyl chitosan carrying ricinoleic functions as an emulsifier for azadirachtin. *Carbohydr. Polym.* 2012, *88* (2), 576–582. https://doi.org/10.1016/j.carbpol.2012.01.002

Feng, J.; Dou, J.; Wu, Z.; Yin, D.; Wu, W. Controlled release of biological control agents for preventing aflatoxin contamination from starch-alginate beads. *Molecules* 2019, *24* (10), 1858.

Fernández-Pérez, M.; Flores-Céspedes, F.; González-Pradas, E.; Villafranca-Sánchez, M.; Pérez-García, S.; Garrido-Herrera, F. J. Use of activated bentonites in controlled-release formulations of atrazine. *J. Agric. Food Chem.* 2004, *52* (12), 3888–3893. https://doi.org/10.1021/jf030833j

França, D.; Medina, Â. F.; Messa, L. L.; Souza, C. F.; Faez, R. Chitosan spray-dried microcapsule and microsphere as fertilizer host for swellable—controlled release materials. *Carbohydr. Polym.* 2018, *196*, 47–55. https://doi.org/10.1016/j.carbpol.2018.05.014

Fu, J.; Wang, C.; Chen, X.; Huang, Z.; Chen, D. Classification research and types of slow controlled release fertilizers (SRFs) used—a review. *Commun. Soil Sci. Plant Anal.* 2018, *25*, 2219–2230. https://doi.org/10.1080/00103624.2018.1499757.

Galán-Jiménez Md. C.; Mishael, Y. G.; Nir, S.; Morillo, E.; Undabeytia, T. Factors affecting the design of slow release formulations of herbicides based on clay-surfactant systems. A methodological approach. *PLoS One* 2013, *8* (3), e59060.

Gámiz, B.; Hermosín, M. C.; Cornejo, J.; Celis, R. Hexadimethrine-montmorillonite nanocomposite: Characterization and application as a pesticide adsorbent. *Appl. Surf. Sci.* 2015, *332*, 606–613.

Ganguly, S.; Dana, K.; Mukhopadhyay, T. K.; Parya, T. K.; Ghatak, S. Organophilic nano clay: A comprehensive review. *T. Indian Ceram. Soc.* 2011, *70* (4), 189–206.

Gerstl, Z.; Nasser, A.; Mingelgrin, U. Controlled release of pesticides into soils from claypolymer formulations. *J. Agric. Food Chem.* 1998, *46*, 3797–3802.

Gharekhani, H.; Olad, A.; Hosseinzadeh, F. Iron/NPK agrochemical formulation from superabsorbent nanocomposite based on maize bran and montmorillonite with functions of water uptake and slow-release fertilizer. *New J. Chem.* 2018, *42* (16), 13899–13914. https://doi.org/10.1039/c8nj01947a

Golbashy, M.; Sabahi, H.; Allahdadi, I.; Nazokdast, H.; Hosseini, M. Synthesis of highly intercalated urea-clay nanocomposite via domestic montmorillonite as eco-friendly slow-release fertilizer. *Arch. Agron. Soil Sci.* 2017, *63* (1), 84–95.

Hassink, J. The capacity of soils to preserve organic C and N by their association with clay and silt particles. *Plant Soil* 1997, *191*, 77–87.

He, Z. L.; Calvert, D. V.; Alva, A. K.; Li, Y. C.; Banks, D. J. Clinoptilolite zeolite and cellulose amendments to reduce ammonia volatilization in a calcareous sandy soil. *Plant Soil* 2002, *247* (2), 253–260. https://doi.org/10.1023/A:1021584300322

Hermosin, M. C.; Calderón, M. J.; Aguer, J.-P.; Cornejo, J. Organoclays for controlled release of the herbicide fenuron. *Pest Manage. Sci.* 2001, *57* (9), 803–809. https://doi.org/10.1002/ps.359

Hermosín, M. C.; Celis, R.; Facenda, G.; Carrizosa, M. J.; Ortega-Calvo, J. J.; Cornejo, J. Bioavailability of the herbicide 2,4-D formulated with organoclays. *Soil Biol. Biochem.* 2006, *38* (8), 2117–2124. https://doi.org/10.1016/j.soilbio.2006.01.032

Hernández-Hernández, K. A.; Illescas, J.; Díaz-Nava, Md. C.; Martínez-Gallegos, S.; Muro-Urista, C.; Ortega-Aguilar, R. E.; Rodríguez-Alba, E.; Rivera, E. Preparation of nanocomposites for the removal of phenolic compounds from aqueous solutions. *Appl. Clay Sci.* 2018, *157*, 212–217.

Hocine, O.; Boufatit, A.; Khouider, A. Use of montmorillonite clays as adsorbents of hazardous pollutants. *Desalination* 2004, *167*, 141–145.

Homem, V.; Santos, L. Degradation and removal methods of antibiotics from aqueous matrices—A review. *J. Environ. Manage.* 2011, *92*, 2304–2347.

Hossain, M. B.; Rahman, M. M.; Biswas, J. C.; Miah, M. M. U.; Akhter, S.; Maniruzzaman, M.; Choudhury, A. K.; Kanti, A.; Ahmed, F.; Shiragi, M. H. K.; Kalra, N. Carbon mineralization and carbon dioxide emission from organic matter added soil under different temperature regimes. *Int. J. Recycl. Org. Waste Agricult.* 2017, *6*, 311–319.

Inskeep, W. P.; Baham, J. Adsorption of Cd (II) and Cu (II) by Na-montmorillonite at low surface coverage. *Soil Sci. Soc. Am. J.* 1983, *47*, 660–665.

Jagadamma, S.; Lal, R. Distribution of organic carbon in physical fractions of soils as affected by agricultural management. *Biol. Fertil. Soils* 2010, *46*, 543–554.

Jiang, L.; Mo, J.; Kong, Z.; Qin, Y.; Dai, L.; Wang, Y.; Ma, L. Effects of organobentonites on imidacloprid release from alginate-based formulation. *Appl. Clay Sci.* 2015, *105–106*, 52–59. https://doi.org/10.1016/j.clay.2014.12.023.

Johannes, A.; Matter, A.; Schulin, R.; Weisskopf, P.; Baveye, P. C.; Boivin, P. Optimal organic carbon values for soil structure quality of arable soils. Does clay content matter? *Geoderma* 2017, *302*, 14–21.

Jorda-Beneyto, M.; Ortuño, N.; Devis, A.; Aucejo, S.; Puerto, M.; Gutiérrez-Praenab, D.; Houtmanb, J.; Pichardo, S.; Maisanaba, S.; Jos, A. Use of nanoclay platelets in food packaging materials: Technical and cytotoxicity approach. *Food Addit. Contam. A, Chem. Anal. Control Expo. Risk Assess.* 2014, *31* (3), 354–363.

Karbout, N.; Moussa, M.; Gasmi, I.; Bousnina, H. Effect of clay amendment on physical and chemical characteristics of sandy soil in arid areas: The case of ground south-eastern Tunisian. *Appl. Sci. Rep.* 2015, *11* (2), 43–48.

Keeble, B. R. BSc MBBS MRCGP (1988) The Brundtland report: 'Our common future', medicine and war, *4* (1), 17–25, doi: 10.1080/07488008808408783

Kenawy, E.-R.; Azaam, M. M.; El-nshar, E. M. Preparation of carboxymethyl cellulose-g-poly (acrylamide)/montmorillonite superabsorbent composite as a slow-release urea fertilizer. *Polym. Adv. Technol.* 2018, *29* (7), 2072–2079. https://doi.org/10.1002/pat.4315

Khatem, R.; Celis, R.; Hermosín, M. C. Cationic and anionic clay nanoformulations of imazamox for minimizing environmental risk. *Appl. Clay Sci.* 2019, *168*, 106–115. https://doi.org/10.1016/j.clay.2018.10.014

Kiem, R.; Kogel-Knabner, I. Refractory organic carbon in particle-size fractions of arable soils II: organic carbon in relation to mineral surface area and iron oxides in fractions <6 μm. *Org. Geochem.* 2002, *33*, 1699–1713.

Kim, K. S.; Park, M.; Lim, W. T.; Komarneni, S. Massive intercalation of urea in montmorillonite. *Soil Sci. Soc. Am. J.* 2011, *75* (6), 2361–2366. https://doi.org/10.2136/sssaj2010.0453

Kim, S. G.; Lofgren, E. A.; Jabarin, S. A. Dispersion of nanoclays with poly (ethylene terephthalate) by melt blending and solid state polymerization. *J. Appl. Polym. Sci.* 2013, *127*(3), 2201–2212.

Kleber, M.; Eusterhues, K.; Keiluweit, M.; Mikutta, C.; Mikutta, R.; Nico, P. S. Mineral–organic associations: Formation, properties, and relevance in soil environments. *Adv. Agron.* 2015, *130*, 1–140.

Kögel-Knabner, I.; Kleber, M. Mineralogical, physicochemical, and microbiological controls on soil organic matter stabilization and turnover. In: *Handbook of Soil Sciences Resource Management and Environmental Impacts*, (2nd edn), Huang, P. M., Li, Y., Sumner, M. E., Eds. CRC Press, Taylor and Francis Group: Boca Raton, FL, 2011, p. 830.

Konta, J. Clay and man: Clay raw materials in the service of man. *Appl. Clay Sci.* 1995, *10* (4), 275–335. https://doi.org/10.1016/0169-1317(95)00029-4

Kulkarni, R. V.; Boppana, R.; Krishna Mohan, G.; Mutalik, S.; Kalyane, N. V. pH-Responsive interpenetrating network hydrogel beads of poly(acrylamide)-*g*-carrageenan and sodium alginate for intestinal targeted drug delivery: Synthesis, in vitro and in vivo evaluation. *J. Colloid Interface Sci.* 2012, *367* (1), 509–517. https://doi.org/10.1016/j.jcis.2011.10.025

Kumar, S.; Nehra, M.; Dilbaghi, N.; Marrazza, G.; Hassan, A. A.; Kim, K. H. Nano-based smart pesticide formulations: emerging opportunities for agriculture. *J. Control. Release* 2019, *294*, 131–153. https://doi.org/10.1016/j.jconrel.2018.12.012

Kumar, V.; Singh, A.; Das, T. K.; Sarkar, D. J.; Singh, S. B.; Dhaka, R.; Kumar, A. Release behavior and bioefficacy of imazethapyr formulations based on biopolymeric hydrogels. *J. Environ. Sci. Heal. B* 2017, *52* (6), 402–409. https://doi.org/10.1080/03601234.2017.1293446

Kumararaja, P.; Manjaiah, K. M.; Datta, S. C.; Ahammed Shabeer T. P.; Sarkar, B. Chitosan-g-poly(acrylic acid)-bentonite composite: A potential immobilizing agent of heavy metals in soil. *Cellulose* 2018, *25*, 3985–3999.

Kumararaja, P.; Manjaiah, K. M.; Datta, S. C.; Ahammed Shabeer, T. P. Potential of bentonite clay for heavy metal immobilization in soil. *Clay Research,* 2014, *33* (2), 19–32.

Kumararaja, P.; Manjaiah, K. M.; Datta, S. C.; Sarkar, B. Remediation of metal contaminated soil by aluminium pillared bentonite: Synthesis, characterisation, equilibrium study and plant growth experiment. *Appl. Clay Sci.* 2017, *137*, 115–122.

Kumararaja, P.; Shabeer T. P. A.; Manjaiah K. M. Effect of bentonite on heavy metal uptake by amaranth (Amaranthus blitum cv. Pusa Kirti) grown on metal contaminated soil. *Indian J. Hort.* 2016, *73* (2), 224–228

Lagaly, G. Pesticide-clay interactions and formulations. *Appl. Clay Sci.* 2001, *8*, 205–209.

Li, J.; Yao, J.; Li, Y.; Shao, Y. controlled release and retarded leaching of pesticides by encapsulating in carboxymethyl chitosan /bentonite composite gel. *J. Environ. Sci. Heal. Part B* 2012, *47* (8), 795–803. https://doi.org/10.1080/03601234.2012.676421

Li, X.; Zhao, Y;, Xi, B.; Meng, X.; Gong, B.; Li, R.; Peng, X.; Liu, H. Decolorization of Methyl Orange by a new clay-supported nanoscale zero-valent iron: Synergetic effect, efficiency optimization and mechanism. *J. Environ. Sci.* 2017, *52*, 8–17.

Li, Z.; Wang, L.; Meng, J.; Liu, X.; Xu J.; Wang, F.; Brookes P. Zeolite-supported nanoscale zero-valent iron: New findings onsimultaneous adsorption of Cd(II), Pb(II), and As(III) in aqueous solution and soil. *J. Hazard. Mater.* 2018, *344*, 1–11.

Liang, R.; Liu, M.; Wu, L. Controlled release NPK compound fertilizer with the function of water retention. *React. Funct. Polym.* 2007, *67* (9), 769–779. https://doi.org/10.1016/j.reactfunctpolym.2006.12.007.

Luo, Z. K.; Wang, E.; Sun, O. J. Soil carbon change and its responses to agricultural practices in Australian agro-ecosystems: A review and synthesis. *Geoderma* 2010, *155*, 211–223.

MacKown, C. T.; Tucker, T. C. Ammonium nitrogen movement in a coarse-textured soil amended with zeolite. *Soil Sci. Soc. Am. J.* 1985, *49* (1), 235–238. https://doi.org/10.2136/sssaj1985.03615995004900010048x

Madusanka, N.; Sandaruwan, C.; Kottegoda, N.; Sirisena, D.; Munaweera, I.; De Alwis, A.; Karunaratne, V.; Amaratunga, G. A. J. Urea–hydroxyapatite-montmorillonite nanohybrid composites as slow release nitrogen compositions. *Appl. Clay Sci.* 2017, *150*, 303–308. https://doi.org/10.1016/j.clay.2017.09.039

Majeed, K.; Jawaid, M.; Hassan A.; Bakar, A. A.; Abdul Khalil, H. P. S.; Salema, A. A.; Inuwa, I. Potential materials for food packaging from nanoclay/natural fibres filled hybrid composites. *Mater. Des.* 2013, *46*, 391–410.

Mandal, B. K.; Kumar, K. M.; Vankayala, R.; Mukherjee, A.; Kumar, K. S.; Reddy, P. S.; Hegde, M. S.; Sreedhar, B. Synthesis of zero valent iron nanoparticles and application to removal of arsenic(III) and arsenic(V) from water. *Indian J. Chem. Soc.* 2012, *89*, 1215–1221.

Manjaiah, K. M.; Mukhopadhyay, R.; Narayanan, N.; Sarkar, B. Clay amendments for environmental clean-up. In: *Soil Amendments for Sustainability: Challenges and Perspectives*, Rakshit, A.; Sarkar, B.; Abhilash, P. Eds. CRC Press, Taylor and Francis: Boca Raton, FL, 2018, p. 404.

Manjaiah, K. M.; Mukhopadhyay, R.; Paul, R.; Datta, S. C.; Kumararaja, P.; Sarkar, B. Clay minerals and zeolites for environmentally sustainable agriculture. In: *Modified Clay and Zeolite Nanocomposite Materials Environmental and Pharmaceutical Applications Micro and Nano Technologies,* 1st edn. Mercurio, M.; Sarkar, B.; Langella, A. Eds. Elsevier: Netherlands, 2019, p. 362.

Margulies, L.; Stern, Th.; Rubin, B.; Ruzo, L. O. Photostabilization of trifluralin adsorbed on a clay matrix. *J. Agric. Food Chem.* 1992, *40*, 152–155.

Martens, D. A.; Reedy, T. E.; Lewis, D. T. Soil organic carbon content and composition of 130-year crop, pasture and forest land-use managements. *Glob. Chang. Biol.* 2004, *10*, 65–78

Messa, L. L.; Froes, J. D.; Souza, C. F.; Faez, R. Híbridos de quitosana-argila para encapsulamento e liberação sustentada do fertilizante nitrato de potássio. *Quim. Nova* 2016, *39* (10), 1215–1220. https://doi.org/10.21577/0100-4042.20160133.

Messa, L. L.; Souza, C. F.; Faez, R. Spray-dried potassium nitrate-containing chitosan/montmorillonite microparticles as potential enhanced efficiency fertilizer. *Polym. Test.* 2020, *81*, 106196. https://doi.org/10.1016/j.polymertesting.2019.106196

Mikutta, R.; Mikutta C.; Kalbitz, K.; Scheel, T.; Kaiser, K.; Jahn, R. Biodegradation of forest floor organic matter bound to minerals via different binding mechanisms. *Geochim. Cosmochim. Acta* 2007, *71*, 2569–2590.

Mishael, Y. G.; Undabeytia, T.; Rabinovitz, O.; Rubin, B.; Nir, S. Sulfosulfuron incorporated in micelles adsorbed on montmorillonite for slow release formulations. *J. Agric. Food Chem.* 2003, *51* (8), 2253–2259. https://doi.org/10.1021/jf0261497.

Moni, C.; Chabbi, A.; Nunan, N.; Rumpel, C.; Chenu C. Spatial dependance of organic carbon–metal relationships: A multi-scale statistical analysis, from horizon to field. *Geoderma* 2010, *158*(3–4), 120–127.

Mouzdahir, Y.; Elmchaouri, A.; Mahboub, R.; Gil, A.; Korili, S. A. Equilibrium modeling for the adsorption of methylene blue from aqueous solutions on activated clay minerals. *Desalination* 2010, *250*, 335–338.

Mukhopadhyay, R.; Adhikari, T.; Sarkar, B.; Barman, A.; Paul, R.; Patra, A. K.; Sharma, P. C.; Kumar, P. Fe-exchanged nano-bentonite outperforms Fe_3O_4 nanoparticles in removing nitrate and bicarbonate from wastewater. *J. Hazard. Mater.* 2019b, *376*, 141–152.

Mukhopadhyay, R.; Bhaduri, D.; Sarkar, B.; Rusmin, R.; Hou, D.; Khanam, R.; Sarkar, S.; Biswas, J. K.; Vithanage, M.; Bhatnagar, A.; Ok. Y. S. Clay–polymer nanocomposites: Progress and challenges for use in sustainable water treatment. *J. Hazard. Mater.* 2020, *383*, 121–125.

Mukhopadhyay, R.; Manjaiah, K. M.; Datta, S. C.; Sarkar, B. Comparison of properties and aquatic arsenic removal potentials of organically modified smectite adsorbents. *J. Hazard. Mater.* 2019a, *377*, 124–131.

Mukhopadhyay, R.; Manjaiah, K. M.; Datta, S. C.; Yadav, R. K. Effect of bentonite on arsenic uptake by beet leaf cultivar Pusa Bharti grown on contaminated soil. *Indian J. Hortic.* 2017a, *74*, 546–551.

Mukhopadhyay, R.; Manjaiah, K. M.; Datta, S. C.; Yadav, R. K.; Sarkar, B. Inorganically modified clay minerals: Preparation, characterization, and arsenic adsorption in contaminated water and soil. *Appl. Clay Sci.* 2017b, *147*, 1–10.

Murray, H. H. Traditional and new applications for kaolin, smectite, and palygorskite: A general overview. *Appl. Clay Sci.* 2000, *17* (5–6), 207–221. https://doi.org/10.1016/S0169-1317(00)00016-8

Nennemann, A.; Mishael, Y.; Nir, S.; Rubin, B.; Polubesova, T.; Bergaya, F.; Van Damme, H.; Lagaly, G. Clay-based formulations of metolachlor with reduced leaching. *Appl. Clay Sci.* 2001, *18* (5–6), 265–275. https://doi.org/10.1016/S0169-1317(01)00032-1

Ni, B.; Liu, M.; Lu, S.; Xie, L.; Zhang, X.; Wang, Y. Novel slow-release multielement compound fertilizer with hydroscopicity and moisture preservation. *Ind. Eng. Chem. Res.* 2010, *49*, 4546–4552.

Ni, X.; Yuejin, W.; Zhengyan, W.; Lin, W.; Guannan, Q.; Lixiang, Y. A novel slow-release urea fertiliser: Physical and chemical analysis of its structure and study of its release mechanism. *Biosyst. Eng.* 2013, *115* (3), 274–282. https://doi.org/10.1016/j.biosystemseng.2013.04.001

Nuruzzaman, M.; Rahman, M. M.; Liu, Y.; Naidu, R. Nanoencapsulation, nano-guard for pesticides: A new window for safe application. *J. Agric. Food Chem.* 2016, *64* (7), 1447–1483. https://doi.org/10.1021/acs.jafc.5b05214

Oades, J. M.; Waters, A. G. Aggregate hierarchy in soils. *Aust. J. Soil Res.* 1991, *29*, 815–828.O'Day, P. A.; Vlassopoulos, D. Mineral-based amendments for remediation. *Elements* 2010, *6*, 375–381.

Oste, L. A.; Lexmond, T. M.; Van Riemsdijk, W. H. Metal immobilization in soils using synthetic zeolites. *J. Environ. Qual.* 2002, *31*, 813–821.

Ovesen, R. G.; Nielsen, J.; Hansen, H. C. B. Biomedicine in the environment: Sorption of the cyclotide Kalata B2 to montmorillonite, goethite and humic acid. *Environ. Toxicol. Chem.* 2011, *30* (8), 1785–1792.

Padidar A.; Jalalian, A.; Abdouss, M.; Najafi, P., Honarjoo, N.; Fallahzade, J.; Effects of nanoclay on some physical properties of sandy soil and wind erosion. *Inter. J. Soil Sci.* 2016, *11*, 9–13.

Pan, X. L.; Deng, C. N.; Zhang, D. Y.; Wang, J. L.; Mu, G. J.; Chen, Y. Toxic effects of amoxicillin on the photosystem II of synechocystis sp. characterized by a variety of in vivo chlorophyll fluorescence tests. *Aquat. Toxicol.* 2008, *89*, 207–213.

Pandey, P.; Giriraj; De, N. Halloysite nanoclay polymer composite: Synthesis, characterization and effect on water retention behaviour of soil. *Chem. Sci. Int. J.* 2018, *23*(3), 1–11.

Park, Y.; Sun, Z.; Ayoko, G. A.; Frost, R. L. Removal of herbicides from aqueous solutions by modified forms of montmorillonite. *J. Colloid Interface Sci.* 2014, *415*, 127–132. https://doi.org/10.1016/j.jcis.2013.10.024.

Pereira, E. I.; Minussi, F. B.; Da Cruz, C. C. T.; Bernardi, A. C. C.; Ribeiro, C. Urea-montmorillonite-extruded nanocomposites: A novel slow-release material. *J. Agric. Food Chem.* 2012, *60* (21), 5267–5272. https://doi.org/10.1021/jf3001229

Pérez-de-Luque, A.; Hermosín, M. C. Nanotechnology and its use in agriculture. In *Bio-Nanotechnology: A Revolution in Food, Biomedical and Health Sciences*; Bagchi Debasis, Bagchi Manashi, Moriyama Hiroyoshi, Fereidoon, S., Ed. Blackwell Publishing Ltd.: Oxford, UK, 2013, pp. 383–398. https://doi.org/10.1002/9781118451915.ch20

Ping-Xiao; Zong, W. Study on structural characteristics of pillared clay modified phosphate fertilizers and its increase efficiency mechanism. *J. Zhejiang Univ. Sci. B*, 2005, *6*, 195–201.

Prabhu, Y. T.; Rao K. V.; Kumari, B. S.; Kumar, V. S. S.; Pavani, T. Synthesis of Fe_3O_4 nanoparticles and its antibacterial application. *Int. Nano Lett.* 2015, *5*, 85–92.

Purakayastha, T. J.; Das, R.; Kumari, S.; Shivay, Y. S.; Biswas, S.; Kumar D.; Chakrabarti, B. CImpact of continuous organic manuring on mechanisms and processes of the stabilisation of soil organic C under rice–wheat cropping system. *Soil Res.* 2019, *58*(1), 73–83.

Radian, A.; Mishael, Y. G. Characterizing and designing polycation-clay nanocomposites as a basis for imazapyr controlled release formulations. *Environ. Sci. Tech.* 2008, *42*, 1511–1516.

Rashidzadeh, A.; Olad, A. Slow-released NPK fertilizer encapsulated by NaAlg-g-Poly(AA-Co-AAm)/MMT superabsorbent nanocomposite. *Carbohydr. Polym.* 2014, *114*, 269–278. https://doi.org/10.1016/j.carbpol.2014.08.010

Rashidzadeh, A.; Olad, A.; Hejazi, M. J. Controlled release systems based on intercalated paraquat onto montmorillonite and clinoptilolite clays encapsulated with sodium alginate. *Adv. Polym. Technol.* 2017, *36* (2), 177–185. https://doi.org/10.1002/adv.21597

Rashidzadeh, A.; Olad, A.; Salari, D.; Reyhanitabar, A. On the preparation and swelling properties of hydrogel nanocomposite based on sodium alginate-g-poly (acrylic acid-co-acrylamide)/clinoptilolite and its application as slow release fertilizer. *J. Polym. Res.* 2014, *21* (2), 1–15. https://doi.org/10.1007/s10965-013-0344-9

Ravier, I.; Haouisee, E.; Clément, M.; Seux, R.; Briand, O. Field experiments for the evaluation of pesticide spray-drift on arable crops. *Pest Manage. Sci.* 2005, *61*, 728–736.

Ray, S. S. *Clay-Containing Polymer Nanocomposites. From Fundamentals to Real Applications.* 1st edn. Elsevier: Oxford, UK, 2013.

Rehab, A.; Akelah, A.; El-Gamal, M. M. Controlled-release systems based on the intercalation of polymeric metribuzin onto montmorillonite. *J. Polym. Sci. Part A Polym. Chem.* 2002, *40* (14), 2513–2525. https://doi.org/10.1002/pola.10326

Ren, C. J.; Chen, J.; Deng, J.; Zhao, F. Z.; Han, X. H.; Yang, G. H.; Tong, X. G.; Feng, Y. Z.; Shelton, S.; Ren, G. X. Response of microbial diversity to C:N:P stoichiometry in fine root and microbial biomass following afforestation. *Biol. Fertil. Soils* 2017, *53*, 457–468.

Robert, M.; Chenu, C. Interactions between soil minerals and microorganisms. In *Soil Biochemistry,* Stotzky, G.; Bollag, J. M., Eds. Marcel Dekker Inc.: New York, 1992, vol. 7, p. 432.

Sahoo, S.; Manjaiah, K. M.; Datta, S. C.; Ahmed Shabeer, T. P.; Kumar, J. Kinetics of metribuzin release from bentonite-polymer composites in water. *J. Environ. Sci. Heal. Part B* 2014, *49* (8), 591–600. https://doi.org/10.1080/03601234.2014.911578

Sahoo, S.; Manjaiah, K. M.; Datta, S. C.; Kumar, R. Polyacrylamide and starch grafted polyacrylamide based nanoclay polymer composites for controlled release of nitrogen. *Clay Res.* 2016, *35*, 16–24.

Said, A.; Zhang, Q.; Qu, J.; Liu, Y.; Lei, Z.; Hu, H.; Xu, Z. Mechanochemical activation of phlogopite to directly produce slow-release potassium fertilizer. *Appl. Clay Sci.* 2018, *165*, 77–81. https://doi.org/10.1016/j.clay.2018.08.006

Saidy, A.; Smernik, R.; Baldock, J.; Kaiser, K.; Sanderman, J. The sorption of organic carbon onto differing clay minerals in the presence and absence of hydrous iron oxide. *Geoderma* 2013, *209*, 15–21.

Sallam, A.; Al-Zahrani, M.; Al-Wabel, M.; Al-Farraj, A.; Usman, A. Removal of Cr (VI) and toxic ions from aqueous solutions and tannery wastewater using polymer clay composites. *Sustainability* 2017, *9*, 1993.

Sánchez-Jiménez, N.; Sevilla, M. T.; Cuevas, J.; Rodríguez, M.; Procopio, J. R. Interaction of organic contaminants with natural clay type geosorbents: Potential use as geologic barrier in urban landfill. *J. Environ. Manage.* 2012, *95*, S182–S187.

Sanchez-Martin, M. J.; Rodriguez-Cruz, M. S.; Andrades, M. S.; Sanchez-Camazano, M. Efficiency of different clay minerals modified with a cationic surfactant in the adsorption of pesticides: Influence of clay type and pesticide hydrophobicity. *Appl. Clay Sci.* 2006, *31*, 216–228.

Sánchez-Verdejo, T.; Undabeytia, T.; Nir, S.; Maqueda, C.; Morillo, E. Environmentally friendly slow release formulations of alachlor based on clay-phosphatidylcholine. *Environ. Sci. Technol.* 2008, *42* (15), 5779–5784. https://doi.org/10.1021/es800743p

Sandeep, S.; Manjaiah, K. M. Radiocesium sorption on soils and clay minerals: effect of oxalates and sodium tetraphenylboron. *Clay Res.* 2009, *28*, 41–67.

Santos, B. R. Dos; Bacalhau, F. B.; Pereira, T. D. S.; Souza, C. F.; Faez, R. Chitosan-montmorillonite microspheres: a sustainable fertilizer delivery system. *Carbohydr. Polym.* 2015, *127*, 340–346. https://doi.org/10.1016/j.carbpol.2015.03.064

Santos, M. C. D.; Ribeiro, M. R.; Mermut, A. R. Submicroscopy of clay microaggregates in an oxisol from Pernambuco, Brazil. *Soil Sci. Soc. Am. J.* 1989, *53* (6), 1895–1901.

Sarkar, B.; Naidu, R.; Rahman, M. M.; Megharaj, M.; Xi, Y. Organoclays reduce arsenic bioavailability and bioaccessibility in contaminated soils. *J. Soils Sedim.* 2012b, *12* (5), 704–712.

Sarkar, B.; Rusmin, R.; Ugochukwu, U. C.; Mukhopadhyay, R.; Manjaiah, K. M. Modified clay minerals for environmental applications. In: *Modified Clay and Zeolite Nanocomposite Materials Environmental and Pharmaceutical Applications Micro and Nano Technologies*, 1st edn., Mercurio, M.; Sarkar, B.; Langella, A. Eds. Elsevier: Amsterdam, The Netherlands, 2019, p. 362.

Sarkar, B.; Singh, M.; Mandal, S.; Churchman, G. J.; Nanthi S. *Clay minerals—organic matter interactions in relation to carbon stabilization in soils.* In: *The Future of Soil Carbon Its Conservation and Formation*; Garcia, C., Nannipieri, P., Hernandez, T., Eds. Academic Press: Cambridge, MA, 2018, pp. 71–86.

Sarkar, B.; Xi, Y.; Megharaj, M.; Krishnamurti, G. S. R.; Bowman, M.; Rose, H., Naidu, R. Bioreactive organoclay: A new technology for environmental remediation. *Crit. Rev. Environ. Sci. Technol.* 2012a, *42* (5), 435–488.

Sarkar, B.; Xi, Y.; Megharaj, M.; Krishnamurti, G. S. R.; Rajarathnam, D.; Naidu, R. Remediation of hexavalent chromium through adsorption by bentonite based Arquad® 2HT-75 organoclays. *J. Hazard. Mater.* 2010, *183* (13), 87–97.

Sarkar, D. J.; Singh, A.; Mandal, P.; Kumar, A.; Parmar, B. S. Synthesis and characterization of poly (CMC-g-Cl-PAam/Zeolite) superabsorbent composites for controlled delivery of zinc micronutrient: Swelling and release behavior. *Polym. Plast. Technol. Eng.* 2015, *54* (4), 357–367.

Sarkar, S.; Datta, S. C.; Biswas, D. R. Synthesis and characterization of nanoclay–polymer composites from soil clay with respect to their water-holding capacities and nutrient-release behaviour. *J. Appl. Polym. Sci.* 2014, *131*, 39951.

Saurabh, K.; Manjaiah, K. M.; Datta, S. C.; Thekkumpurath, A. S.; Kumar, R. Nanoclay polymer composites loaded with urea and nitrification inhibitors for controlling nitrification in soil. *Arch. Agron. Soil Sci.* 2019, *65* (4), 478–491.

Schmidt, M. W. I.; Torn, M. S.; Abiven, S.; Dittmar, T.; Guggenberger, G.; Janssens, I. A.; Kleber, M.; Kögel-Knabner, I.; Lehmann, J.; Manning, D. A. C.; Nannipieri, P.; Rasse,

D. P.; Weiner, S.; Trumbore, S. E. Persistence of soil organic matter as an ecosystem property. *Nature* 2011, *478*, 49–56.

Schneider, M. P. W.; Scheel, T.; Mikutta, R.; Hees, P. V.; Kaiser, K.; Kalbitz, K. Sorptive stabilization of organic matter by amorphous Al hydroxide. *Geochim. Cosmochim. Acta* 2010, *74* (5), 1606–1619.

Schoonheydt, R. A.; Pinnavaia, T.; Lagaly, G.; Gangas, N. Pillared clays and pillared layered solids. *Pure Appl. Chem.* 1999, *71*, 2367–2371.

Schwesig, D.; Kalbitz, K.; Matzner, E. Effects of aluminium on the mineralization of dissolved organic carbon derived from forest floors. *Eur. J. Soil Sci.* 2003, *54* (2), 311–322.

Seliman, A. F.; Lasheen, Y. F.; Youssief, M. A. E., Abo-Aly, M. M.; Shehata, F. A. Removal of some radionuclides from contaminated solution using natural clay: Bentonite. *J. Radioanal. Nucl. Chem.* 2014, *300*, 969–979.

Setia, R.; Rengasamy, P.; Marschner, P. Effect of mono- and divalent cations on sorption of water-extractable organic carbon and microbial activity. *Biol. Fertil. Soils* 2014, *50*, 727–734.

Shabtai, I. A.; Mishael, Y. G. Polycyclodextrin–clay composites: Regenerable dual-site sorbents for bisphenol a removal from treated wastewater. *ACS Appl. Mater. Interfaces* 2018, *10* (32), 27088–27097.

Shaviv, A. Advances in controlled release of fertilizers. *Adv. Agron.* 2000, *71*, 1–49. https://doi.org/10.1016/S0065-2113(01)71011-5

Shi, W.; Shao, H.; Li, H.; Shao, M.; Du, S. Progress in the remediation of hazardous heavy metal-polluted soils by natural zeolite. *J. Hazard. Mater.* 2009, *170* (1), 1–6.

Shokri, E.; Yegani, R.; Pourabbas, B.; Ghofrani, B. Evaluation of modified montmorillonite with di-cationic surfactants as efficient and environmentally friendly adsorbents for arsenic removal from contaminated water. *Water Sci. Technol. Water Supply* 2018, *18* (2), 460–472.

Singh, B.; Sharma, D. K.; Gupta, A. A Study towards release dynamics of thiram fungicide from starch-alginate beads to control environmental and health hazards. *J. Hazard. Mater.* 2009, *161* (1), 208–216. https://doi.org/10.1016/j.jhazmat.2008.03.074

Singh, M.; Sarkar, B.; Biswas, B.; Bolan, N. S.; Churchman, G. J. Relationship between soil clay mineralogy and carbon protection capacity as influenced by temperature and moisture. *Soil Biol. Biochem.* 2017, *109*, 95–106.

Singh, M.; Sarkar, B.; Biswas, B.; Churchman, J.; Bolan, N. S. Adsorption desorption behavior of dissolved organic carbon by soil clay fractions of varying mineralogy. *Geoderma* 2016, *280*, 47–56.

Singh, M.; Sarkar, B.; Sarkar, S.; Churchman, J;, Bolan, N;, Mandal, S.; Menon, M.; Purakayastha, T. J.; Beerling, D. J. Stabilization of soil organic carbon as influenced by clay mineralogy. *Adv. Agron.* 2018, *148*, 33–48.

Six, J.; Conant, R. T.; Paul, E. A.; Paustian, K. Stabilization mechanisms of soil organic matter: Implications for C-saturation of soils. *Plant Soil* 2002, *241*, 155–176.

Smith, P.; Fang, C.; Dawson, J. J. C.; Moncrieff, J. B. Impact of global warming on soil organic carbon. *Adv. Agron.* 2008, *97*, 1–43.

Soliemanzadeh, A.; Fekri, M. The application of green tea extract to prepare bentonite-supported nanoscale zero-valent iron and its performance on removal of Cr (VI): Effect of relative parameters and soil experiments. *Microporous Mesoporous Mater.* 2017, *239*, 60–69.

Solihin; Zhang, Q.; Tongamp, W.; Saito, F. Mechanochemical synthesis of kaolin–KH_2PO_4 and kaolin–$NH_4H_2PO_4$ complexes for application as slow release fertilizer. *Powder Technol.* 2011, *12*, 354–358.

Souza, J. de L.; Chiaregato, C. G.; Faez, R. Green composite based on PHB and montmorillonite for KNO_3 and NPK delivery system. *J. Polym. Environ.* 2018, *26* (2), 670–679. https://doi.org/10.1007/s10924-017-0979-4

Souza, J. L.; de Campos, A.; França, D.; Faez, R. PHB and montmorillonite clay composites as KNO^3 and NPK support for a controlled release. *J. Polym. Environ.* 2019, *27* (9), 2089–2097. https://doi.org/10.1007/s10924-019-01498-9

Srinivasan, R. Advances in application of natural clay and its composites in removal of biological, organic, and inorganic contaminants from drinking water. *Adv. Mater. Sci. Eng.* 2011, *2011*, 1–17.

Stockmann, U.; Adams, M. A.; Crawford, J. W.; Field, D. J.; Henakaarchchi, N.; Jenkins, M.; Minasny, B.; McBratney, A. B.; Courcelles, Vd. Rd.; Singh, K.; Wheeler, I.; Ab- bott, L.; Angers, D. A.; Baldock, J.; Bird, M.; Brookes, P. C.; Chenu, C.; Jastrow, J. D.; Lal, R.; Lehmann, J.; O'Donnell, A. G.; Parton, W. J.; Whitehead, D.; Zimmermann, M. The knowns, known unknowns and unknowns of sequestration of soil organic carbon. *Agric. Ecosyst. Environ.* 2013, *164*, 80–99.

Tahir, S., Marschner, P. Clay addition to sandy soil reduces nutrient leaching—effect of clay concentration and ped size Shermeen and Petra. *Commun. Soil Sci. Plant Anal.* 2017, *48* (15), 1813–1821.

Tahir, S., Marschner, P. Clay amendment to sandy soil—Effect of clay concentration and ped size on nutrient dynamics after residue addition. *J. Soils Sediments* 2016, *16*, 2072–2080.

Takei, T.; Yoshida, M.; Hatate, Y.; Shiomori, K.; Kiyoyama, S. Preparation of polylactide/poly (ε-caprolactone) microspheres enclosing acetamiprid and evaluation of release behaviour. *Polymer Bull.* 2008, *61*, 391–397.

Tan, D.; Yuan, P.; Annabi-Bergaya, F.; Liu, D.; He, H. Methoxy-modified kaolinite as a novel carrier for high-capacity loading and controlled-release of the herbicide Amitrole. *Sci. Rep.* 2015, *5* (1), 1–6. https://doi.org/10.1038/srep08870

Tangkoonboribun, R.; Rauysoongnern, S.; Rambo, P. V.; Tusman, B. Effect of organic and clay material amendment on physical properties of degraded sandy soil for sugarcane production. *Sugar Tech.* 2006, *8*, 44–48.

Tiller, K. G., Soil contamination issues: Past, present and future, a personal perspective. In: *Contaminants and the Soil Environment in the Australasia-Pacific Region.* Naidu, R.; Kookana, R. S.; Oliver, D. P.; Rogers, S.; McLaughlin, M. J. Eds. Kluwer: Dordrecht, 1996, p. 718.

Tomich, T. P.; Chomitz, K.; Francisco, H.; Izac, A. M. N.; Murdiyarso, D.; Ratner, B. D., Thomas, D. E.; Noordwijk, M. Policy analysis and environmental problems at different scales: Asking the right questions. *Agric. Ecosyst. Environ.* 2004, *104*, 5–18.

Toor, M.; Jin, B.; Dai, S.; Vimonses, V. Activating natural bentonite as a cost effective adsorbent for removal of Congo-red in wastewater. *J. Ind. Eng. Chem.* 2015, *21*, 653–661.

Uddin, M. K. A review on the adsorption of heavy metals by clay minerals, with special focus on the past decade. *Chem. Eng. J.* 2017, 308, 438–462. https://doi.org/10.1016/j.cej.2016.09.029

Ugochukwu, U. C.; Jones, M. D.; Head, I. M.; Manning, D. A. C.; Fialips, C. I. Biodegradation and adsorption of crude oil hydrocarbons supported on "homoionic" montmorillonite clay minerals, *Appl. Clay Sci.,* 2014a, *87*, 81–86.

Ugochukwu, U. C.; Jones, M. D.; Head, I. M.; Manning, A. C.; Fialips, C. I. Effect of acid activated clay minerals on biodegradation of crude oil hydrocarbons. *Int. Biodegrad. Biodeterior.* 2014b, *88*, 185–191.

Ugochukwu, U. C.; Manning, A. C.; Fialips, C. I. Effect of interlayer cations of montmorillonite on the biodegradation and adsorption of crude oil polycyclic aromatic compounds. *J. Environ. Manage.* 2014c, *142*, 30–35.

Undabeytia, T.; Nir, S.; Rubin, B. Organo-clay formulations of the hydrophobic herbicide norflurazon yield reduced leaching. *J. Agric. Food Chem.* 2000, *48* (10), 4767–4773. https://doi.org/10.1021/jf9907945

Undabeytia, T.; Nir, S.; Rytwo, G.; Morillo, E.; Maqueda, C. Modeling adsorption desorption processes of Cd on montmorillonite. *Clays Clay Miner.* 1998, *46*, 423–428.

Undabeytia, T.; Nir, S.; Rytwo, G.; Serban, C.; Morillo, E.; Maqueda, C. Modeling adsorption-desorption processes of Cu on edge and planar sites of montmorillonite. *Environ. Sci. Technol.* 2002, *36*, 2677–2683.

Vanaamudan, A.; Sudhakar, P. P. Equilibrium, kinetics and thermodynamic study on adsorption of reactive blue-21 and reactive red-141 by chitosan-organically modified nanoclay (Cloisite 30B) nano-bio composite. *J. Taiwan Inst. Chem. Eng.* 2015, *55*, 145–151.

Vane, C. H.; Kim, A. W.; Beriro, D. J.; Cave, M. R.; Knights, K.; Moss-Hayes, V.; Nathanail, P. C. Polycyclic aromatic hydrocarbons (PAH) and polychlorinated biphenyls (PCB) in urban soils of Greater London, UK. *Appl. Geochem.* 2014, *51*, 303–314.

Verma, M. K.; Pandey, P.; De, N. Characterization of water retention and release capacity of innovative nano clay polymer composite superabsorbent. *J. Pharmacogn. Phytochem.* 2017, *6*, 42–48.

Walker, G. W.; Kookana, R. S.; Smith, N. E.; Kah, M.; Doolette, C. L.; Reeves, P. T.; Lovell, W.; Anderson, D. J.; Turney, T. W.; Navarro, D. A. Ecological risk assessment of nano-enabled pesticides: a perspective on problem formulation. *J. Agric. Food Chem.* 2018, *66* (26), 6480–6486. https://doi.org/10.1021/acs.jafc.7b02373.

Wang, G.; Wang, S.; Sun, W.; Sun, Z.; Zheng, S. Synthesis of a novel illite-carbon nanocomposite adsorbent for removal of Cr (VI) from wastewater. *J. Environ. Sci.* 2017b, *57*, 62–71.

Wang, L.; Yu, G.; Li, J.; Feng, Y.; Peng, Y.; Zhao, X.; Tang, Y.; Zhang, Q. Stretchable hydrophobic modified alginate double-network nanocomposite hydrogels for sustained release of water-insoluble pesticides. *J. Clean. Prod.* 2019a, *226*, 122–132. https://doi.org/10.1016/j.jclepro.2019.03.341

Wang, P.; Wang, J.; Hui, Z.; Dong, Y.; Zhang, Y. The role of iron oxides in the preservation of soil organic matter under long-term fertilization. *J. Soils Sediments* 2019b, *19*, 588–598.

Wang, S.; Jia, Z.; Zhou, X.; Zhou, D.; Chen, M.; Xie, D.; Luo, Y.; Jia, D. Preparation of a biodegradable poly(vinyl alcohol)-starch composite film and its application in pesticide controlled release. *J. Appl. Polym. Sci.* 2017a, *134* (28), 45051. https://doi.org/10.1002/app.45051

Watte, L.; Koekkoek, E. J. W.; Buurman, P.; Van Der Plicht, J.; Wattel, E.; Van Breemen, N. Mean residence time of soil organic matter associated with kaolinite and smectite. *Eur. J. Soil Sci.* 2003, *54*, 1–10.

Wattel-Koekkoek, E J. W.; Buurman, P. Mean residence time of kaolinite and smectite-bound organic matter in mozambiquan soils. *Soil Sci. Soc. Am. J.* 2004. *68*, 154–161.

Wen, Y.; Liu, W.; Deng, W.; He, X.; Yu, G. Impact of agricultural fertilization practices on organo-mineral associations in four long-term field experiments: Implications for soil C sequestration. *Sci. Total Environ.* 2019, *651*, 591–600.

Weng, X.; Cai, W.; Lin, S.; Chen, Z. Degradation mechanism of amoxicillin using clay supported nanoscale zerovalent iron. *Appl. Clay Sci.* 2017, *147*, 137–142.

Wilpiszewska, K.; Spychaj, T.; Paździoch, W. Carboxymethyl starch/montmorillonite composite microparticles: Properties and controlled release of Isoproturon. *Carbohydr. Polym.* 2016, *136*, 101–106. https://doi.org/10.1016/j.carbpol.2015.09.021.

Wiseman, C. L. S.; Puttmann, W. Interactions between mineral phases in the preservation of soil organic matter. *Geoderma* 2006, *134*, 109–118.

Wu, J.; Wei, Y.; Lin, J.; Lin, S. Study on starch-graft-acrylamide/mineral powder superabsorbent composite. *Polymer* 2003, *44* (21), 6513–6520. https://doi.org/10.1016/S0032-3861 (03)00728-6

Xia, Y.; Rubino, M.; Auras, R. Interaction of nanoclay-reinforced packaging nanocomposites with food simulants and compost environments. *Adv. Food Nutr. Res.* 2019, *88*, 275–298.

Xiang, Y.; Ru, X.; Shi, J.; Song, J.; Zhao, H.; Liu, Y.; Guo, D.; Lu, X. Preparation and properties of a novel semi-IPN slow-release fertilizer with the function of water retention. *J. Agric. Food Chem.* 2017, *65* (50), 10851–10858. https://doi.org/10.1021/acs.jafc.7b03827

Xu, R.; Mao, J.; Peng, N.; Luo, X.; Chang, C. Chitin/clay microspheres with hierarchical architecture for highly efficient removal of organic dyes. *Carbohydr. Polym.* 2018, *188*, 143–150.

Xu, X.; Shi, Z.; Li, D.; Rey, A.; Ruan, H.; Craine, J. M.; Liang, J.; Zhou, J.; Luo, Y. Soil properties control decomposition of soil organic carbon: Results from data assimilation analysis. *Geoderma* 2016, *262*, 235–242.

Yan, H.; Chen, X.; Feng, Y.; Xiang, F.; Li, J.; Shi, Z.; Wang, X.; Lin, Q. Modification of montmorillonite by ball-milling method for immobilization and delivery of acetamiprid based on alginate/exfoliated montmorillonite nanocomposite. *Polym. Bull.* 2016, *73* (4), 1185–1206. https://doi.org/10.1007/s00289-015-1542-x

Yang, J.; Yu, K.; Liu, C. Chromium immobilization in soil using quaternary ammonium cations modified montmorillonite: Characterization and mechanism. *J. Hazard. Mater.* 2017b, *321*, 73–80.

Yang, Y.; Liu, B.; Yu, L.; Zhou, Z.; Ni, X.; Tao, L.; Wu, Y. Nitrogen loss and rice profits with matrix-based slow-release urea. *Nutr. Cycl. Agroecosystems* 2018a, *110* (2), 213–225. https://doi.org/10.1007/s10705-017-9892-4

Yang, Y.; Ni, X.; Zhou, Z.; Yu, L.; Liu, B.; Yang, Y.; Wu, Y. Performance of matrix-based slow-release urea in reducing nitrogen loss and improving maize yields and profits. *Field Crops Res.* 2017a, *212*, 73–81. https://doi.org/10.1016/j.fcr.2017.07.005

Yang, Y.; Yu, L.; Ni, X.; Yang, Y.; Liu, B.; Wang, Q.; Tao, L.; Wu, Y. Reducing nitrogen loss and increasing wheat profits with low-cost, matrix-based, slow-release urea. *Agron. J.* 2018b, *110* (1), 380–388. https://doi.org/10.2134/agronj2017.06.0351

Ye, R.; Parajuli, B.; Sigua, G. Subsurface clay soil application improved aggregate stability, nitrogen availability, and organic carbon preservation in degraded ultisols with cover crop mixtures. *Soil Sci. Soc. Am. J.* 2019, *83*, 597–604.

Yuan, G. D.; Theng, B. K. G., Churchman, G. J.; Gates W. P. Clays and clay minerals for pollution control. *Dev. Clay Sci.* 2013, *5*, 587–644.

Yuan, G. D.; Wada, S. I. Allophane and imogolite nanoparticles in soil and their environmental applications. In: *Nature's Nanostructures,* Barnard, A. S.; Guo, H. B. Eds. CRC Press, Taylor & Francis group, Pan Stanford Publishing Pte Ltd: Singapore, 2012, p. 545.

Yuan, W.; Solihin; Zhang, Q.; Kano, J.; Saito, F. Mechanochemical formation of KSi-Ca-O compound as a slow-release fertilizer. *Powder Technol.* 2014, *260*, 22–26.

Zaghouane-Boudiaf, H.; Boutahala, M. Preparation and characterization of organo-montmorillonites. application in adsorption of the 2,4,5-trichlorophenol from aqueous solution. *Adv. Powder Technol.* 2011, *22* (6), 735–740. https://doi.org/10.1016/j.apt.2010.10.014

Zhang, M.; Pu, J. Mineral materials as feasible amendments to stabilize heavy metals in polluted urban soils. *J. Environ. Sci.* 2011, *23* (4) 607–615.

Zhao, J.; Chen, S.; Hu, R.; Li, Y. Aggregate stability and size distribution of red soils under different land uses integrally regulated by soil organic matter, and iron and aluminum oxides. *Soil Tillage Res.* 2017, *167*, 73–79.

Zhong, Z.; Chen, Z.; Xu, Y., Ren, C., Yang, G.; Han, X.; Ren, G.; Feng, Y. relationship between soil organic carbon stocks and clay content under different climatic conditions in central China. *Forests* 2018, *9*, 598.

Zhou, T.; Wang, Y.; Huang, S.; Zhao, Y. Synthesis composite hydrogels from inorganic–organic hybrids based on leftover rice for environment-friendly controlled-release urea fertilizers. *Sci. Total Environ.* 2018, *615*, 422–430. https://doi.org/10.1016/j.scitotenv.2017.09.084

CHAPTER 2

Clay–Enzyme Interactions and Their Implications

RANJAN PAUL* and SONALIKA SAHOO

Division of Soil Resource Studies, ICAR National Bureau of Soil Survey and Land Use Planning, Amravati Road, Nagpur, Maharashtra 440033, India

Corresponding author. E-mail: ranjan.reliance@gmail.com

ABSTRACT

Soil enzymes are responsible for the maintenance of physicochemical properties, fertility, nutrient cycling, and overall soil health. The persistence and stability of the enzyme in soils are generally governed by their association with clays and other colloids. Enzyme molecules interact with mineral surfaces through a variety of physical and chemical bonds where enzymes's functional groups play a crucial role. The adsorption of enzymes on surfaces of clay mineral results in significant change in enzyme activity and kinetics, since catalysis occurs in the mineral solution interface. The specific activity of enzymes such as glucose oxidase, arylsulfatase, urease, acid phosphatase, invertase, amylase, tyrosinase, glucosidase, and catalase is inhibited due to adsorption to clays. There are exceptions to this general trend, and adsorption of enzymes on clays leads to an increase in activity of acid and alkaline phosphatase, glucosidase, xyloxidase, etc. Apart from mineralization of organic substances, clay–enzyme interactions are of great importance in forming micro- and macro-aggregates, which improve the soil structure. The clay–enzyme interaction has a significant role in the sequestration of organic carbon in soils and maintaining soil health. If the mechanism of enzyme adsorption to soil clays is understood well, this may help in developing strategies to apply clay–enzyme complex as biofertilizer in the near future for supplying nutrients from organic sources as per the need of plants.

2.1 INTRODUCTION

Enzymes produced from plant, animal, and microbial sources are considered catalysts of chemical transformation in soil and sediment environments. Enzymes hydrolyze large and complex polymeric organic compounds into small monomers, which afterward passed through cell walls and helped in the growth and respiration of microbial cells. Therefore, enzyme activity within the soil system may be the rate-limiting step in governing the degradation and mineralization of soil organic matter (SOM) for release of carbon and other nutrients necessary for higher plants and animals for survival. Enzyme activity in soil was demonstrated 10–12 decades ago, and about 50–60 soil enzymes representing all major groups have been identified in various soils (Paul and Clark, 1996). The persistence and stability of enzymes in soils are generally governed by their association with clays, since the highest enzyme activity was observed in clay size fractions (McLaren and Packer, 1970). In this regard, Mclaren (1975) inferred soil as "system of humus and clay immobilized enzymes." Enzymes may be categorized according to their location as "intracellular" and "extracellular," which are leaked or lysed from dead cells or actively secreted by living bacteria and fungi cells (Burns, 1982; Skujins, 1976). The adsorption of enzymes by clay mineral surfaces results in a significant change in properties such as activity, optimum pH and thermal range of activity, stability, and kinetics, since catalysis occurs in the mineral solution interface. Therefore, in the natural soil environment, an extracellular enzyme cannot show its "ideal" activity as showing while present in free form in a buffered solution (Boyd and Mortland, 1990). The benefits of adsorbed enzymes lie on the fact that an enzyme is easily separated from the medium after reacting with a substrate in the solution and then can be stabilized and reused sometimes. Thus, longevity of the adsorbed enzyme increased compared to the free enzyme. Pioneer work on enzyme–clay interaction started with Durand (1963), who reported lowering down of activity of uricase (an heptic enzyme) after clay adsorption, which resulted in a lower rate of production of allantoin from uric acid. Boyd and Mortland (1990) and Naidja et al. (2000) also did classical research work on mechanisms behind the clay–enzyme interaction. This chapter highlights the clay–enzyme interaction, which takes place or believed to take place in soil environment, and their effects on enzyme activity and kinetics with a focus on its implications.

2.2 SOIL ENZYMES

Soil enzymes are vital constituents of soils, responsible for various nutrient cycling in the soil, maintenance of soil fertility, soil physicochemical properties, and overall soil quality. Soil enzymes are essentially secreted by plant roots, dead and living microbes, and soil macrofauna. Soil enzymes can be categorized into intracellular soil enzymes (produce and function inside living cells) and extracellular soil enzymes (secreted by living cells but function outside the cells). As stabilized soil enzymes constitute about half of active soil enzymes pool (http://soilquality.org/home.html), enzymes that are stored in the soil and function extracellularly are considered as soil enzymes (Shi, 2010). Soil enzymes are generally associated with the soil constituents and accumulate in soil aqueous phase or form complexes with clay, humus, or organic matter and clay–humus complexes (Ladd, 1978; Burns, 1982).

Soil enzymes can be catagorized based on substrate dependency into two groups: constitutive and inducible. Constitutive enzymes are not substrate dependent and maintain a nearly constant level inside cells at most of the times (e.g., pyrophosphatase). Inducible enzymes are either absent or present in minute quantity initially, but when certain substrate is present, its concentration increases (e.g., Amidase). Soil enzymes are classified into different classes based on the biochemical reactions as follows:

1. *Oxidoreductases*: enzymes involved in oxidation–reduction reaction (catalase, dehydrogenase, and peroxidase).
2. *Transferases*: enzymes participate in the transfer of a group of atoms from donor to an acceptor molecule (rhodonese and aminotransferases).
3. *Hydrolases*: enzymes involved in the hydrolytic splitting of bonds (urease, phosphatase, and cellulose).
4. *Lysates*: enzymes involved in the splitting of bonds other than hydrolysis or oxidation.
5. *Isomerases*: enzymes involved in isomerization reaction.
6. *Ligases*: enzymes involved in the formation of bonds by the splitting ATP (acetyl-CoA carboxylase).

2.3   IMPORTANT SOIL ENZYMES AND THEIR IMPORTANCE IN SOIL QUALITY AND HEALTH

2.3.1   AMYLASE

Amylase is a group of enzymes responsible for starch hydrolysis (Ross, 1976). Amylase is commonly found in both soils and plants with various properties and activities (Ladd and Butler, 1969, 1972). Soil amylase breaks down polysaccharides like starch to glucose (Singaram and Kamalakumari, 2000). Amylases are of two types: α-amylase and β-amylase (Pazur, 1965; Thoma et al., 1971). The α-amylase is synthesized by soil macro- and microorganisms and also by plants. It breakdown the starch molecules by hydrolyzing the α-(1-4) glycosidic bonds (Pazur, 1965). β-amylase is mostly secreted by plants, and it removes glucose disaccharide from "the nonreducing end of the starch" (Thoma et al., 1971). β-amylase is usually extracellular and inducible (Alexander, 1977). Soil amylase activity in soil is affected by substrate quality, type of vegetation, management practices, fungal and bacterial populations, temperature, moisture soil types, and soil pH (Ross and Roberts, 1970; Pancholy and Rice, 1973; Ross, 1975; Sinsabaugh and Linkins, 1987; Joshi et al., 1993). Amylases contribute little toward soil carbon cycling, as soil organic inputs contain a small amount of starch.

2.3.2   ARYLSULFATASE

Arylsulfate catalyzes the breakdown of aromatic sulfate esters ($RO-SO_3$) to release SO_4^{2-} (Elsgaard et al., 2002). Arylsulfatase is found both as intra- and extracellular enzyme. Klose and Tabatabai (1999) reported that extracellular and live cells of microorganisms contribute 45% and 55% of the total determined pool of arylsulfatase, respectively. Soil organic compound contains a certain amount of sulfur as aromatic sulfate esters. Arylsulfatase is responsible for the release of plant-available sulfate by catalyzing the oxidation of soluble organic matter intracellularly and hydrolyzing the aromatic sulfate esters extracellularly (Dodgson et al., 1982). Therefore, the activity of arylsulfatase reflects the trans-formations of organic sulfur in the soil ecosystem (Lipińska et al., 2014).

2.3.3   β-GLUCOSIDASE

β-Glucosidase breaks glycosidic bonds in the glucosides produced by the decomposition of plant residues by cellulases and amylases (Martinez and Tabatabai, 1997). The glucose produced by β-glucosidase action on glucosides is an essential energy source for soil microbes (Esen, 1993). In soil environment, it is an ubiquitous and major enzyme (Tabatabai, 1994). β-Glucosidase is very sensitive to change in soil pH, soil management practices, and soil heavy metal content (Deng and Tabatabai, 1995; Bergstrom et al., 1998; Leirós et al., 1999; Acosta-Martínez and Tabatabai, 2000; Ndiaye et al., 2000; Madejón et al., 2001). Therefore, many researchers had suggested to use this enzyme as the indicator of change in soil biological activity caused by soil acidification, past biological activity, and to measure the SOM stabilization capacity (Ndiaye et al., 2000; Madejón et al., 2001).

2.3.4 CELLULASE

Cellulases catalyze the decomposition of cellulose into monosaccharides such as β-glucose or shorter polysaccharides or oligosaccharides (Deng and Tabatabai, 1994). Complete degradation of cellulose requires cellulases, which basically consist of three types of enzymes: endo-glucosidase or endo-1,4-β-glucanase, exo-glucosidase or exo-1,4-glucanase, and β-glucosidase. Endo-glucosidase breaks the cellulose chains arbitrarily and exo-1,4- glucanase breaks the cellulose by removing oligosaccharides from the nonreducing end of carbohydrate chain. In contrast, the breakdown of cellulose to glucose is done by β-D-glucosidase by hydrolyzing water-soluble cellodextrins and cellobiose (Alef and Nannipieri, 1995). Plant debris is the major contributor of cellulases, while soil fungi and bacteria contribute a little amount toward the total cellulases in soil (Richmond, 1991). As cellulase decomposes the abundantly found organic compound in the soil, that is, cellulose, it plays an important role in soil carbon (C) cycling (Eriksson et al., 1990). Production of cellulase as extracellular enzymes in soil has been found to be affected by various soil properties such as soil moisture content, soil temperature, and soil pH (Rubidge, 1977; Srinivasulu and Rangaswamy, 2006), oxygen content, and the trace elements from some pesticides (Petker and Rai, 1992; Arinze and Yubedee, 2000). The quality, quantity, composition, and location of the substrate in the soil profile also affect the cellulose production by microbes in soil (Gomah, 1980; Linkins et al., 1984; Hope and Burns, 1987) and water. Soil fungal and bacterial population can have a positive effect on soil cellulase activity (Joshi et al., 1993).

2.3.5 DEHYDROGENASE

Dehydrogenase is a respiratory enzyme commonly found in all the soil microorganisms. It involves the transfer of protons and electrons between substrates and acceptors during the respiration pathways of soil microbes. Dehydrogenase activity is used as a comparative index of biological activities in soil (Burns, 1978; Garcia and Herna'ndez, 1997; Gu et al., 2009) and may act as a soil quality indicator. This enzyme does not found extracellularly in the soil as it produced and contributed by only living cells (Nannipieri et al., 1990; Yuan and Yue, 2012). Soil dehydrogenase activity is affected by the soil type, soil redox state, and soil temperature (Glinski and Stepniewski, 1985; Kandeler, 1996; Brzezinska et al., 1998). Dehydrogenase enzyme is often used to study the effect of management practices, type of pesticides and their concentration, and the presence of trace elements and other pollutants on soil quality (Pitchel and Hayes, 1990; Wilke, 1991; Frank and Malkomes, 1993; McCarthy et al., 1994).

2.3.6 PHOSPHATASE

Phosphatase enzymes hydrolyze phosphate ester bond in organic matter and consequently release phosphate. Phosphatases are a group of enzymes consisting of "phosphomonoesterases, phosphodiesterases, triphosphoric monoester hydrolases, and enzymes acting on phosphoryl-containing anhydrides and P–N bonds" (Hinsinger et al., 2018). Phosphomonoesterases consist of phosphoprotein phosphatases, acid and alkaline phosphomonoesterase, phytases, and nucleotidases. Acid phosphatase and alkaline phosphatase are most extensively studied soil phosphatase and are mostly dominant in acid and alkaline soils, respectively (Tabatabai, 1994). Alkaline phosphatase is produced mainly from soil microbes and animals, while acid phosphatase is secreted by soil microbes, plants, and animals (Tisserant et al., 1993). These enzymes increase plant-available P by releasing P from organic sources (Speir and Ross, 1978). Crop species, variety, and crop management practices significantly affect acid phosphatase secretion by plant roots (Wright and Reddy, 2001; Ndakidemi, 2006). Soil properties such as soil pH, moisture content, and temperature along with quality and quantity of leachate input by plants and soil microbe interaction affect the soil phosphatases activity (Speir and Ross, 1978). Phosphatase activity is considered as a soil fertility indicator (Dick and Tabatai, 1992; Dick et al., 2000).

2.3.7 SOIL PROTEASE

Proteases are extensively present in soils with a multitude of actions (Hayano, 1986). Proteases consist of a group of enzymes, which hydrolyze peptide bonds in proteins to release short peptide-like polypeptides and oligopeptides, which further degraded to amino acids (Handa et al., 2000). These enzymes are secreted by plants, soil microorganisms, and soil animals and found to be associated with the soil colloidal fractions (organic and inorganic colloids) (Nannipieri et al., 1996). So, generally, protease is extracellular (Burns, 1982). Humocarbohydrate complex in soil partially contribute toward protease activities in soil (Batistic et al., 1980). Soil microbes use amino acids (released by protein hydrolysis) as nitrogen sources and mineralized these to ammonia. The released ammonia adds toward plant-available nitrogen (N). Therefore, nitrogen mineralization and plant-available N in soils are dependent on soil protease activity (Moreno et al., 2003; Stevenson, 1986).

2.3.8 UREASE

Urease enzymes "hydrolyze urea into ammonium and carbon dioxide" (Fazekasova, 2012), and during the urea hydrolysis process, soil pH increases. Urease enzyme is ubiquitous and is secreted by most of the soil microbes and plant roots (Burns, 1986). It acts both as an extra- and intracellular enzyme (Mobley and Hausinger, 1989). It increases the ammoniacal nitrogen concentration after urea fertilization, and soil nitrogen dynamics is prominently affected by the urease enzyme (Byrnes and Amberger, 1989). Thus, urease activity in soil is used as an index of N mineralization (Nannipieri et al., 2012). Cropping history, SOM content, soil depth, soil temperature, the presence of heavy metals, and agricultural management practices such as application soil amendments influence urease activity in the soils (Tabatabai, 1977; Bremner and Mulvaney, 1978; Yang et al., 2006). Various reports suggested that extracellular urease stabilized on soil colloids contributes more toward total urease activity in soil. Extracellular urease associated with soil organomineral complexes showed better stability as compared to urease present in the soil solution (Burns, 1986). Nannipieri et al. (1978) extracted humus–urease complexes from soil and found them resistant against proteolytic attack and extreme temperature, while intracellular urease (extracted from microorganisms or plants) showed rapid degradation by soil proteolytic enzymes (Zantua and Bremner, 1977).

2.3.9 INVERTASE

Invertase comes under the hydrolase enzyme group, and it converts sucrose to glucose and fructose. Invertase is predominantly present in plants, soil microorganisms, and animals (Alef and Nannipieri, 1995). It decomposes carbohydrate polymers to release simpler sugars and facilitate carbon transformations. It also increases soluble nutrients content in the soil by mediating in complex SOM decomposition. Therefore, soil invertase enzyme activity can also be used as a soil C cycling index (Sardans et al., 2008).

2.3.10 LACCASE

Laccases are multicopper phenoloxidases. These enzymes are found in fungi, bacteria, insects, and some higher plants. Laccases have low substrate specificity. They can oxidize various organopollutants such as polychlorinated biphenyls, polycyclic aromatic hydrocarbons, pesticides and synthetic dyes, and inorganic compounds such as iron–cyanide complexes and iodine (Sinsabaugh, 2010; Eichlerová et al., 2012). Laccases, along with other soil extracellular enzymes, decompose lignin and other polyphenols added to soils. Therefore, laccases also play an essential role in the soil C cycling (Eichlerová et al., 2012).

2.3.11 CHITINASE

Chitinase enzymes catalyze the hydrolysis of chitin. For complete degradation of chitin, a compound consists of N-acetyl-β-glucosaminidase, chitobiase, and chitinase are required. N-acetyl-β- glucosaminidase is essential in the mineralization of N from chitin; therefore, it is used as an indicator of soil chitinase activity (Olander and Vitousek, 2000). The major fraction of soil chitinases is mostly produced by fungi, and a little fraction is contributed by bacteria (Gooday, 1994).

2.4 MECHANISMS OF ENZYME AND CLAY INTERACTION

Variable charges of clay mineral surface become positive or negative depending on the pH of the soil solution (e.g., kaolinite and noncrystalline amorphous minerals that have a pH-dependent charge) (Brady and

Weil, 2008). Enzymes interact with charged clay mineral surfaces through different functional groups (C=O, O–H, NH_2, COOH, etc.) present within the enzyme structure (Theng, 1982). These functional groups are hydrophilic, hydrophobic, and negatively, positively, or neutrally charged and change their shape and configuration in response to change in solution pH, ionic strength, etc. (Skujins, 1976; Stotzky, 1986). Therefore, mechanisms of enzyme–clay interaction are different and complex. Enzyme properties, such as optimum pH for activation, stability, activity, and kinetics, are highly influenced by the adsorption process. These changes are due to influences of surface properties of clay and due to the fact that catalysis occurs on the mineral solution interface (Boyd and Mortland, 1990). Different types of binding mechanisms such as electrostatic interaction, van der Waals interaction, H bonding, ion–dipole interaction, and so on are important (Mortland, 1970) (see Table 2.1).

2.5 INFLUENCE OF CLAY–ENZYME INTERACTION ON ACTIVITY AND KINETIC PROPERTIES OF ENZYME

Along with activity, kinetic parameters are used to provide complete information regarding an enzyme's behavior, which is typically measured at a single saturating substrate concentration (German et al., 2011). Kinetic parameters are generally expressed by Michaelis–Menten parameters (V_{max} and K_m). The V_{max} value describes the maximum velocity of conversion of substrate to the product when all the enzyme active sites are saturated with the substrate. The K_m value represents the substrate concentration at which half of V_{max} is achieved. K_m is the measure of the affinity of the enzyme for the substrate. Higher K_m indicates lower affinity and vice versa. V_{max} and K_m are useful parameters as they are independent of the enzyme concentration used (Schnell and Maini, 2003). The activities of clay-adsorbed enzymes are lower than those of free enzymes in homogeneous solution (Mortland and Gieseking, 1952). A list of the changes in the specific activity of some soil enzymes due to adsorption to various specimen or soil-derived clay minerals are presented in Table 2.2. In some cases, the specific activity of free and adsorbed enzymes was calculated from kinetic parameter (V_{max}) using the Michaelis–Menten equation (Makboul and Ottow, 1979a, 1979b, 1979c; Burns, 1986; Ahn et al., 2007; Calabi-Floody et al., 2012) and then compared. The general observation is that, specific activity of enzymes such as glucose oxidase, arylsulfatase, urease, acid phosphatase, invertase, amylase, tyrosinase, glucosidase, and catalase are inhibited due to adsorption

TABLE 2.1   Bonding Mechanisms of Clay–Enzyme Complexes

Nature of Bonding Mechanism	Specific Conditions
Electrostatic attraction	Charged amiono acid molecule of enzyme is adsorbed electrostatically to the negative charge sites of clay. These reactions are characterized by high degree of reversibility.
Protonation of organic molecules at clay surface	Adsorption of enzyme molecules at the clay surface through protonation. The sources of the protons for such a reaction are exchangeable. H^+, water, and other cationic species present in the vicinity of clay srface.
Covalent bond	Covalent bond between two molecules takes place by mutual sharing of an electron between them by two adjacent atoms. One or both the electrons might be contributed by one atom. Covalent bonding usually takes place between atoms of same electronegativity or between atoms of not differing much in electronegativity.
Hydrogen bonding	The molecules that do not possess any net electrical charge are electrostatically attracted to each other or to charged molecules surfaces through a dipole. Hydrogen bonding is such kind of dipole interaction and is forms when an H atom is bonded to two or more other atoms.
Ion–dipole interaction	Dipoles formed in the enzyme molecule is attracted to ions and ion–dipole forces developed. Such kind of bonding mechanism takes place between nonionic but polar organic molecules of enzyme and ionic COOH or OH groups of silicate surface.
Water bridging	Polar organic molecules are linked to an exchangeable metal cation by bridging a water molecule.
van der Waals forces	Two completely neutral molecule are held together by "instantaneous dipole" due to imbalances in electron distribution. Such type of bonding forces are weak and short range ordered. Forces of this type of interaction decrease very sharply with increase in distance between the interacting species. This type of bonding is significant for higher molecular weight organic compounds.
Entropy effects	Adsorption of some organic molecules from enzyme solution on clay minerals is favored if there is a positive entropy change in the system. Positive entropy change is created by displacing numerous water molecules from clay surface by enzyme molecules and lowering of free energy due to conformational changes (Russel, 1973).

Source: Mortland (1970).

to clays. The decrease in activity varies from 50% to 100%. The inhibitory effect is more pronounced in case of swelling category 2:1 type layer silicates than nonswelling types (1:1) and oxide–hydroxides. There are of exceptions to this general trend, and it is observed that adsorption of enzymes on clays does not always lead to decreased activities. Makboul and Ottow (1979c)

TABLE 2.2 Changes in Activity of Enzyme due to Adsorption to Clay Minerals

Type of Clay Minerals	Mineral Name	Enzyme Used	Change in Activity (%)	Reference
2:1 type layer silicates	Ca-Montmorillonite	Glucose oxidase	−77	Ross and McNeilly (1972)
	Ca-Montmorillonite	Glucose oxidase	−57 to −96	Morgan and Corke (1976)
	Ca-Montmorillonite	Arylsulfatase	−52	Hughes and Simpson (1978)
	Montmorillonite	Urease	−52	Makboul and Ottow (1979a)
	Montmorillonite	Acid phosphatase	−57	Makboul and Ottow (1979b)
	Ca-Montmorillonite	Alkaline phosphatase	+8	Makboul and Ottow (1979c)
	Montmorillonite	Endopeptidases	−66	Haska (1981)
	Montmorillonite	Invertase	−100	Burns (1986)
	Montmorillonite	α Amylase	−96	Burns (1986)
	Montmorillonite	β Amylase	−100	Burns (1986)
	Na-Montmorillonite	Glucosidase	−35 to −100	Quiquampoix (1987a)
	Montmorillonite	Glucosidase	0 to −82	Quiquampoix (1987b)
	Bentonite	Tyrosinase	−84 to −100	Claus and Filip (1988)
	Bentonite	Laccase	−11 to −100	Claus and Filip (1988)
	$Al(OH)_x$-Montmorillonite	Invertase	−89 to −95	Gianfreda et al. (1991)
	Na-Montmorillonite	Invertase	−88 to −96	Gianfreda et al. (1991)
	$Al(OH)_x$-Montmorillonite	Urease	−49 to −67	Gianfreda et al. (1992)
	Na-Montmorillonite	Urease	−41	Gianfreda et al. (1992)
	Montmorillonite complex	Peroxidase	0	Gianfreda and Bollag (1994)
	Montmorillonite	Laccase	0	Gianfreda and Bollag (1994)
	Montmorillonite	Acid phosphatase	−68	Gianfreda and Bollag (1994)

TABLE 2.2 *(Continued)*

Type of Clay Minerals	Mineral Name	Enzyme Used	Change in Activity (%)	Reference
	Al(OH)$_x$-coated montmorillonite	Tyrosinase	−24 to −62	Naidja et al. (1997)
	Na-montmorillonite	Phosphatase	−80	Rao et al. (2000)
	Ca-Montmorillonite	Catalase	−81 to −99	Calamai et al. (2000)
	Al(OH)$_x$-Montmorillonite	Phosphatase	−42	Rao et al. (2000)
	Na-Montmorillonite	Peroxidase	−88 to −99	Lozzi et al (2001)
	Ca-Montmorillonite	Peroxidase	0 to −69	Lozzi et al (2001)
	Montmorillonite	Protease	−100	Tietjen and Wetzel (2003)
	Montmorillonite	Glucosidase	+50	Tietjen and Wetzel (2003)
	Montmorillonite	Alkaline phosphatase	−67	Tietjen and Wetzel (2003)
	Montmorillonite	Xyloxidase	+200	Tietjen and Wetzel (2003)
	Smectite clays	β-Glucosidase	0 to −4	Serefoglou et al. (2008)
	Montmorillonite nanoclay	Acid phosphatase	+4 to +48	Calabi-Floody et al. (2012)
	TiO$_2$–montmorillonite complex	Laccase	−20	Wang (2013)
	Nanoclays dominated by Smectite	Acid phosphatase	−60	Paul et al. (2016)
	Nanoclays dominated by Smectite	Alkaline phosphatase	+800	Paul et al. (2016)
	Nanoclays dominated by Mica	Acid phosphatase	−44	Paul et al. (2016)
	Nanoclays dominated by mica	Alkaline phosphatase	+200	Paul et al. (2016)
	Illite	Catalase	0	Ross and McNeilly (1972)
	Illite	Glucose oxidase	−21	Ross and McNeilly (1972)
	Illite	Urease	−44	Makboul and Ottow (1979a)

TABLE 2.2 *(Continued)*

Type of Clay Minerals	Mineral Name	Enzyme Used	Change in Activity (%)	Reference
	Illite	Acid phosphatase	−56	Makboul and Ottow (1979b)
	Illite	Alkaline phosphatase	−42	Makboul and Ottow (1979c)
	Illite	Invertase	−91	Burns (1986)
	Illite	α-Amylase	−83	Burns (1986)
	Illite	β-Amylase	−99	Burns (1986)
	Muscovite	Invertase	−66	Burns (1986)
	Muscovite	α-Amylase	−4	Burns (1986)
	Muscovite	β-Amylase	−99	Burns (1986)
1:1 type layer silicates	Kaolin	Catalase	+2	Ross and McNeilly (1972)
	Kaolinite	Chitinase	−95	Skujins et al. (1974)
	Kaolinite	Arylsulfatase	−18	Hughes and Simpson (1978)
	Kaolinite	Urease	−40	Makboul and Ottow (1979a)
	Kaolinite	Acid phosphatase	−70	Makboul and Ottow (1979b)
	Kaolinite	Alkaline phosphatase	−4	Makboul and Ottow (1979c)
	Kaolinite	Invertase	−55	Burns (1986)
	Kaolinite	α-Amylase	−99	Burns (1986)
	Kaolinite	β-Amylase	−100	Burns (1986)
	Kaolinite	Glucosidase	0 to −88	Quiquampoix (1987a)
	Kaolinite	Glucosidase	−13 to −100	Quiquampoix (1987a)
	Kaolinite	Tyrosinase	0 to −75	Claus and Filip (1988)
	Kaolinite	Laccase	−30 to −83	Claus and Filip (1988)
	Kaolinite	Acid phosphatase	−64	Gianfreda and Bollag (1994)
	Kaolinite	Laccase	−14	Gianfreda and Bollag (1994)

TABLE 2.2   *(Continued)*

Type of Clay Minerals	Mineral Name	Enzyme Used	Change in Activity (%)	Reference
	Kaolinite	Acid phosphatase	−43	Huang et al. (2005)
	Nanoclays dominated by Kaolinite	Acid phosphatase	−80	Paul et al. (2016)
	Nanoclays dominated by Kaolinite	Alkaline phosphatase	+100	Paul et al. (2016)
Oxides and hydroxides	Allophane	Glucose oxidase	−52	Ross and McNeilly (1972)
	Allophane	Invertase	−80	Burns (1986)
	Allophane	α-Amylase	−43	Burns (1986)
	Allophane	β-Amylase	−88	Burns (1986)
	Goethite	Glucosidase	0 to +40	Quiquampoix (1987a)
	Al(OH)$_x$	Invertase	−94 to −99	Gianfreda et al. (1991)
	Al-hydroxide	Acid phosphatase	−55	Rao et al. (2000)
	Goethite	Acid phosphatase	−32	Huang et al. (2005)
	Al-hydroxide	Tyrosinase	−11	Ahn et al. (2007)
	Allophanic nanoclays	Phytase	+70	Menezes-Blackburn (2011)
	Allophanic nanoclay	Acid phosphatase	+4 to +48	Calabi-Floody et al. (2012)

reported increased and slightly reduced activity of alkaline phosphatase in the presence of Ca-montmorillonite and kaolinite, respectively. Ross and McNeilly (1972) observed no change in activity of catalase when illite and kaolinite were used as support material. Similarly, Gianfreda and Bollag (1994) observed no change in the activity of laccase and peroxidase upon adsorption to montmorillonite. Quiquampoix (1987a) reported increased activity of glucosidase due to adsorption to goethite. Tietjen and Wetzel (2003) observed 1.5 and 2 times increase in the specific activity of glucosidase and xyloxidase, respectively, on montmorillonite support. Serefoglou et al. (2008) found no change in ß-glucosidase activity when immobilized to smectite clays. Menezes-Blackburn (2011) observed a 1.5 times increase in the activity of phytase when adsorbed to nanosize allophane minerals. Calabi-Floody et al. (2012) evaluated soil-derived minerals instead of any specimen mineral as support material for acid phosphatase immobilization. They reported 50% increase in enzyme activity when adsorbed to soil-derived nanosize montmorillonite and allophane. Recently, Paul et al. (2016) observed enhanced catalytic activity of nanoclay bound (smectite, mica, and kaolinite dominated) alkaline phosphatase relative to free enzyme.

Kinetic properties of free enzymes changed due to adsorption to clay surfaces and the Michaelis–Menten parameters can be applied to enzyme–substrate reactions for the adsorbed enzyme also as in case of free enzyme. The general observation is that, V_{max} values either decreased or unchanged for clay-adsorbed enzymes and the K_m values may be higher or smaller than free enzymes (Burns, 1986). Makboul and Ottow (1979a) reported a 50%–60% decrease in V_{max} of urease compared to free enzyme when adsorbed into montmorillonite, kaolinite, and illite. The same pattern of change was observed by Makboul and Ottow (1979b) when acid phosphatase was adsorbed to the montmorillonite, kaolinite, and illite. Presence of illite decreased V_{max} of alkaline phosphatase (Makboul and Ottow, 1979c). Gianfreda (1991) reported a decrease in V_{max} of invertase when adsorbed to montmorillonite and its Al complexes. Huang et al. (1995) observed reduction in V_{max} of acid phosphatase when adsorbed to montmorillonite and its various interlayered complexes. Recently, Paul et al. (2016) reported that highest V_{max} value was observed for free acid phosphatase (48.1 µg p-nitrophenol mg^{-1} min^{-1}) followed by complexes with separated nano-clays from Inceptisol (27.1), Vertisol (18.6), and Alfisol (10.7). Thus, when compared with the free form, adsorbed enzymes often show lower reaction velocities and altered substrate affinities. Although, Sundaram and Crook (1971) observed an increase in V_{max} of urease when purified kaolinite was used as support material. Similarly, Makboul and Ottow (1979c) reported an increase in V_{max} of alkaline phosphatase from 2553 to 2778 µg PNF ×10 mL^{-1} h^{-1} in the presence of montmorillonite but decreased slightly to 2469 µg PNF ×10 mL^{-1} h^{-1} in the presence of kaolinite. This kind of opposite pattern suggests that adsorption to montmorillonite favored hydrolysis of the enzyme–substrate complex. The V_{max} value of acid phosphatase was higher than the free enzyme when immobilized to soil-derived allophanic clays (Rosas et al., 2008). Calabi Floody (2012) also showed immobilization on allophanic or montmorillonite materials improved V_{max} (28%–38%) of acid phosphatase.

Sundaram and Crook (1971) reported a slight increase in K_m of urease when adsorbed to kaolinite. The value of K_m was increased 1.5–2 times when urease and acid phosphatase were adsorbed into montmorillonite, kaolinite, and illite (Makboul and Ottow, 1979a, 1979b). Huang et al. (1995) observed increase in K_m of acid phosphatase when adsorbed to montmorillonite and its various interlayered complexes. On the contrary, the presence of illite decreased K_m values of alkaline phosphatase (Makboul and Ottow, 1979c). Similarly, Rosas et al. (2008) reported a decrease in K_m as compared to the free acid phosphatase, when immobilized to allophanic clays. The K_m value for free acid phosphatase was 0.243 mM, whereas complexes with nanoclays separated from Inceptisol (0.343 mM) and Vertisol (0.276 mM) have higher and Alfisol (0.050 mM) had lower K_m than free enzyme (Paul et al., 2016). The change in K_m values after adsorption is governed by the distribution pattern of substrate and enzyme in the vicinity of the clay surface (Theng, 1982). If the charges of clay–enzyme complex and substrate are opposite, the concentration of substrate is more near the adsorption surface than the bulk solution. Therefore, K_m values will be decreased. However, if the charges of the substrate and clay–enzyme complex are similar, the K_m values will be increased. If the surfaces are uncharged, the K_m values will also be unaffected by the process. The other factors that affect change in K_m values are hinderence created by steric forces during movement of the substrate to the active sites of enzyme, and conformational changes take place in the enzyme, restriction in the diffusion of larger substrate molecules, and so on (Boyd and Mortland, 1990).

2.6   IMPLICATIONS OF CLAY–ENZYME INTERACTION

Besides playing key biochemical functions in the decomposition of SOM for making nutrients available to the plants and supporting microbial life, clay–enzyme interactions in the soil are of fundamental importance in explaining diverse phenomena. Role of clay-immobilized form of enzyme becomes more critical during the periods when enzyme-producing plants, animals, and microbial sources are less or not at all active due to stressed condition prevailing in the soil environment (Stursova and Sinsabaugh, 2008). Decomposition of carbonaceous substances in the soil such as starch, cellulose, amines, and other aromatic compounds, phenolic contents, polyphenols, and so on is governed by enzymes such as amylase, cellulase, β-glucosidase, phenol oxidase, dehydrogenase, laccase, chitinase, and so on (Das and Verma, 2010). Complexes of these enzymes with clay minerals present in soil have a very important role in maintaining their stability and activity for a longer period during hydrolysis (Burns, 1986). Enzymes such as urease and protease play a very important role for making nitrogen bioavailable for plants and microbes, the activity of which is affected by nature and type of clay minerals (Makboul and Ottow, 1979a; Gianfreda et al., 1992). Similarly, acid and alkaline phosphatases are mainly present in soils as extracellular enzymes. They are predominant in clay-sized fraction, as these have a very high affinity for colloidal soil components (McLaren and Packer, 1970). Improvement in catalytic properties of acid phosphatase due to its adsorption to mineral surfaces has been reported by some researchers (Rosas et al., 2008; Calabi et al., 2012). Immobilized acid and alkaline phosphatases can be used to hasten P mineralization from various P-rich organic sources (Calabi et al., 2012; Menezes-Blackburn et al., 2013; Paul et al., 2016). Arylsulfatases are very widespread and are responsible for hydrolysis of sulfate esters in the soil (Das and Verma, 2010). Although its activity was inhibited in the presence of kaolinite and montmorillonite, the extent of inhibition was more in the case of montmorillonite having a higher total surface area than kaolinite (Hughes and Simpson, 1978). Clay minerals stabilize the soil structure by adsorbing enzymes (or other organic molecules) into surface, occluding within nanofabric structures and forming micro- and macroaggregates (Das and Verma, 2010; Singh et al., 2018). Clay–enzyme interactions are of great importance in explaining the sequestration of organic carbon in soils, coastal and marine sediments (Mayer, 1994; Torn et al., 1997), and the global carbon balance, which regulate atmospheric CO_2 concentrations. Complexation with soil clays stabilizes the enzymes against other proteolytic enzyme attack, and this has been shown experimentally (Jastrow and Miller, 1997; Kaiser and Guggenberger, 2000; Zimmerman et al., 2004).

2.7 CONCLUSIONS

The clay–enzyme complexes as prepared in the laboratory and their interaction studies would provide a model, which resembles soil environment where such clay and enzyme interact with each other to contribute to different soil functions. The kind and type of minerals have a substantial impact on the activity and kinetics of various enzymes, as evidenced from research results cited above. In most of the cases, the activity of enzymes has been decreased compared to free enzyme when immobilized to clay support, but there were some promising results where reverse trends were observed. In addition, through the kinetics and stability studies, the mechanism of enzyme adsorption to soil clays could be understood well. Sometimes, this may help in developing strategies to apply clay–enzyme complex as biofertilizer in the near future for supplying nutrients from organic sources as per the need of plants.

KEYWORDS

- soil enzymes
- clay mineral
- enzyme activity
- kinetics

REFERENCES

Acosta-Martı́nez, V.; Tabatabai, M. A. Enzyme activities in a limed agricultural soil. Biol. Fertil. Soils 2000, 31, 85–91.

Ahn, M.Y.; Zimmerman, A.R.; Martinez, C.E.; Archibald, D.D.; Bollag, J.M.; Dec, J. Characteristics of Trametes villosa laccase adsorbed on aluminum hydroxide. Enzyme. Microb. Technol. 2007, 41, 141–148.

Alef, K.; Nannipieri, P. Cellulase activity. In: Methods in Applied Soil Microbiology and Biochemistry; Alef, K.; Nannipieri, P., Eds.; Academic: San Diego, CA, 1995, pp. 345–349.

Alexander, M. Introduction to Soil Microbiology. Wiley: New York, 1977, pp. 50–150.

Arinze, A.E.; Yubedee, A.G. Effect of fungicides on Fusarium grain rot and enzyme production in maize (Zea mays L.). Glob. J. Appl. Sci. 2000, 6, 629–634.

Baldrian, P. Wood-inhabiting ligninolytic basidiomycetes in soils: Ecology and constraints for applicability in bioremediation. Fungal Ecol. 2008, 1, 4–12.

Batistic, L.; Sarkar, J.M.; Mayaudon, J. Extraction, purification and properties of soil hydrolases. Soil Biol. Biochem. 1980, 12, 59–63.

Bergstrom, D.W.; Monreal, C.M.; King, D.J. Sensitivity of soil enzyme activities to conservation practices. Soil Sci. Soc. Am. J. 1998, 62, 1286–1295.

Blanchette, R.A.; Ander, P., Eds.; Microbial and Enzymatic Degradation of Wood and Wood Components. Springer: New York, NY, 1990, pp. 89–180.

Boyd, S.A.; Mortland, M.M. Enzyme interactions with clays and clay-organic matter complexes. In: Soil Biochemistry. Routledge: London, 1990, p. 28.

Brady, N.C.; Weil, R.R.; Weil, R.R. The Nature and Properties of Soils, vol. 13. Prentice-Hall: Upper Saddle River, NJ, 2008, pp. 662–710.

Bremner, J.M.; Mulvaney, R.L. Urease activity in soils. In: Soil Enzymes; Bums, R. G., Ed.; Academic: London, UK, 1978, pp. 149–196.

Brzezinska, M.; Stepniewska, Z.; Stepniewski, W. Soil oxygen status and dehdrogenase activity. Soil Biol. Biochem. 1998, 30(13), 1783–1790.

Burns, R. G. Enzyme activity in soil: Location and a possible role in microbial ecology. Soil Biol. Biochem. 1982, 14, 423–427.

Burns, R.G. Enzyme activity in soil: Some theoretical and practical considerations. In: Soil Enzymes; Bums, R.G., Ed.; Academic: London, UK, 1978, pp. 295–340.

Burns, R.G. Interaction of enzymes with soil mineral and organic colloids. In: Interactions of Soil Minerals With Natural Organics and Microbes; Huang, P.M.; Schnitzer, M., Eds.; Soil Science Society of America: Madison, 1986, pp. 429–452.

Calabi-Floody, M.; Velásquez, G.; Gianfreda, L.; Saggar, S.; Bolan, N.; Rumpel, C.; Mora, M. L. Improving bioavailability of phosphorous from cattle dung by using phosphatase immobilized on natural clay and nanoclay. Chemosphere 2012, 89(6), 648–655.

Calamai, L.; Lozzi, I.; Stotzky, G.; Fusi, P.; Ristori, G.G. Interaction of catalase with montmorillonite homoionic to cations with different hydrophobicity: Effect on enzymatic activity and microbial utilization. Soil Biol. Biochem. 2000, 32, 815–823.

Claus, H.; Filip, Z. Behavior of phenoloxidases in the presence of clays and other soil-related adsorbents. Appl. Microbiol. Biotechnol. 1988, 28, 506–511.

Das, S.K.; Varma, A. Role of enzymes in maintaining soil health. In: Soil Enzymology. Springer: Berlin, Germany, 2010, pp. 25–42.

Deng, S.P.; Tabatabai, M.A. Cellulase activity of soils. Soil Biol. Biochem. 1994, 26, 1347–1354.

Deng, S.P.; Tabatabai, M.A. Cellulase activity of soils: Effect of trace elements. Soil Biol. Biochem. 1995, 27, 977–979.

Dick, W.A.; Cheng, L.; Wang, P. Soil acid and alkaline phosphatase activity as pH adjustment indicators. Soil Biol. Biochem. 2000, 32, 1915–1919.

Dick, W.A.; Tabatabai, M.A. Potential uses of soil enzymes. In: Soil Microbial Ecology: Applications in Agricultural and Environmental Management; Metting, F.B. Jr., Ed.; Marcel Dekker: New York, NY, 1992, pp. 95–127.

Dodgson, K.S.; White, G.; Fitzgerald, J.W. Sulphatase Enzyme of Microbial Origin, Vol. I. CRC Press: Boca Raton, FL, 1982, pp. 156–159.

Durand, G. Microbiologie des sols-sur la degradation des bases puriques et pyrimidiques dans le sol – fixation de ces composes par les argiles-etude en fonction du pH et de la concentration. Comptes Ren. Hebd. des Seances de1 Acad. Des. Sci. 1963, 256 (19) 4126.

Eichlerová, I.; Šnajdr, J.; Baldrian, P. Laccase activity in soils: Considerations for the measurement of enzyme activity. Chemosphere 2012, 88(10), 1154–1160.

Elsgaard, L.; Andersen, G. H.; Eriksen, J. Measurement of arylsulphatase activity in agricultural soils using a simplified assay. Soil Biol. Biochem. 2002, 34, 79–82.

Eriksson, K.E.L.; Blancbette, R.A.; Ander, P. Biodegration of cellulose. In: Microbial and Enzymatic Degradation of Wood and Wood Components; Eriksson, K.E.L.; Blancbette, R.A.; Ander, P., Eds.; Springer: New York, 1990, pp. 89–180.

Esen, A. β-Glucosidases-biochemistry and molecular biology, ACS Symposium Series, 533. American Chemical Society, Washington, DC, 1993, pp. 9–17.

Fazekasova, D. Evaluation of Soil Quality Parameters Development in Terms of Sustainable Land Use: Sustainable Development Authoritative and Leading Edge Content for Environmental Management. In Tech: Rijeka, 2012.

Frank, T.; Malkomes, H.P. Influence of temperature on microbial activities and their reaction to the herbicide Goltix in different soils under laboratory conditions. Zentralblatt für Mikrobiol. 1993, 148, 403–412.

Garcia, C.; Herna´ndez, T. Biological and biochemical indicators in derelict soils subject to erosion. Soil Biol. Biochem. 1997, 29, 171–177.

German, D.P.; Chacon, S.S.; Allison, S.D. Substrate concentration and enzyme allocation can affect rates of microbial decomposition. Ecology 2011, 92(7), 1471–1480.

Gianfreda, L.; Bollag, J.M. Effect of soils on the behavior of immobilized enzymes. Soil Sci. Soc. Am. J 1994, 58, 1672–1681.

Gianfreda, L.; Rao, M.A.; Violante, A. Invertase (beta-fructosidase)—effects of montmoril-lonite, Al-hydroxide and Al (OH) X-montmorillonite complex on activity and kinetic-properties. Soil Biol. Biochem. 1991, 23, 581–587.

Gianfreda, L.; Rao, M. A.; Violante, A. Adsorption, activity and kinetic-properties of urease on montmorillonite, aluminum hydroxide and Al (OH) X-montmorillonite complexes. Soil Biol. Biochem. 1992, 24, 51–58.

Glinski, J.; Stepniewski, W. Soil Aeration and its Role for Plants. CRC Press: Boca Raton, FL, 1985.

Gomah, A.M. CM-cellulase activity in soil as affected by addition of organic material, temperature, storage and drying and wetting cycles. Zeitschrift fuer Pflanzenernaehrung und Bodenkunde 1980, 143, 349–356.

Gooday, G.W. Physiology and microbial degradation of chitin and chitosan. In: Biochemistry of Microbial Degradation; Ratledge, C., Ed.; Kluwer Academic Publishers: Dordrecht, The Netherlands, 1994, pp. 279–312.

Gu, Y.; Wang, P.; Kong, C. Urease, invertase, dehydrogenase and polyphenol activities in paddy soils influenced by allelophatic rice variety. Eur. J. Soil Biol. 2009, 45, 436–441.

Handa, S.K.; Agnihothri, M.P.; Kulshresta, G. Effect of pesticides on soil fertility. In: Pesticide Residue Analysis and Significance. Research Periodicals and Publishing House: New Delhi, 2000, pp. 184–198.

Haska, G. Activity of bacteriolytic enzymes adsorbed to clays. Microb. Ecol. 1981, 7, 331–341.

Hayano, K. Cellulase complex in tomato field soil; Introduction localization and some properties. Soil Biol. Biochem. 1986, 18, 215–219.

Hope, C.F.A.; Burns, R.G. Activity, origins and location of cellulases in a silt loam soil. Biol. Fert. Soils 1987, 5, 164–170.

Huang, Q.Y.; Liang, W.; Cai, P. Adsorption, desorption and activities of acid phosphatase on various colloidal particles from an Ultisol. Colloid Surf B 2005, 45, 209–214.

Huang, Q.; Shindo, H.; Goh, T. B. Adsorption, activities and kinetics of acid phosphatase as. Adsorption 1995, 159 (4).

Hughes, J.D.; Simpson, G.H. Arylsulfatase-clay interactions. 2. Effect of kaolinite and montmorillonite on arylsulfatase activity. Aust. J. Soil. Res 1978, 16, 35–40.

Jastrow, J. D.; Miller, R. M. Soil Aggregate Stabilization and Carbon Sequestration: Feedbacks Through Organo-Mineral Associations. CRC Press: Boca Raton, FL, 1997.

Joshi, S.R.; Sharma, G.D.; Mishra, R.R. Microbial enzyme activities related to litter decomposition near a highway in a sub-tropical forest of North east India. Soil Biol. Biochem. 1993, 25, 1763–1770.

Kaiser, K.; Guggenberger, G. The role of DOM sorption to mineral surfaces in the preservation of organic matter in soils. Org. Geochem. 2000, 31, 711–725.

Kandeler, E. Nitrate. Methods in Soil Biology; Schinner, F.; Öhlinger, R.; Kandeler, E.; Margesin, R., Eds.; Springer: Berlin, Germany, 1996, pp. 408–410.

Klose, S.; Tabatabai, M.A. Arylsulphatase activity of microbial biomass in soils. Soil Sci. Soc. Am. J. 1999, 63, 569–574.

Ladd, J.N. Origin and range of enzymes in soil. In: Soil Enzymes; Burns, R.G., Ed.; Academic: New York, 1978, pp. 51–96.

Ladd, J.N.; Butler, J.H.A. Inhibition and stimulation of proteolytic enzyme activities by soil humic acids. Austr. J. Soil Res. 1969, 7, 253–261.

Ladd, J.N.; Butler, J.H.A. Short-term assays of soil proteolytic enzyme activities using proteins and peptide derivatives as substrates. Soil Biol. Biochem.1972, 4, 19–30.

Leirós, M.C.; Trasar-Cepeda, C.; García-Fernández, F.; Gil-Sotres, F. Defining the validity of a biochemical index of soil quality. Biol. Fertil. Soils 1999, 30,140–146.

Linkins, A.E.; Melillo, J.M.; Sinsabaugh, R.L. Factors affecting cellulase activity in terrestrial and aquatic ecosystems. In: Current Perspectives in Microbial Ecology; Klug, M.J.; Reddy, C.A., Eds.; American Society of Microbiology: Washington, 1984, pp. 572–579.

Lipińska, A.; Kucharski, J.; Wyszkowska, J. Activity of arylsulphatase in soil contaminated with polycyclic aromatic hydrocarbons. Water Air Soil Pollut. 2014, 225(9), 2097.

Lozzi, I.; Calamai, L.; Fusi, P.; Bosetto, M.; Stotzky, G. Interaction of horseradish peroxidase with montmorillonite homoionic to Naþ and Ca2þ: Effects on enzymatic activity and microbial degradation. Soil Biol. Biochem. 2001, 33,1021–1028.

Madejón, E.; Burgos, P.; López, R.; Cabrera, F. Soil enzymatic response to addition of heavy metals with organic residues. Biol. Fertil. Soils 2001, 34, 144–150.

Makboul, H.E.; Ottow, J.C.G. Clay minerals and the Michaelis constant of urease. Soil Biol. Biochem. 1979a, 11(6), 683–686.

Makboul, H.E.; Ottow, J.C.G. Michaelis constant (K_m) of acid phosphatase as affected by montmorillonite, illite, and kaolinite clay-minerals. Microb. Ecol. 1979b, 5, 207–213.

Makboul, H.E.; Ottow, J.C.G. Alkaline-phosphatase activity and Michaelis constant in the presence of different clay-minerals. Soil Sci. 1979c, 128, 129–135.

Martinez, C.E.; Tabatabai, M.A. Decomposition of biotechnology by-products in soils. J. Environ. Qual. 1997, 26, 625–632.

Mayer, L.M. Relationships between mineral surfaces and organic carbon concentrations in soils and sediments. Chem. Geol. 1994, 114, 347–363.

McCarthy, G.W.; Siddaramappa, R.; Reight, R.J.; Coddling, E.E.; Gao, G. Evaluation of coal combustion by products as soil liming materials: Their influence on soil pH and enzyme activities. Biol. Fert. Soils. 1994, 17, 167–172.

McLaren, A.D.; Packer, L. Some aspects of enzyme reactions in heterogeneous systems. Adv Enzymol Relat. Areas Mol. Biol. 1970, 33, 245–308.

McLaren, A. D. Soil as a system of humus and clay immobilized enzymes. Chemica Scripta 1975, 8, 97–99.

Menezes-Blackburn, D.; Jorquera, M.; Gianfreda, L.; Rao, M.; Greiner, R.; Garrido, E.; De, la, Luz, Mora, M. Activity stabilization of Aspergillus niger and Escherichia coli phytases immobilized on allophanic synthetic compounds and montmorillonite nanoclays. Biores. Tech. 2011, 102(20), 9360–9367.

Menezes-Blackburn, D.; Jorquera, M. A.; Greiner, R.; Gianfreda, L.; De, la, Luz, Mora, M. Phytases and phytase-labile organic phosphorus in manures and soils. Crit. Rev. Env. Sci. Tech. 2013, 43(9), 916–954.

Mobley, H.L.T.; Hausinger, R.P.; Microbial urease: Significance, regulation and molecular characterization. Microbiol. Rev. 1989, 53, 85–108.

Moreno, J.L.; Garcia, C.; Hernandez, T. Toxic effect of cadmium and nickel on soil enzymes and the influence of adding sewage sludge. Eur. J. Soil Sci. 2003, 54, 377–386.

Morgan, H. W.; Corke, C.T. Adsorption, desorption, and activity of glucose oxidase on selected clay species. Can J. Microbiol 1976, 22, 684–693.

Mortland, M. M. Clay-organic complexes and interactions. Adv. Agron. 1970, 22(75), 117.

Mortland, M. M.; Gieseking, J. E. The influence of clay minerals on the enzymatic hydrolysis of organic phosphorus compounds 1. Soil Sci. Soc. Am. J 1952, 16(1), 10–13.

Naidja, A.; Huang, P.M.; Bollag, J.M. Activity of tyrosinase immobilized on hydroxy aluminum montmorillonite complexes. J. Mol. Catal. A: Chem. 1997, 115, 305–316.

Naidja, A.; Huang. P.M.; Bollag, J.M. Enzyme-clay interactions and their impact on transformations of natural and anthropogenic organic compounds in soil. J. Environ. Qual. 2000, 29, 677–691.

Nannipieri, P.; Ceccanti, B.; Cervelli, S.; Sequi, P. Stability and kinetic properties of humus-urease complexes. Soil Biol. Biochem. 1978, 10, 143–147.

Nannipieri, P.; Ceccanti, B.; Grego, S. Ecological significance of biological activity in soil. In: Soil Biochemistry, Vol. 6. Bollag, J.M.; Stotzky, G., Eds.; Marcel Dekker: New York, 1990, pp. 293–355.

Nannipieri, P.; Sequi, P.; Fusi, P. Humus and enzyme activity. In: Humic Substances in Terrestrial Ecosystems; Piccolo, A., Ed.; Elsevier: New York, 1996, pp. 293–328.

Nannipieri, P.; Landi, L.; Giagnoni, L.; Renella, G. Past, present and future in soil enzymology. In: Soil Enzymology in the Recycling of Organic Wastes and Environmental Restoration, Environmental Science and Engineering; Trasar-Cepeda, C.; Hernandez, T.; Garcia, C.; Rad, C.; Gonzalez-Carcedo, S., Eds.; Springer-Verlag: Berlin, 2012, pp. 1–17.

Ndakidemi, P.A. Manipulating legume/cereal mixtures to optimize the above and below ground interactions in the traditional African cropping systems. Afr. J. Biotechnol. 2006, 5 (25), 2526–2533.

Ndiaye, E.L.; Sandeno, J.M.; McGrath, D.; Dick, R.P. Integrative biological indicators for detecting change in soil quality. Am. J. Altern. Agric. 2000, 15, 26–36.

Olander, L. P.; Vitousek, P. M. Regulation of soil phosphatase and chitinase activity by N and P availability. Biogeochem. 2000, 49(2), 175–191.

Pancholy, S.K.; Rice, E.L. Soil enzymes in relation to old field succession; Amylase, cellulose, invertase, dehydrogenase and urease. Soil Sci. Soc. Am. J. 1973, 37, 47–50.

Paul, E.A.; Clark, F.E. Soil Microbiology and Biochemistry, Academic Press: New York, 1996.

Paul, R., Interaction of Soil nanoclay with phosphatase in relation to P mineralization (Doctoral dissertation, Division of Soil Science and Agricultural Chemistry Indian Agricultural Research Institute, New Delhi-1) 2016.

Pazur, J.H. Enzymes in the synthesis and hydrolysis of starch. In: Starch: Chemistry and Technology. Vol. 1 Fundamental Aspects; Whistler, R.; Paschall, E.F., Eds.; Academic Press: New York, 1965, pp. 133–175.

Petker, A.S.; Rai, P.K/ Effect of fungicides on activity, secretion of some extra cellular enzymes and growth of Alternaria alternata. Indian J. Appl. Pure Biol. 1992, 7, 57–59.

Pitchel, J.R.; Hayes, J.M. Influence of fly ash on soil microbial activity and populations. J. Environ. Qual. 1990, 19, 593–597.

Quiquampoix, H. A stepwise approach to the understanding of extracellular enzymeactivity in soil.1. Effect of electrostatic interactions on the conformation of a beta-D-glucosidase adsorbed on different mineral surfaces. Biochimie 1987a, 69, 753–763.

Quiquampoix, H. A stepwise approach to the understanding of extracellular enzymeactivity in soil. 2. Competitive effects on the adsorption of a beta-D-glucosidase in mixed mineral or organo mineral systems. Biochimie 1987b, 69, 765–771.

Rao, M. A.; Violante, A.; Gianfreda, L. Interaction of acid phosphatase with clays, organic molecules and organo-mineral complexes: Kinetics and stability. Soil Biol. Biochem. 2000, 32, 1007–1014.

Richmond, P.A. Occurrence and functions of native cellulose. In: Biosynthesis and Biodegradation of Cellulose; Haigler, C.H.; Weimer, P.J., Eds.; Dekker: New York, 1991, pp. 5–23.

Rosas, A.; Mora, M.D.; Jara, A.A; Lopez, R., Rao, M.A.; Gianfreda, L. Catalytic behaviour of acid phosphatase immobilized on natural supports in the presence of manganese or molybdenum. Geoderma 2008, 145, 77–83.

Ross, D.J.; McNeilly, B.A. Some influences of different soils and clay minerals on the activity of glucose. Soil Biol. Biochem. 1972, 4, 9–18.

Ross, D.J. Invertase and amylase activities in ryegrass and white clover plants and their relationships with activities in soils under pasture. Soil Biol. Biochem. 1976, 8, 351–356.

Ross, D.J. Studies on a climosequence of soils in tussock grasslands-5. Invertase and amylase activities of topsoils and their relationships with other properties. NZ. J. Sci. 1975, 18, 511–518.

Ross, D.J.; Roberts, H.S. Enzyme activities and oxygen uptakes of soils under pasture in temperature and rainfall sequences. J. Soil Sci. 1970, 21, 368–381.

Rubidge, T. The effect of moisture content and incubation temperature upon the potential cellulase activity of John Innes no. 1 soil (ISSN. 0020-6164). Int. Biodeterior. Bul. 1977, 13, 39–44.

Russel, W. E. Soil Conditions and Plant Growth. Long Man Group: London, 1973.

Sardans, J.; Peñuelas, J.; Estiarte, M. Changes in soil enzymes related to C and N cycle and in soil C and N content under prolonged warming and drought in a Mediterranean shrubland. Appl. Soil Ecol. 2008, 39, 223–235.

Schnell, S. A.; Maini, P. K. A century of enzyme kinetics. Should we believe in the K_m and V_{max} estimates? Oxford University Research Article 2003.

Serefoglou, E.; Litina, K.; Gournis, D.; Kalogeris, E.; Tzialla, A.A; Pavlidis, I.V.; Stamatis, H.; Maccallini, E.; Lubomska, M.; Rudolf, P. Smectite clays as solid supports for immobiliza-tion of beta-glucosidase: Synthesis, characterization, and biochemical properties. Chem. Mater. 2008, 20, 4106–4115.

Singaram, P. and Kamala, K. Effect of continuous application of different levels of fertilizers with farm yard manure on enzyme dynamics of soil. Madras Agricul. J. 2000, 87 (4-6), 364–365.

Singh, M.; Sarkar, B.; Sarkar, S.; Churchman, J.; Bolan, N.; Mandal, S.; Beerling, D. J. Stabilization of soil organic carbon as influenced by clay mineralogy. In: Advances in Agronomy. vol. 148, Academic Press: Cambridge, MA, 2018, pp. 33–84.

Sinsabaugh, R.L. Phenol oxidase, peroxidase and organic matter dynamics of soil. Soil Biol. Biochem. 2014, 2, 391–404.

Sinsabaugh, R.L.; Linkins, A.E. Inhibition of the Trichoderma viride cellulase complex by leaf litter extracts. Soil Biol. Biochem. 1987, 19, 719–725.

Skujins, J.; Pukite, A.; McLaren, A.D. Adsorption and activity of chitinase on kaolinite. Soil Biol. Biochem. 1974, 6, 179–182.

Skujins, J. Extracellular enzymes in soils. CRC Crit. Rev. Microbiol. 1976, 4, 383–421.

Speir, T.W.; Ross, D.J. Soil phosphatase and sulphatase. In: Soil Enzymes; Burns, R.G., Ed.; Academic Press: London, UK, 1978, p. 380.

Srinivasulu, M.; Rangaswamy, V. Activities of invertase and cellulase as influenced by the application of tridemorph and captan to groundnut (Arachis hypogaea) soil. Afr. J. Biotechnol. 2006, 5, 175–180.

Stevenson, F.J. Cycles of Soil-Carbon, Nitrogen, Phosphorus, Sulfur, Micronutrients; Wiley Int. Science Publ., John Wiley & Sons: New York, 1986.

Stotzky, G. Influence of soil mineral colloids on metabolic processes, growth, adhesion, and ecology of microbes and viruses. Interact. Soil Miner. Nat., Org. Microbes 1986, 17, 305–428.

Stursova, M.; Sinsabaugh, R. L. Stabilization of oxidative enzymes in desert soil may limit organic matter accumulation. Soil Biol. Biochem. 2008, 40(2), 550–553.

Sundaram, P. V.; Crook, E. M. Preparation and properties of solid-supported urease. Can. J. Biochem. 1971, 49(12), 1388–1394.

Tabatabai, M.A. Effect of trace elements on urease activity in soils. Soil Biol. Biochem. 1977, 9, 9–13.

Tabatabai, M.A. Soil enzymes. In: Methods of Soil Analysis, Part 2. Microbiological and Biochemical Properties; Weaver, R.W.; Angle, J.S.; Bottomley, P.S., Eds.; SSSA book series no. 5. Soil Science Society America: Madison, WI, 1994, pp. 775–833.

Theng, B. K. G. Clay-polymer interactions: Summary and perspectives. Clay Clay Miner. 1982, 30(1), 1–10.

Thoma, J.A.; Spradlin, J.E.; Dygert, S. Plant and animal amylases. In: The Enzymes, Vol. 5, Boyer, P.D., Ed.; International Society of Soil-Science: The Netherlands, 1971 pp. 115–189.

Tietjen, T.; Wetzel, R.G. Extracellular enzyme-clay mineral complexes: Enzyme adsorption, alteration of enzyme activity, and protection from photodegradation. Aquat. Ecol. 2003, 37, 331–339.

Tisserant, B.; Gianinazzi-Pearson, V.; Gianinazzi, S.; Gollotte, A. In planta histochemical staining of fungal alkaline phosphatase activity for analysis of efficient arbuscular mycorrhizal infections. Mycol. Res. 1993, 97, 245–250.

Torn, M.S.; Trumbore, S.E.; Chadwick, O.A.; Vitousek, P.M.; Hendricks, D.M. Mineral control of soil organic carbon storage and turnover. Nature 1997, 389, 170–173.

Wang, Q.; Peng, L.; Li, G.; Zhang, P.; Li, D.; Huang, F.; Wei, Q. Activity of laccase immobilized on TiO2-montmorillonite complexes. Int. J. Mol. Sci. 2013, 14(6), 12520–12532.

Wilke, B.M. Effect of single and successive additions of cadmium, nickel and zinc on carbon dioxide evolution and dehydrogenase activity in a sandy Luvisol. Biol. Fert. Soils 1991, 11, 34–37.

Wright, A.L.; Reddy, K.R. Phosphorus loading effects on extracellular enzyme activity in everglades wetland soil. Soil Sci. Soc. Am. J. 2001, 65, 588–595.

Yang, Z.; Liu, S.; Zheng, D.; Feng, S. Effects of cadmium, zinc and lead on soil enzyme activities. J. Environ. Sci. 2006, 18, 1135–1141

Yuan, B.; Yue, D. Soil microbial and enzymatic activities across a chronosequence of Chinese pine plantation development on the loess plateau of China. Pedosphere 2012, 22, 1–12.

Zantua, M.I.; Bremner, J.M. Stability of urease in soils. Soil Biol. Biochem. 1977, 9, 135–140.

Zimmerman, A.R.; Chorover, J.; Goyne, K. W.; Brantley, S. L. Protection of mesopore-adsorbed organic matter from enzymatic degradation. Environ. Sci. Technol. 2004, 38, 4542–4548.

CHAPTER 3

Short-Range Order (SRO) Minerals and Their Implications in Sustainable Agriculture by Exerting Influence on Carbon Sequestration and Fixation of Phosphorus and Potassium

DIBYENDU CHATTERJEE[1,2*], S. C. DATTA[1], K. M. MANJAIAH[1] and NINTU MANDAL[3]

[1]ICAR Indian Agricultural Research Institute, New Delhi 110012, India

[2]ICAR National Rice Research Institute, Cuttack, Odisha 753006, India

[3]Bihar Agricultural University, Bhagalpur, Bihar, India

*Corresponding author. E-mail: dibyenducha@gmail.com

ABSTRACT

Amorphous oxide (free or combined) of Si, Al, and Fe present in soil is also known as short-range order (SRO) mineral. The examples of such minerals are proto-imogolite allophane, imogolite, allophane, ferrihydrite, and hisingerite. Dispersion, flocculation, aggregation, infiltration, erosion, and landscape stability are different properties of soil that are affected by SRO minerals. Physical characteristics such as high water-holding capacity, low bulk densities, large void ratios, and anomalous compaction behaviors are controlled by SRO minerals. Different chemical properties such as sequestration of carbon, cation and anion exchange capacity, phosphate and potassium fixation capacity, and high pH-low base saturation are the characteristics of these minerals. In this chapter, we discuss how these minerals are formed and its implications on C sequestration and P and K fixation capacities.

3.1 INTRODUCTION

The work on short-range order (SRO) minerals was started long back in the 1970s (Murti et al., 1976). Free and combined oxides of Si, Al, and Fe in clay size inorganic colloidal fraction (<2 μm) of soils which are amorphous to X-rays are called amorphous ferri-aluminosilicates (AFASs), most recently known as SRO minerals (Murti et al., 1976). The examples of such minerals are proto-imogolite allophane, imogolite, allophane, ferrihydrite, and hisingerite. Due to the absence of sharp peak during the X-ray diffraction analysis, these minerals are "X-ray amorphous" (Konta, 2009). Electron microscopy, infrared spectroscopy, nuclear magnetic resonance spectroscopy, and thermal analysis interpreted certain structural order of these materials, but only over a "short range" in one direction. Interestingly, the structures of SRO minerals are irregular with small particle size, which give broad and low intensity peaks in X-ray diffractograms. Hence, these minerals are described as "SRO" instead of "amorphous." These minerals are generally of two types—silicates and nonsilicates. Silicates include all ferri-aluminosilicate minerals, while the nonsilicates include amorphous iron and aluminum oxides, hydroxides, etc.

Imogolite is very important among all ferri-aluminosilicate minerals in soils due its high specific surface area and it is reactive toward anions and cations (Wada, 1989). The chemical formula of imogolite is $(HO)_3Al_2O_3SiOH$ and these are naturally present in hydrated aluminosilicate polymer (Cradwick et al., 1972). This mineral is characterized by transmission electron microscopy as bundles of long tubes. Some researchers argued that imogolite has long-range order in one dimension along the tubes for which it is not considered as SRO minerals (Lowe, 1995).

Allophane, containing silica, alumina, and water in chemical combination, is also a SRO mineral (Parfitt, 1990). The major part of volcanic-derived soil contained allophane. In nonvolcanically derived soils, these minerals are also found in clay size fraction (Spark, 1995). The low bulk density and large quantity of organic matter accumulating capacity of volcanic ash soils are believed to be due to the presence of allophane. The $SiO_2:Al_2O_3$ ratio of allophane is in the range of 0.84–2. The tetrahedral and octahedral coordinates are occupied by the aluminum in allophane. The Al-rich allophane is a spherical particle that has diameter of 3–6 nm. These spherical particles contain micropores of 0.3–2.0 nm in diameter. These micropores act as "defect sites" in the mineral moiety. Empirical composition of allophane is $Al_2Si_2O_5 \cdot nH_2O$; however, type of allophane is determined by variation in Si:Al ratio. These are typically three types (Lowe, 1995): (1)

Al-rich allophane (Al:Si $\geq$ 2.0) such as proto-imogolite allophane, (2) Si-rich allophane (Al:Si $\approx$ 1.0) such as halloysite-like allophane, and (3) stream deposit allophane (Al:Si $\approx$ 0.9–1.8) such as hydrous feldspathoid allophane.

Ferrihydrite, another important SRO minerals, is basically a poorly crystalline hydrated ferric oxhydroxides (Carlson and Schwertmann, 1981). Ferrihydrite acts as a precursor of more crystalline iron minerals such as goethite, lepidocrocite, and hematite. However, the formation of these crystalline minerals from ferrihydrite is impeded when it is coprecipitated with silica in Si-rich environment (Vempati and Loeppert, 1989). Stoichiometric composition of ferrihydrite is FeO(OH), which is varied in between $Fe_5O_3(OH)_9$ and $Fe_4(OH)_{12}$ (Eggleton and Fitzpatrick, 1988).

3.2 FORMATION OF SRO MINERALS WITHIN RHIZOSPHERE WHERE PHOSPHORUS AND POTASSIUM SUPPLY IS RESTRICTED

3.2.1 RELEASE OF LOW-MOLECULAR-WEIGHT ORGANIC ACIDS (LMWOAS) IN RHIZOSPHERE

Rhizosphere soil can be defined as the soil, which adheres to or influenced by the root and can be removed by shaking with sterile water (Ryan et al., 2009). The rhizosphere zone is different from the bulk or nonrhizosphere soil on the account of nutrients and microbial activities facilitated by root exudates. Plants have developed various strategies to convert the rhizosphere to lessen the impact of these environmental stresses. Understanding these strategies, propose various ways in which the rhizosphere can be modified to improve crop health and productivity. Plant roots have various abilities to synthesize, accumulate, and secrete several arrays of compound. It releases more than 200 carbon compound in the form of exudates, which is termed as "rhizodeposition" (Kuzyakov and Domanski, 2000). These compounds are classified into two groups on the basis of molecular weight of which the first group is the high-molecular-weight compounds such as mucilage and ectoenzymes and second group is the low-molecular-weight compounds such as amino acids, organic acids, sugar, phenolics, vitamins, phytosiderophores, and flavonoids (Bertin et al., 2003). Root exudates can be characterized by the release of the carbon-containing compounds derived from the internal metabolic processes of the plant such as respiration, while the external secretion processes are mainly meant for nutrient acquisition (Uren, 2000).

The transformation of the layer silicates to SROs is sometimes mediated by LMWOAs in rhizosphere. Several aliphatic acids are present in the rhizosphere soils such as citric, oxalic, formic, acetic, succinic, malonic, malic, maleic, lactic, aconitic, and fumaric acids that are coming in the group of LMWOAs (Van Hees et al., 1996). Several of these LMWOAs are produced during the decomposition of organic matter in the soils (Fox and Comerford, 1990), root exudates (Chatterjee et al., 2014b), and leaf washing (Ström et al., 1994). In the soil solution, LMWOA concentrations have been reported in the range of 1–1000 mM (Van Hees et al., 1996). The content of oxalic, malonic, fumaric, and succinic acids in bulk soils (45.4 mg kg^{-1} dry weight of soil) was always lower than those in rhizosphere soils (66.4–100.1 mg kg^{-1} dry weight of soil) (Chiang et al., 2006). Although the concentration of organic acids in soil solution is usually low (10^{-3} to 4×10^{-4} mol L^{-1}), but larger quantities are reported in the rhizosphere soils (Vance et al., 1996). However, the constituent and quantity of root exudates may differ with age of plant, type of stress, and plant types.

3.2.2   TRANSFORMATION OF CRYSTALLINE LAYER SILICATE TO SRO MINERALS

Stress condition induces plants to release LMWOAs in root exudates that may attack the layer silicates and simultaneously do not allow the oxides of silicon, aluminum, and iron to combine to form layer silicates again by hindering its proton exclusion mechanism, which cannot be possible under proton-rich (acid) environment (Figure 3.1). Thus, under stress condition, noncrystallinity is increased and in rhizosphere soils after growing crops which were reflected by increase in the content of SRO minerals (Datta et al., 2009; Chatterjee et al., 2014b, 2015a).

LMWOAs form strong complexes with aluminum and iron (McColl and Pohlman, 1986). In nonacidified podzolized soils, about 20%–40% of aluminum and 15%–25% of iron are complexed with LMWOAs (Van Hees et al., 2000). In addition, LMWOA serves as a means of translocation of elements in the soil profile and contributes to the movement of aluminum and iron from top of the profile toward bottom layers (Lundström and Giesler, 1995). One eventual process of the immobilization of Al and Fe bound to these acids is biodegradation of the complex (Lundström and Giesler, 1995). Aluminum and iron, thus, released during the organic complex degradation can then coprecipitate with silicon. Gustafsson et al. (1995) and Karltun et al. (2000) reported the presence of "imogolite-type materials" in podzolized "B" horizons, which formed in the above process.

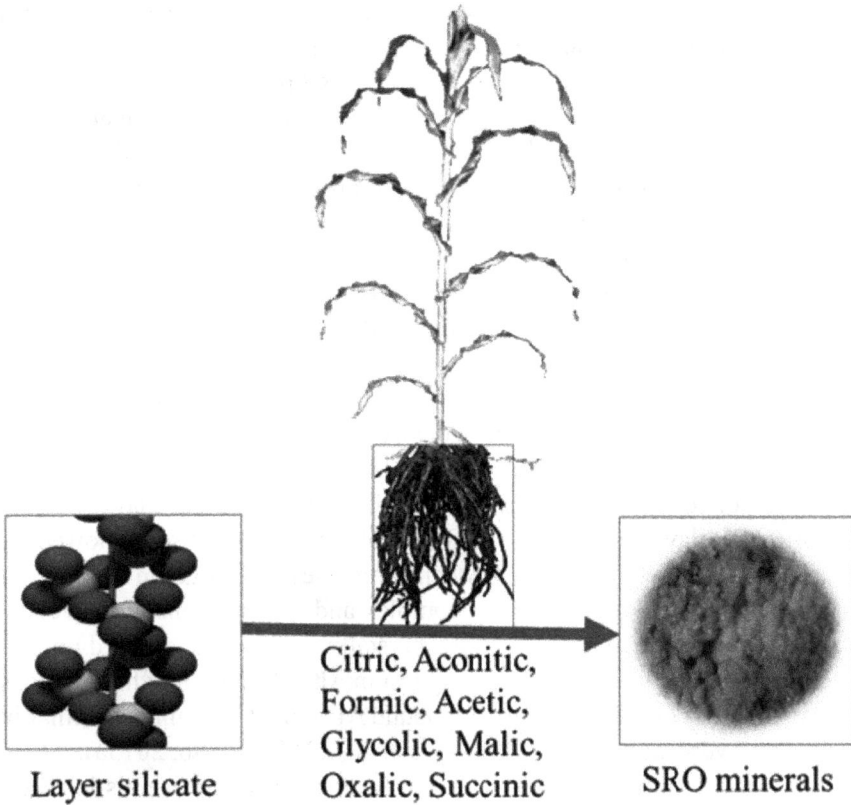

FIGURE 3.1   Schematic display of transformation of layer silicates to SROs in rhizosphere.

The SRO minerals can be made artificially in the laboratory. Violante and Huang (1993) reported synthesized noncrystalline materials and reported its transformation to pseudoboehmite. In a laboratory study, considerable amount of Si, Al, Fe, K, and P is released when citric acid is passed through a filled soil column (Chatterjee et al., 2015b, 2016). These released Si, Al, and Fe recombined to form SRO minerals after the degradation of LMWOAs. Khademi and Arocena (2008) studied that palygorskite and sepiolite were formed from kaolinite in the rhizosphere soils. They observed that transformation of sepiolite in the rhizosphere of selected agricultural crops was linked to the ability of plants to extract magnesium (Mg) from soil. Sometimes, bacterial species actively participated in the formation of silicate minerals through outer sphere complex between the silicate anions and amino group (Urrutia and Beveridge, 1994). The formation of SRO aluminosilicates was found in the neosynthesis of mineral product in the presence of bacterial surface and organic ligand (Urrutia and Beveridge, 1995).

In nature, allophane and imogolite are formed preferentially at pH > 4.9. In more acid conditions, Al–humus complexes tend to be dominant (Shoji and Fujiwara, 1984). When pH is lower than 4.9, antiallophenic effect occurs due to formation of Al–humus complex. In absence of Al, the formation of allophane or imogolite is hindered. It has been further suggested that the allophane precipitation is impeded by competition between orthosilicic acid and humic substances for soluble aluminum (Shoji et al., 1993). The formation of noncrystalline material and organic matter accumulation in volcanic soils is called "andosolization" (Dachaufour, 1977). Unlike podzolization, aluminum, iron, and organic matter do not translocate significantly to the subsurface horizons (Ugolini and Dahlgren, 2002). In a research with maize crop, it was observed that the highest AFAS was found in Inceptisol followed by Vertisol and Alfisol for colloidal clay, while in noncolloidal clay the highest AFAS was observed in Vertisol followed by Alfisol and Inceptisol (Table 3.1). Mean allophane plus imogolite was highest in Inceptisol colloidal clay–humus complex and Vertisol noncolloidal clay–humus complex, whereas ferrihydrite, crystalline Fe, poorly ordered Fe, and organically bound iron were found highest in Alfisol. In both of the fractions of clay–humus complex, organically bound Al content was almost similar in Alfisol and Vertisol samples (Chatterjee et al., 2014a, 2014b, 2015a).

The interaction between biochemical compounds and minerals in rhizosphere region results in precipitation of SRO Al and Fe oxides. All LMWOAs may interact with hydrolytic products of Al and Fe and form organomineral complexes (Vance et al., 1996). Organic substances such as humic acid (HA) and fulvic acid (FA) and many LMWOAs form stable complexes with aluminum, disrupt the crystallization of aluminum hydroxides, and promote the formation of SRO aluminum hydroxides (Colombo et al., 2004). The surface properties of Al transformation products are largely affected by the organic acids. The formation and transformation of SRO Fe oxides are greatly influenced by the interactions of soil minerals with organic substances and microorganisms (Cornell and Schwertmann, 2003). The redox cycling, hydrolytic reactions, crystallization of iron precipitation products, and the subsequent formation of pedogenic oxides of iron were also influenced by the organic substances and microorganisms (Huang and Wang, 1997). Allophane-like precipitates are formed by coprecipitation of the hydroxides of silicon, aluminum, and iron hydroxides through neutralization (Nogami, 2004). Imogolite may form from the surface of weathered plagioclase, which

TABLE 3.1 Content of SRO Minerals (%) in Soil–clay and Clay–humus Complexes Averaged over Fertility and Rhizospheric Status

SRO Minerals (%)	Formula	Colloidal			Noncolloidal		
		Alfisol	Inceptisol	Vertisol	Alfisol	Inceptisol	Vertisol
AFAS	$(\%Si_nO_2 + \%Al_{n2}O_3 + \%Fe_{n2}O_3)/0.9$	13.25	18.06	16.25	9.23	7.23	12.87
Allophane + Imogolite	$\% Si_o \times 7.1$	1.73	4.15	2.60	1.70	5.00	5.43
Ferrihydrite	$\% Fe_o \times 1.7$	0.66	0.53	0.40	0.72	0.47	0.52
Crystalline Fe	$Fe_d - Fe_o$ (%)	7.39	1.79	2.08	8.35	2.06	2.40
Poorly ordered Fe	$Fe_o - Fe_p$ (%)	0.25	0.18	0.11	0.26	0.13	0.18
Organically bound Al	Al_p (%)	1.42	1.20	1.46	1.33	1.25	1.33
Organically bound Fe	Fe_p (%)	0.14	0.13	0.12	0.16	0.15	0.13

n = NaOH extractable, o = oxalate extractable, d = citrate-bicarbonate-dithionite (CBD) extractable, p = pyrophosphate extractable (modified after Chatterjee et al., 2014a)

initially occurs as small tiny bumps and later on develops into a fibrous imogolite (Tazaki, 1979). Kawano and Obokata (2007) reported that basic amino acids (lysine, histidine, and arginine) may interact more strongly with the negatively charged amorphous silica than other nonbasic amino acids.

Lilienfein et al. (2003) studied soil formation along with organic matter accumulation in California. They reported an increase in allophane at the rate of 0.14 g kg−1 year−1 and concentration of allophane was 68 times higher than older soils. Besides, in the same soil, concentration of ferrihydrite was increased by 2.3 times. Carbon and nitrogen storage was increased with soil age over the first ~600 years and after that the accretion rates were lower. Sarkar et al. (1979) observed that the amount of precipitation products of noncrystalline aluminum or iron extracted from rhizosphere soil was greater than those from nonrhizosphere soil and significant positive correlations were showed between extractable Al and Fe and extractable C suggesting chemical associations between these organic and inorganic components. Yagasaki et al. (2006) researched on the two volcanic ash soils and found that these soils are rich in soil organic matter (SOM) and organically bound Al. They found SRO aluminosilicates such as imogolite and proto-imogolite were plenty in these soils. However, the soil horizon A was found undersaturated with respect to imogolite and proto-imogolite. This suggests that low disintegration rate of these minerals, possibly due to transformation of the physicochemical characteristics of their surfaces, caused by complexation with SOM. Johnson-Maynard et al. (1997) evaluated the chemical and mineralogical conversion of allophanic soils under forest and bracken fern coverage. The undisturbed forest soils contained more of SRO aluminum–iron minerals, while the bracken fern-influenced soils contained more of metal–humus complexes. Successional communities of the bracken fern are responsible for a shift from allophanic to nonallophanic properties in these soils, probably due to bracken fern associated with increased levels of soil organic carbon (SOC) and a subsequent enhancement in formation of Al–humus complexes. The SRO material is slightly less (20%–37%) in temperate region compared to these in tropical soils (29%–40%) (Tan et al., 1970).

3.3 CARBON SEQUESTRATION PROPERTY OF SRO MINERALS

3.3.1 SRO MINERALS AS A SINK

The atmospheric concentration of carbon dioxide and methane is increasing very fast (Chatterjee and Saha, 2018). Protecting organic materials that present in soil are one of the main ways to curb the emission of greenhouse gases (Chatterjee et al., 2019). Through the interactions with highly reactive SROs, the SOC can be stabilized in the soil (Chatterjee et al., 2013). They also have a strong tendency to get adsorbed in humic substances, microbial biomass, and LMWOAs. Thus, soils contained high quantity of SROs (e.g., volcanic ash soils, rhizospheric soils) usually have much higher carbon content than other soils in the same environmental condition.

Plant roots release carbon-rich compounds into the deeper soil layer that can be sequestered. For example, 20%–30% of total assimilated carbon is transferred into the soil by the cereal roots of which about one-third is lost in the form of CO_2 by root respiration and microbial utilization (Gregory and Atwell, 1991). The amount of belowground carbon released by cereals and grasses is about 1500 and 2200 kg C ha^{-1}, respectively, during the vegetative period (Bowen and Rovira, 1999). During cropping, more humus has been removed from interlayer position in rhizosphere soil as compared to nonrhizosphere in Alfisol and from interlayer position of nonrhizosphere soil in Vertisol (Mandal and Datta, 2005). In rhizosphere soil, interlayer position of the smectite mineral expands in which the humus can enter and smectite–kaolinite layer can be formed. In the rhizosphere samples of Vertisol, clay–organic complex is more stable than that of Alfisol.

SOC pools in soil are twice more than atmospheric pools and three times more than terrestrial organic pools (Lal, 2003). This pool is a large C store under changing climate scenario. Egli et al. (2007) reported that concentration of SOC in the topsoil, the SOC stocks in the profiles, the humus fractions such as HA and FA, functional groups and organic substances, imogolite-type materials, and oxyhydroxides were strongly related to the altitude. The Histosols soil order accumulate higher quantity of carbon (204.5 kg m^{-2} up to 100 cm depth) compared to Andisols soil order (30.6 kg m^{-2} up to 100 cm depth) (Eswaran et al., 1993). The capacity of soil to preserve organic matter is directly proportional to stability of SOM, its resistance to oxidation (Righi et al., 1995), and the presence of some natural recalcitrant compounds such as aliphatic macromolecules (suberans, lipids, algaenans, cutans), sporopollenins, charcoal, biochar, and lignins (Munda et al., 2018). Higher SOM was found in Andosols than Ultisols, which is related to transformation of stable fractions to the more labile ones in Ultisols (Aguilera et al., 1997). Barbera et al. (2008) found that the calcitrant organic fraction was aliphatic compounds and did not greatly interact with the kaolinite, smectite, or poorly Fe or Al crystalline phases. In contrast, a strong relationship with poorly crystalline oxyhydroxides and kaolinite was found with the oxidizable fraction.

Allophane minerals containing soil such as Andosols exhibit high concentrations of SOC compared to other kinds of the soil (Vacca et al., 2009). These minerals expressed different chemical properties such as the ability to sequester carbon, cation and anion exchange capacity (pH dependent), large fixation capacity of phosphate, adsorption, and high pH-low base saturation relationship (Wada and Harward, 1974). Large amount of organic carbon has been found in Andosols (Óskarsson et al., 2004). Chevallier et al. (2008) found that the transformation and decomposition of organic matter into the CO_2 are less when the soil contains more allophane. The SOC in allophanic soils is too stable to be mineralized by microorganisms. These stable SOC contributes to higher C accumulation in allophanic soils (Boudot et al., 1986). In many soils, the type of soil is more important for soil C storage than the total amount of clay (Torn et al., 1997; Percival et al., 2000). The protective effect to SOM in allophanic soils was thought to result from: (i) the flocculation of initially soluble organo–Al complexes and (ii) the binding of organic molecules within either the intertubular spaces of imogolite threads or the interspherular spaces of allophane aggregates (Boudot, 1992). Woignier et al. (2007) reported that nitrogen and carbon content in soils are also dependent on the allophane content. Basile-Doelsch et al. (2005) found that buried horizons contained much of poorly crystalline material such as proto-imogolite and proto-imogolite allophane and stored large amounts of organic matter which turned over very slowly. They also studied that organic matter is not only chelated poorly crystalline material, but it may also act as limiting factor in the polymerization of mineral phases between proto-imogolite and proto-imogolite allophane. Several reports indicate that SRO minerals have capable to adsorbing organic molecules due to its large specific surface area (Parfitt et al., 1997). Basile-Doelsch et al. (2007) observed that the organomineral complexes of imogolite-type materials formed through the largest proportion (82.6%) of organic matter in the horizon of volcanic ash soil were associated with SRO minerals. Chatterjee et al. (2013) noted significantly positive correlation of total organic carbon with AFAS and allophane in coarse and fine clay–humus complex, respectively. Significantly positive correlation was also found between nonlabile carbon (NLC) and AFAS in both the clay–humus complex, while NLC and allophane content in coarse clay–humus complex. Fine clay–humus is the most important fraction in terms of total amount of sequestered carbon. The role of Al and Fe compounds is more prominent in carbon accumulation in coarse clay–humus fraction of Alfisol. Rhizosphere sequestered more carbon than nonrhizosphere and among the soils, the most potential of sequestering carbon was found in Alfisol. They also used an index of carbon sequestration that can be determined through NLC/labile carbon (LC) ratio. Higher NLC/ LC ratio refers to higher carbon sequestration in soil. In rhizospheric soils, NLC/LC ratio was >1, which expressed that out of the total organic carbon, majority was nonlabile, that is, less susceptible to microbial degradation. Rhizospheric soils (average NLC/LC ratio is 1.31, >1) had higher NLC/LC ratio than nonrhizospheric soils (average 0.38, <1) (Figure 3.2).

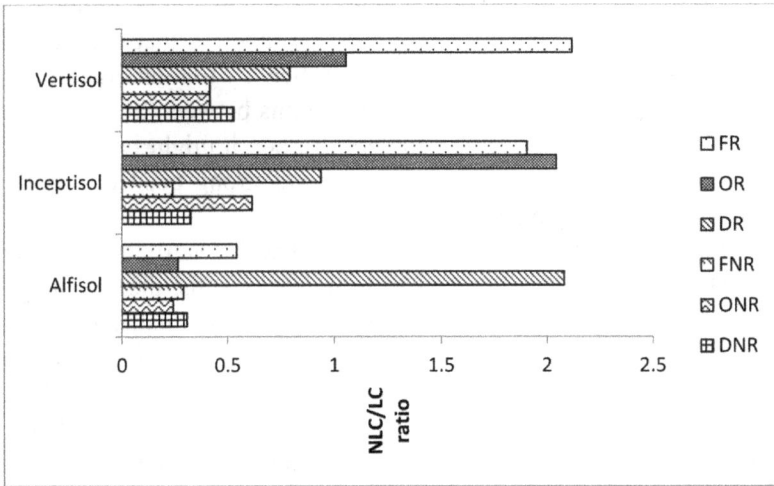

(A)

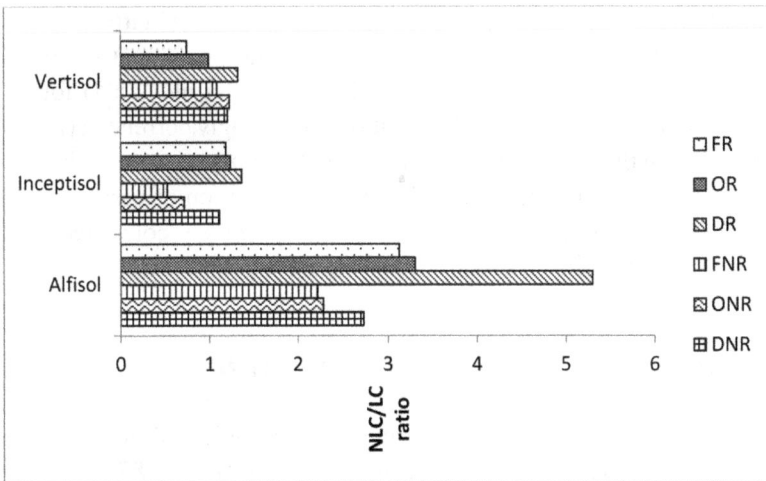

(B)

FIGURE 3.2 NLC/LC ratio of (A) colloidal and (B) noncolloidal clay–humus complex (modified after Chatterjee et al., 2013).

In contrast to the previous findings, Egli et al. (2008) observed that imogolite-type material significantly not contributed to the stabilization of SOM. Similarly, acid ammonium acetate extracted Al was highly correlated with SOC ($R^2 = 0.88$) in Inceptisols; however, the allophane content did not show strong correlation with SOC ($R^2 = 0.14$) (Matus et al., 2006). The possible reasons of such findings may be related to the mechanism of SOC stabilization in andic soils is noncrystalline Al hydroxides and insoluble organically Al complex rather than allophane (or imogolite) (Boudot, 1992; Percival et al., 2000). Tonneijck et al. (2009) reported that organomineral complexes were not found to contribute significantly to total organic carbon stabilization as proved by negatively correlations between SOC and clay ($r = -0.71$) or allophane ($r = -0.86$) content. Rather, it might be stabilized by organometallic complexes as evidenced by the strong positive correlation between SOC and exchangeable Al ($r = +0.87$).

Yonebayashi and Hattori (1988, 1989) evaluated that in Andisols, hydrogen/carbon (H/C) ratio is less as compared to nonandic soils, thereby suggested that there is a greater amount of condensation of humic matter in andic soil than in other soil. Lilienfein et al. (2003) expressed that dissolved organic carbon adsorption has been increased with increasing SRO content. The SRO minerals are helpful to stabilize microbial biomass and metabolites (Saggar et al., 1996). Torn et al. (1997) showed that an increase in C stability in Andisols as compared to Inceptisols and other soil order was due to presence of higher quantity of SRO minerals in Andisols. The organic matter decay rate is more in nonallophanic soils as compared to the allophanic soils (Martin et al., 1982) whereas the allopathic humus complex is the final phase that regulates the formation of soil (Zunino and Borie, 1985). Matus et al. (2008) found that allophane content (estimated by Al/Si ratio) was strongly correlated ($R^2 = 0.82$) with the SOC in the fine silt plus clay and that Al–humus together with C in the finest particles explained ($R^2 > 60$) the largest proportion of variation of SOC across the studied soils. SOM content is often positively correlated with the clay content of the soils (Bosatta and Ågren, 1997).

3.3.2 MECHANISM OF CARBON SEQUESTRATION

A major part of the SOM in predominantly inorganic soil is usually found in silt and clay size fractions (Catroux and Schnitzer, 1987). The carbon protection mechanism can be broadly classified into two groups (Chevallier et al., 2008): (1) physical protection in mesopore and (2) chemical protection through binding with SRO or aluminum and iron oxides.

SRO is a major constituent of volcanic ash soils. SOM stabilization in the volcanic ash soils is generally related to the formation of organomineral (SRO–SOM) complexes (Basile-Doelsch et al., 2005; Mikutta et al., 2005; Wiseman and Püttmann, 2006) and formation of organometallic (Al/ Fe–SOM) complexes (Wiseman and Püttmann, 2006; Egli et al., 2008). For instance, the sequestration of SOM is facilitated by the poorly crystalline illite with high surface area (Bock and Mayer, 2000). However, these protection mechanism methods are still debatable (Dahlgren et al., 2004). In the preservation of SOC, sorption plays an important role (Wiseman and Püttmann, 2006). The valence unsatisfied hydroxyl groups of Al, Si, and Fe hydroxides produce organomineral complex, which have the ability to absorb ions and compounds of opposite charge (Shoji et al., 1993; Mikutta et al., 2005) and the relationship of SOM to clay particle (C) through polyvalent cations (P) that can be represented as $[(C-P-SOM)_x]_y$, where x and y depend upon the primary clay particles size (Manjaiah et al., 2010). Additionally, ligand exchange, coulombic attraction, van der Waals forces, polyvalent cation bridging with or without mediated by hydration water, hydrogen bonding, and water bridging mechanisms of interaction are also involved in this process (Theng and Tate, 1989). Organomineral and organometallic complexes reserve C, not only because of the toxic effect of aluminum that has an effect on soil microbes, but also because of the limited access to enzymes, which protect the mineralization of organic compounds. In Andisols, the mean residence time or organic C is very high and the turnover time is very low (Saggar et al., 1996; Torn et al., 1997; Basile-Doelsch et al., 2005). Conte et al. (2003) reported that due to abundance of carboxyl groups in andic humic soils, these groups form stable complexes with aluminum that is present in allophane. The andic soil showed a lower degree of organic matter (OM) transformation, which is due to the stable complex formation by virtue of larger number of carboxyl and methoxyl groups in humic matter. Increase in carbonyl and carboxyl groups and reduction of hydroxyl, phenolic, and methoxy groups are observed with accumulation of humic substances in andic soil (Yonebayashi and Hattori, 1989).

The SOC can also be protected physically from enzymatic degradation by entering within mineral mesopores (2–50 nm in diameter) (Mayer et al., 2004). There are some evidences that the fractal structure may be involved in the C sequestration. Respiration of microorganisms in mesopores is affected due to low SOM availability and low oxygen diffusion into it because of its tortuous porosity. Allophane may also be considered as a natural gel that shows C sequestering property (Woignier et al., 2006). In the aggregates, the SOM tends to be irregularly distributed, occurring as blebs and smears intimately associated with clays (Ransom et al., 1997; Curry et al., 2007). The incorporation of SOM resulted in significant decrease in permeability of soil as part of pore space that is occupied by SOM and subsequently this created almost impervious microporosity (Curry et al., 2007). The physical association justified greater preservation potential of SOM by inhibiting enzymatic breakdown; the enzyme-controlled degradation becomes diffusion limited and much of the SOM occupies pore space inaccessible to the reactants (Rothman and Forney, 2007).

3.4 ROLE OF SRO MINERALS ON PHOSPHORUS AND POTASSIUM FIXATION

3.4.1 PHOSPHORUS FIXATION

Phosphorus (P) is the second most important nutrient for the development of plants and it is limiting crop production in many areas of the world (Holford, 1997). Availability of P to crops is very important to make agriculture more efficient and sustainable at the global level (Vance, 2001). Only a small portion of P is available by the plants mainly due to chemical precipitation and physical adsorption of P by the soil (Barrow, 1980). Increase the phosphorus concentration in the soil solution through LMWOAs is well-known today (Bolan et al., 1994). The LMWOAs have been attributed to desorption and solubilization of P from adsorbent. The effect of LMWOAs on P adsorption/desorption may occur through the involvement of three main mechanisms: (1) competition for P adsorption sites, (2) dissolution of adsorbents, and (3) changes in the surface charge of adsorbents (Traina et al., 1986). In the allophanic soil, the active surface of allophane and imogolite is considered for P fixation sites. Chemisorption and displacement of structural silicon are included in the fixation mechanisms in these soils. The competition of oxalate with phosphorus on oxides or the competition of FA and phosphate on gibbsite and goethite is well documented (Violante et al., 1991). Soil pH is a major factor that influences the competition between LMWOAs and P. At low pH, phosphate adsorption is lower than at high pH. Mucilages are also involved in plant nutrition of phosphorus (Nagarajah et al., 1970) by decreasing the adsorption of phosphate on goethite (Grimal et al., 2001).

The allophane strongly retains phosphate and organic matter due to its inherent chemical characteristics such as high surface area and reactivity (Takahasihi et al., 1993). In a study, it was observed that phosphate sorption on the solids was hindered in the presence of tartrate and its efficiency increased by increasing the molar ratio of initial tartrate/phosphate. Furthermore, phosphate sorption by titrate in the acid system was higher as compared to alkaline systems (Cristofaro et al., 2000). The hindrance of P sorption was much more effective when equimolar mixture of tartrate and oxalate was used as compared to the tartrate alone under the same organic ligand concentrations. Such results can be explained as more sites with high affinity for both the organic ligands were occupied by tartrate and oxalate than by tartrate alone (He et al., 1999). In a study, it was observed that total phosphorus content of Andosols (993–3469 mg kg^{-1}) was greater than those of the fully developed Ultisols (733–1108 mg kg^{-1}) (Escudey et al., 2001). Saigusa and Matsuyama (1998) revealed that the presence of allophane in the clay fraction of Andosols are the main reason for such unique properties such as thick humus horizon, high fixation of P, high water-holding capacity, and low bulk density. Allophane content explained about 72% of the phosphate sorption capacity of European volcanic ash soils (Füleky, 2004). However, the allophane is more reactive and had better involvement in P fixation as compared to imoglite (Henmi et al., 1982). The phosphate retention in allophenic soils is mainly due to the ability of amorphous aluminum silicate soils to adsorb phosphate strongly through the ligand-exchange mechanism (Parfitt, 1980). Matsuyama et al. (1998) found that nonallophanic Andosols also adsorbed more phosphate per unit of acid-oxalate extractable aluminum, sometimes it is even higher than allophanic Andosols. It suggested that Al from humus complex is more active in the P adsorption than that from allophane. The allophane is a natural nanoclay, which can be used as low cost and environmentally friendly material for effective removal of P from a wide range of concentrations and hence it is used for the eutrophic water remediation and the treatment of P-rich effluent and sewage (Yuan and Wu, 2007).

Oxalate-soluble Al was reported to explain about 73% of the phosphate sorption ability of European volcanic soils (Füleky, 2010). Gunjigake and Wada (1980) also indicated that high reactive forms of Al in Andisols: (1) allophane and imogolite, (2) allophane-like components, (3) Al associated with humic substances, and (4) aluminum present in the interlayers of expandable phyllosilicates is involved in fixing phosphate. Córdova et al. (1996) reported that high P fixing capacity is found in the soil in which Al–humus complex is found in sufficient quantity and in this type of soil, a large amount of P application is necessary for the nutritional management of crops. SOM influences P fixation by blocking of P fixing sites. Allophanic soil of Japan showed a different trend in P fixation based on their SOM contents (Imai et al., 1981). It was observed that the adsorption of phosphate was decreased by increasing the content of organic matter and increased with the increasing content of allophanic materials (Pigna and Violante, 2003). Chatterjee et al. (2014a) presented that phosphorus and potassium fixation occurs on the SRO minerals. However, some other factors such as active aluminum and iron and crystalline phyllosilicates may also working simultaneously in the natural system. Not only soil components but also some management methods such as application of fertilizer and rhizospheric effect (due to differential presence of SRO minerals) can change the P and K fixation capacity of the soil–humus complex. They also reported that allophane content (for noncolloidal clay–humus complex), kaolinite content, ferrihydrite, crystalline iron, and poorly ordered iron showed positive correlation with P fixation capacity in both colloidal and noncolloidal clay–humus fractions.

3.4.2 POTASSIUM FIXATION

Potassium (K) is an essential nutrient that is needed for plant growth and is a dynamic ion in soil systems; its importance is well recognized in the agricultural sector (Martin and Sparks, 1985; Das et al., 2018). Among the various forms of K such as exchangeable and solution K equilibrates rapidly, whereas nonexchangeable K equilibrates very slowly with the exchangeable and solution forms. The transfer of potassium from minerals to any of the other three forms is extremely slow; these states of potassium are essentially unavailable to crops (Havlin et al., 1999). The total removal of K from the soil is always greater than the decrease in water soluble and exchangeable soil K. The nonexchangeable K fraction is released when the levels of soil solution and exchangeable K are reduced by plant uptake and leaching (Niebes et al., 1993). Soils that are rich in vermiculite and mica can have large amounts of nonexchangeable K, whereas soils containing kaolinite, quartz, and other siliceous minerals contain less available and exchangeable K (Martin and Sparks, 1985). A considerable improvement in nonexchangeable K-release was reported in manured plots (Srinivasarao et al., 2010).

The amorphous minerals K-fixing capacity was in the range up to 2–6 mEq/100 g. However, K-fixation capacity and composition were not related. Such a relationship would not be expected, if potassium is being fixed by channels in amorphous mineral acting as molecular sieves (Van Reeuwijk and de Villiers, 1968). In natural system, Chatterjee et al. (2014a) observed that mean K fixation was highest in depleted fertility status, Vertisol and rhizosphere soils for both colloidal and noncolloidal clays. They also showed a positive significant correlation between AFAS content for the K fixation in both colloidal and noncolloidal clay–humus complex (Chatterjee et al., 2014a). Li et al. (2010) reported that in rapeseed–rice rotation after one cycle, the amount and rate of K fixation was more in rhizosphere soil than in nonrhizosphere soil. The K added treatments had significantly less amount and rate of K fixation compared to the K skipped treatments. Generally, K^+ adsorption mechanism is almost similar to NH_4^+ adsorption mechanism. Similar to K fixation by SROs, a significant correlation between NH_4^+ adsorption potential and the amount of poorly crystallized illite is also found (Liu et al., 2008). Hinsinger et al. (1993) observed that phlogopite transforms into vermiculite when the K^+ concentration decreases below a threshold of about 80 mmol m^{-3} in the rhizosphere of annual ryegrass and rape within several weeks or months.

There have been several reports in which soil volcanic materials are formed which have a strong K^+ or NH_4^+ retention (Escudey et al., 1997). However, no common mechanism has yet been identified to define this high K selectivity. However, high values of K selectivity of soils containing SRO minerals can be explained by the presence of: (1) halloysite–smectite mixed-layer clays, (2) the presence of 2:1 layer silicate contaminants, (3) the presence of trace concentrations of zeolites (e.g., clinoptilolite), and (4) the existence of a high-charge halloysite with a strong affinity for K (Takahashi et al., 2001).

3.5 IMPLICATION OF SROS ON SUSTAINABLE AGRICULTURE

SRO minerals play a very important role in soil management in relation to dispersion, flocculation, aggregation, infiltration, erosion, and landscape stability. The SRO minerals control some physical properties such as the high water retention capacity, slippery but nonsticky consistency, high Atterberg limits, larger values for liquid limits and plastic limits, low bulk densities, high pores ratios, and inconsistent compaction behaviors. The chemical properties which considered the ability to sequester carbon, pH-dependent cation and anion exchange capacity, large fixation capacity of phosphate, adsorption, and high pH-low base saturation relationship are also shown by these minerals (Wada and Harward, 1974). Separation of SROs without destroying their associations with crystalline materials and organic matter is practically impossible. Therefore, the chemical composition and the characteristic properties ascribed to allophane are difficult to evaluate in a quantitative way.

KEYWORDS

- allophane
- carbon sequestration
- imogolite
- K fixation
- P fixation
- short-range order (SRO) minerals

REFERENCES

Aguilera, S. M., Borie, G., Peirano, P. and Galindo, G. (1997). Organic matter in volcanic soils in Chile: chemical and biochemical characterization. *Commun. Soil Sci. Plant Anal.*, **28**: 899–912.

Barbera, V., Raimondi, S., Egli, M. and Plötze, M. (2008). The influence of weathering processes on labile and stable organic matter in Mediterranean volcanic soils. *Geoderma*, **143**: 191–205. doi: 10.1016/j.geoderma.2007.11.002.

Barrow, N. J. (1980). Evaluation and utilization of residual phosphorus in soils. In: *The Role of Phosphorus in Agriculture*. Khasawneh, F. E. and Sample, E. C. (Eds.), American Society of Agronomy, Madison, WI, USA, 333–359.

Basile-Doelsch, I., Amundson, R., Stone, W. E. E., Borschneck, D., Bottero, J. Y., Moustier, S., Masin, F. and Colin, F. (2007). Mineral control of carbon pools in a volcanic soil horizon. *Geoderma*, **137**: 477–489. doi: 10.1016/j.geoderma.2006.10.006.

Basile-Doelsch, I., Amundson, R., Stone, W. E. E., Masiello, C. A., Bottero, J. Y., Colin, F., Masin, F., Borschneck, D. and Meunier, J. D. (2005). Mineralogical control of organic carbon dynamics in a volcanic ash soil on La Réunion. *Eur. J. Soil Sci.*, **56**: 689–703. doi: 10.1111/j.1365–2389.2005.00703.x.

Bertin, C., Yang, X. and Weston, L. A. (2003). The role of root exudates and allelochemicals in the rhizosphere. *Plant Soil*, **256**: 67–83.

Bock, M. J. and Mayer, L. M. (2000). Mesodensity organo-clay associations in a near-shore sediment. *Mar. Geol.*, **163**: 65–75.

Bolan, N. S., Naidu, R., Mahimairaja, S. and Baskaran, S. (1994). Influence of low-molecular-weight organic acids on the solubilization of phosphates. *Biol. Fertil. Soils*, **18**: 311–319.

Bosatta, E. and Ågren, G. (1997). Theoretical analyses of soil texture effects on organic matter dynamics. *Soil Biol. Biochem.*, **29**: 1633–1638.

Boudot, J. P. (1992). Relative efficiency of complexed aluminium noncrystalline Al hydroxide, allophane and imogolite in retarding the biodegradation of citric acid. *Geoderma*, **52**: 29–39. doi: 10.1016/0016–7061(92)90073-G.

Boudot, J. P., Hadj, B. A. B. and Chrone, T. (1986). Carbon mineralization in Andosols and aluminium-rich highlands soils. *Soil Biol. Biochem.*, **18**: 457–461.

Bowen, G. D. and Rovira, A. D. (1999). The rhizosphere and its management to improve plant growth. *Adv. Agron.*, **66**: 1–102.

Carlson, L. and Schwertmann, U. (1981). Natural ferrihydrites in surface deposits from Finland and their association with silica. *Geochim. Cosmochim. Acta*, **45**: 421–429.

Catroux, G. and Schnitzer, M. (1987). Chemical, spectroscopic and biological characteristics of the organic matter in particle size fractions separated from an aquoll. *Soil Sci. Soc. Am. J.*, **51**: 1200–1207.

Chatterjee, D., Kuotsu, R., Ao, M., Saha, S., Ray, S. K. and Ngachan, S. V. (2019). Does rise in temperature effect adversely on soil fertility, carbon fractions, microbial biomass and enzymatic activities under different land use? *Curr. Sci.*, **116**(12): 2044–2054 (accepted on March 12, 2019).

Chatterjee, D. and Saha, S. (2018). Response of soil properties and soil microbial communities to the projected climate change. In: Advances in Crop Environment Interaction. Bal, S. K., Mukherjee, J., Choudhury, B. U. and Dhawan, A. K. (Eds.), Springer, Singapore, 87–136.

Chatterjee, D., Datta, S. C. and Manjaiah, K. M. (2013). Clay carbon pools and their relationship with short-range order minerals: avenues to mitigate climate change? *Curr. Sci.*, **105**(10): 1404–1410.

Chatterjee, D., Datta, S. C. and Manjaiah, K. M. (2014a). Fractions, uptake and fixation capacity of phosphorus and potassium in three contrasting soil orders. *J. Soil Sci. Plant Nutr.*, **14**(3): 640–656. doi: 10.4067/S0718-95162014005000051.

Chatterjee, D., Datta, S. C. and Manjaiah, K. M. (2014b). Transformation of short-range order minerals in maize (*Zea mays* L.) rhizosphere. *Plant Soil Env*, **60**(6): 241–248.

Chatterjee, D., Datta, S. C. and Manjaiah, K. M. (2015a). Characterization of citric acid-induced transformation of short-range order minerals in Alfisol, Inceptisol and Vertisol of India. *Eur. J. Mineral.*, **27**: 551–557. doi: 10.1127/ejm/2015/0027-2446.

Chatterjee, D., Datta, S. C. and Manjaiah, K. M. (2015b). Effect of citric acid treatment on release of phosphorus, aluminum and iron from three dissimilar soils of India. *Arch. Agron. Soil Sci.*, **61**(1): 105–117. doi: 10.1080/ 03650340.2014.919449.

Chatterjee, D., Datta, S. C. and Manjaiah, K. M. (2016). Citric acid-induced potassium and silicon release in Alfisols, Vertisols and Inceptisols of India. *Proc. Natl. Acad. Sci. India Sect. B Biol. Sci.*, **86**(2):429–439. doi: 10.1007/s40011–014-0464-y.

Chevallier, T., Woignier, T., Toucet, J., Blanchart, E. and Dieudonne, P. (2008). Fractal structure in natural gels: effect on carbon sequestration in volcanic soils. *J. Sol-Gel Sci. Technol.*, **48**: 231–238. doi: 10.1007/s10971-008-1795-z.

Chiang, K. Y., Wang, Y. N., Wang, M. K. and Chiang, P. N. (2006). Low-molecular-weight organic acids and metal speciation in rhizosphere and bulk soils of a temperate rain forest in Chitou, Taiwan. *Taiwan J. For. Sci.*, **21**: 327–337.

Colombo, C., Ricciardella, M., Cerce, A. D., Maiuro, L. and Violante, A. (2004). Effect of tannate, pH, sample preparation, aging and temperature on the formation and nature of Al oxyhydroxides. *Clays Clay Miner.*, **52**: 721–733.

Conte, P., Spaccini, R., Chiarella, M. and Piccolo, A. (2003). Chemical properties of humic substances in soils of an Italian volcanic system. *Geoderma*, **117**: 243–250.

Córdova, J., Valverde, F. and Espinosa, J. (1996). Phosphorus residual effect in Andisols cultivated with potatoes. *Better Crops Int.*, **10**: 6–8.

Cornell, R. M. and Schwertmann, U. (2003). *The Iron Oxides*, VCH, Weinheim, Germany.

Cradwick, P. D. G., Farmer, V. C., Russell, J. D., Masson, C. R., Wada, K. and Yoshinaga, N. (1972). Imogolite, a hydrated aluminum silicate of tubular structure. *Nat. Phys. Sci.*, **240**: 187–189.

Cristofaro, A. D., He, J. Z., Zhou, D. H. and Violante, A. (2000). Adsorption of phosphate and tartrate on hydroxy-aluminum–oxalate precipitates. *Soil Sci. Soc. Am. J.*, **64**: 1347–1355.

Curry, K. J., Bennett, R. H., Mayer, L. M., Curry, A., Abril, M., Biesiot, P. M. and Hulbert, M. H. (2007). Direct visualization of clay microfabric signatures driving organic matter preservation in fine-grained sediment. *Geochim. Cosmochim. Acta*, **71**: 1709–1720.

Dachaufour, P. (1977). *Pédologie. I. Pédogenese et Classification.* Masson, Paris.

Dahlgren, R., Saigusa, M. and Ugolini, F. (2004). The nature, properties and management of volcanic soils. *Adv. Agron.*, **82**: 113–182.

Das, D., Nayak, A. K., Thilagam, V. K., Chatterjee, D., Shahid, M., Tripathi, R., Mohanty, S., Kumar, A., Lal, B., Gautam, P. and Panda, B. B. (2018). Measuring potassium fractions is not sufficient to assess the long-term impact of fertilization and manuring on soil's potassium supplying capacity. *J. Soils Sedim.*, **18**(5): 1806–1820. https://doi.org/10.1007/s11368-018-1922-6.

Datta, S. C., Takkar, P. N. and Verma, U. K. (2009). Effect of partial removal of adsorbed humus on kinetics of potassium and silica release by tartaric acid from clay–humus complex from two dissimilar soil profiles. *Aust. J. Soil Res.*, **47**: 715–724.

Eggleton, R. A. and Fitzpatrick, R. W. (1988). New data and a revised structural model for ferrihydrite. *Clays Clay Miner.*, **36**: 111–124.

Egli, M., Nater, M., Mirabella, A., Raimondi, S., Plötz, M. and Alioth, L. (2008). Clay minerals, oxyhydroxide formation, element leaching and humus development in volcanic soils. *Geoderma*, **143**: 101–114. doi: 10.1016/j.geoderma.2007.10.020.

Egli, M., Alioth, L., Mirabella, A., Raimondi, S., Nater, M. and Verel, R. (2007). Effect of climate and vegetation on soil organic carbon, humus fractions, allophanes, imogolite, kaolinite, and oxyhydroxides in volcanic soils of Etna (Sicily). *Soil Sci.*, **172**: 673–691.

Escudey, M., Diaz, P., Foerster, J. E. and Galindo, G. (1997). Adsorbed ion activity coefficients in K–Ca exchange on soil fractions derived volcanic materials. *Aust. J. Soil Res.*, **35**: 123–130.

Escudey, M., Galindo, G., Förster, J. E., Briceño, M., Diaz, P. and Chang, A. (2001). Chemical forms of phosphorus of volcanic ash-derived soils in Chile. *Commun. Soil Sci. Plant Anal.*, **32**: 601–616.

Eswaran, H., van den Berg, E. and Reich, P. (1993). Organic carbon in soils of the world. *Soil Sci. Soc. Am. J.*, **57**: 192–194.

Füleky, G. (2004). Phosphate sorption of European volcanic soils. In: Arnalds, O. and Oskarsson, H. Volcanic Soil Resources in Europe, Rala Report no 214, COST action 622 final meeting Abstracts, pp. 100–101.

Füleky, G. (2010). Phosphate sorption capacity of European volcanic soils. *Agrokémia és Talajtan*, **59**: 77–84.

Gregory, P. J. and Atwell, B. J. (1991). The fate of carbon in pulse labeled crops of barley and wheat. *Plant Soil*, **136**: 205–213.

Grimal, J. Y., Frossard, E. and Morel, J. L. (2001). Maize root mucilage decreases adsorption of phosphate on goethite. *Biol. Fertil. Soils*, **33**: 226–230.

Gunjigake, N. and Wada, K. (1980). Effects of phosphorus concentration and pH on phosphate retention by active aluminum and iron of Andosols. *Soil Sci.*, **132**: 347–352.

Gustafsson, J. P., Bhattacharya, P., Bain, D. C., Fraser, A. R. and McHardy, W. J. (1995). Podzolization mechanisms and the synthesis of imogolite in northern Scandinavia. *Geoderma*, **66**: 167–184.

Havlin, J. L., Beaton, J. D., Tisdale, S. L. and Nelson, W. L. (1999). *Soil Fertility and Fertilizers*. Prentice-Hall International (UK) Limited, London.

He, J. Z., Cristofaro, A. D. and Violante, A. (1999). Comparison of adsorption of phosphate, tartrate, and oxalate on hydroxy aluminum montmorillonite complexes. *Clays Clay Miner.*, **47**: 226–233.

Hemni, T., Nakai, M., Nakata, T. and Yoshinaga, N. (1982). Removal of Phosphate by Utilizing Clays of Volcanic Ash Soil and Weathered Pumice. Memoirs of College of Agriculture, Ehime University, pp. 17–24.

Hinsinger, P., Elsass, F., Jaillard, B. and Robert, M. (1993). Root-induced irreversible transformation of a trioctahedral mica in the rhizosphere of rape. *J. Soil Sci.*, **44**: 535–545.

Holford, I. C. R. (1997). Soil phosphorus: its measurement and its uptake by plants. *Aust. J. Agric. Res.*, **35**: 227–239.

Hsu, P. H. (1963). Effect of initial pH, phosphate and silicate on the determination of aluminum with aluminon. *Soil Sci.*, **96**: 230–238.

Huang, P. M. and Wang, M. K. (1997). Formation chemistry and selected surface properties of iron oxides. In: *Advances in Geoecology: Soil and Environments*. Auerswald, K., Stanjek, H. and Bigham, J. (Eds.), Catena, Reiskirchen, Germany, 241–270.

Imai, H., Goulding, K. W. T. and Talibudeen, O. (1981). Phosphate adsorption in allophanic soils. *J. Soil Sci.*, **32**: 555–570.

Johnson-Maynard, J. L., McDaniel, P. A., Falen A. L. and Ferguson, D. E. (1997). Chemical and mineralogical conversion of Andisols following invasion by bracken fern. *Soil Sci. Soc. Am. J.*, **61**: 549–555.

Karltun, E., Bain, D., Gustafsson, J. P., Mannerkoski, H., Murad, E., Wagner, U., Fraser, T., McHardy, B., Melkerud, P. A. and Starr, M. (2000). Surface reactivity of poorly ordered minerals in podzol B horizon. *Geoderma*, **94**: 263–286.

Kawano, M. and Obokata, S. (2007). The effect of amino acids on the dissolution rates of amorphous silica in near-neutral solution. *Clays Clay Miner.*, **55**: 361–368.

Khademi, H. and Arocena, J. M. (2008). Kaolinite formation from palygorskite and sepiolite in rhizosphere soils. *Clays Clay Miner.*, **56**: 429–436.

Konta, J. (2009). Phyllosilicates in the sediment-forming processes: weathering, erosion, transportation, and deposition. *Acta Geodyn. Geomater.*, **6**: 13–43.

Krishnamurti, G. S. R., Sarma, V. A. K. and Rengasamy, P. (1976). Amorphous ferri-aluminosilicates in some tropical ferruginous soils. *Clay Miner.*, **11**: 137–147.

Kuzyakov, Y. and Domanski, G. (2000). Carbon input by plants into the soil. *J. Plant Nutr. Soil Sci.*, **163**: 421–431. doi: 10.1002/1522-2624(200008) 163:4<421::AID-JPLN421>3.0.CO;2-R.

Lal, R. (2003). Global potential of soil carbon sequestration to mitigate the greenhouse effect. *Crit. Rev. Plant Sci.*, **22**: 151–184.

Li, X., Lu, J., Wu, L., Chen, F. and Malhi, S. S. (2010). Potassium fixation and release characteristics in rhizosphere and nonrhizosphere soils for a rapeseed–rice cropping sequence. *Commun. Soil Sci. Plant Anal.*, **41**: 865–877.

Lilienfein, J., Qualls, R. G., Uselman, S. M. and Bridgham, S. D. (2003). Soil formation and organic matter accretion in a young andesitic chronosequence at Mt. Shasta, California. *Geoderma*, **116**: 249–264.

Lowe, D. J. (1995). Teaching clays: from ashes to allophane. In: *Clays: Controlling the Environment*. Churchman, G. J., Fitzpatrick, R. W. and Eggleton, R. A. (Eds.), *Proceedings 10th International Clay Conference*, Adelaide, Australia (1993). CSIRO Publishing, Melbourne, 19–23.

Lundström, U. S. and Giesler, R. (1995). Use of aluminum species composition in soil solution as an indicator of acidification. *Ecol. Bull.*, **44**: 114–122.

Mandal, D. and Datta, S. C. (2005). Clay-organic complexation and mineral transformation in the rhizosphere during cropping: a study through differential X-ray diffraction (DXRD) and fitting of XRD profile. *Clay Res.*, **24**: 169–181.

Manjaiah, K. M., Kumar, S., Sachdev, M. S., Sachdev, P. and Datta, S. C. (2010). Study of clay–organic complexes. *Curr. Sci.*, **98**: 915–921. doi: 533, 35400019165191.0150.

Martin, H. W. and Sparks, D. L. (1985). On the behavior of nonexchangeable potassium in soils. *Commun. Soil Sci. Plant Anal.*, **16**: 133–162.

Martin, J. P., Zunino, H., Peirano, P., Caiozzi, M. and Haider, K. (1982). Decomposition of C^{14}-labeled lignins, model humic acid polymers, and fungal melanins in allophanic soils. *Soil Biol. Biochem.*, **14**: 289–293.

Matsuyama, M., Kudo, K. and Saigusa, M. (1998). Active aluminum of cultivated Andosols and related soil chemical properties in Japan. *Bull. Fac. Agr. Life Sci.*, **1**: 30–36.

Matus, F., Amigo, X. and Kristiansen, S. M. (2006). Aluminum stabilization controls organic carbon levels in Chilean volcanic soils. *Geoderma*, **132**: 158–168. doi: 10.1016/j.geoderma.2005.05.005.

Matus, F., Garrido, E., Sepúlveda, N., Cárcamo, I., Panichini, M. and Zagal, E. (2008). Relationship between extractable Al and organic C in volcanic soils of Chile. *Geoderma*, **148**: 180–188.

Mayer, L., Schick, L., Hardy, K., Wagai, R. and McCarthy, J. (2004). Organic matter content of small mesopores in sediments and soils. *Geochim. Cosmochim. Acta*, **68**: 3863–3872.

McColl, J. G. and Pohlman, A. A. (1986). Soluble organics and their chelating influence on aluminum and other metal dissolution in forest soils. *Water Air Soil Pollut.*, **31**: 917–927.

Mikutta, R., Kleber, M. and Jahn, R. (2005). Poorly crystalline minerals protect organic carbon in clay subfractions from acid subsoil horizons. *Geoderma*, **128**: 106–115. doi: 10.1016/j.geoderma.2004.12.018.

Munda, S., Bhaduri, D., Mohanty, S., Chatterjee, D., Tripathi, R., Shahid, M., Kumar, U., Bhattacharya, P., Kumar, A., Adak, T. and Jangde, H. K. (2018). Dynamics of soil organic carbon mineralization and C fractions in paddy soil on application of rice husk biochar. *Biomass Bioenergy*. **115**: 1–9.

Nagarajah, S., Posner, A. M. and Quirk, J. P. (1970). Competitive adsorptions of phosphate with polygalacturonate and other organic anions on kaolinite and oxide surfaces. *Nature*, **228**: 83–84.

Niebes, J. F., Dufey, J. E., Jaillard, B. and Hinsinger, P. (1993). Release of non-exchangeable potassium from different size fractions of two highly K-fertilized soils in the rhizosphere of rape (*Brassica napus* cv Drakkar). *Plant Soil*, **155**: 403–406.

Nogami, K. (2004). Relationship in chemical composition between mother solution and allophane-like aluminosilicate precipitate through neutralization of acid hydrothermal water by seawater. *Earth Planets Space*, **56**: 457–462.

Óskarsson, H., Arnalds, Ó., Gudmundsson, J. and Gudbergsson, G. (2004). Organic carbon in Icelandic Andosols: geographical variation and impact of erosion. *Catena*, **56**: 225–238.

Parfitt, R. L., Theng, B. K. G., Whitton, J. S. and Shepherd, T. G. (1997). Effects of clay minerals and land use on organic matter pools. *Geoderma*, **75**: 1–12. doi: 10.1016/ S0016-7061(96)00079-1.

Parfitt, R. L. (1980). Chemical properties of variable charge soils. In: *Soils with Variable Charge*. Theng, B. K. G. (Ed.), New Zealand Soil Bureau Publishing, Lower Hutt, New Zealand, 167–194.

Parfitt, R. L. (1990). Allophane in New Zealand—a review. *Aust. J. Soil Res.*, **28**: 343–360.

Percival, H. J., Parfitt, R. L. and Scott, N. A. (2000). Factors controlling soil carbon levels in New Zealand grassland: is clay content important? *Soil Sci. Soc. Am. J.*, **64**: 1623–1630. doi: 10.2136/sssaj2000.6451623x.

Pigna, M. and Violante, A. (2003). Adsorption of sulfate and phosphate on Andisols. *Commun. Soil Sci. Plant Anal.*, **34**: 2099–2113.

Ransom, B., Bennett, R. H., Baerwald, R. and Shea, K. (1997). TEM study of in situ organic matter on continental margins: occurrence and the "monolayer" hypothesis. *Mar. Geol.*, **138**: 1–9.

Righi, D., Dinel, H., Shulten, H. R. and Schnitzer, M. (1995). Characterization of clay–organic matter complexes to oxidation by peroxide. *Eur. J. Soil Sci.*, **46**: 423–429.

Rothman, D. H. and Forney, D. C. (2007). Physical model for the decay and preservation of marine organic carbon. *Science*, **316**: 1325–1328.

Ryan, P. R., Dessaux, Y., Thomashow, L. S. and Weller, D. M. (2009). Rhizosphere engineering and management for sustainable agriculture. *Plant Soil*, **321**: 363–383.

Saggar, S., Parshotam, A., Sparling, G. P., Feltham, C. W. and Hart, P. B. S. (1996). ^{14}C-labeled ryegrass turnover and residence times in soils varying in clay content and mineralogy. *Soil Biol. Biochem.*, **28**: 1677–1686.

Saigusa, M. and Matsuyama, N. (1998). Active aluminum of cultivated Andosols and related soil chemical properties in Japan. *Tohoku. J. Agr. Res.*, **48**: 75–83.

Sarkar, A. N., Jenkins, D. A. and Jones, R. G. W. (1979). Modifications to mechanical and mineralogical composition of soil within the rhizosphere. In: *The Soil–Root Interface*. Harley, J. L. and Russel, R. S. (Eds.), Academic Press, San Diego, pp. 125–136.

Shoji, S. and Fujiwara, Y. (1984). Active aluminum and iron in the humus horizons of Andosols from northeastern Japan: their forms, properties, and significance in clay weathering. *Soil Sci.*, **137**: 216–226.

Shoji, S., Nanzyo, M. and Dahlgren, R. (1993). Volcanic ash soils—genesis, properties and utilization. *Dev. Soil Sci.*, **21**: 1–288.

Sparks, D. L. (1995). *Environmental Soil Chemistry.* Academic Press, San Diego.

Srinivasarao, C. H., Vittal, K. P. R., Kundu, S., Gajbhiye, P. N. and Babu, M. V. (2010). Continuous cropping, fertilization, and organic manure application effects on potassium in an Alfisol under arid conditions. *Commun. Soil Sci. Plant Anal.*, **41**: 783–796.

Ström, L., Olsson, T. and Tyler, G. (1994). Differences between calcifuge and acidifuge plants in root exudation of low-molecular organic acids. *Plant Soil*, **167**: 239–245.

Takahashi, T., Dahlgren, R. and van Sustcren, P. (1993). Clay mineralogy and chemistry of soils formed in volcanic materials in xeric moisture regime of northern California. *Geoderma*, **59**: 131–150.

Tan, K. H., Perkins, H. F. and McCreery, R. A. (1970). The nature and composition of amorphous material and free oxides in some temperate region and tropical soils. *Commun. Soil Sci. Plant Anal.*, **1**: 227–238.

Tazaki, K. (1979). Scanning electron microscopic study of imogolite formation from plagioclase. *Clays Clay Miner.*, **27**: 209–212.

Theng, B. K. G. and Tate, K. R. (1989). Interactions of clays with soil organic constituents. *Clay Res.*, **8**: 1–10.

Tonneijck, F., Jansen, B., Nierop, L., Verstraten, K. and Sevink, J. (2009). Toward understanding of carbon stabilization mechanisms in volcanic ash soils in Andean ecosystems. *Earth Env. Sci.*, **6**: 042033. doi: 10.1088/1755-1307/6/4/042033.

Torn, M. S., Trumbore, S. E., Chadwick, O. A., Vitousek, P. M. and Hendricks, D. M. (1997). Mineral control of soil organic carbon storage and turnover. *Nature*, **398**: 170–173.

Traina, S. J., Sposito, G., Hesterberg, D. and Kafkafi, U. (1986). Effects of pH and organic acids on orthophosphate solubility in an acidic, montmorillonitic soil. *Soil Sci. Soc. Am. J.*, **50**: 45–52.

Tu, S. X., Guo, Z. F. and Sun, J. H. (2007). Effect of oxalic acid on potassium release from typical Chinese soils and minerals. *Pedosphere*, **17**: 457–466.

Ugolini, F. C. and Dahlgren R. A. (2002). Soil development in volcanic ash. *Global Env. Res.*, **6**: 69–82.

Uren, N. C. (2000). Types, amounts, and possible functions of compounds released into the rhizosphere by soil-grown plants. In: *The Rhizosphere: Biochemistry and Organic Substances at the Soil–plant Interface*. Pinton, R., Varanini, Z. and Nannipieri, P. (Eds.), Marcel Dekker, Inc., New York, 19–40.

Urrutia, M. M. and Beveridge, T. J. (1994). Formation of fine-grained silicate minerals and metal precipitates by a bacterial surface (*Bacillus subtilis*). *Chem. Geol.*, **116**: 261–280.

Urrutia, M. M. and Beveridge, T. J. (1995). Formation of short-range ordered aluminosilicates in the presence of a bacterial surface (*Bacillus subtilis*) and organic ligands. *Geoderma*, **65**: 149–165.

Vacca, S., Capra, G. F., Coppola, E., Rubino, M., Madrau, S., Colella, A., Langella, A. and Buondonno, A. (2009). From andic nonallophanic to nonandic allophanic Inceptisols on alkaline basalt in Mediterranean climate: a toposequence study in the Marghine district (Sardinia, Italy). *Geoderma*, **151**: 157–167. doi: 10.1016/j.geoderma.2009.03.024.

Van Hees, P. A. W. and Lundström, U. S. (2000). Equilibrium models of aluminum and iron complexation with different organic acids in soil solution. *Geoderma*, **94**: 201–221.

Van Hees, P. A. W., Andersson, A. M. and Lundström, U. S. (1996). Separation of organic low molecular weight aluminum complexes in soil solution by liquid chromatography. *Chemosphere*, **33**: 1951–1966.

Van Reeuwijk, L. P. and de Villiers, J. M. (1968). Potassium fixation by amorphous aluminosilicates gel. *Proc. Soil Sci. Soc. Am.*, **32**: 238–240.

Vance, C. (2001). Symbiotic nitrogen fixation and phosphorus acquisition: plant nutrition in a world of declining renewable resources. *Plant Physiol.*, **127**: 390–397.

Vance, G. E, Stevenson, E. J. and Sikora, E. J. (1996). Environmental chemistry of aluminum-organic complexes. In: *The Environmental Chemistry of Aluminum*, Sposito, G. (Ed.), CRC Press, Lewis Publisher, Boca Raton, Florida, 169–220.

Vempati, R. K. and Loeppert, R. H. (1989). Influence of structural and adsorbed Si on the transformation of synthetic ferrihydrite. *Clays Clay Miner.*, **37**: 273–279.

Violante, A. and Huang, P. M. (1993). Formation mechanism of aluminum hydroxide polymorphs *Clays Clay Miner.*, **41**: 590–597.

Violante, A., Colombo, C. and Buondonno, A. (1991). Competitive adsorption of phosphate and oxalate by aluminum oxides. *Soil Sci. Soc. Am. J.*, **55**: 65–70.

Wada, K. (1989). Allophane and imogolite. In: *Minerals in Soil Environments*, Dixon, J. B. and Weed, S. B. (Eds.), 2nd Edition. SSSA Book Ser., SSSA, Madison, WI, Ch. 21. 1051–1087.

Wada, K. and Harward, M. E. (1974). Amorphous clay constituents of soil. *Adv. Agron.*, **26**: 211–260.

Wiseman, C. L. S. and Püttmann, W. (2006). Interactions between mineral phases in the preservation of soil organic matter. *Geoderma*, **134**: 109–118.

Woignier, T., Pochet, G., Doumenc, H., Dieudonné, P. and Duffours, L. (2007). Allophane: a natural gel in volcanic soils with interesting environmental properties. *J. Sol-Gel Sci. Technol.*, **41**: 25–30. doi: 10.1007/s10971-006-7593-6.

Woignier, T., Primera, J. and Hashmy, A. (2006). Application of the DLCA model to "natural" gels: the allophanic soils. *J. Sol-Gel Sci. Technol.*, **40**: 201–207.

Yagasaki, Y., Mulder, J. and Okazaki, M. (2006). The role of soil organic matter and short-range ordered aluminosilicates in controlling the activity of aluminum in soil solutions of volcanic ash soils. *Geoderma*, **137**: 40–57.

Yonebayashi, K. and Hattori, T. (1988). Chemical and biological studies on environmental humic acids: I. composition of elemental and functional groups of humic acids. *Soil Sci. Plant Nutr.*, **34**: 571–584.

Yonebayashi, K. and Hattori, T. (1989). Chemical and biological studies on environmental humic acids: II. ^{1}H-NMR and IR spectra of humic acids. *Soil Sci. Plant Nutr.*, **35**: 383–392.

Yuan, G. and Wu, L. (2007). Allophane nanoclay for the removal of phosphorus in water and wastewater. *Sci. Tech. Adv. Mater.*, **8**: 60–62.

Zunino, H. and Borie, F. (1985). Materia orgánica y procesos biológicos en suelos alofánicos. In: *Suelos Volcánicos de Chile*. Tosso, J. (Ed.), INIA, Santiago, Chile.

PART II
Potential Utilization of Microbes for Stressed Agriculture

PART II

CHAPTER 4

Rhizospheric Microbes for Nutrient Fortification in Crop Plants

SHALOO VERMA, DIKCHHA SINGH, MALAY KR DAS,
GYAN PRAKASH SRIVASTAVA, MURUGAN KUMAR, and
HILLOL CHAKDAR*

*ICAR-National Bureau of Agriculturally Important Microorganisms
(NBAIM), Kushmaur, Mau 275 103, Uttar Pradesh, India*

*Corresponding author. E-mail: hillol.chakdar@gmail.com

ABSTRACT

In contrast to the macronutrients, elements like Fe, Zn, Mn, Cu, Se, Ca, etc., are required in very low amount for both plants and humans but undoubtedly they are indispensable for essential cellular functions. Availability of these micronutrients in soil may be limited due to diverse physicochemical factors and even if they are accumulated in sufficient quantity in plants, their bioavailability for humans may be restricted by antinutritional factors and losses during processes like milling, polishing. The plant rhizosphere represents a dynamic and metabolically active niche densely populated by diverse microorganisms. Endowed with versatile biological functions, rhizosphere inhabiting microorganisms can increase the plant availability of micronutrients through solubilization, mineralization, chelation, etc. Further, such microorganisms can improve the root architecture which can increase plants' nutrient use efficiency and also modulate the ion transporters to enhance the uptake of micronutrients. Production of phytase through rhizospheric microorganisms can help to dephytinize edible plant parts/foods to enhance micronutrient bioavailability. Microorganism offers a sustainable solution to complement agronomic and genetic approaches to enrich crop plants with micronutrients and thereby combating the problem of "hidden hunger."

4.1 INTRODUCTION

Micronutrients are low in the majority of the staple crops like rice, wheat, potato, etc., on which majority of the global population are dependent. Fe and Zn are the two most important micronutrients whose bioavailability in soil, crop plant, food as well as in human are limited. Due to their involvement in multitude of biological functions, deficiency of Fe and Zn is widely distributed especially in developing nations. Apart from inherent low bioavailability of Fe and Zn in cereals, postharvest processes like polishing, milling, and pearling also lower the amount of these micronutrients (Borg et al., 2009). Moreover, antinutritional factors like phytates, fibers, tannins, etc., may further lower the availability of these minerals (White and Broadley 2005; Brinch-Pedersen et al., 2007; Pfeiffer and McClafferty 2007). It is estimated that more than two billion people are iron deficient while more than a billion are at risk of zinc deficiency (Tulchinsky 2010; Joy et al., 2015a). The application of specific-micronutrient bearing chemical fertilizers had been tried but was not effective as they form complexes in soil which cannot be taken up by plants efficiently. Among other interventions, dietary supplementation or diversification had been used in the past but had limited success. Biofortification has now emerged as an efficient strategy to sustainably enhance the bioavailable content of micronutrients in staple food crops. A number of strategies like conventional and molecular breeding, genetic modification, agronomic, and/or soil management have been applied to increase the micronutrient availability in food crops. As most of the micronutrients available in soil are in inaccessible form, so to make them available for plants, breeding strategies may not work always due to the fact that a variety/line may not achieve its genetic potential when the nutrients are not present in the bioavailable form in soil. Soil microorganisms which remain in close association to the plant roots perform key functions in biogeochemical cycling. Due to their versatile biological properties and ecological services rendered in plant nutrition and sustenance, they hold considerable promise for the fortification of nutrients. The application of microorganisms as a part of soil management can make the unavailable nutrients available to plants and can also modulate the specific transporters for enhanced uptake. In this chapter, we will discuss the potential of rhizospheric microorganisms for fortifying micronutrients in crop plants and their edible parts. Although not widely used this approach hold the realistic potential to supplement other traditional approaches for combating the "hidden hunger" caused due to micronutrient deficiency.

4.2   RHIZOSPHERIC MICROORGANISMS: AN OVERVIEW

The region of soil surrounding the plant roots where the highest microbiological activities are found is generally known as rhizosphere (Figure 4.1). Secretion of root exudates containing numerous sugars, amino acids, vitamins, and other chemical signals make the root adhering region a conducive niche for colonization and proliferation. The composition of root exudates strongly influences the nature and number of colonizers (Nautiyal and Dion 2008;   Zhalnina et al., 2018). In return to nutritional factors from plants, microorganisms also render multiple services to plants. The services or functions of rhizospheric microbes may broadly be divided into two major categories: (1) *Direct*: for example, solubilization/mineralization of fixed form of nutrients, fixation of nitrogen, production of phytohromones, chelation of nutrients, etc., through which plant growth and nutrition are directly influenced and (2) *Indirect*: for example, production of antibiotics, cyanides, lytic enzymes, antimicrobial peptide, high affinity chelators, etc., which can suppress the growth of pathogens and indirectly help in plants sustenance. Rhizospheric microorganism has been well-studied and used for increasing the plant availability of nutrients like nitrogen, phosphorus, potassium, and as biocontrol agents for multiple pathogens. *Rhizobium, Bacillus, Paenibacillus, Azotobacter, Azospirillum, Pseudomonas, Enterobacter, Pantoea, Streptomyces*, etc., are some well-known genera which are abundantly present in rhizosphere of various plants and exert versatile plant beneficial functions (Backer et al., 2018). Besides nitrogen fixers, nutrient solubilizers/mineralizers have received huge research focus globally due to their ability to increase availability of nutrients like P, K, through organic acid production or secretion of phosphatases. The organic acid producing microorganisms can solubilize the insoluble nutrients in rhizosphere and make them available to plants. Not only for P or K, but this mechanism also works for Zn and availability of Zn in soil solution can be greatly increased. Similarly, siderogenic microbes producing chelators can complex with nutrients present in low amount and help them to be available for plants. Production of phytohormones like auxin by microorganisms can help as an additional trigger for root growth and development of plants which in turn help is achieving better nutrient use efficiency. The mechanisms of plant growth promotion have been discussed in detail by recent reviews by (Backer et al., 2018; Mhatre et al., 2019). With the growing concern of deteriorating soil health and fertility due to applications of chemical fertilizers and pesticides, biofertilizers and biopesticides based on rhizospheric microorganisms offer a sustainable solution to reduce the use of agrochemicals and thereby restoring the soil health. Obviously, the focus of earlier as well as majority of the present works related to characterization and application of rhizospheric microbes was on macronutrients, but studies report their potential application in enhancing micronutrient uptake and accumulation in plants. The following sections of this chapter discuss various mechanisms through which microorganisms can help to achieve micronutrient fortification in crop plants.

4.3 ROLE OF RHIZOSPHERIC MICROORGANISMS IN NUTRIENT FORTIFICATION OF CROP PLANTS

As discussed in the previous section, rhizospheric microbes exert versatile plant beneficial functions with great variety of mechanisms. In case of microbe-mediated micronutrient fortification, only a few of them are studied and understood (Figure 4.1). Globally, reports strongly suggest that microorganisms can effectively enrich the crop plants with various micronutrients (Table 4.1). This section discusses those selected mechanisms.

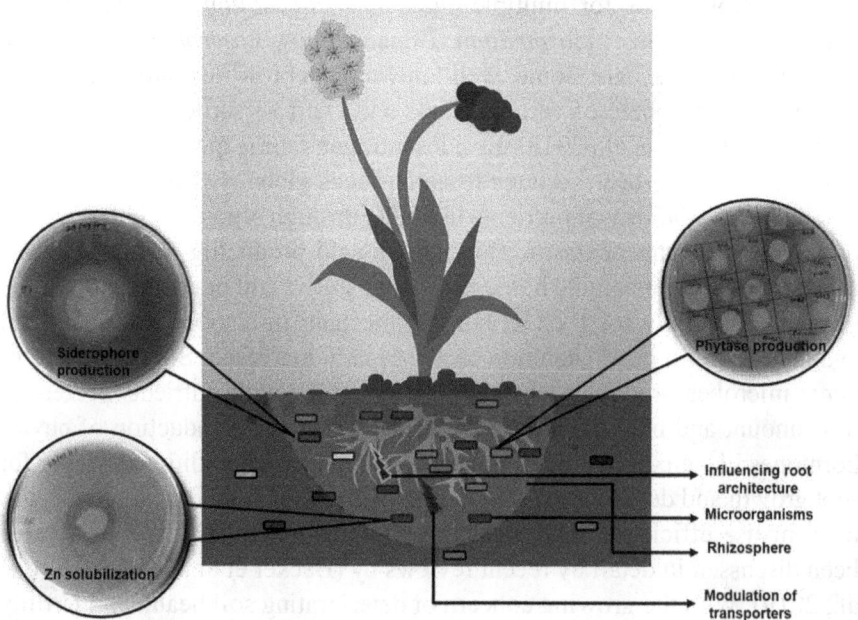

FIGURE 4.1 Diagrammatic representation of rhizosphere and potential roles played by the rhizosphere inhabiting microorganisms for enriching micronutrients in crop plants.

TABLE 4.1 Some Recent Works Reporting Enrichment of Various Micronutrients in Crop Plants using Microbes Including Rhizospheric Microorganisms

Sr. No.	Crop	Inoculant	Isolation Source of Inoculant	Enhancement Level of Micronutrients	Reference
1.	Wheat	*Bacillus subtilis* ZM63 and *B. aryabhattai* ZM31	Maize	68% and 78% increase in grain Zn and Fe, respectively	Mumtaz et al., 2018
2.		*B. subtilis* and AM fungi	Agricultural soil	62.5% and 38% increase in Fe and Zn, respectively	Yadav et al., 2020
3.		AM fungi, *G. versiforme*, and *Funneliformis mosseae*	Maize	>1.85-fold increase in Selenium	Luo et al., 2019
4.		*B. subtilis, Arthrobacter* sp.	Wheat	2-fold increase of Zn content in grain	Singh et al., 2017
5.	Chickpea	*Enterobacter ludwigii*	Rice	Increased 18% Fe, 23% Zn, 19% Cu, 2% Mn, and 22% Ca	Gopalakrishnan et al., 2016
6.		*Enterobacter* sp. MN17	Agricultural soil	14% increase in grain Zn	Ullah et al., 2020
7.		*Streptomyces* sp. CAI-21 and *Streptomyces* sp. MMA-32	–	>35% increase in seed Fe	Sathya et al., 2016
8.	Pigeon pea	*Acinetobacter tandoii*	Rice	Increased 12% Fe, 5% Zn, 8% copper (Cu), 39% Mn, and 11% Ca	Gopalakrishnan et al., 2016
9.	Mung bean	*Pantoea dispersa* MPJ9 and *Pseudomonas putida* MPJ6	Mung bean	3.4-fold and 2.8-fold increase in Fe content, respectively	Patel et al., 2018
10.	Soybean	*Paraburkholderia megapolitana*	Castor bean	7.4-fold increase in Se content	Trivedi et al., 2019
11.		*Bradyrhizobium japonicum* and *Streptomyces griseoflavus*	Soybean and Sweet pea root respectively	Increase in Ca content by 1.3-fold in flowering stage, 1.95-fold in pod-fill stage, 1.47-fold in seed-fill stage, 1.18-fold in maturity	Htwe et al., 2018

TABLE 4.1 *(Continued)*

Sr. No.	Crop	Inoculant	Isolation Source of Inoculant	Enhancement Level of Micronutrients	Reference
12.	Sorghum	*Pseudomonas putida* P159 and *Pseudomonas fluorescens* T17–24	Agricultural soil	25.4% and 20.9% increase in Mn content	Abbaszadeh-Dahaji et al., 2019
13.	Maize	*Bacillus* sp. AZ6	Rhizosphere of maize	Zn contents of grain and shoot by 46% and 52%	Hussain et al., 2019
14.		*Azospirillum brasilense* (HM053)	–	4-fold higher iron content	Scott et al., 2020
15.	Quinoa	*Burkholderia phytofirmans* PsJN (with 1% biochar)	Agricultural soil	71% increase in Fe concentration	Naveed et al., 2020

4.3.1 SOLUBILIZATION OF NUTRIENTS BY MICROORGANISMS

In soil, plant nutrients are found in different forms in which some are insoluble but nutritionally important for both plant as well as humans (Joy et al., 2015b; Ku et al., 2019). In soil, micronutrients either rapidly reacts with other compounds to form precipitates or forms complexes which make them unavailable for plants (Dimkpa and Bindraban 2016). The uptake of micronutrient, such as Fe, Cu, Mn, and others, is preceded after their conversion to more soluble ionic forms (Dimkpa and Bindraban 2016). It is worthy to mention that rhizospheric microorganisms do not rely on any single strategy to solubilize the trace elements into bioavailable forms. Under nutrient-starved conditions, root exudates of plants stimulate rhizospheric microbes to produce wide range of organic acids (Meena et al., 2017). For example, the presence of glucose in ZnO liquid medium stimulated the production of acids like gluconic, malonic, oxalic, etc., by Zn solubilizing *Pseudomonas*, *Plantibacter*, *Streptomyces*, *Stenotrophomonas*, and *Curtobacterium* isolated from wheat rhizosphere. On the other hand, Zn solubilization may be mediated via proton extrusion (in *Plantibacter*) or complexation with organic acids (in *Curtobacterium*) in absence of glucose (Costerousse et al., 2018). Microorganism producing organic acids have the ability to bind metal ions in soil solution through complexation reaction. A number of factors like soil pH, number and position of carboxylic groups, chemical forms of the binding heavy metals, etc., influence the stability of metal ion-organic acid complexes (Jones 1998). Mostly, gluconic and keto-gluconic acid have been reported to be associated with Zn solubilization (Fasim et al., 2002; Saravanan et al., 2007b; Saravanan et al., 2007a; Sunithakumari et al., 2016). However, production of citric acid, malic acid, oxalic acid, tartaric acid, formic acid, and acetic acid has also been reported to be associated with Zn solubilization (Martino et al., 2003; Li et al., 2010; Sah et al., 2017). Increased zinc uptake by plants can lead to enhancement in grain zinc accumulation (Ramesh et al. (2014) implicated organic acids for the reduction of pH of soybean rhizosphere and enhanced uptake and accumulation of zinc by Zn and P solubilizing *Enterobacter cloaceae* MDSR9. Rhizobacterial isolates from wheat and sugarcane, such as *Pseudomonas fragi*, *Pantoea dispersa*, *Pantoea agglomerans*, *Enterobacter cloacae*, and *Rhizobium* sp. can solubilize Zn when tested on $ZnCO_3$ containing medium and help in increase Zn content and growth promotion in wheat (Kamran et al., 2017).

4.4   CHELATION OF NUTRIENTS

The rhizosphere is the habitat of a diverse range of microorganisms of which many of them are known for their beneficial role in the bioavailability of metal through chelation (Mishra et al., 2017; Kumar et al., 2017). Diverse microorganisms like bacteria, actinomycetes, fungi, and certain algae produce siderophores, a low molecular weight (>10 kDa) secondary metabolite that can specifically chelate ions like Fe^{3+} (Dimkpa and Bindraban, 2016; Khan et al. 2018). Iron is one of the most abundant element on earth and is essential for almost all living beings including plants (Shirvani and Nourbakhsh, 2010). The presence of Fe in insoluble minerals Goethite (FeOOH) or hematite (Fe_2O_3) limit the bioavailability of Fe for plants (Osorio Vega, 2007). Under this iron-limiting condition, certain microorganisms produce siderophores which have the high affinity to form complexes with Fe^{3+} (1:1) and can uptake of the complexes (Rashid et al., 2016). In addition to iron, siderophores can also bind with other metals such as Cu^{2+}, Mn^{2+}, Zn^{2+}, and many others which also have nutritional importance in plant growth and development (Braud et al., 2009; Meena et al., 2017). After transforming unavailable forms of micronutrient-bearing minerals into available forms, micronutrient can be taken up directly by plants as micronutrient–siderophore complexes or root-mediated processes, such as chelate degradation followed by release of ions or through ligand exchange reaction (Holmén and Casey, 1996; Radzki et al., 2013; Ahmed and Holmström, 2014). For example, iron-chelated bacterial siderophores from *Chryseobacterium* spp. and *Pseudomonas fluorescens* under iron-limited conditions resulted in better plant growth as compared to iron sufficient conditions (Radzki et al., 2013). Patel et al., (2018) reported that catecholate-type siderophore producing bacterial isolates such as *Pantoea dispersa* and *Pseudomonas putida*, enhanced iron content of the mung bean with 89.9% and 85.3%, respectively, siderophore activity. Root inoculation with *Pseudomonas putida* improved iron content in maize (Jumadi et al., 2019). Sathya et al. (2016) reported that siderophore producing actinobacteria promoted mineral density of chickpea seed by 24%–28% for Fe and 25%–28% for Zn. Application of different rhizobacterial strains *Pseudomonas putida*, *Pseudomonas fluorescens*, and *Azospirillum lipoferum* to the field grown rice plants showed increase in iron uptake by plants as well as the translocation of iron into the grains (Sharma et al., 2013). Enhancement of selenium content in wheat grain by co-inoculation of selenobacteria and arbuscular mycorrhizal fungi (AMF) has been also been reported by Durán et al., 2013.

4.5 MODULATION FOR MICRONUTRIENT TRANSPORTERS BY RHIZOSPHERIC MICROBES

Although chelation and solubilization helps to convert micronutrients from unavailable to available form for plant uptake, various rhizospheric microorganisms help to modulate plant nutrient transporter system. Certain transporter proteins like ZIP, IRT, ZRT, NRAMP, COPT, and MOPT are known to be involved to transport micronutrients from soil to roots. AhNRAMP1 involved in the uptake of Mn and Zn in root and shoot of pea was found to be upregulated by using *P. fluorescens* and *P. putida* (Wang et al., 2019; Ghanbari Zarmehri et al., 2013). Inoculation of *Azospirillum* and mycorrhizae upregulated AtNRAMP1 which enhanced the affinity of Mn transporter in the root in *A. thaliana* during Mn starvation (Cailliatte et al., 2010). AtNRAMP1, AtNRAMP3, and AtNRAMP4 upregulate the transport of Fe in *A. thaliana* by *B. subtilis* (Vivek Kumar et al., 2016; Stacey et al., 2008). HvZIP13 was upregulated in Barley through inoculation of AMF (*Rhizophagus irregularis*) in Zn deficient condition. *Enterobacter cloacae strain ZSB14 are upregulated the OsZIP1 and OsZIP5 in root and shoot of Rice (Watts-Williams and Cavagnaro, 2018)*. ZAT (Zn transporter in *A. thaliana*) gene was also upregulated in *A. thaliana* by *Azospirillum brasilense*. *Paenibacillus polymyxa* BFKC01 and *B. subtilis* GB03, significantly enhanced Fe uptake with upregulation of FIT1 (responsible for the expression of IRT1 and FRO2) in *Arabidopsis* (Zhang et al., 2009). *B. cereus, B. licheniformis*, and *B. pichinotyi* showed great potential to increase Se uptake in wheat plants (Yasin et al., 2015) and (Naznin et al., 2014). *Bacillus amyloliquefaciens* releases volatile organic compounds for the activation of Fe and Se uptake in *Arabidopsis* (Wang et al., 2017). Cu accumulation in rice roots was increased in *Rhizobium* inoculated rice plant as compared to control due to the upregulation of OsCOPT2 (Adak et al., 2016; Perea-García et al., 2013). Inoculation of *Providencia sp.* in wheat plant, recorded 45.5% increase in Zn accumulation in roots due to enhanced ZIP expression (Rana et al., 2012; Guerinot, 2000).

4.6 INFLUENCING ROOT SYSTEM ARCHITECTURE

Root system architecture influences the rhizosphere through spatial and temporal distribution of roots in soil. Roots explore the deep soil domain for nutrient availability enhancing acquisition of micronutrients. Due to heterogenous distribution of nutrient in the soil, it was proved that plants sense and direct root growth with the help of rhizospheric microbes (McNear Jr, 2013). Hence, different types of root phenotypes are observed under different condition presenting different architectural traits. Some root architectural traits include: (1) higher number of lateral roots and root hairs, (2) increased root surface area, (3) development of lateral roots close to root apices, (4) greater root penetration into the strong soils to improve access to soil resources, (5) greater uptake of nutrients during transpiration by diffusion (Miwa and Fujiwara, 2010). Top soil foraging was appropriate root architecture for Mn and Cu as it is immobile in soil and concentrated in top soil. Since Fe, Zn, Mn, and Cu are restricted in alkaline soil, their mobility in soil is also restricted, hence, their acquisition can be improved through extensive root system in mineral-rich patches (White and Broadley, 2011) and (White and Greenwood, 2013). The mass and volume of *T. caerulescens* (Pennycress member of cruciferae family) roots was higher with increase in its root surface area than in the axenic controls due to bacterial inoculation enhancing Zn acquisition (Schwartz et al., 1999). *Arabidopsis thaliana* inoculation with *Azospirillum brasilense* increased the number of lateral roots and root hairs, and elevated the internal auxin concentration in the plant (Spaepen et al., 2014). Selenium volatization rate was four times higher than control plant in mustard with enhanced root hair production under inoculation of rhizobacterial strain BJ2 and BJ15 which increased Se concentration in shoot and roots (de Souza et al., 1999). Root length, average root diameter, root surface area, or total root volume of wheat plant inoculated with *A. brasilense* and *P. fluorescens* was found to be better than uninoculated plant (Combes-Meynet et al., 2011). Singh et al. (2017) observed that inoculation of endophytic bacteria viz. *Bacillus subtilis* and *Arthrobacter* sp. resulted in significant improvement in root length, surface area, volume and diameter which might have helped the wheat plants to accumulate more zinc in grains.

4.7   PHYTASE PRODUCING RHIZOSPHERIC MICROORGANISMS

Phytic acid or phytate (mixed cation salts of phytic acid) is the group of organic phosphorus (P) that are synthesized by plants stored in seeds during the ripening period (Jorquera et al., 2008). Phytate accounts for up to 80% of the total P in seeds (Taliman et al., 2019; Raboy and Dickinson, 1987). Phytic acids have a lower binding capability for metals like Fe, Zn, Ca, Mg, and Cu, which makes micronutrients unavailable form for absorption (Gupta et al., 2015; Jorquera et al., 2008). Monogastric animal including human do not contain phytase enzyme in their digestive tract and fail to process the phytates present in seeds (Sparvoli and Cominelli, 2015). Removal of phytic acids through microbial activity can be of help to augment the micronutrient availability. Microorganisms especially fungi like *Aspergillus, Sporotrichium, Pichia,* etc., have been used for dephytinization of foods (Ranjan et al., 2015; Sapna, 2014; Kaur; and Satyanarayana, 2010). Diverse microorganisms produce diverse phytases like Cysteine phytases (CPhy), Histidine acid phosphatases (HAP), beta propeller phytases (BPP), and Purple acid phosphatases. HAPs are more predominant in filamentous fungi while BPPs are most prevalent in bacteria (Jorquera et al., 2008; Singh and Satyanarayana, 2015). Diverse phytase producing bacteria like *Bacillus, Paenibacillus, Serratia,* etc., have been reported from rhizosphere of different plants (Gulati et al., 2007; Jorquera et al., 2011; Shedova et al., 2008; Sanguin et al., 2016). Despite the potential of rhizospheric microorganisms for phytase production, their use in dephytinization is very limited and reports are also scanty. For example, Amritha et al. (2018) reported that dephytinization of seed coat matter of Finger Millet (*Eleusinecoracana*) by *Lactobacillus pentosus* CFR3 improved zinc bioavailability.

4.8 MYCORRHIZA AND OTHER FUNGUS FOR NUTRIENT FORTIFICATION OF CROP PLANTS

AMF are mutualistic fungi forming a symbiotic association with wild and cultivated plants belonging to angiosperms, gymnosperms, pteridophytes, and bryophytes (Wang and Qiu, 2006; Andrew Smith and Smith, 2011). AMF spread in soil and forms extramatrical mycelium, a fungal hyphae penetrate root cortical cells to form arbuscules to exchange nutrients and carbon, playing a significant role in escalating uptake of immobile micronutrients (Fe, Zn, Cu, Mn) from soil having low diffusion coefficient (Lehmann et al., 2014). AMF helps in the establishment of alternative nutrient assimilation pathways in plants by extra- and intraradical hyphae, arbuscules, and the root apoplast interface via the mycorrhizal pathway. Extraradical hyphae help in increasing the surface area and explorable soil volume (Parniske, 2008; Smith and Read, 2008) and enhancing absorption and transport of Cu, Mn, and Fe to its associated plants (Hart and Forsythe, 2012; González-Guerrero et al., 2010). AMF's ability to reach, colonize, and nutrient acquisition by solubilization, excreting chelating compounds, producing ectoenzyme, and mobilization in plant (Suman et al., 2016; Verma et al., 2017; Mishra et al. 2015; Upadhayay et al., 2019) growing under insufficient phytoavailable nutrients depends on edaphic factors such as soil texture, soil pH, clay, and organic matter content, cation exchange capacity, soil P, and nutrient concentration (Chagnon et al., 2013; Lehmann and Rillig 2015) present in soil. Lehmann and Rillig, (2015) analyzed 233 studies and found a significant and positive impact of AMF on crop plants for Cu, Fe, and Mn nutrition in its edible parts. Pellegrino and Bedini (2014) found that the foreign AMF *Funneliformis mosseae* and *R. irregularis* symbiosis in a leguminous crop plant chickpea (*Cicer arietinum* L.) as single and dual species inoculum resulted in higher plant growth, nutrient uptake, grain Fe and Zn accumulation representing a valid biofertilization and biofortification strategy. Intraradical colonization of *Glomus intraradices* increased up to 82% in wild tobacco (*Nicotiana rustica* L.) on addition of varying concentration of Zn (Audet and Charest, 2006). Similar result was observed in maize roots inoculated with *G. intraradices* (Seres et al., 2006; Subramanian et al., 2013) and in wheat grain yield and Zn content (Pellegrino et al., 2015). Up to 101% and 75% increased Zn content in modern and old variety of *Triticum turgidum* L. subsp. *durum* (Desf.) Husn, respectively, upon inoculation with *R. intraradices* along with high content of Fe, Cu, Mn (Ercoli et al., 2017). Mycorrhiza ectomycorrhizal fungi, AM fungi, ericoid mycorrhizal fungi, arbutoid mycorrhizal fungi, monotropoid mycorrhizal fungi, and orchid mycorrhizal fungi (Wang and Qiu, 2006) show acidification of rhizospheric region by different organic acid, phenolic compounds, and siderophore production making them a demanding metal chelating agent for Fe, Cu, Mn, and other metals (White and Broadley, 2009) and vesicular arbuscular mycorrhizae root colonization shows increased spatial Zn availability (Thompson et al., 2013). Upadhayay et al., (2019) reviewed the effect of mycorrhiza on micronutrient uptake and reported that AMF successfully helped in modulating the root architecture and helped the colonized root to grow under micronutrient depletion zone up to a distance and could take up the nutrient from the soil where it is available. Durán et al. (2013) observed enhanced Se levels in wheat grains upon co-inoculation of several selenobacteria and the mycorrhizal fungus *Glomus claroideum*. In another study, Durán et al. (2016) found increased Se content in lettuce plant upon co-inoculation of endophytic selenobacteia and AM fungus *Rhizophagus intraradices* and helped to improve tolerance against drought stress. Hart et al. (2015) observed that *R. irregularis* and *Funneliformis mossae* upon root colonization in tomato showed increased Cu concentration in its fruit with high carotenoids and antioxidants contents. Watts-Williams et al. (2015) investigated that 24% of Zn in the shoots of the AM plants, tomato (*Solanum lycopersicum* L.) inoculated with *R. irregularis*, increased growing under low Zn soil.

Piriformospora indica, a Basidiomycota, belongs Sebacinaceae family from Hymenomycetes class, shows similarities to AMF (Hibbett et al., 2007). This endophytic fungus help plants to survive under biotic and abiotic stress condition by promoting growth, nutrient uptake, early flowering, and seed production (Yadav et al., 2010; Das et al., 2012). They colonize in the cortex of roots and rhizospheric zone in an endophytic manner by the formation of pear shaped inter- and intracellular chlamydospores but it does not invade the endodermis and aerial part of plants (Varma et al., 2012). They help to extract, mobilize, and translocate Zn, Fe, Cu, and Mn to host plant (Gosal et al., 2013; Gosal et al., 2010; Gosal et al., 2007). Padash et al. (2016) investigated the effect of *P. indica* and Zn treatment in *Lactuca sativa* and found that a significant increase in plant growth parameters (shoot fresh weight, shoot dry weight, shoot height, and leaf number per plant), chlorophyll content and leaf Zn concentrations (7.6-fold) upon *P. indica* inoculation thus helped in Zn enrichment in lettuce plants. M. Kumar et al., (2009) found similar results upon the interaction between maize and *P. indica*.

4.9   FUTURE PROSPECTS

Deficiency of micronutrients (especially Fe and Zn) is a grave concern for the entire world, especially for the Asian and African countries. Efforts are being made to enrich the foods including the staple crops with micronutrients like Fe and Zn through agronomic or genetic biofortification. Agronomic biofortification is not any permanent solution to this problem, while genetic methods including breeding strategies are time and cost intensive. Rhizospheric microorganisms due to their huge metabolic diversity, known role in biogeochemical cycling and intricate interaction with plant roots and soil can be a better choice to mobilize micronutrients to the plants. Application of microorganisms for micronutrient fortification should be carried out after thorough examination of interaction among potential microbes, crop genotypes with varying accumulation pattern and soils with differing micronutrient status. In future, emphasis should also be given toward characterization of rhizospheric microorganisms of plants which are known to be hyperaccumulators of various micronutrients. Further, attention should also be given to improve the available rhizomicrobial strains for increased nutrient solubilization or chelation ability. However, the application of microorganism cannot be the sole solution for combating the hidden hunger. Integration of genetic as well as agronomic biofortification should be explored to work out a viable, economical, and sustainable option for biofortification.

KEYWORDS

- **fortification**
- **microbes**
- **phytase**
- **rhizosphere**

REFERENCES

Abbaszadeh-Dahaji, P.; Masalehi, F.; Akhgar, A. Improved growth and nutrition of Sorghum (Sorghum bicolor) plants in a low-fertility calcareous soil treated with plant growth–promoting rhizobacteria and Fe-EDTA. *J Soil Sci Plant Nutr.* 2020, 20(1): 31–42.

Adak, A.; Prasanna, R.; Babu, S.; Bidyarani, N.; Verma, S.; Pal, M. et al. Micronutrient enrichment mediated by plant–microbe interactions and rice cultivation practices. *J Plant Nutr.* 2016, 39(9): 1216–1232.

Ahmed, E.; Holmström, S.J.M. Siderophores in environmental research: Roles and applications. *Microb Biotechnol.* 2014, 7(3): 196–208.

Amritha, G.K.; Dharmaraj, U.; Halami, P.M.; Venkateswaran, G. Dephytinization of seed coat matter of finger millet (*Eleusine coracana*) by *Lactobacillus pentosus* CFR3 to improve zinc bioavailability. *LWT.* 2018, 87: 562–566.

Andrew, S.F.; Smith, S.E. What is the significance of the arbuscular mycorrhizal colonisation of many economically important crop plants? *Plant Soil.* 2011, 348: 63–79.

Audet, P.; Charest, C. Effects of AM colonization on "wild tobacco" plants grown in zinc-contaminated soil. *Mycorrhiza.* 2006, 16(4): 277–283.

Backer, R.; Rokem, J.S.; Ilangumaran, G.; Lamont, J.; Praslickova, D.; Ricci, E.; et al. Plant growth-promoting rhizobacteria: context, mechanisms of action, and roadmap to commercialization of biostimulants for sustainable agriculture. *Front Plant Sci.* 2018, 9: 1473.

Borg, S.; Brinch-Pedersen, H.; Tauris, B.; Holm, P.B. Iron transport, deposition, and bioavailability in the wheat and barley grain. *Plant Soil.* 2009, 325(1): 15–24.

Braud, A.; Hannauer, M.; Mislin, G.L.A.; Schalk, I.J. The *Pseudomonas aeruginosa* pyochelin-iron uptake pathway and its metal specificity. *J Bacteriol.* 2009, 191(11): 3517–325.

Brinch-Pedersen, H.; Borg, S.; Tauris, B.; Holm, P.B. Molecular genetic approaches to increasing mineral availability and vitamin content of cereals. *J Cereal Sci.* 2007, 46(3): 308–326.

Cailliatte, R.; Schikora, A.; Briat, J.F.; Mari, S.; Curie, C. High-affinity manganese uptake by the metal transporter NRAMP1 is essential for *Arabidopsis* growth in low manganese conditions. *Plant Cell.* 2010, 22(3): 904–917.

Chagnon, P.L.; Bradley R.L.; Maherali, H.; Klironomos, J.N. A trait-based framework to understand life history of mycorrhizal fungi. *Trends Plant Sci.* 2013, 18(9): 484–491.

Combes-Meynet, E.; Pothier, J.F.; Moënne-Loccoz, Y.; Prigent-Combaret, C. The Pseudomonas secondary metabolite 2,4-diacetylphloroglucinol is a signal inducing rhizoplane expression of Azospirillum genes involved in plant-growth promotion. *Mol Plant-Microbe Interact.* 2011, 24(2): 271–284.

Costerousse, B.; Schönholzer-Mauclaire, L.; Frossard, E.; Thonar, C. Identification of heterotrophic zinc mobilization processes among bacterial strains isolated from wheat rhizosphere (*Triticum aestivum* L.). *Appl Environ Microbiol*. 2018, 84(1): e01715– e01717.

Das, A.; Kamal, S.; Shakil, N.A.; Sherameti, I.; Oelmüller, R.; Dua, M, et al. The root endophyte fungus *Piriformospora indica* leads to early flowering, higher biomass and altered secondary metabolites of the medicinal plant, *Coleus forskohlii*. *Plant Signal Behav*. 2012, 7(1): 103–112.

Dimkpa, C.O.; Bindraban, P.S. Micronutrients fortification for efficient agronomic production. *Agron Sustain Dev*. 2016, 36: 1–26.

Durán, P.; Acuña, J.J.; Armada, E.; López-Castillo, O.M.; Cornejo, P.; Mora, M.L. et al. Inoculation with selenobacteria and arbuscular mycorrhizal fungi to enhance selenium content in lettuce plants and improve tolerance against drought stress. *J Soil Sci Plant Nutr*. 2016, 16(1): 211–225.

Durán, P.; Acuña, J.J.; Jorquera, M.A.; Azcón, R.; Borie, F.; Cornejo, P. et al. Enhanced selenium content in wheat grain by co-inoculation of selenobacteria and arbuscular mycorrhizal fungi: A preliminary study as a potential Se biofortification strategy. *J Cereal Sci*. 2013, 57(3): 275–280.

Ercoli, L.; Schüßler, A.; Arduini, I.; Pellegrino, E. Strong increase of durum wheat iron and zinc content by field-inoculation with arbuscular mycorrhizal fungi at different soil nitrogen availabilities. *Plant Soil*. 2017, 419(1–2): 153–167.

Fasim, F.; Ahmed, N.; Parsons, R.; Gadd, G.M. Solubilization of zinc salts by a bacterium isolated from the air environment of a tannery. *FEMS Microbiol Lett*. 2002, 213(1): 1–6.

Ghanbari, Z.S.; Moosavi, S.G.; Zabihi, H.R.; Seghateslami, M.J. The effect of plant growth promoting rhizobacteria (PGPR) and zinc fertilizer on forage yield of maize under water deficit stress conditions. *Technol J Eng Appl Sci*. 2013, 3: 3281–3290.

González-Guerrero, M.; Benabdellah, K.; Valderas, A.; Azcón-Aguilar, C.; Ferrol, N. GintABC1 encodes a putative ABC transporter of the MRP subfamily induced by Cu, Cd, and oxidative stress in *Glomus intraradices*. *Mycorrhiza*. 2010, 20(2): 137–146.

Gopalakrishnan, S.; Vadlamudi, S.; Samineni, S.; Sameer Kumar, C.V. Plant growth-promotion and biofortification of chickpea and pigeonpea through inoculation of biocontrol potential bacteria, isolated from organic soils. *Springerplus*. 2016.

Gosal, S.K.; Kalia, A.; Varma, A. Piriformospora indica: perspectives and retrospectives. In: *Piriformospora indica*. Springer; 2013. p. 53–77.

Gosal, S.K.; Karlupia, A.; Gosal, S.S.; Chhibba, I.M.; Varma, A. Biotization with Piriformospora indica and Pseudomonas fluorescens improves survival rate, nutrient acquisition, field performance and saponin content of micropropagated Chlorophytum sp. *Indian J Biotechnol*. 2010, 9(3): 289–297.

Gosal, S.K.; Kumar, L.; Kalia, A.; Chouhan, R.; Varma, A. Role of Piriformospora indica as biofertilizer for promoting growth and micronutrient uptake in *Dendrocalamus strictus* seedlings. *J Bamboo Ratt*. 2007, 6: 223–228.

Guerinot, M. Lou. The ZIP family of metal transporters. *Biochim Biophys Acta*. 2000, 1465 (1–2): 190–8.

Gulati, H.K.; Chadha, B.S.; Saini, H.S. Production and characterization of thermostable alkaline phytase from Bacillus laevolacticus isolated from rhizosphere soil. *J Ind Microbiol Biotechnol*. 2007, 34(1): 91–98.

Gupta, R.K.; Gangoliya, S.S.; Singh, N.K. Reduction of phytic acid and enhancement of bioavailable micronutrients in food grains. *J Food Sci Technol*. 2015, 52(2): 676–684.

Hart, M.; Ehret, D.L.; Krumbein, A.; Leung, C.; Murch, S.; Turi, C. et al. Inoculation with arbuscular mycorrhizal fungi improves the nutritional value of tomatoes. *Mycorrhiza.* 2015, 25(5): 359–76.

Hart, M.M.; Forsythe, J.A. Using arbuscular mycorrhizal fungi to improve the nutrient quality of crops; nutritional benefits in addition to phosphorus. *Sci Hortic (Amsterdam).* 2012.

Hibbett, D.S.; Binder, M.; Bischoff, J.F.; Blackwell, M.; Cannon, P.F.; Eriksson, O.E. et al. A higher-level phylogenetic classification of the Fungi. *Mycol Res.* 2007, 11(5): 509–547.

Holmén, B.A.; Casey, W.H. Hydroxamate ligands, surface chemistry, and the mechanism of ligand-promoted dissolution of goethite [α-FeOOH (s)]. *Geochim Cosmochim Acta.* 1996, 60(22): 4403–4416.

Htwe, A.Z.; Moh, S.M.; Moe, K.; Yamakawa, T. Effects of co-inoculation of Bradyrhizobium japonicum SAY3-7 and Streptomyces griseoflavus P4 on plant growth, nodulation, nitrogen fixation, nutrient uptake, and yield of soybean in a field condition. *Soil Sci Plant Nutr.* 2018, 64(2): 222–229.

Hussain, A.; Zahir, Z.A.; Ditta, A.; Tahir, M.U.; Ahmad, M.; Mumtaz, M.Z., et al. Production and implication of bio-activated organic fertilizer enriched with zinc-solubilizing bacteria to boost up maize (*Zea mays* L.) production and biofortification under two cropping seasons. *Agronomy.* 2020, 10(1): 39.

Jones, D.L. Organic acids in the rhizosphere—a critical review. *Plant Soil.* 1998, 205(1): 25–44.

Jorquera, M.; Martinez, O.; Maruyama, F.; Marschner, P.; de la Luz Mora, M. Current and future biotechnological applications of bacterial phytases and phytase-producing bacteria. *Microbes Environ.* 2008, 23(3): 182–191.

Jorquera, M.A., Crowley, D.E, Marschner, P.; Greiner, R.; Fernández, M.T.; Romero, D., et al. Identification of β-propeller phytase-encoding genes in culturable Paenibacillus and Bacillus spp. from the rhizosphere of pasture plants on volcanic soils. *FEMS Microbiol Ecol.* 2011, 75(1): 163–172.

Joy, E.J.M.; Kumssa, D.B.; Broadley, M.R.; Watts, M.J.; Young, S.D.; Chilimba, A.D.C., et al. Dietary mineral supplies in Malawi: spatial and socioeconomic assessment. *BMC Nutr* [Internet]. 2015a, 1(1): 42. Available from: http://bmcnutr.biomedcentral.com/articles/10.1186/s40795-015-0036-4

Joy, E.J.M.; Stein, A.J.;, Young, S.D.; Ander, E.L.; Watts, M.J;, Broadley, M.R. Zinc-enriched fertilisers as a potential public health intervention in Africa. *Plant Soil.* 2015b, 389: 1–24.

Jumadi, S.S.; Djide, M.N.; Mallongi, A. Root inoculation with pseudomonas putida IFO 14796 for improving iron contents in maize grain. *Science,* 2019, 8(1): 1–5.

Kamran, S.; Shahid, I.; Baig, D.N.; Rizwan, M.; Malik, K.A.; Mehnaz, S. Contribution of zinc solubilizing bacteria in growth promotion and zinc content of wheat. *Front Microbiol.* 2017, 8: 2593.

Kaur, P.; Satyanarayana, T. Improvement in cell bound phytase activity of Pichia anomala by permeabilization and applicability of permeabilized cells in soymilk dephytinization. *J Appl Microbiol.* 2010, 108(6): 2041–2409.

Khan, A.; Singh, P.; Srivastava A. Synthesis, nature and utility of universal iron chelator–siderophore: A review. *Microbiol Res.* 2018, 212: 103–111.

Ku, Y.S.; Rehman, H.M.; Lam, H.M.. Possible roles of rhizospheric and endophytic microbes to provide a safe and affordable means of crop biofortification. *Agronomy.* 2019, 9(11): 764.

Kumar, V.; Kumar, M.; Shrivastava, N.; Bisht, S.; Sharma, S.; Varma, A. Interaction among rhizospheric microbes, soil, and plant roots: Influence on micronutrient uptake and bioavailability. In: Plant, Soil and Microbes. Springer; 2016. p. 169–185.

Kumar, V.; Menon, S.; Agarwal, H.; Gopalakrishnan, D. Characterization and optimization of bacterium isolated from soil samples for the production of siderophores. *Resour Technol.* 2017, 3(4): 434–439.

Kumar, M.; Yadav, V.; Tuteja, N.; Johri, A.K. Antioxidant enzyme activities in maize plants colonized with Piriformospora indica. *Microbiology.* 2009, 155(3): 780–790.

Lehmann, A.; Rillig, M.C. Arbuscular mycorrhizal contribution to copper, manganese and iron nutrient concentrations in crops–A meta-analysis. *Soil Biol Biochem.* 2015, 81: 147–158.

Lehmann, A.; Veresoglou, S.D.; Leifheit, E.F.; Rillig, M.C. Arbuscular mycorrhizal influence on zinc nutrition in crop plants—a meta-analysis. *Soil Biol Biochem.* 2014, 69: 123–131.

Li, W.C.; Ye, Z.H.; Wong, M.H. Metal mobilization and production of short-chain organic acids by rhizosphere bacteria associated with a Cd/Zn hyperaccumulating plant, *Sedum alfredii. Plant Soil.* 2010, 326(1–2): 453–467.

Luo, W.; Li, J.; Ma, X.; Niu, H.; Hou, S.; Wu, F. Effect of arbuscular mycorrhizal fungi on uptake of selenate, selenite, and selenomethionine by roots of winter wheat. *Plant Soil.* 2019, 438(1–2): 71–83.

Martino, E.; Perotto, S.; Parsons, R.; Gadd, G.M. Solubilization of insoluble inorganic zinc compounds by ericoid mycorrhizal fungi derived from heavy metal polluted sites. *Soil Biol Biochem.* 2003, 35(1): 133–141.

McNear Jr, D.H. The rhizosphere-roots, soil and everything in between. *Nat Educ Knowl.* 2013, 4(3): 1.

Meena, V.S.; Meena, S.K.; Verma, J.P.; Kumar, A.; Aeron, A.; Mishra, P.K.; et al. Plant beneficial rhizospheric microorganism (PBRM) strategies to improve nutrients use efficiency: a review. *Ecol Eng.* 2017, 107: 8–32.

Mhatre, P.H,; Karthik, C.; Kadirvelu, K.; Divya, K.L; Venkatasalam, E.P.; Srinivasan, S.; et al. Plant growth promoting rhizobacteria (PGPR): A potential alternative tool for nematodes bio-control. *Biocatal Agric Biotechnol.* 2019;17: 119–28.

Mishra, J.; Singh, R.; Arora, N.K. Alleviation of heavy metal stress in plants and remediation of soil by rhizosphere microorganisms. *Front Microbiol* 8: 1706.

Mishra, S.; Singh, A.; Keswani, C.; Saxena, A.; Sarma, B.K.; Singh, H.B. Harnessing plant-microbe interactions for enhanced protection against phytopathogens. In: *Plant Microbes Symbiosis: Applied Facets.* Springer; 2015. pp. 111–125.

Miwa, K.; Fujiwara, T. Boron transport in plants: co-ordinated regulation of transporters. *Ann Bot.* 2010, 105(7): 1103–1108.

Mumtaz, M.Z.; Ahmad, M.; Jamil, M.; Hussain, T. Zinc solubilizing Bacillus spp. potential candidates for biofortification in maize. *Microbiol Res.* 2017, 202: 51–60.

Nautiyal, C.S.; Dion, P. Molecular mechanisms of plant and microbe coexistence. Springer; 2008.

Naveed, M.; Ramzan, N.; Mustafa, A.; Samad, A.; Niamat, B.; Yaseen, M. et al. Alleviation of salinity induced oxidative stress in Chenopodium quinoa by Fe biofortification and biochar—endophyte interaction. *Agronomy.* 2020, 10(2): 168.

Naznin, H.A.; Kiyohara, D.; Kimura, M.; Miyazawa, M.; Shimizu, M.; Hyakumachi, M. Systemic resistance induced by volatile organic compounds emitted by plant growth-promoting fungi in *Arabidopsis* thaliana. *PLoS One.* 2014, 9(1): e86882.

Osorio Vega, N.W. A review on beneficial effects of rhizosphere bacteria on soil nutrient availability and plant nutrient uptake. *Rev Fac Nac Agron Medellín.* 2007, 60(1): 3621–3643.

Padash, A.; Shahabivand, S.; Behtash, F.; Aghaee A. A practicable method for zinc enrichment in lettuce leaves by the endophyte fungus Piriformospora indica under increasing zinc supply. *Sci Hortic* (Amsterdam). 2016, 213: 367–372.

Parniske, M. Arbuscular mycorrhiza: The mother of plant root endosymbioParniske, M. Arbuscular mycorrhiza: The mother of plant root endosymbioses. *Nat Rev Microbiol.* 2008, 6: 763–775.

Patel, P.; Trivedi, G.; Saraf, M. Iron biofortification in mungbean using siderophore producing plant growth promoting bacteria. *Environ Sustain.* 2018, 1(4): 357–365.

Pellegrino, E.; Bedini, S. Enhancing ecosystem services in sustainable agriculture: biofertilization and biofortification of chickpea (*Cicer arietinum* L.) by arbuscular mycorrhizal fungi. *Soil Biol Biochem.* 2014, 68: 429–439.

Pellegrino, E.; Opik, M.; Bonari, E.; Ercoli, L. Responses of wheat to arbuscular mycorrhizal fungi: a meta-analysis of field studies from 1975 to 2013. *Soil Biol Biochem.* 2015, 84: 210–217.

Perea-García, A.; Garcia-Molina, A.; Andrés-Colás, N.; Vera-Sirera, F.; Pérez-Amador, M.A.; Puig, S.; et al. *Arabidopsis* copper transport protein COPT2 participates in the cross talk between iron deficiency responses and low-phosphate signaling. *Plant Physiol.* 2013, 162(1): 180–194.

Pfeiffer, W.H.; McClafferty, B. HarvestPlus: Breeding crops for better nutrition. In: *Crop Science.* 2007.

Raboy, V.; Dickinson, D.B. The timing and rate of phytic acid accumulation in developing soybean seeds. *Plant Physiol.* 1987, 85(3): 841–844.

Radzki, W.; Mañero, F.J.G.;Algar. E.; García, J.A.L.; García-Villaraco, A.; Solano, B.R. Bacterial siderophores efficiently provide iron to iron-starved tomato plants in hydroponics culture. *Antonie Van Leeuwenhoek.* 2013, 104(3): 321–330.

Ramesh, A.; Sharma, S.K.; Sharma, M.P.; Yadav, N.; Joshi, O.P. Inoculation of zinc solubilizing Bacillus aryabhattai strains for improved growth, mobilization and biofortification of zinc in soybean and wheat cultivated in Vertisols of central India. *Appl Soil Ecol.* 2014, 73: 87–96.

Rana, A.; Saharan, B.; Nain, L.; Prasanna, R.; Shivay, Y.S. Enhancing micronutrient uptake and yield of wheat through bacterial PGPR consortia. *Soil Sci Plant Nutr.* 2012, 58(5): 573–82.

Ranjan, B.; Singh, B.; Satyanarayana, T. Characteristics of recombinant phytase (rSt-Phy) of the thermophilic mold Sporotrichum thermophile and its applicability in dephytinizing foods. *Appl Biochem Biotechnol.* 2015, 177(8): 1753–66.

Rashid, M.I.; Mujawar, L.H.; Shahzad, T.; Almeelbi, T.; Ismail, I.M.I.; Oves, M. Bacteria and fungi can contribute to nutrients bioavailability and aggregate formation in degraded soils. *Microbiol Res.* 2016, 183: 26–41.

Sah, S.; Singh, N.; Singh, R. Iron acquisition in maize (*Zea mays* L.) using Pseudomonas siderophore. *Biotech.* 2017, 7(2): 121.

Sanguin, H.; Wilson, N.L.; Kertesz, M.A. Assessment of functional diversity and structure of phytate-hydrolysing bacterial community in Lolium perenne rhizosphere. *Plant Soil.* 2016, 401(1–2): 151–167.

Sapna, S.B. Phytase production by Aspergillus oryzae in solid state fermentation and its applicability in dephytinization of wheat bran. *Appl Biochem Biotechnol.* 2014, 173(7): 1885–1895.

Saravanan, V.S.; Kalaiarasan, P.; Madhaiyan, M.; Thangaraju M. Solubilization of insoluble zinc compounds by *Gluconacetobacter diazotrophicus* and the detrimental action of zinc ion (Zn2+) and zinc chelates on root knot nematode Meloidogyne incognita. *Lett Appl Microbiol.* 2007a; 44(3): 235–241.

Saravanan, V.S.; Madhaiyan, M.; Thangaraju, M. Solubilization of zinc compounds by the diazotrophic, plant growth promoting bacterium *Gluconacetobacter diazotrophicus*. *Chemosphere.* 2007b, 66(9): 1794–1798.

Sathya, A.; Vijayabharathi, R.; Srinivas, V.; Gopalakrishnan, S. Plant growth-promoting actinobacteria on chickpea seed mineral density: an upcoming complementary tool for sustainable biofortification strategy. *Biotech.* 2016, 6(2): 138.

Schwartz, C.; Morel, J.L.; Saumier, S.; Whiting, S.N.; Baker, A.J.M. Root development of the zinc-hyperaccumulator plant Thlaspi caerulescens as affected by metal origin, content and localization in soil. *Plant Soil.* 1999, 208(1): 103–115.

Scott, S.; Housh, A.; Powell, G.; Anstaett, A.; Gerheart, A.; Benoit, M. et al. Crop Yield, Ferritin and Fe (II) boosted by Azospirillum brasilense (HM053) in Corn. *Agronomy.* 2020, 10(3): 394.

Seres, A.; Bakonyi,G.; Posta, K. Zn uptake by maize under the influence of AM-fungi and Collembola Folsomia candida. *Ecol Res.* 2006, 21(5): 692.

Sharma, A.; Shankhdhar, D.; Shankhdhar, S.C. Enhancing grain iron content of rice by the application of plant growth promoting rhizobacteria. *Plant, Soil Environ.* 2013, 59(2): 89–94.

Shedova, E.; Lipasova, V.; Velikodvorskaya, G.; Ovadis, M.; Chernin. L.; Khmel, I. Phytase activity and its regulation in a rhizospheric strain of *Serratia plymuthica*. *Folia Microbiol (Praha).* 2008, 53(2): 110–114.

Shirvani, M.; Nourbakhsh, F. Desferrioxamine-B adsorption to and iron dissolution from palygorskite and sepiolite. *Appl Clay Sci.* 2010, 48(3): 393–397.

Singh, D.; Rajawat, M.V.S.; Kaushik, R.; Prasanna, R.; Saxena, A.K. Beneficial role of endophytes in biofortification of Zn in wheat genotypes varying in nutrient use efficiency grown in soils sufficient and deficient in Zn. *Plant Soil.* 2017, 416: 107–116.

Singh, B.; Satyanarayana, T. Fungal phytases: characteristics and amelioration of nutritional quality and growth of non-ruminants. *J Anim Physiol Anim Nutr (Berl).* 2015, 99(4): 646–660.

Smith, S.E.; Read, D.J. *Mycorrhizal Symbiosis.* 3rd. Academic Press, New York, ISBN. 2008, 440026354: 605.

de Souza, M.P.; Chu, D.; Zhao, M.; Zayed, A.M.; Ruzin, S.E.; Schichnes, D.; et al. Rhizosphere bacteria enhance selenium accumulation and volatilization by Indian mustard. *Plant Physiol.* 1999, 19(2): 565–574.

Spaepen, S.; Bossuyt, S.; Engelen, K.; Marchal, K.; Vanderleyden, J. Phenotypical and molecular responses of *Arabidopsis thaliana* roots as a result of inoculation with the auxin-producing bacterium A zospirillum brasilense. *New Phytol.* 2014, 201(3): 850–861.

Sparvoli, F.; Cominelli, E. Seed biofortification and phytic acid reduction: a conflict of interest for the plant? *Plants.* 2015, 4(4): 728–55.

Stacey. M.G.; Patel. A.; McClain, W.E.; Mathieu, M.; Remley, M.; Rogers, E.E.; et al. The *Arabidopsis* AtOPT3 protein functions in metal homeostasis and movement of iron to developing seeds. *Plant Physiol.* 2008, 146(2): 589–601.

Subramanian, K.S.; Balakrishnan, N.; Senthil, N. Mycorrhizal symbiosis to increase the grain micronutrient content in maize. *Aust J Crop Sci.* 2013, 7(7): 900–910.

Suman, A.; Yadav, A.N.; Verma, P. Endophytic microbes in crops: diversity and beneficial impact for sustainable agriculture. In: *Microbial Inoculants in Sustainable Agricultural Productivity.* Springer; 2016. pp. 117–143.

Sunithakumari, K.; Padma Devi, S.N.; Vasandha, S. Zinc solubilizing bacterial isolates from the agricultural fields of Coimbatore, Tamil Nadu, India. *Curr Sci.* 2016, 110(2): 196–205.

Taliman, N.A.; Dong, Q.; Echigo, K.; Raboy, V.; Saneoka, H. Effect of phosphorus fertilization on the growth, photosynthesis, nitrogen fixation, mineral accumulation, seed yield, and seed quality of a soybean low-phytate line. *Plants.* 2019, 8(5): 119.

Thompson, J.P.; Clewett, T.G.; Fiske, M.L. Field inoculation with arbuscular-mycorrhizal fungi overcomes phosphorus and zinc deficiencies of linseed (*Linum usitatissimum*) in a vertisol subject to long-fallow disorder. *Plant Soil.* 2013, 371: 117–137.

Trivedi, G.; Patel, P.; Saraf, M. Synergistic effect of endophytic selenobacteria on biofortification and growth of Glycine max under drought stress. *South African J Bot.* 2019, https://doi.org/10.1016/j.sajb.2019.10.001.

Tulchinsky, T.H. Micronutrient deficiency conditions: Global health issues. *Public Health Rev.* 2010, 32(1): 243.

Ullah, A.; Farooq, M.; Hussain, M. Improving the productivity, profitability and grain quality of *kabuli* chickpea with co-application of zinc and endophyte bacteria *Enterobacter* sp. MN17. *Arch. Agron. Soil Sci.* 2020, 66: 897–912.

Upadhayay, V.K.; Singh, J.; Khan, A.; Lohani, S.; Singh, A.V. Mycorrhizal mediated micro-nutrients transportation in food based plants: a biofortification strategy. In: *Mycorrhizo-sphere and Pedogenesis.* Springer; 2019, pp. 1–24.

Varma, A.; Bakshi, M.; Lou, B.; Hartmann, A.; Oelmueller, R. Piriformospora indica: a novel plant growth-promoting mycorrhizal fungus. *Agric Res.* 2012, 1(2): 117–131.

Verma, P.; Yadav, A.N.; Kumar, V.; Singh, D.P.; Saxena, A.K. Beneficial plant-microbes interactions: biodiversity of microbes from diverse extreme environments and its impact for crop improvement. In: *Plant–Microbe Interactions in Agro-Ecological Perspectives.* Springer; 2017. pp. 543–580.

Wang, N.; Qiu, W.; Dai, J.; Guo, X.; Lu, Q.; Wang, T.; et al. AhNRAMP1 enhances manganese and zinc uptake in plants. *Front Plant Sci.* 2019, 10: 415.

Wang, B.; Qiu, Y.L. Phylogenetic distribution and evolution of mycorrhizas in land plants. *Mycorrhiza.* 2006, 16(5): 299–363.

Wang, J.; Zhou, C.; Xiao, X.; Xie, Y.; Zhu, L.; Ma, Z. Enhanced iron and selenium uptake in plants by volatile emissions of *Bacillus amyloliquefaciens* (BF06). *Appl Sci.* 2017, 7(1): 85.

Watts-Williams, S.J.; Cavagnaro, T.R. Arbuscular mycorrhizal fungi increase grain zinc concentration and modify the expression of root ZIP transporter genes in a modern barley (*Hordeum vulgare*) cultivar. *Plant Sci.* 2018, 274: 163–170.

Watts-Williams, S.J.; Smith, F.A.; McLaughlin, M.J.; Patti, A.F.; Cavagnaro, T.R. How important is the mycorrhizal pathway for plant Zn uptake? *Plant Soil.* 2015, 390(1–2): 157–166.

White, P.J.; Broadley, M.R. Biofortifying crops with essential mineral elements. Vol. 10, *Trend Plant Sci.* 2005. p. 586–593.

White, P.J.; Broadley, M.R. Biofortification of crops with seven mineral elements often lacking in human diets—iron, zinc, copper, calcium, magnesium, selenium and iodine. *New Phytologist.* 2009, 182(1): 49–84.

White, P.J.; Broadley, M.R. Physiological limits to zinc biofortification of edible crops. Front Plant Sci [Internet]. 2011, 2. Available from: http://journal.frontiersin.org/article/10.3389/fpls.2011.00080/abstract

White, P.J.; Greenwood, D.J. Properties and management of cationic elements for crop growth. *Soil Cond Plant Growth.* 2013, 12: 160–194.

Yadav, V.; Kumar, M.; Deep, D.K.; Kumar, H.; Sharma, R.; Tripathi, T. et al. A phosphate transporter from the root endophytic fungus Piriformospora indica plays a role in phosphate transport to the host plant. *J Biol Chem*. 2010, 285(34): 26532–26544.

Yadav, R.; Ror, P.; Rathore, P.; Ramakrishna, W. Bacteria from native soil in combination with arbuscular mycorrhizal fungi augment wheat yield and biofortification. *Plant Physiol Biochem*. 2020, 150: 222–233.

Yasin, M.; El-Mehdawi, A.F.; Anwar, A.; Pilon-Smits, E.A.H.; Faisal, M. Microbial-enhanced selenium and iron biofortification of wheat (*Triticum aestivum* L.)-applications in phytoremediation and biofortification. *Int J Phytorem*. 2015, 17(4): 341–347.

Zhalnina, K.; Louie, K.B.; Hao, Z.; Mansoori, N.; da Rocha, U.N.; Shi, S.; et al. Dynamic root exudate chemistry and microbial substrate preferences drive patterns in rhizosphere microbial community assembly. *Nat Microbiol*. 2018, 3(4): 470–480.

Zhang, H.; Sun, Y.; Xie, X.; Kim, M.; Dowd, S.E.; Paré, P.W. A soil bacterium regulates plant acquisition of iron via deficiency-inducible mechanisms. *Plant J*. 2009, 58(4): 568–577.

CHAPTER 5

Managing Abiotic Stressed Agriculture through Microbes

CHIRSMITA[1], BHOLA NATH SAHA[2], SHYAMASHREE ROY[3], and RAJIV RAKSHIT[1*]

[1]Department of Soil Science and Agricultural Chemistry, Bihar Agricultural University, Sabour 813210, Bhagalpur, Bihar, India

[2]Department of Soil Science and Agricultural Chemistry, Dr. Kalam Agricultural College, Affiliated to Bihar Agricultural University, Kishanganj 855007, Bihar, India

[3]Department of Agronomy, RRS (OAZ), Majhian, Uttar Banga Krishi Viswavidyalaya, Dakshin Dinajpur 733133, West Bengal, India

*Corresponding author. E-mail: rajiv.ssaciari@gmail.com

ABSTRACT

Agriculture is a dynamic system that always seeks constant requirement of energy to maintain its stable state usually known as homeostasis. Any interference in this state of homeostasis may be considered as biological stress. Likewise, plants are more affected by abiotic stresses like heat, drought, nutrient deficiency, salinity, and heavy metal stress due to the devastating consequences which can be seen across the world, leading to hampering the sustainable agriculture. Under these stressed conditions, microbes could play significant roles to combat the ill effects generated due to this stressed environment and subsequently leading to improving plant growth vis-à-vis soil health. Plant-growth-promoting rhizobacteria (PGPR) influences or changes the properties of rhizospheric soils through the production of polysaccharides, biofilms, releasing osmoprotectants, and sometimes heat-shock proteins. PGPR produces proline and glycine betaine under the salt stress condition which is utilized as osmoprotectants. Mechanisms like displacement of sorption equilibrium, microbial-mediated nutrient transfer, and biomass turnover are considered as the indirect pathway through microbes increase the transfer of nutrients ions into soil solution under nutrient-deficient condition. PGPR mitigates these impacts through other mechanisms like producing antioxidants, releasing of cytokinins, and affecting the ethylene precursor 1-aminocyclopropane-1-carboxylate deaminase. Microbes alleviate heavy metals through like expulsion of metals, bioaccumulation, biotransformation, and metal adsorption to evade the toxicity of metals in soils. Inoculation of these microorganisms can alleviate stresses in plants, and hence opens up an emerging avenue in sustainable agriculture.

5.1 INTRODUCTION

A number of incidences of abiotic stresses-affecting productivity in crops are being witnessed across the globe. Abiotic stress means the negative impact of nonliving factors on living entity under a specific environment and this includes drought, salinity, chilling, freezing, and other environmental extremes. Among them, the most common factor that limits the growth of crops is the low water supply. Water stress in its broadest sense includes both drought and salinity stress. An approximately one-fifth of the irrigated lands in the world is affected by salinity (Morton et al., 2019). These factors are believed to have a distressing impact on growth and yield of plants under field conditions (Suzuki et al., 2014). Stresses of this nature have been the primary cause of crop failure worldwide, affecting or declining crop yields by 50% (Wang et al., 2007). Among these stresses, the dips in crop yields are due to high temperature (20%), low temperature (7%), drought (9%), and other forms of stresses (4%) (Minhas et al., 2017). Moreover, only 9% of the area is conducive for crop production, while 91% is under stress in the world, and hence could lead to a huge impact on the exchequer of various nations. Now, it is a great challenge to develop efficient and low-cost methods to cope up with these abiotic stresses. A number of tolerant crop varieties are developed along with new resource management practices to handle these stresses. Recently, studies indicate the use of microorganisms to deal with this abiotic stresses. These microorganisms colonize the rhizosphere/ endorhizosphere of plants and improve the growth and development of plants through various primary and secondary mechanisms (Saxena et al., 2005). With the time, the role of microbes in the management of biotic and abiotic stresses is gaining importance. Induced systemic tolerance has been proposed to define the role of plant-growth promoting rhizobacteria (PGPR) that means bringing some induced physical and chemical changes, which increase the tolerance capacity of plants against these abiotic stresses. The PGPR could also play an important role in the maintenance of soil health under stressed environments. The application of organic materials or residues acts as a soil conditioner for the beneficial microbes and enhances the plant–microbe associations, and hence increases the yield of crops. Hence, a triple combination of plants, tolerant microbes, and organic amendment represents an association through which higher yields can obtain even under stressed agriculture situations. This triple association offers a congenial environment to the proliferation of microbes that in turn enhances the growth of plants in a disturbed agroecosystem.

5.2   RESPONSE OF PLANTS TO ABIOTIC STRESSES

Naturally, plants often experience a number of abiotic stresses. Climate change is an established phenomenon with a number of literatures available across the world. The rise of carbon dioxide in the atmosphere and temperature extremes has shown their individual as well combined ill effects on the plants. The uneven distribution of rainfall leads to drought in many places. During drought, the available water to plants steadily increases, and hence causing death of the plants prematurely. Under such situations, plant growth arrest due to the reduction in metabolic demands. Drought primarily causes a decline in grain weight and plant height. By contrast, heat stress causes aborted spikes, and hence there is a decline in the number of grains. Heat stress affects the pollen development, whereas drought affects the development of pistils in flowers. Even, an increase in temperature across the world has become a great concern, which has an ill-effect on the growth as well as productivity in agricultural crops. Heat stress basically affects the germination of seeds and photosynthetic efficiency. Unlike in temperate countries, plants usually encounter cold stress. Cold stress in plants mainly affects the cellular function as well as the postharvest quality of crops. So, in the process of adaptation to cold stress, plant acquires its tolerance mechanism known as acclimation. There are places where evaporation rates are higher than precipitation, and hence cause salinity. Salinity is a global threat and reduces the crop growth by inducing ion toxicity and osmotic stress. Under salt stress, there is wilting of plants and reduced uptake of minerals due to differences in the osmotic pressure in soil solution and in plant cells. Moreover, salinity deteriorates cell expansion and membrane functions. The dependence on chemical fertilizers and with rapid industrialization, there is an accumulation of toxic metals in the environment causing ill effects on the soil-plant continuum. As per the criteria of essentiality, some of the metals are important in the crucial biological processes of plants. However, few other heavy metals like arsenic, cadmium, chromium, etc., can lead to toxic symptoms in plants only when their concentration reaches beyond supraoptimal values (Tiwari and Lata, 2018). Heavy metals cause morphological abnormalities, and metabolic disorders are leading to the production of reactive oxygen species and disruption of the redox homeostasis of cells. This change in the redox status is known to be a chief cause of toxicity in plants (Shahid et al., 2015).

5.3   GENERAL MECHANISMS OF PGPR TOWARD STRESS

The unique properties of microbes can be utilized in tolerance to abiotic stresses. Microbial diversity can also be harnessed to tackle the problems due to abiotic stress. The role of exopolysaccharides from PGPR on promoting plant resistance was reported by a number of investigators (Wafaa et al., 2014). PGPR influences or changes the properties of rhizospheric soils through the production of polysaccharides, biofilms, releasing osmoprotectants, and sometimes heat-shock proteins. These microorganisms per se can lessen these stresses in crops. The PGPR assists the plant growth through mechanisms like resource acquisition, mineralization of nutrients, phytohormone production, and synergism with plants to mitigate the stress. They can also protect plants through inhibition of growth of deleterious microbes. The cell wall of fungus containing amino acids, carboxyl, hydroxyl, and other functional groups acts as binding agents for metal in contaminated soils. In saline soils, inoculation of microorganisms accelerates the binding of sodium ions, and hence reduces the same for plant uptake (Kohler et al., 2009). Proline, an amino acid, accumulates in many plant species in response to environmental stress. Proline acts as an osmoprotective where salt stress is a problem (Szabados and Savoure, 2009). In leguminous plants, trehalose metabolism is a signaling phenomenon of adaptation to abiotic stress (Goel et al., 2018). The secretion of siderophores by microorganisms can also help in the uptake of trace metals like iron, besides microbial communities' releases cations from minerals under nutrient deficient conditions. Some cold tolerant PGPR synthesize cold-shock proteins and cold acclimation proteins under low temperature conditions. This protein helps to combat the low temperature impact on plants. Another important role of rhizobacteria is the synthesis of phytohormones includes auxin, the most common being the indole acetic acid, cytokinins, and gibberellins. Diverse bacterial genera such as *Acinetobacter, Arthrobacter, Bacillus, Corynebacterium, Delftia, Duganella, Exiguobacterium, Kocuria, Lysinibacillus, Methylobacterium, Micrococcus, Paenibacillus, Pantoea, Pseudomonas, Psychrobacter, Serratia,* and *Stenotrophomonas* have been reported to produce these hormones (Nath and Nath, 2018). During drought, high amount of ethylene seems inhibitory for the growth of plants. PGPR possess an enzyme 1-aminocyclopropane-1-carboxylate (ACC) deaminase enzyme that converts the precursor of ethylene to α-ketobutyrate and ammonium, and hence lowering the amount of ethylene (Mishra et al., 2017). The mechanisms through which microbes facilitate plants growth are mentioned in Figure 5.1.

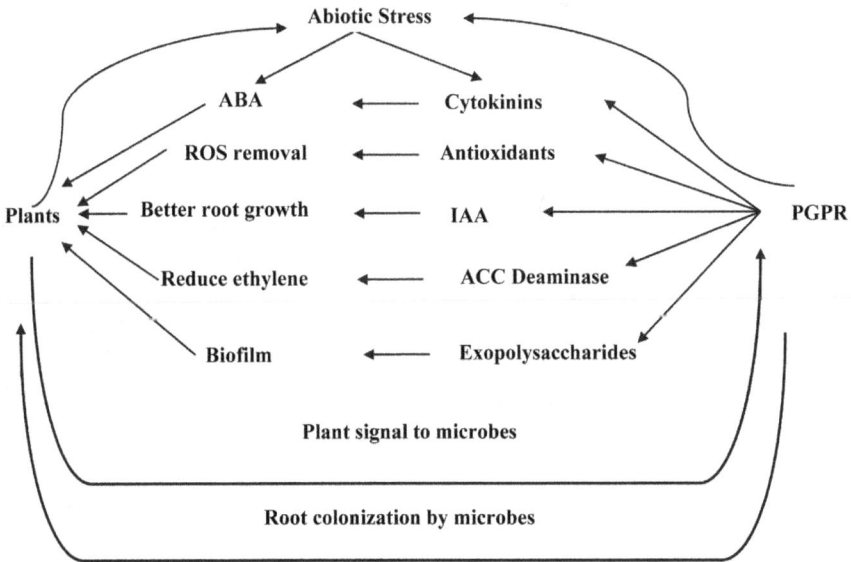

FIGURE 5.1 Mechanisms behind combating the stresses through microbes (modified from Grover et al., 2011).

5.4 ROLE OF MICROBES UNDER SALINITY STRESS

Salinity toxicity is a widespread problem across many areas of the world and considered as one of the obstacles for agricultural production.

Worldwide, over 20% of farmland is affected by salts and almost 10% of irrigated lands are being under saline stress (Ruan et al., 2010) which tunes to a monetary loss of around US$12 billion (Flowers et al., 2010). Plants developed an array of the regulatory system to respond to salinity stress. PGPR is found to enhance the growth of many crops like bean, tomato, pepper, and lettuce under saline condition (Salwan et al., 2019). Rhizobacteria with 1-aminocyclopropane-1-carboxylate deaminase (ACCD) activity improves the root tolerance capacity against salts (Qin et al., 2016). Representative-halotolerant organisms like *Bacillus* and *Pseudomonas* have greater ACCD activity and been explored for their ability to grow under higher sodium chloride concentration of 2%–11% (Singh and Jha, 2016). PGPR uses proline, glycine betaine, and potassium ions as major osmolytes to combat salt stress (Csonka, 1989). A greenhouse experiment was conducted to test the effects of *Piriformospora indica* and *Azospirillum* strain under increasing salinity levels on wheat seedling growth and proline accumulation (Zarea et al., 2012). The result indicated a positive influence of the organisms on salinity tolerance and was related to the better water status and proline accumulation in wheat seedlings inoculated with those organisms. Regulation of sodium ions maintains normal growth in saline stress. A number of PGPRs are found to enhance the Na^+ export and K^+ imports in plants. Biofilm formation under salt stress is an important strategy adopted by microorganisms for their survival in rhizospheric zone. With this idea, Kasim et al. (2016) studied the role of biofilm producing PGPR on salt tolerance of barley. A *Bacillus amyloliquefaciens* strain was identified responsible for the tolerance of barley to salinity. This strain was able to synthesize plant growth hormones like indole acetic acid (IAA), gibberellins, etc., and improved the uptake of nutrients from soils (Wang et al., 2009). AM fungus accumulates soluble sugars in the roots and thus AM infected plants are more resistant to osmotic stress induced during salt stress. Some microbes are capable of producing plant growth is speculated a key factor responsible for salt tolerance (Kang et al. 2014). *Sinorhizobium meliloti* strain produces IAA and enhances the growth of *Medicago truncatula* under salinity stress (Bianco and Defez 2009). *Bacillus licheniformis* collected from Kutch desert was able to solubilize phosphorus and improved the growth of *Arachis hypogaea* under 50 mM NaCl concentration (Goswamia et al. 2014). Under salt stress, halo-tolerant PGPR may play a critical role in the combating physiological abnormalities observed in plants. Halotolerant PGPR improves the morphological parameters and productivity of plants through antioxidant production, extra polysaccharides (EPS) formation, and osmotic adjustments. There is an ample scope to utilize these characters of halotolerant for producing biofertilizers and its use in sustainable agricultural production in saline belts. Microbial communities responsible for specific improvement in plant growth under saline condition are shown in Table 5.1.

TABLE 5.1 Role of Microbial Strains in Crops Under Saline Conditions

Strain	Crops	Character/s Improved	Author, year
Halomonas Bacillus	Alfalfa	Binding of salts	Kearl et al., 2019
Micrococcus yunnanensis	Sugarbeet	Decrease in ethylene production, increase in photosynthetic activity	Zhou et al., 2017
Pseudomonas sp.	Cucumber	Improved plant growth	Yuan et al., 2016
Arthrobacter pascens	Maize	Increase in osmolytes and antioxidant enzymes	Ullah and Bano, 2015
Enterococcus sp.	*Vigna radiata* L.	Less uptake of sodium	Panwar et al., 2016
Ochrobactrum sp.	*Arachis hypogaea* L.	Production of ACC deaminase and IAA	Paulucci et al., 2015
Arthrobacter woluwensis, Microbacterium oxydans, Arthrobacter aurescens, Bacillus megaterium, and *Bacillus aryabhattai*	Soybean	Increased production of IAA, GA, and siderophores and increased phosphate solubilization	Khan et al., 2019
Bacillus sp., *Z. halotolerans, S. succinus, B. gibsonii, O. oncorhynchi, Halomonas* sp. and *Thalassobacillus* sp.	Wheat	IAA production, nitrogen fixation, phosphorus solubilization, osmolytes accumulation	Orhan, 2016
Halobacillus sp. *Halobacillus dabanensis*	Rice	IAA, nitrogen fixation	Rima et al., 2018
Curtobacterium flaccumfaciens	Barley	Inducing systemic resistance	Cardinale et al., 2015

5.5 ROLE OF MICROBES UNDER DROUGHT STRESS

Scarcity of water is a harsh environmental limitation to crop productivity. By 2050, drought is expected to become a serious problem for more than 50% arable land (Kasim et al., 2013). PGPR could be utilized for its important role in the alleviation of drought stress in crop plants. Beneficial microorganisms settle in the rhizosphere and impart tolerance by producing EPS, releasing ACC deaminase, accumulating osmolytes, and regulating stress genes under drought situation. Changes in the structure of plant-associated microbial communities in the rhizosphere and their variation across the various zones of the rhizosphere impart a great resistance toward drought stress. PGPR secretes high molecular weight compounds known as EPS representing 40–95% of the bacterial weight (Ojuederie et al., 2019). Generally, like *Pseudomonas, Bacillus,* and *Acinetobacter* are EPS producers which lead to withstand harsh drought situation. In foxtail millet, inoculation of *Pseudomonas fluorescens* and *Enterobacter hormaechei* leads to better growth under drought stress due to EPS production by the bioinoculants (Niu et al., 2018). *Achromobacter piechaudii* capable of producing ACC deaminase alleviated the oxidative stress in tomato and pepper and significantly increased the yields of both the crops (Mayank et al., 2004). *Capsicum annum* L. inoculated with bacteria from desert areas exhibited a tolerance to water scarcity. Inoculation with bacterial isolates enhanced the growth of roots leading to increased water uptake under drought situation (Marasco et al., 2013). Even application of AM fungi individually or with *Azospirillum brasilense* in rice increased proline and soluble sugars to cope with water-limited conditions. Under such situation, the photosystem-II performance gets improved, and hence better growth and yield of rice (Sanchez et al., 2011). Khan et al., 2019 reported that the culmination of PGPR along with growth regulators maintains the photosynthetic efficiency of chickpea by inducing soluble sugar storage, and hence increase the tolerance capacity to drought. Auxin regulates most of the biological process in plants; and IAA production is a direct mechanism of PGPR in plants which can change the architecture of roots, number of root tips, and surface area found to be effective in water and mineral uptake by plants under drought condition (Vacheron et al., 2013). In addition, microbes release volatile organic compounds (VOCs) which are low molecular weight compounds that evaporate at normal temperature and pressure; these compounds are responsible for induced-systemic tolerance. A wide range of bacteria is known to produce VOCs, responsible for inducing resistance in plants. *Pseudomonas chlororaphis* when colonizes the plants, it prevents the loss of water by closing stomata due to release of 2,3-butanediol considered as one of the important VOC (Cho et al., 2008). Volatiles dimethylhexadecylamine produced by certain microorganisms improved the root number, length, density, and architecture to increase the water uptake during drought conditions (Sharifi and Ryu, 2018). Microbial communities, which are responsible for specific improvement in plants growth under the drought condition, are shown in Table 5.2.

TABLE 5.2 Role of Microbial Strains in Crops Under Drought Conditions

Strain	Crops	Character/s improved	Author, year
Bacillus cereus, Planomicrobium chinense	Wheat	Antioxidant enzymes, proline production, EPS production	Khan and Bano, 2019
Bacillus licheniformis	Pepper	Production of ACC deaminase	Lim and Kim, 2013
Burkholderia vietnamiensis, Rhizobium tropici,	Poplar	Increased biomass, decreased stomatal conductance	Khan et al., 2016
Pseudomonas sp*., Bacillus* sp. and *Artrobacter* sp.	Rice	Increase antioxidant activity	Gusain et al., 2015
Sphingomonas sp.	Soybean	Increased production of sugar and amino acids	Asaf et al., 2017
Bacillus sp*., Enterobacter* sp.	Wheat, Maize	Changes in root architecture	Jochum et al., 2019
Gluconacetobacter diazotrophicus	Sugarcane	Activation of drought stress responsive	Vargas et al., 2014
Bacillus amyloliquefaciens	Grapes	Secreted melatonin, reduced H_2O_2	Jiao et al., 2016

5.6 ROLE OF MICROBES UNDER NUTRIENT DEFICIENCY STRESS

Besides the role of soil microorganisms in influencing several abiotic as well as biotic stresses in plants, they also involved in managing nutrient deficiency tolerance directly or indirectly. Plants have developed numerous mechanisms to overcome nutrient deficiency limitations. There are several mechanisms of plant nutrients uptake affected by soil microorganisms like root area and volume expansion, root hair branching and development. Mechanism like displacement of sorption equilibrium, microbial-mediated nutrient transfer and biomass turnover are considered as the indirect pathway through microbes increase the transfer of ions into soil solution. As soil-borne microorganisms are innumerable in their counts in the rhizosphere, they play an enormous role through their activity as well as metabolic activities to combat abiotic stresses. Plant–microbe interactions comprise mutualistic associations that amend local and systemic mechanisms in plants offering defense under adverse conditions (Meena et al., 2017). Under nutrients-limiting conditions, growth is reduced and plants modify their physiology and morphology to cope with scarce resources. These modifications depend on the specific nutrient mobility in soil (Hodge, 2004). The first response to nutrient deficiency appears to be related to stresses rather than nutrient deprivation (Hammond et al., 2003). Indeed, plants have the ability to remobilize major nutrients from the pool present in the cell. Rhizospheric microorganisms possess the capability to mineralize the bound or fixed form of nutrients in the soil organic matter and causes decomposition of cellulose, hemicellulose, and lignin; solubilize the bound phosphorus; causes oxidation–reduction, or mineralization of sulfur compounds. Some rhizosphere bacteria carry out fixation of atmospheric inert nitrogen into ammonia using the enzyme nitrogenase. Rhizobacteria from root nodules of leguminous plants, involves in nitrogen fixation, and delivers it to the plants. These affiliations have given data about the mutualistic relationships since plants have created constitutive and inducible defense mechanisms to keep away from destructive communications. Biological nitrogen fixation is one of the major contributions of the rhizosphere bacteria that are beneficial for soil and plant health, besides nutrient mobilization, solubilization and plant growth promotion through secondary metabolites. Phosphate-solubilizing microorganisms include bacteria like Pseudomonas, Bacillus, nitrogen-fixing cyanobacteria, other free-living diazotrophs and actinomycetes and fungi like Aspergillus, *Penicillium, Paecilomyces,* etc. Arbuscular mycorrhizal fungi facilitate phosphate nutrient transfer to the higher plants. The presence of these microorganisms in the rhizosphere facilitates the activity 10 times more than when they are present as free-living saprophytes because of the easy availability of organic soluble compounds in root exudates for their growth. The principal method which increases phosphate availability is the release of acids by microbes that dissolve minerals, releasing soluble forms of phosphorus. The capacity of organic acids excreted by soil bacteria to solubilize soil minerals is exemplified by studies of 2-ketogluconic acid productions by Gram-negative bacteria growing in glucose media. Additionally, alteration in pH of soils due to acidification of root environment mobilizes soil minerals. Solubilization of a variety of metals contained in natural and synthetic insoluble salts and minerals has been demonstrated. Siderophores and other metal chelators are synthesized by a variety of bacteria, including common soil organisms. Rhizosphere microorganisms that produce chelating agents (e.g., siderophore production) increase the solubility of iron and manganese compounds, and thus iron and manganese may be more available to plants. It has also been shown that microorganisms on roots significantly increase the uptake rates of calcium by the roots. This increase may be due to the high concentrations of CO_2 in the rhizosphere produced by microorganisms, which increase the solubility, and thus the availability of calcium. Translocation of various radiolabeled-organic compounds and heavy metals along mycelial filaments has also been demonstrated. Several investigators have reported the response of trees to AM fungal inoculation. Some fungi form mycorrhizal association with higher plants and help in the nutrient uptake of phosphorus and survival and under drought conditions. The mycorrhizal fungi form both ectotrophic as well as endotrophic associations. About 4000 fungal species belonging primarily to Basidiomycota and fewer to the Ascomycota are known to form ectomycorrhiza. The ectotrophic fungi are *Boletus, Lactarius, Amanita, Pisolithus,* and *Elaphomyces.* Endotrophic mycorrhizal fungi (VAM–AM fungi) include *Glomus, Gigaspora,* Acaulospora, Endogene. Several heterotrophic fungi secrete acidic substances during their growth which helps in solubilization of insoluble bound phosphates in soil. Members of the genera *Aspergillus, Trichoderma, Paecillomyces, Penicillium,* and *Curvularia* have been found to solubilize insoluble bound phosphates compounds in soil.

5.7 ROLE OF MICROBES UNDER HEAVY METALS STRESS

Microbes play an important role in maintaining the fertility of the soils through numerous interactions with physicochemical components of soils. There are strategies like expulsion of metals, bioaccumulation, biotransformation, and metal adsorption through which microbes are able to evade the toxicity of metals in soils. Through these mechanisms, PGPR as bioinoculant can significantly improve the growth of plants facing heavy metal stress (Ahemad, 2014). Soil microorganisms can affect the mobility of metals and reduce their availability to plants through acidification, siderophores production, and by changing redox potential (Burd et al., 2000). PGPR that is capable of producing siderophore solubilizes unavailable form of iron that can be assimilated through roots-mediated mechanisms. The biosorption mechanism also contributed to reduce the impact of heavy metals as microbial cells can adsorb heavy metals through their metabolic processes (Khan et al., 2007). *Azotobacter chroococcum* produced siderophores and ACC deaminase under heavy metal stress of Cu and Pb, which promotes the yield and growth of maize plants (Rizvi and Khan, 2018). PGPRs including *Pseudomonas, Streptomyces,* and *Methylobacterium* are reported to have metal tolerance capacity (Sessitsch et al., 2013). Pandey and co-workers in 2013 reported cadmium tolerant *Ochrobactrum* sp. That helps in the remediation of Cd and improved the growth and yield of rice. In spite of this conventional practice, the use of genetically modified bacteria is gaining consideration, but limited to laboratory trials (Gupta and Singh, 2017). Some genetically modified organisms like *Pseudomonas putida, Mesorhizobium huakuii,* and *Enterobacter cloacae* are known to have metal tolerance capacity (Ullah et al., 2015). Microbial EPS released in the rhizosphere possess a number of anionic functional groups, and thus helps to remove the heavy metals through biosorption (Ayangbenro and Babalola, 2017). EPS production sometimes leads to the formation of biofilm which acts as a protective sheath over microbes and is known to sequester heavy metals. The use of PGPR for remediating heavy metals still lacks commercialization because in some areas, the accumulation of metals in tissue downturns the remediation process in heavily contaminated areas (Ma et al., 2011). Microbial communities responsible for the specific improvement in plant growth under heavy metal stressed conditions are shown in Table 5.3.

TABLE 5.3   Microbial Strains in Crops Used Under Heavy Metal Stressed Conditions

Strain	Crops	Metal Ameliorated	Author, year
Bacillus amyloliquefaciens	Rice	Copper	Shahzad et al., 2019
Kocuria sp.	Brassica nigra	Copper	Hansda and Kumar, 2017
Bacillus cereus	Wheat	Copper and Chromium	Hassan et al., 2017
Klebsiella pneumoniae	Rice	Cadmium	Pramanik et al., 2017
Ralstonia eutropha	Sunflower	Zinc	Marques et al., 2013
Sinorhizobium meliloti and *Agrobacterium tumefaciens*	*Medicago lupulina*	Copper and Zinc	Jian et al., 2019
Pseudomonas aeruginosa and *Burkholderia, Gladioli*	Tomato	Cadmium	Khanna et al., 2019
Pseudomonas and *Bacillus* sp.	Spinach	Cadmium, Lead, and Zinc	Shilev et al., 2019

5.8   ROLE OF MICROBES UNDER TEMPERATURE STRESS

Under changing the climate scenarios, adverse temperatures like heat as well as cold need to mitigate for enhancing the crop productivity. Among different environmental stresses in crop plants, temperature stress is one of the limiting factors that causing a great degree of damage to plants. During unfavorable soil temperature as well as climatic variations of temperature, several physiological, morphological, and biochemical changes occur in plant leaves and roots to adapt themselves or to modify the environmental stresses. These adaptations include the ability of root systems to excrete a wide variety of molecules altering the plants–microbes associations. Microorganisms could play an important role in increasing the tolerance to heat and cold stresses in agricultural plants. PGPR mitigates temperature stresses (cold and heat) through the production of biofilm and EPS. PGPR mitigates these impacts through other mechanisms like producing antioxidants, release of cytokinins, and affecting the ethylene precursor ACC deaminase. AM fungi and other endophytic bacteria also tend to mitigate this type of cold and heat stresses (Chakraborty and Saha, 2019). Thus, the deleterious effect of heat stress in plants productivity in high-temperature regions could be mitigated by using promising species of PGPR. Therefore, inoculation of PGPR in existing cereal-based cropping systems with the inclusion of deep-rooted cultivars may be one of the strategies for mitigating the stress. These beneficial PGPRs colonize the rhizosphere or endorhizosphere and promote the growth of plants through various mechanisms. Rhizospheric soils are colonized with diverse temperature (cold or heat) tolerant microbes. In temperate areas, a microbe tolerant to cold colonizes the rhizosphere and improves the plant growth properties under low temperature. The PGPRs in cereals under low soil temperature and in soils deficient in phosphorus (Yadav et al., 2014), and in pulses under low soil temperature have been reported (Meena et al., 2015) with an efficient biofertilizer ability. The PGPRs isolated from rhizospheric soil like *Brevundimonas terrae, Arthrobacter nicotianae,* and *Pseudomonas cedrina* are adapted for low temperature and are known to have multifunction plant growth promoting ability. Several authors have also reported that the bacteria isolated from low-temperature regions express their antimicrobial activity (Javani et al., 2015). As temperature plays a significant role in regulating biochemistry and physiology of soil microorganisms under unsuitable temperature conditions or in changing climatic variations, there exists a wide variation in variability in their tolerance toward temperature stress. Under such conditions, microbes develop process to protect their membrane, proteins and nucleic acids. Microbes even alter their genetic expression and enhance the protein and related enzymes to combat this temperature stress. Gene expression of heat and cold tolerant protein and enzyme is enhanced under these conditions. Some bacteria and fungi accumulate trehalose during heat stress, which protects microbes from thermal injury (Li et al., 2009). Trehalose is also responsible for reducing the denaturation of enzymes and proteins under stress due to its important property of tolerating freezing and desiccation. *Serratia nematodiphila* increases the growth of black pepper under low-temperature stress by the production of gibberellins (Kang et al., 2015). In grapes, *Burkholderia phytofirmans* modulated the metabolism of carbohydrates to reduce frost injuries under low-temperature stress (Fernandez et al., 2012). Subramanian et al., 2015 in their study with tomato inoculated with *Pseudomonas vancouverensis* OB155 and *P. frederiksbergensis* OS261 showed increased expression of cold-tolerant genes and antioxidant activity in tomato plant tissue.

5.9 CONCLUSION

Abiotic stresses are responsible for the decrease in yield across many parts of the world. Microbes associated with crop plants and soil confers a great resistance to stresses of abiotic nature. Microorganisms follow a number of mechanisms to withstand these stresses. One alternative could be developing stress-tolerant plants through changes in the genetic makeup, but harnessing microbes is one of the effective environment friendly options available within a short time. Potential microbes are well-identified for a range of crops under diverse stress conditions, but their evaluation on the field seems necessary for its applications in wider zones.

KEYWORDS

- abiotic stress
- crop plants
- PGPRs
- stress tolerance

REFERENCES

Ahemad, M. Remediation of metalliferous soils through the heavy metal resistant plant growth promoting bacteria: Paradigms and prospects. *Arab. J. Chem.* 2019, 12, 1365–1377.

Asaf, S.; Khan, A.L.; Khan, M.A.; Imran, Q.M.; Yun, B.W.; Lee, I.J. Osmoprotective functions conferred to soybean plants via inoculation with *Sphingomonas* sp LK11 and exogenous trehalose. *Microbiol. Res.* 2017, 205, 135–145.

Ayangbenro, A.; Babalola, O. A new strategy for heavy metal polluted environments: a review of microbial biosorbents. *Int. J. Environ. Res. Public Health.* 2017, 14, 94.

Bianco, C.; Defez, R. Medicago truncatula improves salt tolerance when nodulated by an indole-3-acetic acid overproducing *Sinorhizobium meliloti* strain. *J. Exp. Bot.* 2009, 60, 3097–3107.

Burd, G.I.; Dixon, D.G.; Glick, B.R. Plant growth promoting bacteria that decrease heavy metal toxicity in plants. *Can. J. Microbiol.* 2000, 46, 237–245.

Cardinale, M.; Ratering, S.; Suarez, C.; Montoya, A.M.Z.; Plaum, R.G.; Schnell, S. Paradox of plant growth promotion potential of rhizobacteria and their actual promotion effect on growth of barley (*Hordeum vulgare* L.) under salt stress. *Microbiol. Res.* 2015, 181, 22–32.

Chakraborty, S.; Saha, N. Role of microorganisms in abiotic stress management. *Int. J. Sci. Environ. Technol.* 2019, 8 (5), 1028–1039.

Cho, S.M.; Kang, B.R.; Han, S.E.; Anderson, A.J.; Park, J.Y.; Lee, Y.H.; Cho, B.H.; Yang, K.Y.; Ryu, C.M.; Kim, Y.C. 2R,3R-Butanediol, a bacterial volatile produced by *Pseudomonas chlororaphis* O6, is involved in induction of systemic tolerance to drought in Arabidopsis thaliana. *Mol. Plant Microbe Interact.* 2008, 21, 1067–1075.

Csonka, L.N. Physiological and genetic responses of bacteria to osmotic stress. *Microbiol. Rev.* 1989, 53, 121–147.

Dardanelli, M.S. Arachis hypogaea PGPR isolated from Argentine soil modifies its lipids components in response to temperature and salinity. *Microbiol. Res.* 2015, 173, 1–9.

Fernandez, O.; Theocharis, A.; Bordiec, S.; Feil, R.; Jacquens, L.; Clement, C. Burkholderia phytofirmans PsJN acclimates grapevine to cold by modulating carbohydrate metabolism. *Mol. Plant Microbe Interact.* 2012, 25, 496–504.

Flowers, T.J.; Galal, H.K.; Bromham, L. Evolution of halophytes: multiple origins of salt tolerance in land plants. *Funct. Plant Biol.* 2010, 37, 604–612.

Goel, R.; Suyal, D.C.; Kumar, V.; Jain, L.; Soni, R. Stress-tolerant beneficial microbes for sustainable agricultural production. In *Microorganisms for Green Revolution* (Switzerland: Springer), 2018, pp. 141–159.

Goswamia, D.; Dhandhukiab, P.; Patela, P.; Thakker, J.N. Screening of PGPR from saline desert of Kutch: growth promotion in *Arachis hypogea* by *Bacillus licheniformis* A2. *Microbiol. Res.* 2014, 169, 66–75.

Grover, M.; Ali, S.Z.; Sandhya, V.; Rasul, A.; Venkateswarlu, B. Role of microorganisms in adaptation of agriculture crops to abiotic stresses. *World J. Microbiol. Biotechnol.* 2011, 27, 1231–1240.

Gupta, S.; Singh, D. Role of genetically modified microorganisms in heavy metal bioremediation. in *Advances in Environmental Biotechnology.* eds R. Kumar; A. Sharma; and S. Ahluwalia (Singapore: Springer), 2017, 197–214.

Gusain, Y.S.; Singh, U.S.; Sharma, A.K. Bacterial mediated amelioration of drought stress in drought tolerant and susceptible cultivar of rice (*Oryza sativa* L.). *Afr. J. Biotechnol.* 2015, 149, 764–773.

Hammond, J.P.; Bennett, M.J.; Bowen, H.C.; Broadley, M.R.; Eastwood, D.C.; May, S.T.; Rahn, C.; Swarup, R.; Woolaway, K.E.; White, P.I. Changes in gene expression in *Arabidopsis* shoots during phosphate starvation and the potential for developing smart plants. *Plant Physiol.* 2003, 132, 578–96.

Hansda, A.; Kumar, V. Cu-resistant Kocuria sp. CRB15: a potential PGPR isolated from the dry tailing of Rakha copper mine. 3*Biotech.* 2017, 7, 132.

Hassan, T.U.; Bano, A.; Naz, I. Alleviation of heavy metals toxicity by the application of plant growth promoting rhizobacteria and effects on wheat grown in saline sodic field. *Int. J. Phytoremed.* 2017, 19, 522–529.

Hodge, A. The plastic plant: root responses to heterogeneous supplies of nutrients. *New Phytol.* 2004, 162, 9–24.

Javani, S. Four psychrophilic bacteria from Antarctica extracellularly biosynthesize at low temperature highly stable silver nanoparticles with outstanding antimicrobial activity. Colloids Surf. *Physicochem. Eng. Aspects*. 2015, 483, 60–69.

Jian, L.; Bai, X.; Zhang, H.; Song, X.; Li, Z. Promotion of growth and metal accumulation of alfalfa by coinoculation with Sinorhizobium and Agrobacterium under copper and zinc stress. 2019. DOI: 10.7717/peerj.6875.

Jiao, J.; Ma, Y.; Chen, S.; Liu, C.; Song, Y.; Qin, Y.; Liu, Y. Melatonin producing endophytic bacteria from grapevine roots promote the abiotic stress-induced production of endogenous melatonin in their hosts. *Front. Plant Sci*. 2016, 7, 1387.

Jochum, M.D.; McWilliam, K.L.; Borrego, E.J.; Kolomiets, M.V.; Niu, G.; Pierson, E.A.; Jo, Y.K. Bioprospecting plant growth-promoting rhizobacteria that mitigate drought stress in grasses. *Front. Microbiol*. 2019, 10, 2106.

Kang, S.M.; Khan, A.L.; Waqas, M.; You, Y.H.; Hamayun, M.; Joo, G.J. Gibberellin-producing *Serratia nematodiphila* PEJ1011 ameliorates low temperature stress in *Capsicum annuum* L. *Eur. J. Soil Biol*. 2015, 68, 85–93.

Kang, S.M.; Khan, A.L.; Waqas, M.; You, Y.H.; Kimd, J.H.; Kimc, J.G.; Hamayune, M.; Lee, I.J. Plant growth-promoting rhizobacteria reduce adverse effects of salinity and osmotic stress by regulating phytohormones and antioxidants in *Cucumiss ativus. J. Plant Interact*. 2014, 9(1), 673–682.

Kasim, W.A.; Osman, M.E.; Omar, M.N.; Abd El-Daim, I.A.; Bejai, S.; Meijer, J. Control of drought stress in wheat using plant growth promoting bacteria. *J. Plant Growth Regul*. 2013, 32, 122–130.

Kasim, W.A.; Gaafar, R.M.; AbouAli, R.M.; Omar, M.N.; Hewait, H.M. Effect of biofilm forming plant growth promoting rhizobacteria on salinity tolerance in barley. *Ann. Agric. Sc*. 2016, 61(2), 217–227.

Kearl, J.; McNary, C.; Lowman, J.S.; Mei, C.; Aanderud, Z.T.; Smith, S.T.; West, J.; Colton, E.; Hamson, M.; Nielsen, B.L. Salt-tolerant halophyte rhizosphere bacteria stimulate growth of Alfalfa in salty soil. *Front. Microbiol*. 2019, 10, 1849.

Khan, N.; Bano, A. Exopolysaccharide producing rhizobacteria and their impact on growth and drought tolerance of wheat grown under rainfed conditions. *PLoS One*. 2019, 14(9), 2019 e0222302.

Khan, Z.; Rho, H.; Firrincieli, A.; Hung, S.H.; Luna, V.; Masciarelli, O.; Doty, S.L. Growth enhancement and drought tolerance of hybrid poplar upon inoculation with endophyte consortia. *Curr. Plant Biol*. 2016, 6, 38– 47.

Khan, M.A.; Asaf, S.; Khan, A.L.; Adhikari, A.; Jan, R.; Ali, S.; Imran, M.; Kim, K.M.; Lee, I.J. Halotolerant rhizobacterial strains mitigate the adverse effects of NaCl stress in soybean seedlings. *BioMed. Res. Int*. 2019, Article ID 9530963, 15 pages.

Khan, M.S.; Zaidi, A.; Wani, P.A. Role of phosphatesolubilizing microorganisms in sustainable agriculture—a review. *Agron. Sustain Dev.* 2007, 27, 29–43.

Khan, N.; Bano, A.; Rahman, M.A.; Guo, J.; Kang, Z.; Babar, M.A. Comparative physiological and metabolic analysis reveals a complex mechanism involved in drought tolerance in chickpea (*Cicer arietinum* L.) induced by PGPR and PGRs. *Sci. Rep*. 2019, 9, 2097.

Khanna, K.; Jamwal, V.L.; Gandhi, S.G.; Ohri, P.; Bharadwaj, R. Metal resistant PGPR lowered Cd uptake and expression of metal transporter genes with improved growth and photosynthetic pigments in *Lycopersicon esculentum* under metal toxicity. *Sci. Rep*. 2019, 9, 5855. | https://doi.org/10.1038/s41598-019-41899-3

Kohler, J.; Caravaca, F.; Carrasco, L.; Roldán, A. Contribution of *Pseudomonas mendocina* and *Glomus intraradices* to aggregates stabilization and promotion of biological properties in rhizosphere soil of lettuce plants under field conditions. *Soil Use Manage.* 2006, 22, 298–304.

Kohler, J.; Hernandez, J.A.; Caravaca, F.; Roldan, A. Induction of antioxidant enzymes is involved in the greater effectiveness of a PGPR versus AM fungi with respect to increasing the tolerance of lettuce to severe salt stress. *Environ. Exp. Bot.* 2009, 65, 245–252.

Li, H.; Ding, X.; Wang, C.; Ke, H.; Wu, Z.; Wang, Y. Control of tomato yellow leaf curl virus disease by *Enterobacter asburiae* BQ9 as a result of priming plant resistance in tomatoes. *Turk. J. Biol.* 2016, 40, 150–159.

Lim, J.H.; Kim, S.D. Induction of drought stress resistance by multi-functional PGPR Bacillus licheniformis K11 in Pepper. *Plant Pathol. J.* 29 (2), 201–208.

Ma, Y.; Prasad, M.N.V.; Rajkumar, M.; Freitas, H. Plant growth promoting rhizobacteria and endophytes accelerate phytoremediation of metalliferous soils. *Biotechnol. Adv.* 2011, 29, 248–258.

Marasco, R.; Rolli, E.; Vigani, G.; Borin, S.; Sorlini, C.; Ouzari, H.; Zocchi, G.; Daffonchio, D. Are drought-resistance promoting bacteria cross-compatible with different plant models? *Plant Signal. Behav.* 2013, 8 (10), e26741.

Marques, A.P.; Moreira, H.; Franco, A.R.; Rangel, A.O.; Castro, P.M. Inoculating Helianthus annuus (sunflower) grown in zinc and cadmium contaminated soils with plant growth promoting bacteria—effects on phytoremediation strategies. *Chemosphere.* 2013, 92, 74–83.

Mayak, S.; Tirosh, T.; Glick, B.R. Plant growth-promoting bacteria that confer resistance to water stress in tomatoes and peppers. *Plant Sci.* 2004, 166, 525–530.

Meena, K.K.; Sorty, A.M.; Bitla, U.M.; Choudhary, K.; Gupta, P.; Pareek, A.; Singh, D.P.; Prabha, R.; Sahu, P.K.; Gupta, V.K.; Singh, H.B.; Krishanani, K.K.; Minhas, P.S. Abiotic stress responses and microbe-mediated mitigation in plants: the omics strategies. Front. *Plant Sci.* 2017, 8, 172.

Meena, R.K. Isolation of low temperature surviving plant growth–promoting rhizobacteria (PGPR) from pea (*Pisum sativum* L.) and documentation of their plant growth promoting traits. *Biocatal. Agric. Biotechnol.* 2015, 4, 806–811.

Microorganisms for Sustainability 7, pp: 141–159. DOI: 10.1007/978-981-10-7146-1_8.

Minhas, P.S.; Rane, J.; Pasala, R.K. Abiotic stresses in Agriculture: An overview. In *Abiotic Stress Management for Resilient Agriculture.* 2017. DOI: 10.1007/978-981-10-5744-1_1.

Mishra, G.I.; Sapre, S.; Kachare, S. Molecular diversity of 1-aminocyclopropane-1-carboxylate (ACC) deaminase producing PGPR from wheat (*Triticum aestivum* L.) rhizosphere. *Plant Soil.* 2017, 414, 213–227.

Morton, M.J.L.; Awlia, M.; Al-Tamimi, N.; Saade, S.; Pailles, Y.; Negrão, S.; Tester, M. Salt stress under the scalpel—dissecting the genetics of salt tolerance. *Plant J.* 2019, 97, 148–163.

Niu, X.; Song, L.; Xiao, Y.; Ge, W. Drought-tolerant plant growth-promoting rhizobacteria associated with foxtail millet in a semi-arid agroecosystem and their potential in alleviating drought stress. *Front. Microbiol.* 2018, 8, 2580.

Ojuederie, O.B.; Olanrewaju, O.S; Babalola, O.O. Plant growth promoting rhizobacterial mitigation of drought stress in crop plants: implications for sustainable agriculture. *Agronomy.* 2019, 9, 712.

Orhan, Furkan. Alleviation of salt stress by halotolerant and halophilic plant growth-promoting bacteria in wheat (*Triticum aestivum*). *Braz. J. Microbiol*. 2016, 47 (3), 621–627.

Pandey, S.; Ghosh, P.K.; Ghosh, S.; De, T.K.; Maiti, T.K. Role of heavy metal resistant *Ochrobactrum* sp. and *Bacillus* spp. strains in bioremediation of a rice cultivar and their PGPR like activities. *J. Microbiol*. 2013, 51, 11–17.

Panwar, M.; Tewari, R.; Nayyar, H. Native halo-tolerant plant growth promoting rhizobacteria *Enterococcus* and *Pantoea* sp. improve seed yield of Mungbean (*Vigna radiata* L.) under soil salinity by reducing sodium uptake and stress injury. *Physiol. Mol. Biol. Plants*. 2016, 22 (4), 445–459.

Paulucci, N.S.; Gallarato, L.A.; Reguera, Y.B.; Vicario, J.C.; Cesari, A.B.; de Lema, M.B.G.; Qin, Y.; Druzhinina, I.S.; Pan, X.; Yuan, Z. Microbially mediated plant salt tolerance and microbiome-based solutions for saline agriculture. *Biotechnol. Adv.* 2016, 34 (7), 1245–1259.

Pramanik, K.; Mitra, S.; Sarkar, A.; Soren, T.; Maiti, T.K. Characterization of cadmium-resistant Klebsiella pneumoniae MCC 3091 promoted rice seedling growth by alleviating phytotoxicity of cadmium. *Environ. Sci. Pollut. Res*. 2017, 24, 24419–24437.

Rima, F.S.; Biswas, S.; Sarker, P.K. Bacteria endemic to saline coastal belt and their ability to mitigate the effects of salt stress on rice growth and yields. Ann. *Microbiol*. 2018, 68, 525–535.

Rizvi, A.; Khan, M.S. Heavy metal induced oxidative damage and root morphology alterations of maize (*Zea mays* L.) plants and stress mitigation by metal tolerant nitrogen fixing *Azotobacter chroococcum*. *Ecotoxicol. Environ. Saf.* 2018, 157, 9–20

Ruan, C.J.; da Silva, J.A.T.; Mopper, S.; Qin, P.; Lutts, S. Halophyte improvement for a salinized world. *Crit. Rev. Plant Sci*. 2010, 29, 329–359.

Salwan, R.; Sharma, A.; Sharma, V. Microbes mediated plant stress tolerance in saline agricultural ecosystem. *Plant Soil*. 2019. https://doi.org/10.1007/s11104-019-04202-x

Sanchez, M.R., Armada, E., Munoz, Y.; Salamone, E.G.; Aroca, R.; Lozano, J.M.R.; Azcon, R. *Azospirillum* and arbuscular mycorrhizal colonization enhance rice growth and physiological traits under well-watered and drought conditions. *J. Plant Physiol*. 2011, 168, 1031–1037.

Saxena, AK.; LataShende, R.; Pandey, A.K. Culturing of plant growth promoting rhizobacteria. In: Gopi KP, Varma A (eds) Basic research applications of mycorrhizae. I K International Pvt Ltd, New Delhi, 2005, pp. 453–474.

Sessitsch, A.; Kuffner, M.; Kidd, P.; Vangronsveld, J.; Wenzel, W.W.; Fallmann, K. The role of plant-associated bacteria in the mobilization and phytoextraction of trace elements in contaminated soils. *Soil Biol. Biochem*. 2013, 60, 182–194.

Shahid, M.; Khalid, S.; Abbas, G.; Shahid, N.; Nadeem, M.; Sabir, M. Heavy metal stress and crop productivity. In: *Crop Production and Global Environmental Issues*, ed. K. R. Hakeem (Cham: Springer International Publishing). 2015, 1–25.

Shahzad, R.; Bilal, S.; Imran, Md.; Khan, A.L.; Alusaimi, A.A.; Al-Shwey, H.A.; Almahasheer, H.; Rehman, S; Lee, I.N. Amelioration of heavy metal stress by endophytic Bacillus amyloliquefaciens RWL-1 in rice by regulating metabolic changes: potential for bacterial bioremediation. *Biochem. J.*. 2019, 476, 3385–3400.

Sharifi, R.; Ryu, C.M. Revisiting bacterial volatile-mediated plant growth promotion: lessons from the past and objectives for the future. *Ann. Bot*. 2018, 122, 349–358.

Shilev, S.; Babrikova, I.; Babrikov T. Consortium of plant growth-promoting bacteria improves spinach (*Spinacea oleracea* L.) growth under heavy metal stress conditions. *J. Chem. Technol. Biotechnol*. 2019. DOI 10.1002/jctb.6077.

Subramanian, P.; Mageswari, A.; Kim, K.; Lee, Y.; Sa, T. Psychrotolerant endophytic Pseudomonas sp strains OB155 and OS261 induced chilling resistance in tomato plants (*Solanum lycopersicum* mill.) by activation of their antioxidant capacity. *Mol. Plant Microbe Interact.* 2015, 28, 1073–1081.

Suzuki, N.; Rivero, R.M.; Shulaev, V.; Blumwald, E.; Mittler, R. Abiotic and biotic stress combinations. *New Phytologist.* 2014, 203, 32–43.

Szabados, L.; Savoure, A. Proline: a multifunctional amino acid. *Trends Plant Sci.* 2009, 15 (2), 89–97.

Tiwari, S.; Lata, C. Heavy metal stress, signaling, and tolerance due to plant-associated microbes: an overview. *Front. Plant Sci.* 2018, 9, 452.

Ullah, A.; Heng, S.; Munis, M.F.H.; Fahad, S.; Yang, X. Phytoremediation of heavy metals assisted by plant growth promoting (PGP) bacteria: a review. *Environ. Exp. Bot.* 2015, 117, 28–40

Ullah, S.; Bano, A. Isolation of plant-growth-promoting rhizobacteria from rhizospheric soil of halophytes and their impact on maize (*Zea mays* L.) under induced soil salinity. *Can. J. Microbiol.* 2015, 61, 307–313.

Vacheron, J.; Desbrosses, G.; Bou_aud, M.L.; Touraine, B.; Moënne-Loccoz, Y.; Muller, D.; Legendre, L.; Wisniewski-Dyé, F.; Prigent-Combaret, C. Plant growth-promoting rhizobacteria and root system functioning. *Front. Plant Sci.* 2013, 4, 356.

Vargas, L.; Santa Brígida, AB.; MotaFilho, J.P.; de Carvalho, T.G.; Rojas, C.A.; Vaneechoutte, D.; Hemerly, A.S. Drought tolerance conferred to sugarcane by association with Gluconacetobacter diazotrophicus: a transcriptomic view of hormone pathways. 2014, *PLoS One* 9:e114744.

Wafaa, M.H.W.; Hussein, M.M.; Mehanna, H.M.; El-Moneim, D.; Bacteria polysaccharides elicit resistance of wheat against some biotic and abiotic stress. *Int. J. Pharm. Sci. Rev. Res.* 2014, 29(2), 292–298.

Wang, S.A.; Wu, H.J.; Qiao, J.Q.; Ma, L.L.; Liu, J.; Xia, Y.F.; Gao, X.W. Molecular mechanism of plant growth promotion and induced systemic resistance to tobacco mosaic virus by Bacillus spp. *J. Microbiol. Biotechnol.* 2009, 19, 1250–1258.

Wang, W.; Vinocur, B.; Altman, A. Plant responses to drought, salinity and extreme temperatures towards genetic engineering for stress tolerance. *Planta.* 2007, 218, 1–14

Yadav, J. Evaluation of PGPR and different concentration of phosphorus level on plant growth, yield and nutrient content of rice (*Oryza sativa*). *Ecol. Eng.* 2014, 62, 123–128.

Yadav, A.N.; Yadav, N. Stress-adaptive microbes for plant growth promotion and alleviation of drought stress in plants. *Acta Sci. Agric.* 2018, 2.6, 85–88.

Yuan, Z.; Druzhinina, I.S.; Labbé, J.; Redman, R.; Qin, Y.; Rodriguez, R. Specialized microbiome of a halophyte and its role in helping non-host plants to withstand salinity. *Sci. Rep.* 2016, 6, 32467. DOI: 10.1038/srep32467

Zarea, M.J.; Hajinia, S.; Karimi, N.; Goltapeh, E.M.; Rejali, F.; Varma, A. Effect of Piriformospora indica and Azospirillum strains from saline or non-saline soil on mitigation of the effects of NaCl. *Soil Biol. Biochem.*, 2012, 45, 139–146.

Zhou, N.; Zhao, S.; Tian, C.Y. Effect of halotolerant rhizobacteria isolated from halophytes on the growth of sugar beet (*Beta vulgaris* L.) under salt stress. *FEMS Microbiol. Lett.* 2017, 364:fnx091. DOI: 10.1093/femsle/fnx091.

CHAPTER 6

An Overview of Biochar Application on Biological Soil Health Indicators and Greenhouse Gas Emission

IFEOLUWA ADESINA, ARNAB BHOWMIK*, HARMANDEEP SHARMA, and ABOLGHASEM SHAHBAZI

^{1}Department of Natural Resources and Environmental Design, North Carolina AandT State University, Greensboro, NC, USA

*Corresponding author. E-mail: abhowmik@ncat.edu

ABSTRACT

The rate of greenhouse gas (GHG) emission has been on the steady rise and pose a threat to the global climate affecting terrestrial ecosystems. Agricultural management practices are reported to be a major source of GHG emissions. Hence, climate adaptive management strategies intended to reduce GHG emissions are of pivotal importance to maintain healthy soil ecosystem functioning. Biochar, a vital material produced from controlled pyrolysis, when applied to soil has been proposed as an effective means of incorporating climate adaptive soil management strategies by enhancing soil physico-chemical as well as biological properties. This chapter reviews the various effects of biochar on the emission of two selected GHG (carbon dioxide [CO_2] and nitrous oxide [N_2O]) and biological soil health indicators. Biochar amended soil has been documented to not only increase carbon storage in soil but also help in mitigating GHG emissions like CO_2 and N_2O. The aromatic structure of biochar makes it very stable, and its slow-releasing form of carbon in the soil makes it highly essential for the sequestration of carbon on the long-term basis. Biochar properties like pyrolysis temperature, biochar feed stock quality, rate, and method of application influence soil biological properties on incorporation. Biochar addition to soil has many benefits including stimulation of specific soil microorganisms, soil enzymatic activity, and soil respiration which could lead to enhanced biological activity and nutrient cycling or retention.

6.1 INTRODUCTION

6.1.1 EMISSION OF GHG FROM THE AGRICULTURAL SECTOR

Greenhouse gases (GHGs) are specific gases in the upper atmospheric region that are instrumental in the capturing and absorption of heat. Watson et al. (1992) listed the significant GHGs that include methane (CH_4), carbon dioxide (CO_2), nitrous oxide (N_2O), water vapor, chlorofluorocarbons (CFCs), and ozone (O_3). Generally, these GHGs are essential and vital to human lives as the planet would be extremely cold without them, leading to an almost impossible life on this planet (Casper, 2009). Over the years, before the industrial revolution, the world's carbon supply was reasonably stable as natural processes could consume as much carbon as has been released, leading to an equilibrium. However, major anthropogenic activities such as burning of fossil fuels, destruction of natural flora through defor-estation, and intensive agricultural production activities have increased the concentration of GHG in the atmosphere. Agriculture provides us with food and necessary raw materials for various industries but instead of mitigating climate change, certain agricultural management practices have been proven to be a net producer of GHG. This could be directly through conventional farming practices that vastly diminish soil carbon stocks while emitting GHG and indirectly through land use change (Lal, 2004). The agriculture sector contributes to up to about 10%–12% of the total anthropogenic GHG production (EPA, 2014). Since the dawn of agriculture, soils have lost anywhere between 30% and 75% of the native soil organic carbon (SOC) (Lal et al., 2007). With the advent of mechanized farming during the mid-20th century, some of the management practices like use of synthetic nitrogen (N) fertilization, tillage, and mono-cropping have accelerated the depletion and reduction in SOC stocks and resulted in anthropogenic CO_2 and N_2O emissions (Khan et al., 2007; Lal, 2004).

CO_2 has contributed more than any other GHG to climate variability and is referred to as highly significant GHG, particularly because of its abundance and long atmospheric residence time (Lindsey, 2018). According to the United States Environmental Protection Agency, in 2018, up to 65% of CO_2 was emitted from fossil fuel and industrial processing while forestry and other land use contributed about 11%, thus accounting for 76% of world total GHG (USEPA, 2019). Also, agriculture sector in the US accounted for 10% of total GHG produced in 2018 and CO_2 emission was 81.3% of total GHG emissions from human activities. CO_2 emissions in the US increased by about 5.8% between 1990 and 2018 (USEPA, 2019). Overall, the rapid increase in CO_2 far exceeds the amount that can be recycled by nature. Houghton et al. (1983) reported that one of the vital and major contributory sources of atmospheric CO_2 is the decomposition of soil organic matter as a result of land use conversion of natural ecosystems (forest or grassland) to agriculture and soil disturbance due to tillage activities. Other anthropogenic sources that emit CO_2 in the agricultural sector include fossil fuel consumption by farm equipment (Shin and Kim, 2018). Such agricultural activities also involve burning of fuels that occur with the production and use of farm machinery and implements coupled with agricultural inputs and fertilizers. N_2O is a potent GHG which also contributes severely to the fast depletion of the stratospheric O_3 layer in the 21st century (Ravishankara et al., 2009). The average lifetime of N_2O is 114 years in the atmosphere which is around 300 times more potent than CO_2 over a 100 year time frame (IPCC, 2014). On a global scale agriculture contributes to 65%–80% of the total N_2O emissions which result from the increase in synthetic fertilizer usage. Future predictions estimate that agricultural soils will contribute close to 59% of total N_2O emissions by the year 2030. Soils amended with nitrogenous fertilizers, compost and manures are the predominant sources of N_2O emissions from agricultural soils (Smith, 2017). Fertilizers release inorganic nitrogen into the soils, which is subsequently converted to N_2O by the action of soil nitrifying and denitrifying bacteria. On average, about 1% of the applied nitrogen from organic amendments and fertilizers are returned back to the atmosphere (Eggleston et al., 2006). Thus, it can be concluded that the emission rate of N_2O is majorly related to the quantity, type, and method of application of such fertilizers, although weather patterns and soil types are also dependent factors of N_2O emissions. Banerjee et al. (2016) reported that N_2O emissions are also dependent on antecedent soil moisture conditions.

One of the keys to shift agriculture from being a source of GHG to serving as a carbon sink is adaptation of climate adaptive soil health management practices (Bhowmik et al. 2016, 2017a, 2017b). Sustainable agriculture systems could provide ecological and environmental services that promote soil conservation, reduce GHG emissions, and improve soil health. An effective counter measure for the increasing GHG emissions and soil degradation is soil carbon sequestration through biochar amendment. Biochar inputs facilitate stable pools of carbon in soil which not only affect the soil biological properties but also has the potential for climate change mitigation. The objective of this chapter is to provide a literature summary on the effect of different types of biochar on soil biological health properties and GHG emissions (mainly CO_2 and N_2O).

6.2 BIOCHAR

Biochar is a carbon-abundant material that is synthesized from pyrolysis, which is a highly controlled chemical process involving the thermal break-down of biomass like wood, manure, crop residues, and so forth, in the absence of oxygen (Cha et al., 2016; Lehmann et al., 2009). By heating the biomass to a temperature between 300 and 1000 °C, in a condition devoid of oxygen, a stable biochar solid is produced. Pyrolysis changes about 10%–50% of the initial carbon biomass into a stable carbon form, which can last in the soil for several years (Lehmann et al., 2007). Hence, biochar is a stable material that can resist degradation over many years. It is similar in appearance to charcoal, but their usage varies since biochar is a soil amendment that could be used to improve overall soil health. Soil health infers to the capacity of soil to act as a crucial and important living ecosystem continuously, supporting, assisting, and sustaining living organisms over a period of time. Biochar usage as a soil amendment has a long history, especially in non-productive soil and the model of adding biochar to soil dates back to several years when plant growth responses on former charcoal storage sites were observed. Subsequently, a lot of notable research began, and Liebig (1852) concluded that biochar could improve the availability of nutrients in soils. By the 19th century, number of scientific studies on biochar had increased significantly.

Previous evidences established that carbon in biochar exhibit high dominance in the soil, with residence time ranging from about hundreds of years to as much as even thousands of years, which is roughly 10–1000 times more than soil organic matter (Gurwick et al., 2013). Consequently, because carbon is persistently held in the soil, it often leads to reductions in the rate and quantity of CO_2 released from soils. Studies have shown reduced GHG emissions after biochar applications, others had contradictory results (Stewart et al., 2013; Ameloot et al., 2013). Therefore, the impacts of biochar on GHG emission is complex and depend upon the biochar and soil attributes (He et al., 2017). The physico-chemical properties of biochar are controlled by pyrolysis method and feedstock material used which are highly heterogeneous in nature. There are several additional benefits of applying biochar to soils besides their high carbon content. These include increase in soil pH, total nitrogen, nutrient retention, soil moisture retention, total phosphorus, improvement of the soil structure, reduced leaching losses of nitrogen, adsorption of synthetic chemicals including steroid hormones and heavy metals (Atkinson et al., 2010; Cao et al., 2009; Sohi et al., 2010), and better root development (Abiven et al., 2015).

6.2.1 BIOCHAR USAGE: LIMITATIONS

Even though several documented benefits of biochar continue to increase, there are but limited studies which focused on biochar effects on the community composition of soil microorganisms. The impacts of bio-interactions that exist between biochar, microbes, soils, and plant roots have not been adequately defined (Joseph et al., 2010). Although biochar was reported to enhance the growth of soil microbial biomass, and its diverse activities as reported by Smith et al. (2010), most of the biochar-carbon are reportedly unavailable for use by soil microbes (Thies and Rillig, 2012). For example, a short-term experiment to ascertain the impacts and effects of straw and biochar on the emission of CO_2 and microbial community using a ^{13}C-labeled rice straw and its biochar (^{13}C-labeled biochar) showed that more straw-C was utilized by soil microbes than biochar-C (Pan et al., 2016). Additionally, there is limited knowledge on how to produce a good-quality biochar in large proportions as well as determine biochar loading capacity across various soil types.

Currently, the highest amount of carbon that can be added to the soil in biochar form without negatively affecting other soil properties or the environment is still unclear. Hence, to optimize the amount of biochar that could be added to soil, more research should be conducted. More studies should be conducted to ascertain if biochar application differs based on type of soil and crop grown. It is also worthy of great note that all biochar compositions are not the same (Spokas et al., 2012); hence, it is pertinent and necessary to determine most beneficial biochar to use under various climatic and environmental conditions. Additionally, the negative responses of crop yields to biochar application have also been observed (Crane-Droesch et al., 2013), but most often, biochar either elevates or has no effect on crop biomass production. Due to variability in the environmental factors, it is possible that mostly observed short term effects of biochar often recorded in the laboratory may not always yield similar results in the field studies. Hence, biochar should be designed with unique characteristics that best fit the specific environmental or agronomic settings where it will be applied (Novak et al., 2012).

6.2.2 BIOCHAR: ENVIRONMENTAL AND AGRONOMIC IMPLICATIONS

Biochar added to soil has several beneficial agronomic and environmental implications (Kookana et al., 2011). Biochar application improves soil physical characteristics such as particle density, bulk density, water holding capacity, infiltration, hydraulic conductivity, etc. as reported by several studies (Githinji, 2014; Basso et al., 2013; Karhu et al., 2011; Laird et al., 2010). In a study conducted in the United Kingdom using biochar produced from a mixture of wood feedstock of Oak (*Quercus sp.*), sycamore (*Acer pseudoplatanus* L.), Beech (*Fagus sylvatica* L.), and Bird Cherry (*Prunus padus* L.) pyrolyzed at 600 °C increased bulk density, water retention of sandy-loam soils as reported by Ulyett et al. (2014). Similarly, a related study involved nine biochar pyrolyzed from five feedstocks at two different temperatures were incubated with Norfolk loamy sand. All biochar-treated samples exhibited an enhanced soil water storage capacity; even though their effects varied with both feedstock selection and prevailing temperature during the pyrolysis process (Novak et al., 2012). Comparison was also made between two switch-grass biochars pyrolyzed in a temperature range of 250–500 °C using both raw and uncharred switchgrass on moisture storage and changes in bulk density in a loamy sand in England (Novak, 2013). Findings showed that the uncharred switchgrass had similar soil moisture retention capacity as the two switchgrass biochars.

The diverse effects of biochar application on soil pH have also been reported by several researchers. One study analyzed the effects of new and old biochars on the amelioration of acidic soils with different incubation methods (Zhao et al., 2015). Results indicated that new biochar performed better in the reclamation of soil acidity compared to old biochar and stated that although old biochar could be used to alter soil acidity, the efficacy declined with short-term aging (Zhao et al., 2015). Another documented research study reported that the application of biochar consequentially resulted in reduced soil pH even with alkaline biochar material (Liu et al., 2012). On the contrary, some laboratory incubation and field studies revealed that biochars pyrolyzed at low temperatures of 300–600 °C made from various feedstocks led to a gradual increase in soil pH, cation exchange capacity (CEC), aggregation, and abundance and altered community structure of soil microorganisms (Gul et al., 2015).

A laboratory study was conducted with a temperate sandy loam soil to test the effects of wood and straw gasification biochar on soil carbon storage. Result from the 22-months soil incubation study showed that the biochar application enhanced the sequestration of soil carbon, CEC and soil pH (Hansen et al., 2016). The effects of three biochars which include *Miscanthus* straws, woody material, and coffee husks were investigated in relation to chemical characteristics, soil respiration and the kinetic release of bio-available silicon ($CaCl_2$-extractable Si). The findings from this study indicated that biochar from *Miscanthus* straws enhanced soil fertility and carbon accumulation. Furthermore, it was also suggested as a potent and likely source of bioavailable silicon (Houben et al., 2014). Biochar has soil conditioning properties (Novak et al., 2009), and when added to soil from a grass pasture reduce N_2O emissions from ruminant urine as documented by Taghizadeh-Toosi et al. (2011). Biochar can also provide valuable sinks for dissolved contaminants at places or depths where leachates and runoffs are stored or irrigated onto the surface of the soil (Robinson et al., 2007). However, the success is dependent on how much of contaminant the biochar can store before attaining saturation and the potency of the biochar contaminant complex, which is dependent on the source and pyrolysis temperature of the biochar (Gell et al., 2011). A one year laboratory incubation experiment was carried out to establish the effects of biochar produced from both bamboo and rice straws on the enzymatic activity and extractability of heavy metals (i.e. Cd, Zn, Pb, and Cu) (Yang et al., 2016). The results clearly indicated that although biochar type, particle size, and application rate changed the extractability of enzyme activity and heavy metal, rice straw biochar indicated a higher potential for reducing the bioavailability of heavy metals in soils as compared to bamboo biochar. The potential application of biochar for land remediation was also explored in another experiment where biochar generated from pinewood under slow pyrolysis was used. The results confirmed that pinewood biochar was able to adsorb magnesium, lead, chromium, and calcium in solution (Abdel-Fattah et al., 2015). Biochar also helps soil to retain beneficial fungi and other microorganisms (Chen et al., 2011). A study was conducted by Bamminger et al. (2014a) to determine the stability of two biochars which were derived from both the pyrolysis and hydrothermal carbonization of maize in arable soil. Results showed that even though biochar significantly increased the SOC content by 20%–40% in soil samples, the highest respiration rates, enzymatic activity, and increase in microbial activity were observed in soils amended with hydrothermal biochar as compared to the unamended controls. Few research studies have also documented biochar effects on soil microbial activity as biochar can be employed in enhancing soil fertility without having adverse effects on the soil biota (Bamminger et al., 2014b;

Rutigliano et al., 2014; Trupiano et al., 2017). Biochar creates a better habitat by changing the soil environment for microorganisms, although a long-term alteration in microbial communities depends on soil texture and the biochar source (Gul et al., 2015). For example, microorganisms tend to be more responsive to biochar source in sandy as compared to clayey soils. Similarly, biochar derived from crops and manure residue of feedstocks when added to soil has higher microbial abundance as compared to wood-derived biochar amended soils (Gul et al., 2015).

6.3   EFFECTS OF BIOCHAR ON BIOLOGICAL INDICATORS OF SOIL HEALTH

6.3.1   SOC AND ITS FRACTIONS

Leng and Huang (2018) documented that the biological effects and stability of biochar in soils are greatly affected by not only the temperature of the pyrolysis but also by the properties of the biochar rather than just by the soil characteristics. Biochar from sewage sludge reduced the quantity of dissolved mobile carbon and bio-accessible carbon polycyclic aromatic hydrocarbons (PAH) in contaminated soils. Biochar which was produced at 700 °C effectively reduced bio-accessible carbon than biochar produced at a lower temperature of about 500 °C (Zielińska and Oleszczuk, 2016). In contrast, the SOC and CEC of soil were reported to be significantly elevated (Shenbagavalli et al., 2012). This further buttress the fact that biochar has excellent potency to improve soil carbon storage. When oak and bamboo biochar were used to amend a degraded red soil, after a 372-day incuba-tion period, results showed clearly that biochar application elevated the total organic carbon content, enhanced microbial activity, and consequently increased macro aggregation, and overall soil quality (Demisie et al., 2014). In China, a two-year research study was conducted to evaluate the poten-tial effect of biochar's application on accumulation of carbon, stability of organic carbon, microbial biomass carbon, and dissolved organic carbon in soil using wheat stalk biochar. Biochar application significantly enhanced the accumulation of organic carbon and improved the stability of SOC. On the other hand, biochar application decreased the content of dissolved organic carbon in soil and microbial biomass carbon increased at first before gradually declining and finally disappeared as the incubation time period increased (Zhang et al., 2012). In a related study conducted to establish the effect of feedstock and pyrolysis temperature on soil carbon fractions, four biochars (rice straw, pine wood, pig manure, and sewage sludge) were pyrolyzed at different temperatures of 300 °C , 400 °C , 500 °C , 600 °C, and 700 °C, respectively. Results indicated that production of biochar at temperature below 500 °C produced higher dissolved organic carbon with low carbon stability while high pyrolysis temperature resulted in higher carbon content and stability. Biochar from pine wood also had the lowest dissolved organic carbon content out of the four biochars used (Wei et al., 2019). Ma et al. (2012) also analyzed the effect of biochar application on SOC content and its fractions in a gray desert soil in China. Biochar from dried cotton stalks were pyrolyzed at varying temperatures of 450, 600, and 750 °C, respectively. Their results indicated that the total SOC and readily oxidizable carbon content significantly increased with a proportionate increase in the temperature of the pyrolysis and the application rate compared to the control. Water-soluble organic carbon also increased with increasing biochar application rate but was not affected by the pyrolysis temperature while microbial biomass carbon increased irrespective of biochar properties and application rate. In a three-year field experiment conducted on ICAR-National Rice Research Institute in Orissa, India, the dynamics of SOC and carbon mineralization using biochar from rice husk were investigated. Subsequent results from the investigation indicated that the rice husk biochar increased the total SOC and carbon mineralization during the first 16 days of application and then declined gradually as the experiment progressed (Munda et al., 2018).

6.3.2 SOIL ENZYMATIC ACTIVITY AND MICROBIAL COMMUNITY STRUCTURE

When fast pyrolysis hardwood biochar, a process that occurs at higher temperatures with heating rates higher than 200 °C/min and residence time less than 10 s, was applied to a calcareous soil, a noticeable and significant alteration in the total soil microbial community structure was recorded. This led to a physiological stress response in the Gram-negative bacteria and also a sudden reduction in soil NO_3–N as about 85–97% reduction was reported in the treated compared to the control which were sustained over a certain period of time (Ippolito et al., 2014). A laboratory-based study where the effect of poultry litter biochar was added at a rate of 3% indicated that biological properties were considerably improved by the addition of biochar (Lu et al., 2015). Water hyacinth (*Eichorniacrassipes*) biochar added at a dose of 20 g/kg also increased the soil enzymatic activity, that is, acid phosphatase activity, alkaline phosphatase activity, and fluorescein hydrolases by 32%, 22.8%, and 50%, respectively, hence significantly improving the soil biological activity. The active microbial biomass tripled while and soil respiration almost doubled as compared to control as reported by Masto et al. (2013). A six-month study where a Portneuf subsoil was treated with switchgrass (*Panicum virgatum*) biochar pyrolyzed at a temperature of 350 °C to measure the microbial abundance response of five genes involved in nitrogen cycling was conducted. All the genes showed higher abundances in biochar amended treatments compared to the non-treated groups (Ducey et al., 2013). Thus, the results confirmed that biochar derived from activated switchgrass designed for use as a soil conditioner influenced the treated soils microbial communities. Activities of invertase, phosphomonoesterase, urease, β-glucosidase, β-glucosaminidase, and arylsulphatase increased with a proportionate increase in biochar rate when four different biochars including sewage sludge, deinking sewage sludge, *Miscanthus*, and pinewood were incorporated to two tropical soils under proso millet (Paz-Ferreiro et al., 2014). In a 219-day incubation experiment conducted to assess the impact of biochar amendments on soil enzymatic activities, soil was amended with two different biochar samples derived from pig manure pyrolyzed at different temperatures of 300 °C and 500 °C. The results from this study indicated that soil amended with biochar increased phosphomonoesterase, dehydrogenase, and phosphodiesterase activities when compared to control (Gascó et al., 2016). After a 90-day incubation period, a fluvo-aquic soil which was amended with maize biochar showed enhanced soil extracellular enzyme activity involved in carbon and sulfur cycling, except for β-xylosidase which increased at the lower rate of maize biochar amendment and decreased with a higher rate of maize biochar. On the contrary, the activities of urease and l-leucine aminopeptidase considerably increased with an increasing rate of biochar addition suggesting that adding biochar to soil could elevate the enzymatic activities (Wang et al., 2015). Adding of charred and uncharred corn feedstock to two different calcareous soils with both fine and coarse texture significantly triggered the activities of catalase, invertase, cellulose dehydrogenase, and protease when compared to the control non-treatment group (Khadem and Raiesi, 2017). Although the results of the study varied with prevailing temperature of pyrolysis, the rate of addition and soil texture, the findings further indicated that a combination of lower temperature and uncharred biochars feedstock improved enzyme activities in both soils than biochar produced at higher temperatures. Soil samples taken from a vineyard in Austria were amended with acid activated wood biochar, and found increased exoglucanase, β-glucosidase, exochitinase, and protease activities as compared to controls (Ameur et al., 2018). In another greenhouse experiment conducted to ascertain the effects of biochar made from poultry litter on clay loam soil, researchers found that the soil enzyme activities of urease, alkaline phosphatase, β-glucosidase, and arylsulphatase were significantly increased by the biochar amendment as reported by Akça and Namlı (2015).

Biochar amendment also influences soil microbial community structure. A research study investigated the positive and substantial effect of the application of biochar derived from the oak pellet on the microbial abundance in temperate soils (Gomez et al., 2014). In a long-term experimental setting, the addition of biochar at higher rates favored more Gram-negative bacteria compared to Gram-positive bacteria and fungi. These effects could be very important for soil organic matter decomposition, CH_4 emissions, sulfur, and nitrogen cycling. In an incubation study where two arable soils were amended with ^{13}C-depleted biochar from both wheat husk and willow plants, the phospholipid fatty acids (PLFAs) were monitored for 100 days during which any increase in PLFAs was observed and attributed to increase in actinomycetes and Gram-negative bacteria (Watzinger et al., 2014). A short-term incubation experiment where two ^{13}C-labeled biochars derived from wheat or eucalypt shoots were integrated into an acidic arenosol was conducted. Results showed that when compound-specific isotopic analysis (CSIA) of PLFAs were carried out, the amount of the biologically available fraction of both biochars was utilized rapidly within three days by the Gram-positive bacteria (Farrell et al., 2013). The PLFA evaluation clearly showed that biochar derived from a hardwood tree increased fungal biomass at the beginning and gradually declined over time while the ratio of Gram-positive to Gram-negative bacteria increased throughout the treatments (Jindo et al., 2012). A research study was also conducted to assess the relationship between chemical properties and microbial communities using four different biochars (swine manure, fruit peels, *Phragmites australis*, and *Brassica rapa*) applied at different rates (Muhammad et al., 2014). Soil PLFA results from amended soils implied that the biochar types and application rates induced necessary changes to the soil microbial community. While biochar increased actinomycetes, bacteria, fungi, Gram-positive, Gram-negative bacteria, and sulfate reduced in all treatments (at varying rates), protozoa PLFA was only elevated in *Brassica rap and Phragmites australis*. Unamended soils also contained higher concentrations of certain iso:anteiso PLFAs, which are indicators of environmental stress rather than the biochar amended soils. In another laboratory-based research work conducted over a 24-week incubation period, biochar produced from the pyrolytic action on sugar maple wood at a temperature of 500 °C was used to examine the short-term soil microbial changes to biochar amended soil samples. The results indicated that the total soil CO_2 respired was increased in biochar-amended samples compared to controls, whereas PLFA concentration which are specific to both Gram-negative, Gram-positive bacteria, and actinomycetes were initially lower than those recorded in the controls during the first four months. However, its gradual increase was reported over time during the course of the study. An increase in bacteria/fungi ratio coupled with a corresponding lower ratio of Gram-negative/Gram-positive bacteria indicated a major shift in the microbial community in favor of Gram-positive bacteria due to the biochar amendment as observed by Mitchell et al. (2015). Principal component analysis (PCA) of the extracted PLFAs indicated varying microbial community structures while determining the effect of biochar additions on temperate sandy-loamy soil in an incubation experiment conducted by Ameloot et al. (2013). In this study, the Gram-positive and Gram-negative bacteria were more profuse in the biochar treatments compared to the control. Steinbeiss et al. (2009) experimented on two different types of biochar derived from glucose and yeast added to the soil in a greenhouse experiment, and PLFA analysis was conducted on the amended soil at the beginning and after four months of incubation. Results indicated that biochar from yeast promoted and contributed to the growth and abundance of fungi in the soil, while glucose promoted the Gram-negative bacteria. A one-year research study was conducted to examine the effects of different addition rates of 0%, 1%, 5%, 10%, and 20% by mass of fast-pyrolysis biochar derived from wood on the temporal dynamics of PLFA in four different and separate temperate soils (Gomez et al., 2014). Results from this research study indicated that the addition of biochar considerably increased the abundance of microbes in all studied soils and thereby shifted the community composition towards Gram-negative bacteria while reducing the Gram-positive bacteria and fungi. Soil microbial biomass, bacterial diversity, water holding capacity, pH, soil electrical conductivity, respiration rate, and net nitrogen mineralization were all elevated when the soils were incorporated with corn-straw biochar treatments (Xu et al., 2016). Addition of biochar pyrolyzed at 350 °C from *Miscanthus giganteus* to soils at pH 4 and 8, respectively, was found to exhibit about 20% change in biomass when the same biochar pyrolyzed at 700 °C exhibited less than 2% microbial biomass as reported in the findings of Luo et al. (2013). Rutigliano et al. (2014) conducted an experimental study in Italy to determine the impact of biochar on the total microbial soil biomass, their activity, and diversity using soil samples from a wheat crop field amended with wood-derived biochar. Their findings reported that the soil sampled during the initial three months of field application showed a steady elevated microbial activity after which the functional diversity changed. However, bacterial genetic diversity was not affected. No biochar effect was found after 14 months of field biochar application, which could indicate that the positive effect of biochar addition was short-lived even though it enhanced soil microbial activity without any apparent disturbance.

6.3.3   SOIL RESPIRATION

The addition of biochar has been reported to cause temporary alterations in soil respiration rates (Jones et al., 2012). A research conducted over two years with biochar amended with mango (*Mangifera indica* L.) on sandy clay loam soil illustrated that 41% and 18% more carbon was respired in the first and second years, respectively, with the addition of biochar when compared to the control experiment devoid of biochar (Major et al., 2010). In another study conducted in the subtropical forest of China to determine the varying effects of biochar on respiration indicated that amendments using biochar led to an upsurge in respiration by 20.92%. In comparison, a 20.25% increase was noted in the temperate forests. However, a negligible effect was recorded in the sub-tropical rain forests, which suggest that biochar responses to respiration are dependent on numerous factors like soil textures and climatic or environmental conditions (Zhou et al., 2017). On the contrary, biochar addition was reported not to increase soil respiration during the first year of the study due to decrease in the sensitivity of soil respiration to temperature (Q_{10}). Another study also suggested that method of biochar application (surface application or incorporation at a specific depth) impacted soil respiration (He et al., 2016). Chen et al. (2018) reported in their own research that biochar alleviated soil respirations under controlled and regulated laboratory conditions in topsoil samples collected from plots in dry crop land in the Northern plain of China, three years after amendment with the use of single application of biochar. Another laboratory-based study was conducted in China in which soil respiration was examined on Latosol amended with biochar produced from sugarcane pyrolysis at three different temperatures of 300 °C, 500 °C, and 700 °C, respectively. The soil respiration was found to increase with increase in temperature (Deng et al.,

2017). In another short-term incubation experiment designed to determine the impacts of pyrolysis and hydrothermal carbonization on soil respirations, results indicated that hydrothermal carbonization char significantly increased soil respiration likewise supporting the dominance of fungi, while pyrolysis char was documented to not have a significant effect on both respiration and fungi dominance (Lanza et al., 2016). An incubation study was conducted on cancerous soils which had been amended with biochar derived from both cow manure and wheat straw produced under varying pyrolytic conditions to ascertain their effects on two biological soil indicators which include dehydrogenase enzyme activity and soil respiration (Beheshti et al., 2018). They reported that both biological conditions decreased with increase in temperature of pyrolysis (300 °C and 500 °C) irrespective of the source of feedstock. This study further concluded and suggested that several factors which include biochar conditions, short- and long-term effect of biochars and forms of biochars should be considered while amending calcareous soils with biochar. Another study conducted to investigate the effects of biochar on soil respiration in tea plantation in China showed that biochar had the tendency not only to increase the soil respiration rate but also decrease soil sensitivity to respiration during temperature change (Q_{10}).

6.4   EFFECTS OF BIOCHAR ON GHG EMISSIONS

6.4.1   EFFECTS ON N_2O EMISSION

N_2O is an important GHG produced mainly from two soil processes namely nitrification and denitrification. Nitrification process occurs when *Nitrosomonas* converts ammonium into nitrite and then subsequently converted to nitrate during the nitrogen cycling process. Nitrate is then further broken down by microorganisms under oxygen limited conditions to produce nitric oxide (NO), N_2O, and dinitrogen gas (N_2) (Austin, 1967). Reduction in N_2O production is achievable in two ways, either by reducing total nitrogen denitrified with which some amount of N_2O will be emitted from soil through the intermediate reaction or by enhancing its complete reduction to N_2 (Baggs, 2011). For the latter, if more inert N_2 is produced from total N instead of N_2O, then the environmental consequences of N_2O emissions greatly decreases. Biochar application can alter all the important factors which are known to determine both the ratio of $N_2O/(N_2 + N_2O)$ produced as well as the total nitrogen denitrified; by modifying the activities of soil microorganisms (Lehmann et al., 2011), the amount of available

NO_3^-, organic carbon (Prendergast-Miller et al., 2011), soil aeration (Kinney et al., 2012), and soil pH (Enders et al., 2012). In a study conducted to evaluate N_2O emission in soils with low and high organic matter content tested under the availability and unavailability of soil feeding worms, the addition of both Peanut hull and Miscanthus biochar to soils revealed that the presence of biochar significantly mitigated N_2O emission from soil without earthworms compared to control soil with no biochar addition. The study further revealed that the earthworms increased the emission of N_2O about 13 times as compared to treatment without earthworms, even though both types of biochar significantly reduced N_2O emission with the availability of earthworms (Augustenborg et al., 2012). In a long-term incubation study, biochars synthesized from both oak and hickory at a temperature range of 450–500 °C were added to fine loam soil and were documented to significantly reduce N_2O emissions (Rogovska et al., 2011). Another soil incubation experiment where hardwood biochar was incorporated with sandy-loam soil documented that hardwood-based biochar was able to suppress the emission of N_2O to about 98% when compared to the control (Case et al., 2012). Biochar application could be an effective management plan and strategy for improving the soil carbon storage on the long-term basis without negatively increasing N_2O emission rates. In another experimental study by Cheng et al. (2012), the incorporation or addition of wheat straw stimulated NO_3^- and NH_4^+ immobilization rates by 95.2% and 30.2%, respectively, but further suppressed the rate of nitrification by 32.2% while increasing the production of N_2O by about 40% as compared to wheat straw derived biochar amended soil. In an experiment comparing N_2O emissions from field experiment and laboratory incubations of green waste biochar amended soils, N_2O emissions from laboratory incubation was almost twice the amount from a field experiment (Felber et al., 2014). In a laboratory incubation study to measure N_2O emission over 100 days using Chernozemic soil amended with cereal straws and biochar rates of 0.67% and 1.68%, the biochar application largely reduced N_2O emissions, and it was concluded that biochar made from cereal straw significantly reduced N_2O emissions (Wu et al., 2013). Biochar amendment in an aerobic laboratory incubation experiment significantly decreased N_2O while increasing the yield of both the rice and wheat by as much as 12% and 17%, respectively. These increases in yields were fairly ascribed to the nitrogen fertilization on biochar amended soils. It was concluded that despite the fact that biochar amplified the potential of global warming potential with the application of nitrogen based fertilizers, biochar integration largely decreased N_2O emissions while positively improving and promoting crop production as reported by Wang et al. (2012). Another finding also affirmed that the presence of biochars significantly led to a drastic reduction in GHG emissions in nitrogen fertilized silt loam soils by decreasing N_2O flux up to 60% (Zheng et al., 2012).

Even though most of the reported and documented studies reported N_2O alleviation with soils amended with biochar, some studies reported contradictory results. An initial increase in N_2O emission after amending soil with biochar was observed by Singh et al. (2010) which might be due to higher labile nitrogen content of biochar and increased microbial activity. However, this spike eventually reduced over time. In a laboratory-based study, the addition of 10% municipal waste biochar to a clay-loam soil reduced N_2O emission by 89% at 78% water-filled pore space while N_2O emission was significantly increased up to 51% when same soil was re-wetted to 83% water-filled pore space (Yanai et al., 2007). This evidence supported the fact that soil aeration increases N_2O emission by inhibiting the activity of N_2O consumers (Bhowmik et al., 2017c). In another study, it was suggested that the age of the biochar had the potential to influence N_2O emissions with time (Spokas, 2013).

6.4.2   EFFECT ON CO_2 EMISSION

Novak et al. (2009) affirmed that the central concept of adding biochar to soils is to combat and regulate the concentrations of CO_2 in the atmosphere caused by anthropogenic activities. Biochars produced at high temperatures are preferable if the aim of such an amendment is to store carbon in soil while researchers have clearly documented that biochars of low-temperature origin should be opted for immediate soil fertility requirements for plant growth and nutrition. The temperature employed during the biochar pyrolysis greatly influence the CO_2 emission rate. An inverse relation has been established between CO_2 emission and the temperature used in the pyrolytic process (Brewer et al., 2012). Biochar produced at low temperatures of ≤600 °C extensively increased CO_2 emissions while biochar produced at higher temperature >600 °C significantly reduced CO_2 emission (Song et al., 2016). Studies on CO_2 emissions conducted by Smith et al. (2010) and Zimmerman et al. (2011) indicated that the preliminary increase in the CO_2 level witnessed could be attributed to both biotic and abiotic factors caused by the mineralization of the labile carbon which is embedded within the biochar and also increased microbial activity leading to an initial upsurge of soil organic matter decomposition. Several studies involving CO_2 emissions in soils amended by biochar have documented an increase in CO_2 whereas, some reported a negligible effect. However, it is worth noting that these effects usually increase with increasing the rate of applied biochar.

Results from an experimental study designed to determine the effects of the application of biochar on of GHG fluxes indicated that the application of biochar significantly elevated CO_2 in soil by about 19% (He et al., 2017). Other findings indicates that the duration of the research study, biochar pH, soil pH, soil texture, biochar feedstock, and application rates are among many of the factors that affect the emission of CO_2 from biochar amended soil. In a study intended to compare the effect of biochar amendment in laboratory incubations, upland fields, and paddy fields, it was reported that application of biochar significantly led to the increase in the CO_2 emissions by 28% increase in the laboratory experiment and 5% in the field experiment under similar conditions. In the same study, biochar derived from husk notably decreased the emission of CO_2, while, on the contrary biochar from wood, straw, and poultry manure considerably increased CO_2 emission (Wang et al., 2016). This research study also indicated that there is a considerable effect of sampling and experimental condition on the emission of CO_2, as CO_2 emissions from the soils amended with biochar were more in the laboratory incubation setting than those witnessed in the field under similar conditions (Wang et al., 2016). Other findings also documented reductions in emissions of CO_2 from silt loam soils when amended with wood chip biochar compared to the non-amended controls (Spokas et al., 2009).

Results from a 46-day incubation study conducted with soils collected from China amended with biochar from *Radix isatidis*, a common medicinal herb, pyrolyzed at different temperatures of 300, 500, and 700 °C revealed that biochar pyrolyzed at 700 °C was able to suppress soil CO_2 emission while biochar produced at lower temperatures of 300 °C and 500 °C resulted into a noticeable increase in CO_2 emitted from the soil (Yuan et al., 2014). A study investigating the effect of biochar on CO_2 release, organic carbon accumulation, and aggregation of soil using corn straw biochar pyrolyzed at 600 °C concluded that after seven months of incubation, biochar was able to enhance the formation of soil aggregates while significantly suppressing CO_2 emission (Hua et al., 2014). In a study conducted in China, soils amended with maize stover and a maize stover-derived biochar for three consecutive maize growing seasons in the field increased labile organic carbon in soil but the biochar was able to significantly suppress CO_2 emissions as compared to stover residues added to soil (Yang et al., 2017). In a short laboratory study investigating the effect of raw rice straw and its derived biochar on mitigating GHG, rice straw biochar effectively suppressed CO_2 emission on all the five soils selected for this experiment while direct rice straw amended had increased CO_2 emissions by 4 to 34 times (Li et al., 2013). The effects of biochar amendment on CO_2 emissions from paddy fields under water-saving irrigation and one-time application of biochar under flooding irrigation were studied by Yang et al. (2018). While biochar successfully increased rice yields, water use efficiency and suppressed CO_2 emissions by 2.22% from paddy fields under water-saving irrigation, CO_2 emissions under flooding irrigation was significantly higher. In a three-week laboratory study, filter cake, a by-product of sugarcane processing, and its biochar pyrolyzed at 575 °C were incorporated into a highly weathered tropical soil. The findings of the experimental study showed that filter cake derived biochar significantly increased the capacity of the soil to retain water, nutrient availability, and suppressed CO_2 emissions compared to its raw residue (Eykelbosh et al., 2014). While investigating CO_2 emission following the application of rice husk derived biochar to cultivated grassland soils. When biochar produced from *Conocarpus erectus* L. and its raw woody waste were applied to soil at 0, 30, and 50 g kg^{-1}, respectively, applications of both *Conocarpus erectus* L. and its biochar decreased CO_2 emission compared to the control, but the least CO_2 emission was recorded in the soil amended only with biochar (El-Mahrouky et al., 2015). Biochar produced from saccharification residue of rice straw enhanced the growth of CO_2-fixing bacteria and suppressed CO_2 emissions up to about 37.13% in comparison to biochar produced directly from rice straw (Hu et al., 2019). In this study, total SOC was also increased up to 1.8-fold when a mixture of biochar produced from saccharification residue of rice straw and autotrophic bacteria was applied to the soil. In China, four treatments of biochar was added to a bamboo plantation in Luvisol soil at the rate of 0, 5, 10, and 20 t ha^{-1}. The biochar was produced from Moso bamboo (*Phyllostachys* eduis) branches pyrolyzed at a temperature of 500 °C and the results indicated that only biochar added at 5 t ha^{-1} significantly reduced the total CO_2 emissions. Hence, the study suggested that biochar applied at a lower dosage have a better and promising effect in suppressing overall CO_2 emissions and storing more SOC in bamboo forests in subtropical regions of China (Ge et al., 2019).

6.5 CONCLUSION

This chapter builds on the possibilities, impacts, and effects of different types of biochar amendments on improving biological soil health properties and GHG emissions. Reports from the literature suggest that the potential to reduce GHG emissions in most cases depends on the interaction of biochars derived from different biomass material at varying pyrolysis temperatures with soil properties. Biochar derived from high carbonaceous biomass when incorporated to soil significantly reduced soil N_2O emissions. Although some experimental findings documented various contrasting reports, most of it could be attributed to variable factors that affect the potency of biochars, for example, source of the biochar, the rate of application, temperature of pyrolysis, etc. Biochars are valuable and environmentally friendly way of mitigating GHG emissions; however, cost could be an issue. It is worthy to note that several agricultural experiments have reported innumerable beneficial effects of biochars ranging from carbon sequestration potential, enhancement of soil structure, and various soil physico-chemical properties. Biochar also influenced the type of dominant soil microbial groups, soil respiration rates, and enzymatic activity after application to soil. Biochar produced under high pyrolysis temperature did not influence microbial populations and their activity to a high extent but were preferable for long-term carbon storage in soil. Overall, in the quest to alleviate GHG emissions from different soil management practices, it is best to design biochars with unique characteristics that best fit the specific environmental or agronomic conditions where it will be applied.

ACKNOWLEDGMENT

This work was supported by the United States Department of Agriculture—National Institute of Food and Agriculture (Evans Allen NC.X332-5-21-130-1 Accession no. 1023321).

KEYWORDS

- **biochar**
- **GHG emission**
- **soil amendment**
- **PLFA soil health indicator**
- **soil enzyme analysis**
- **biological soil health**

REFERENCES

Abdel-Fattah, T. M.; Mahmoud, M. E.; Ahmed, S. B.; Huff, M. D.; Lee, J. W.; and Kumar, S. Biochar from woody biomass for removing metal contaminants and carbon sequestration. *J. Ind. Eng. Chem.* 2015, *22*, 103–109.

Abiven, S.; Hund, A.; Martinsen, V.; and Cornelissen, G. Biochar amendment increases maize root surface areas and branching: a shovelomics study in Zambia. *Plant Soil.* 2015, *395*(1), 45–55. doi:10.1007/s11104-015-2533-2

Akça, M. O.; and Namlı, A. Effects of poultry litter biochar on soil enzyme activities and tomato, pepper and lettuce plants growth. *Eurasian. J. Soil. Sci.* 2015, *4*(3), 161–168.

Ameloot, N.; De Neve, S.; Jegajeevagan, K.; Yildiz, G.; Buchan, D.; and Funkuin, Y. N. Short-term CO_2 and N_2O emissions and microbial properties of biochar amended sandy loam soils. *Soil Biol. Biochem.* 2013, *57*, 401–410.

Ameur, D.; Zehetner, F.; Johnen, S.; Jöchlinger, L.; Pardeller, G.; Wimmer, B.; Rosenr, F; Dersch, G; Zechmeister-Boltenstern, S.; Mentler, A.; Soja, G.; Keiblinger, K. M. Activated biochar alters activities of carbon and nitrogen acquiring soil enzymes. *Pedobiologia.* 2018, *69*, 1–10. doi:https://doi.org/10.1016/j.pedobi.2018.06.001

Atkinson, C. J.; Fitzgerald, J. D.; and Hipps, N. A. Potential mechanisms for achieving agricultural benefits from biochar application to temperate soils: a review. *Plant Soil.* 2010, *337*(1–2), 1–18.

Augustenborg, C. A.; Hepp, S.; Kammann, C.; Hagan, D.; Schmidt, O.; and Müller, C. Biochar and earthworm effects on soil nitrous oxide and carbon dioxide emissions. *J. Environ. Qual.* 2012, *41*(4), 1203–1209.

Austin, A.T. The chemistry of the higher oxides of nitrogen as related to the manufacture, storage and administration of nitrous oxide. *Br. J. Anaesth.* 1967, *39*(5), 345–350.

Baggs, E. M. Soil microbial sources of nitrous oxide: recent advances in knowledge, emerging challenges and future direction. *Curr. Opin. Environ. Sustainability.* 2011, *3* (5), 321–327.

Bamminger, C.; Marschner, B.; and Jüschke, E. An incubation study on the stability and biological effects of pyrogenic and hydrothermal biochar in two soils. *Eur. J. Soil Sci.* 2014a, *65*(1), 72–82.

Bamminger, C.; Zaiser, N.; Zinsser, P.; Lamers, M.; Kammann, C.; and Marhan, S. Effects of biochar, earthworms, and litter addition on soil microbial activity and abundance in a temperate agricultural soil. *Biol. Fertil. Soils.* 2014b, *50*(8), 1189–1200.

Banerjee, S.; Helgason, B.; Wang, L.; Winsley, T.; Ferrari, B. C.; and Siciliano, S. D. Legacy effects of soil moisture on microbial community structure and N_2O emissions. *Soil Biol. Biochem.* 2016, *95*, 40–50. doi:10.1016/j.soilbio.2015.12.004

Basso, A. S.; Miguez, F. E.; Laird, D. A.; Horton, R.; and Westgate, M. Assessing potential of biochar for increasing water-holding capacity of sandy soils. *Global Change Biol.* 2013, *5*(2), 132–143.

Beheshti, M.; Etesami, H.; and Alikhani, H. A. Effect of different biochars amendment on soil biological indicators in a calcareous soil. *Environ. Sci. Pollut. Res.* 2018, *25*(15), 14752–14761. doi:10.1007/s11356-018-1682-2

Bhowmik, A.; Fortuna A. M.; Cihacek, L. J.; Rahman, S.; Borhan, M. S.;and Carr, P. M. Use of laboratory incubation techniques to estimate greenhouse gas footprints from conventional and no-tillage organic agroecosystems. *Soil Biol. Biochem.* 2017a, *112*, 204–215.

Bhowmik, A.; Fortuna, A. M.; Cihacek, L. J.; Bary, A. I.; Carr, P. M.; and Cogger, C. G. Potential carbon sequestration and nitrogen cycling in long-term organic management systems. *Renewable Agric. Food Syst.* 2017b, *32*(6), 498–510.

Bhowmik, A.; Cloutier, M.; Ball, E.; and Bruns, M.A. Underexplored microbial metabolisms for enhanced nutrient cycling in agricultural soils. *AIMS Microbiol.* 2017c, *3*, 826–845.

Bhowmik, A; Fortuna, A. M.; Cihacek, L. J.; Bary, A. I.; and Cogger, C. G. Use of biological indicators of soil health to estimate reactive nitrogen dynamics in long-term organic vegetable and pasture systems. *Soil Biol. Biochem.* 2016, *103*, 308–319.

Brewer, C. E.; Hu, Y.-Y.; Schmidt-Rohr, K.; Loynachan, T. E.; Laird, D. A.; and Brown, R. C. Extent of pyrolysis impacts on fast pyrolysis biochar properties. *J. Environ. Qual.* 2012, *41*(4), 1115–1122.

Cao, X.; Ma, L.; Gao, B.; and Harris, W. Dairy-manure derived biochar effectively sorbs lead and atrazine. *Environ. Sci. Technol.* 2009, *43*(9), 3285–3291.

Case, S. D.; McNamara, N. P.; Reay, D. S.; Whitaker, J. The effect of biochar addition on N_2O and CO_2 emissions from a sandy loam soil—the role of soil aeration. *Soil Biol. Biochem.* 2012, *51*, 125–134.

Casper, J. K. *GHGes: worldwide impacts*; Infobase Publishing: New York, 2009.

Cha, J. S.; Park, S. H.; Jung, S.-C.; Ryu, C.; Jeon, J.-K.; Shin, M.-C.; and Park, Y.-K. Production and utilization of biochar: a review. *J. Ind. Eng. Chem.* 2016, *40*, 1–15. doi:10.1016/j.jiec.2016.06.002

Chen, H.; Du, Z.; Guo, W.; and Zhang, Q. Effects of biochar amendment on cropland soil bulk density, cation exchange capacity, and particulate organic matter content in the North China Plain. *J. Appl. Ecol.* 2011, *22*(11), 2930–2934.

Chen, J.; Sun, X.; Zheng, J.; Zhang, X.; Liu, X.; Bian, R.; … Pan, G. Biochar amendment changes temperature sensitivity of soil respiration and composition of microbial communities 3 years after incorporation in an organic carbon-poor dry cropland soil. *Biol. Fertil. Soils.* 2018, *54*(2), 175–188. doi:10.1007/s00374-017-1253-6

Cheng, Y,; Cai, Z,; Chang, S. X.; Wang, J.; and Zhang, J. Wheat straw and its biochar have contrasting effects on inorganic N retention and N_2O production in a cultivated Black Chernozem. *Biol. Fertil. Soils.* 2012, *48*, 941–946.

Crane-Droesch, A.; Abiven, S.; Jeffery, S.; and Torn, M. S. Heterogeneous global crop yield response to biochar: a meta-regression analysis. *Environ. Res. Lett.* 2013, *8*(4), 044049. doi:10.1088/1748–9326/8/4/044049

Demisie, W.; Liu, Z.; and Zhang, M. Effect of biochar on carbon fractions and enzyme activity of red soil. *CATENA.* 2014, *121*, 214–221. doi:10.1016/j.catena.2014.05.020

Deng, W.; Van Zwieten, L.; Lin, Z.; Liu, X.; Sarmah, A. K.; and Wang, H. Sugarcane bagasse biochars impact respiration and GHG emissions from a latosol. *J. Soils Sediments.* 2017, *17*(3), 632–640.

Ducey, T. F.; Ippolito, J. A.; Cantrell, K. B.; Novak, J. M.; and Lentz, R. D. Addition of activated switchgrass biochar to an aridic subsoil increases microbial nitrogen cycling gene abundances. *Appl. Soil Ecol.* 2013, *65*, 65–72. doi:10.1016/j.apsoil.2013.01.006

Eggleston, S.; Buendia, L.; Miwa, K.; Ngara, T.; and Tanabe, K. *2006 IPCC guidelines for national GHG inventories*, 2006: Vol. 5: Institute for Global Environmental Strategies Hayama, Japan.

El-Mahrouky, M.; El-Naggar, A. H.; Usman, A. R.; and Al-Wabel, M. Dynamics of CO_2 Emission and Biochemical Properties of a Sandy Calcareous Soil Amended with Conocarpus Waste and Biochar. *Pedosphere.* 2015, *25*(1), 46–56. doi:10.1016/S1002-0160(14)60075-8

Enders, A.; Hanley, K.; Whitman, T.; Joseph, S.; and Lehmann, J . Characterization of biochars to evaluate recalcitrance and agronomic performance. *Bioresour. Technol.* 2012, *114*, 644–653.

Eykelbosh, A. J.; Johnson, M. S.; Santos de Queiroz, E.; Dalmagro, H. J.; and Guimarães Couto, E. Biochar from sugarcane filtercake reduces soil CO_2 emissions relative to raw residue and improves water retention and nutrient availability in a highly-weathered tropical soil. *PloS One.* 2014, *9*(6), e98523–e98523. doi:10.1371/journal.pone.0098523

Farrell, M.; Kuhn, T. K.; Macdonald, L. M.; Maddern, T. M.; Murphy, D. V.; Hall, P. A.; ... Baldock, J. A. Microbial utilisation of biochar-derived carbon. *Sci. Total Environ.* 2013, *465*, 288–297.

Felber, R.; Leifeld, J.; Horák, J.; and Neftel, A. Nitrous oxide emission reduction with greenwaste biochar: comparison of laboratory and field experiments. *Eur. J. Soil Sci.* 2014, *65*(1), 128–138.

Gascó, G.; Paz-Ferreiro, J.; Cely, P.; Plaza, C.; and Méndez, A. Influence of pig manure and its biochar on soil CO_2 emissions and soil enzymes. *Ecol. Eng.* 2016, *95*, 19–24. doi:10.1016/j.ecoleng.2016.06.039

Ge, X.; Cao, Y.; Zhou, B.; Wang, X.; Yang, Z.; and Li, M.-H. Biochar addition increases subsurface soil microbial biomass but has limited effects on soil CO_2 emissions in subtropical moso bamboo plantations. *Appl. Soil Ecol.* 2019, *142*, 155–165. doi:10.1016/j. apsoil.2019.04.021

Gell, K.; van Groenigen, J.; and Cayuela, M. L. Residues of bioenergy production chains as soil amendments: immediate and temporal phytotoxicity. *J. Hazard. Mater.* 2011, *186*(2–3), 2017–2025.

Githinji, L. Effect of biochar application rate on soil physical and hydraulic properties of a sandy loam. *Arch. Agron. Soil Sci.* 2014, *60*(4), 457–470. doi:10.1080/03650340.2013. 821698

Gomez, J.; Denef, K.; Stewart, C.; Zheng, J.; and Cotrufo, M.. Biochar addition rate influences soil microbial abundance and activity in temperate soils. *Eur. J. Soil. Sci.* 2014, *65*(1), 28–39.

Gul, S.; Whalen, J. K.; Thomas, B. W.; Sachdeva, V.; and Deng, H. Physico-chemical properties and microbial responses in biochar-amended soils: Mechanisms and future directions. *Agric., Ecosyst. and Environ.* 2015, *206*, 46–59. doi:10.1016/j.agee.2015.03.015

Gurwick, N. P.; Moore, L. A.; Kelly, C.; and Elias, P. A systematic review of biochar research, with a focus on its stability in situ and its promise as a climate mitigation strategy. *PloS One.* 2013, *8*(9), e75932–e75932. doi:10.1371/journal.pone.0075932

Hansen, V.; Müller-Stöver, D.; Munkholm, L.J.; Peltre, C.;Hauggaard-Nielsen, H. and Jensen, L.S. The effect of straw and wood gasification biochar on carbon sequestration, selected soil fertility indicators and functional groups in soil: an incubation study. *Geoderma.* 2016, *269*, 99–107.

He, X.; Du, Z.; Wang, Y.; Lu, N.; and Zhang, Q. Sensitivity of soil respiration to soil temperature decreased under deep biochar amended soils in temperate croplands. *Appl. Soil Ecol.* 2016, *108*, 204–210. doi:10.1016/j.apsoil.2016.08.018

He, Y.; Zhou, X.; Jiang, L.; Li, M.; Du, Z.; Zhou, G.; . . . Hosseini Bai, S. Effects of biochar application on soil GHG fluxes: a meta-analysis. *Global Change Biol.* 2017, *9*(4), 743–755.

Houben, D.; Sonnet, P.; and Cornelis, J. Biochar from Miscanthus: a potential silicon fertilizer. *Plant Soil.* 2014, *374*(1–2), 871–882.

Houghton, R.; Hobbie, J.; Melillo, J. M.; Moore, B.; Peterson, B.; Shaver, G.; and Woodwell, G. Changes in the carbon content of terrestrial biota and soils between 1860 and 1980: a net release of CO_2 to the atmosphere. *J. Ecol. Mono.* 1983, *53*(3), 235–262.

Hu, J.; Guo, H.; Wang, X.; Gao, M.-t.; Yao, G.; Tsang, Y. F.; Li, J.; Yan, L.; Zhang, S. Utilization of the saccharification residue of rice straw in the preparation of biochar is a novel strategy for reducing CO_2 emissions. *Sci. Total Environ.* 2019, *650*, 1141–1148. doi:10.1016/j.scitotenv.2018.09.099

Hua, L.; Lu, Z.; Ma, H.; and Jin, S. Effect of biochar on carbon dioxide release, organic carbon accumulation, and aggregation of soil. *Environ. Prog. Sustainable Energy.* 2014, *33*(3), 941–946. doi:10.1002/ep.11867

IPCC. Global emissions management. United States Environment Protection Agency. 2014 https://www3.epa.gov/climatechange/ghgemissions/global.html. (accessed May 19, 2020)

Ippolito, J.; Stromberger, M.; Lentz, R.; and Dungan, R. Hardwood biochar influences calcareous soil physicochemical and microbiological status. *J. Environ. Qual.* 2014, *43*(2), 681–689.

Jindo, K.; Sánchez-Monedero, M. A.; Hernández, T.; García, C.; Furukawa, T.; Matsumoto, K.; … Bastida, F. Biochar influences the microbial community structure during manure composting with agricultural wastes. *Sci. Total Environ.* 2012, *416*, 476–481.

Jones, D.; Rousk, J.; Edwards-Jones, G.; DeLuca, T.; and Murphy, D. Biochar-mediated changes in soil quality and plant growth in a three year field trial. *Soil. Biol. Biochem.* 2012, *45*, 113–124.

Joseph, S.; Camps-Arbestain, M.; Lin, Y.; Munroe, P.; Chia, C.; Hook, J.; van Zwieten, L.; Kimber, S.; Cowie, A.; Singh, B. P.; Lehmann, J.; Foidl, N.; Smernik, R. J.; and Amonette, J. E. An investigation into the reactions of biochar in soil. *Aust. J. Soil Res.* 2010, *48*(7), 501–515.

Karhu, K.; Mattila, T.; Bergström, I.; Regina, K. Biochar addition to agricultural soil increased CH4 uptake and water holding capacity—results from a short-term pilot field study. *J. Agric. Ecosyst. Environ.* 2011, *140*(1–2), 309–313.

Khadem, A.; and Raiesi, F. Influence of biochar on potential enzyme activities in two calcareous soils of contrasting texture. *Geoderma,* 2017, *308*, 149–158. doi:10.1016/j.geoderma.2017.08.004

Khan, S.A.; Mulvaney R.L.; Ellsworth T.R.; and Boast C.W. The myth of nitrogen fertilization for soil carbon sequestration. *J. Environ. Qual.* 2007, *36*, 1821.

Kinney, T.; Masiello, C.; Dugan, B.; Hockaday, W.; Dean, M.; and Zygourakis, K. Hydrologic properties of biochars produced at different temperatures. *Biomass Bioenergy.* 2012, *41*, 34–43.

Kookana, R. S.; Sarmah, A. K.; Van Zwieten, L.; Krull, E.; and Singh, B. Biochar application to soil: agronomic and environmental benefits and unintended consequences. *Adv. Agron.* 2011, *112*, 103–143.

Laird, D. A.; Fleming, P.; Davis, D. D.; Horton, R.; Wang, B.; and Karlen, D. L. Impact of biochar amendments on the quality of a typical Midwestern agricultural soil. *Geoderma.* 2010, *158*(3–4), 443–449.

Lal, R. Soil carbon sequestration impacts on global climate change and food security. *Science.* 2004, *304*, 1623–1627.

Lal, R.; Follett R.F; Stewart, B.A.; and Kimble, J.M. Soil carbon sequestration to mitigate climate change and advance food security. *Soil Sci.* 2007, *172*, 943–956.

Lanza, G.; Rebensburg, P.; Kern, J.; Lentzsch, P.; and Wirth, S. Impact of chars and readily available carbon on soil microbial respiration and microbial community composition in a dynamic incubation experiment. *Soil Tillage Res.* 2016, *164*, 18–24. doi:10.1016/j. still.2016.01.005

Lehmann, J.; Rillig, M. C.; Thies, J.; Masiello, C. A.; Hockaday, W. C.; Crowley, *D.* Biochar effects on soil biota—a review. *Soil biol. Biochem.* 2011, *43*(9), 1812–1836.

Lehmann, *J.* Bio-energy in the black. *Front. Ecol. Environ.* 2007, *5*(7), 381–387.

Leng, L.; and Huang, H. An overview of the effect of pyrolysis process parameters on biochar stability. *Bioresour. Technol.* 2018, *270*, 627–642. doi:10.1016/j.biortech.2018.09.030

Li, F.; Cao, X.; Zhao, L.; Yang, F.; Wang, J.; and Wang, S. Short-term effects of raw rice straw and its derived biochar on GHG emission in five typical soils in China. J. *Soil Sci. Plant Nutr.* 2013, *59*(5), 800–811. doi:10.1080/00380768.2013.821391

Lindsey, R. *Climate Change: Atmospheric Carbon Dioxide |National Oceanographic and Atmospheric Administration, News and Features.* August 2018. https://www.climate. gov/news-features/understanding-climate/climate-change-atmospheric-carbon-dioxide (accessed April 29, 2020)

Liu, X.-H.; and Zhang, X.-C. Effect of biochar on pH of alkaline soils in the loess plateau: results from incubation experiments. *Int. J. Agric. Biol.* 2012, *14*(5), 745–750.

Lu, H.; Li, Z.; Fu, S.; Méndez, A.; Gascó, G.; and Paz-Ferreiro, J. Effect of biochar in cadmium availability and soil biological activity in an anthrosol following acid rain deposition and aging. *Water Air Soil Pollut.* 2015, *226*(5), 164.

Luo, Y.; Durenkamp, M.; De Nobili, M.; Lin, Q.; Devonshire, B. J.; and Brookes, P. C. Microbial biomass growth, following incorporation of biochars produced at 350 °C or 700 °C, in a silty-clay loam soil of high and low pH. *Soil Biol. Biochem.* 2013, *57*, 513–523. doi:10.1016/j.soilbio.2012.10.033

Ma, L.; Lv, N.; Ye, J.; Ru, S.; Li, G.; and Hou, Z. Effects of biochar on organic carbon content and fractions of gray desert soil. *J. Eco.-Agric.* 2012, *20*(8), 976–981.

Major, J.; Lehmann, J.; Rondon, M.; and Goodale, C. Fate of soil-applied black carbon: downward migration, leaching and soil respiration. *Global Change Biol.* 2010, *16*(4), 1366–1379.

Masto, R. E.; Kumar, S.; Rout, T. K.; Sarkar, P.; George, J.; Ram, L. C. Biochar from water hyacinth (*Eichornia crassipes*) and its impact on soil biological activity. *Catena.* 2013, *111*, 64–71.

Mitchell, P. J.; Simpson, A. J.; Soong, R.; and Simpson, M. Shifts in microbial community and water-extractable organic matter composition with biochar amendment in a temperate forest soil. *Soil Biol. Biochem.* 2015, *81*, 244–254.

Muhammad, N.; Dai, Z.; Xiao, K.; Meng, J.; Brookes, P. C.; Liu, X.; … Xu, J. Changes in microbial community structure due to biochars generated from different feedstocks and their relationships with soil chemical properties. *Geoderma.* 2014, *226*, 270–278.

Munda, S.; Bhaduri, D.; Mohanty, S.; Chatterjee, D.; Tripathi, R.; Shahid, M.; … Nayak, A. K. Dynamics of soil organic carbon mineralization and C fractions in paddy soil on application of rice husk biochar. *Biomass Bioenergy.* 2018, *115*, 1–9. doi:10.1016/j.biombioe.2018.04.002

Novak, J.; Watts, D. Augmenting soil water storage using uncharred switchgrass and pyrolyzed biochars. *Soil. Use. Manage.* 2013, *29*(1), 98–104.

Novak, J. M.; Busscher, W. J.; Laird, D. L.; Ahmedna, M.; Watts, D. W.; and Niandou, M. Impact of biochar amendment on fertility of a southeastern coastal plain soil. *J. Soil. Sci.* 2009, *174*(2), 105–112.

Novak, J. M.; Busscher, W. J.; Watts, D. W.; Amonette, J. E.; Ippolito, J. A.; Lima, I. M.; ... Ahmedna, M. Biochars impact on soil-moisture storage in an ultisol and two aridisols. *J. Soil. Sci.* 2012, *177*(5), 310–320.

Pan, F.; Li, Y.; Chapman, S. J.; Khan, S.; and Yao, H. Microbial utilization of rice straw and its derived biochar in a paddy soil. *Sci. Total Environ.* 2016, *559*, 15–23.

Paz-Ferreiro, J.; Fu, S.; Méndez, A.; and Gascó, G. Interactive effects of biochar and the earthworm Pontoscolex corethrurus on plant productivity and soil enzyme activities. *J. Soils Sediments.* 2014, *14*(3), 483–494. doi:10.1007/s11368-013-0806-z

Prendergast-Miller, M. T.; Duvall, M.; Sohi, S. P Localisation of nitrate in the rhizosphere of biochar-amended soils. *Soil. Biol. Biochem.* 2011, *43*(11), 2243–2246.

Ravishankara A.R; Daniel J.S; Portmann R.W. Nitrous oxide (N2O): the dominant ozone-depleting substance emitted in the 21st century. *Science.* 2009, *326*(5949), 123–125.

Robinson, B. H.; Green, S.; Chancerel, B.; Mills, T.; and Clothier, B. Poplar for the phytomanagement of boron contaminated sites. *Environ. Pollut.* 2007, *150*(2), 225–233.

Rogovska, N.; Laird, D.; Cruse, R.; Fleming, P.; Parkin, T.; and Meek, D. Impact of biochar on manure carbon stabilization and GHG emissions. *Soil. Sci. Soc. Am. J.* 2011, *75*(3), 871–879.

Rutigliano, F. A.; Romano, M.; Marzaioli, R.; Baglivo, I.; Baronti, S.; Miglietta, F.; and Castaldi, S. Effect of biochar addition on soil microbial community in a wheat crop. *Euro. J. Soil Biol.* 2014, *60*, 9–15. doi:10.1016/j.ejsobi.2013.10.007

Shenbagavalli, S.; and Mahimairaja, S. Characterization and effect of biochar on nitrogen and carbon dynamics in soil. *Int. J. Agric.* 2012, *2*(2), 249–255.

Shin, C. S.; and Kim, K. U. CO_2 emissions by agricultural machines in South Korea. *Appl. Eng. Agric.* 2018, *34*(2), 311–315.

Singh, B. P.; Hatton, B. J.; Singh, B.; Cowie, A. L.; and Kathuria, A. Influence of biochars on nitrous oxide emission and nitrogen leaching from two contrasting soils. *J. Environ. Qual.* 2010, *39*(4), 1224–1235.

Smith, J. L.; Collins, H. P.; Bailey, V. L. The effect of young biochar on soil respiration. *Soil Biol. Biochem.* 2010, *42*(12), 2345–2347.

Smith, K. A. Changing views of nitrous oxide emissions from agricultural soil: key controlling processes and assessment at different spatial scales. *Eur. J. Soil Sci.* 2017, *68*(2), 137–155. doi:10.1111/ejss.12409

Sohi, S. P.; Krull, E.; Lopez-Capel, E.; and Bol, R. A review of biochar and its use and function in soil. *Adv. Agron.* 2010, *105*, 47–82.

Song, X.; Pan, G.; Zhang, C.; Zhang, L.; and Wang, H. Effects of biochar application on fluxes of three biogenic GHGes: a meta-analysis. *Ecosyst. Health Sustainability.* 2016, *2*(2), e01202. doi:10.1002/ehs2.1202

Spokas, K.; Koskinen, W.; Baker, J.; and Reicosky, D. Impacts of woodchip biochar additions on GHG production and sorption/degradation of two herbicides in a Minnesota soil. *Chemosphere.* 2009, *77*(4), 574–581.

Spokas, K. A.; Cantrell, K. B.; Novak, J. M.; Archer, D. W.; Ippolito, J. A.; Collins, H. P.; ... McAloon, A. Biochar: a synthesis of its agronomic impact beyond carbon sequestration. *J. Environ. Qual.* 2012, *41*(4), 973–989.

Spokas, K. A. Impact of biochar field aging on laboratory GHG production potentials. *Global Change Biol.* 2013, *5*(2), 165–176.

Steinbeiss, S.; Gleixner, G.; Antonietti, M. Effect of biochar amendment on soil carbon balance and soil microbial activity. *Soil Biol. Biochem.* 2009, *41*(6), 1301–1310.

Stewart, C. E.; Zheng, J.; Botte, J.; and Cotrufo, M. F. Co-generated fast pyrolysis biochar mitigates green-house gas emissions and increases carbon sequestration in temperate soils. *Global Change Biol.* 2013, *5*(2), 153–164.

Taghizadeh-Toosi, A.; Clough, T. J.; Condron, L. M.; Sherlock, R. R.; Anderson, C. R.; and Craigie, R. A. Biochar incorporation into pasture soil suppresses in situ nitrous oxide emissions from ruminant urine patches. *J. Environ. Qual.* 2011, *40*(2), 468–476.

Thies, J. E.; and Rillig, M. C. Characteristics of biochar: biological properties. In *Biochar for Eenvironmental Management.* 2012, (pp. 117–138): Routledge.

Trupiano, D.; Cocozza, C.; Baronti, S.; Amendola, C.; Vaccari, F. P.; Lustrato, G.; . . . Scippa, G. S. The effects of biochar and its combination with compost on lettuce (Lactuca sativa L.) growth, soil properties, and soil microbial activity and abundance. *Int. J. Agron.* 2017. doi:10.1155/2017/3158207

Ulyett, J.; Sakrabani, R.; Kibblewhite, M.; and Hann, M. Impact of biochar addition on water retention, nitrification and carbon dioxide evolution from two sandy loam soils. *Eur. J. Soil Sci.* 2014, *65*(1), 96–104.

USDA. Soil Health | Natural Resources Conservation Service; 2020. https://www.nrcs.usda.gov/wps/portal/nrcs/main/soils/health/. (Accessed April 29, 2021)

USDA. What is Pyrolysis? | Biomass Pyrolysis Research; 2017. https://www.ars.usda.gov/northeast-area/wyndmoor-pa/eastern-regional-research-center/docs/biomass-pyrolysis-research-1/what-is-pyrolysis/. (Accessed May 10, 2021)

US Environmental Protection Agency | Inventory of US GHG emissions and sinks: 2012, 1990–2010. In: US Environmental Protection Agency Washington, DC.

US Environmental Protection Agency | Global GHG Emissions Data. 2019. https://www.epa.gov/ghgemissions/global-greenhouse-gas-emissions-data. (Accessed May 5, 2021)

Wang, J.; Pan, X.; Liu, Y.; Zhang, X.; and Xiong, Z. Effects of biochar amendment in two soils on GHG emissions and crop production. *Plant Soil.* 2012, *360*(1), 287–298. doi:10.1007/s11104-012-1250-3

Wang, X.; Song, D.; Liang, G.; Zhang, Q.; Ai, C.; and Zhou, W. Maize biochar addition rate influences soil enzyme activity and microbial community composition in a fluvo-aquic soil. *Appl. Soil Ecol.* 2015, *96*, 265–272. doi:10.1016/j.apsoil.2015.08.018

Wang, J. Y.; Xiong, Z. Q.; and Kuzyakov, Y. Biochar stability in soil: meta-analysis of decomposition and priming effects. *Global Change Biol. Bioenergy.* 2016, *8*(3), 512–523.

Watson, R.T.; Meira Filho, L.G.; Sanhueza, E.; and Janetos, A. Greenhouse gases: sources and sinks. In *Climate Change 1992: The Supplementary Report to the IPCC Scientific Assessment*, edited by J.T. Houghton, B.A. Collander, and S.K. Varney. 1992, (pp. 25–46). Cambridge, MA, USA: University Press.

Watzinger, A.; Feichtmair, S.; Kitzler, B.; Zehetner, F.; Kloss, S.; Wimmer, B.; . . . Soja, G. Soil microbial communities responded to biochar application in temperate soils and slowly metabolized ^{13}C-labelled biochar as revealed by 13C PLFA analyses: results from a short-term incubation and pot experiment. *Eur. J. Soil Sci.* 2014, *65*(1), 40–51.

Wei, S.; Zhu, M.; Fan, X.; Song, J.; Peng, P. a.; Li, K.; . . . Song, H. Influence of pyrolysis temperature and feedstock on carbon fractions of biochar produced from pyrolysis of rice straw, pine wood, pig manure and sewage sludge. *Chemosphere.* 2019, *218*, 624–631. doi:10.1016/j.chemosphere.2018.11.177

Wu, F.; Jia, Z.; Wang, S.; Chang, S. X.; and Startsev, A. Contrasting effects of wheat straw and its biochar on GHG emissions and enzyme activities in a Chernozemic soil. *Biol. Fertil. Soils.* 2013, *49*(5), 555–565. doi:10.1007/s00374-012-0745-7

Xu, N.; Tan, G.; Wang, H.; and Gai, X. Effect of biochar additions to soil on nitrogen leaching, microbial biomass and bacterial community structure. *Eur. J. Soil Biol.* 2016, *74*, 1–8.

Yanai, Y.; Toyota, K.; Okazaki, M. Effects of charcoal addition on N_2O emissions from soil resulting from rewetting air-dried soil in short-term laboratory experiments. *J. Soil Sci. Plant Nutr.* 2007, *53*(2), 181–188.

Yang, S.; Jiang, Z.; Sun, X.; Ding, J.; and Xu, J. Effects of Biochar Amendment on CO_2 Emissions from Paddy Fields under Water-Saving Irrigation. *Int. J. Environ. Res. Public Health.* 2018, *15*(11), 2580.

Yang, X.; Liu, J.; McGrouther, K.; Huang, H.; Lu, K.; Guo, X. Effect of biochar on the extractability of heavy metals (Cd, Cu, Pb, and Zn) and enzyme activity in soil. *Environ. Sci. Poll. Res.* 2016, *23*(2), 974–984.

Yang, X.; Meng, J.; Lan, Y.; Chen, W.; Yang, T.; Yuan, J.; … Han, J. Effects of maize stover and its biochar on soil CO_2 emissions and labile organic carbon fractions in Northeast China. *Agric. Ecosyst. Environ.* 2017, *240*, 24–31. doi:10.1016/j.agee.2017.02.001

Yuan, H.; Lu, T.; Wang, Y.; Huang, H.; and Chen, Y. Influence of pyrolysis temperature and holding time on properties of biochar derived from medicinal herb (radix isatidis) residue and its effect on soil CO_2 emission. *J. Anal. Appl. Pyrolysis.* 2014, *110*, 277–284. doi:10.1016/j.jaap.2014.09.016

Zhang, M.; Bayou, W. D.; Tang, H. Effects of biochar's application on active organic carbon fractions in soil. *J. Soil. Water Conserv.* 2012, *26*(2), 127–131.

Zhao, R.; Coles, N.; Kong, Z.; Wu, J. J. S.; and Research, T. (2015). Effects of aged and fresh biochars on soil acidity under different incubation conditions. *Soil Tillage Res.* 2015, *146*, 133–138.

Zheng, J.; Stewart, C. E.; and Cotrufo, M. F.. Biochar and nitrogen fertilizer alters soil nitrogen dynamics and GHG fluxes from two temperate soils. *J. Environ. Qual.* 2012, *41*(5), 1361–1370.

Zhou, G.; Zhou, X.; Zhang, T.; Du, Z.; He, Y.; Wang, X.; … Xu, C. Biochar increased soil respiration in temperate forests but had no effects in subtropical forests. *For. Ecol. Manage.* 2017, *405*, 339–349. doi:10.1016/j.foreco.2017.09.038

Zielińska, A.; and Oleszczuk, P. Bioavailability and bioaccessibility of polycyclic aromatic hydrocarbons (PAHs) in historically contaminated soils after lab incubation with sewage sludge-derived biochars. *Chemosphere.* 2016, *163*, 480–489.

Zimmerman, A. R.; Gao, B.; and Ahn, M. Y. Positive and negative carbon mineralization priming effects among a variety of biochar-amended soils. *Soil Biol. Biochem.* 2011, *43*(6), 1169–1179.

PART III

Resource Conservation Technologies and Prospective Carbon Management Strategies

CHAPTER 7

Resource Conservation Technologies for Sustainable Soil Management

ABIR DEY[1*], PRAVIN K. UPADHYAY[2], MAHESH C. MEENA[1], and
BRAHMA S. DWIVEDI[1]

[1]Division of Soil Science and Agricultural Chemistry,
ICAR-Indian Agricultural Research Institute, New Delhi, Delhi, India

[2]Division of Agronomy, ICAR-Indian Agricultural Research Institute,
New Delhi, Delhi, India

*Corresponding author. E-mail: abirdey21@gmail.com

ABSTRACT

Ushering of Green Revolution in the mid-1960s has shifted Indian agriculture from "traditional animal-based subsistence" to "intensive chemical and machinery-based." This paradigm shift triggered deterioration of soil health and sustainability of natural resources, as reported from many parts of the country. Intensive tillage, removal/burning of crop residues (CRs), extensive mining of plant nutrients, and intensive mono cropping systems are often found to be the root cause of these "second generation problems," namely, subsoil compaction, multi-nutrient deficiencies, soil organic matter depletion, decline in nutrient-supplying capacity of soil, low-nutrient use efficiencies, diminishing soil biodiversity, increased energy cost, etc. Often, a lack of proper residue management technologies leads to residue burning for clearing the field. Added to this, excess fossil fuel combustion for tilling machineries results into further atmospheric pollution. In recent years, conservation agriculture (CA) has emerged as an alternative farming practice to address soil degradation resulting from agricultural practices that deplete the organic matter and nutrient content of the soil, aiming at sustained crop productivity with lower production costs. CA is based mainly on three principles, that is, minimal mechanical soil disturbance, permanent organic soil cover, and diversified crop rotations. A successful implementation of CA requires effective resource conservation technologies (RCTs), including selection of suitable crop rotations, balanced and integrated plant nutrient supply system, precision nutrient management using soil-test or sensor-based intelligent decision support tools, recycling of CRs, improved tillage practices, use of low grade mining bi-products, biofertilizers, organic manures, green manures, etc. It is envisaged that development of such strategies will not only help in sustaining higher crop productivity but also improving soil health and environmental quality. The chapter covers the application details of such RCTs, along with their benefits and bottlenecks for adoption in Indian conditions.

7.1   INTRODUCTION

7.1.1   SOIL SUSTAINABILITY

Soil is fundamental to sustenance and development of human civilization. Soil serves major functions in the ecosystem, namely, (1) producing crops, feed, fiber, fuel, (2) filtering and cleaning of water, (3) regulating greenhouse gas (GHG) emission by acting as a sink to atmospheric carbon (C), and (4) acting as a source of biodiversity etc. As described in World Soil Charter of FAO, "Soil management is sustainable if the supporting, provisioning, regulating, and cultural services provided by soil are maintained or enhanced without significantly impairing either the soil functions that enable those services or biodiversity. The balance between the supporting and provisioning services for plant production and the regulating services the soil provides for water quality and availability and for atmospheric greenhouse gas composition is a particular concern" (FAO, 2017). Sustainable soil management (SSM) has the potential to increase food production, enhance nutrient content of food, and mitigate climate change. The SSM can also serve as a basis for poverty eradication, agricultural, rural, and over-all societal development, and ultimately promotion of food and nutritional security. The main objectives of SSM are to (1) minimalize soil erosion, (2) enrich organic matter, (3) foster soil nutrient balance, (4) mitigate soil contamination, acidification, salinization, and alkalinization, (5) preserve and enhance soil biodiversity, (6) minimize and mitigate soil sealing and compaction, (7) improve soil water management, etc.

7.1.2 POST-GREEN REVOLUTION THREATS TO SOIL SUSTAINABILITY

The post-Green Revolution agriculture faces several problems, namely, decreased soil organic matter (SOM) levels, wide spread occurrence of multi-nutrient deficiencies, soil compaction etc. as a result intensive cultivation (Dwivedi et al., 2012). Unchecked use of irrigation water is a serious concern nowadays due to receding level of groundwater. The fertilizer applications are heavily skewed toward N, ignoring balanced nutrition. Soils of high production zones are the most affected due to nutrient mining, where nutrient replenishments lags far behind than the replenishments. The unbalanced use of fertilizer promotes low-nutrient use efficiencies along with a deterioration of soil health. The tillage-intensive mono-cropping along with puddling and clean cultivation increased subsoil compaction, declined the quality and quantity of SOM in Indian soils (Chauhan et al., 2012; Dey et al., 2018).

In the recent past, soil health gained public interest and the stakeholders acknowledged the fragility of the natural resource. The soil health plays a critical role in sustainability of agricultural system along with productivity and financial profit. The current increment in population pressure along with shrinkage in arable-land resources further emphasized on the quality of the soil. Government of India acknowledged the importance of soil health for enhancing farmers' income. This calls for development of efficient SSM strategies including selection of suitable crop rotations, development of novel fertilizer products, enhancing nutrient use efficiencies, balanced and integrated plant nutrient supply system, recycling of crop residues (CRs), improved tillage practices, use of low grade P, and K minerals. It is envisaged that development of such strategies will not only help in sustaining higher crop productivity but also maintaining a healthy soil and clean environment. CR management remained a major problem for Indian farmers, especially in the Indo-Gangetic plains (IGP), where cereal–cereal cropping system dominates. Retaining a part of CR on field can provide some sustainable solution. The permissible quantity of residues of different crops should be quantified which can be retained depending on cropping systems, soil, and climate without creating operational problems.

7.1.3 RESOURCE CONSERVATION TECHNOLOGIES

Resource conserving technologies (RCTs) are defined as package of practices that improves resource or input-use efficiency. The RCT is a broad term, encompassing a number of innovative package of practices. The RCTs broadly cover the five management aspects, namely tillage, CR, nutrient, crop diversification, and water management. Reduction in the intensity of tillage operations, followed by crop establishment under zero tillage (ZT) or reduced tillage (RT) conditions is beneficial for SOM. Crop establishment on permanent raised beds (PB) is another example of tillage reduction (Humphreys and Roth 2008; Dey et al., 2016; Jat et al., 2019). Specialized seed drills, namely, zero-till seed-cum-fertilizer drill, turbo seeder, happy seeder, etc., are used to sow seeds in midst of retained stubbles or residues of previous crops in the field.

Precision nutrient management (PNM) synchronizes between crop demand and nutrient supply, in turn improving nutrient use efficiency (Dwivedi et al., 2016, 2017). Soil-test-based fertilizer application, site-specific nutrient management (SSNM), in-season real-time N management, integrated nutrient management (INM), decision support tools, etc., are some components of PNM which proved extremely beneficial in several on-farm and on-field trials across India. After launching of Soil Health Card (SHC) scheme, awareness and interest of the farmers in soil testing have been increased, and the service needs to be revamped for better efficiency and reliability (Dwivedi et al., 2017). Leaf color chart (LCC), chlorophyll meter or GreenSeeker (GS) are proved to be instrumental in monitoring in-season crop N demand, and synchronizing it with nutrient supply. "Nutrient Expert' (NE) for rice, maize, and wheat improved yield substantially over farmers' fertilizer practice through rationalization of inputs (Dwivedi et al., 2016). Legume inclusion in the cropping cycles is a proven way of enhancing crop yield and NUE. Studies on diversification of cereal–cereal system through inclusion of legumes as a short duration grain crop, forage crop, substitute crop, or green manure revealed improvement in nutrient availability and a decrease in NO_3–N leaching (Dwivedi et al., 2016; Meena et al., 2018; Parihar et al., 2019). Induced defoliation (ID) in the extra-short duration legumes can also be used as source of additional SOM to soil (Mandal et al., 2013) and improved physical health of soil (Mandal et al., 2019).

Water-use efficiency has to be increased at a large scale and optimize the agricultural productivity so that pressure on natural resources could be minimized (Rockström et al., 2007; Teixeira et al., 2014). Several researchers reported benefits of subsurface drip and precision water management (PWM) technologies in ZT and residue retained systems in terms of higher crop productivity, financial profit, and enhanced soil health along with improvement of water and energy use efficiency (Sidhu et al., 2019).

7.1.4 CONSERVATION AGRICULTURE

Conservation agriculture (CA) is relatively recent agricultural management system gaining popularity at global scale. It is essentially an integration of modern scientific technology with traditional soil husbandry knowledge gained over by many generations of the farmers. The CA is mainly based on three principles, that is, least mechanical soil disturbance, permanent CR soil cover, and diversified crop rotations (Dey et al., 2018; Parihar et al., 2019). According to Food and Agriculture Organization (FAO), "CA aims to conserve, improve and ensure efficient use of natural resources through their integrated management." The CA aims at a sustainable crop production technology which has the potential of maximizing crop yields without the undue exploitation of soil, environmental or ecosystem resources (Hobbs et al., 2008; Erenstein et al., 2012).

The RCTs and CA are often used synonymously, but there is sharp distinction between these two. The CA is comprised of specific sets of RCTs. Under CA, the soil disturbance due to periodic tillage should be limited to 15 cm wide strip or 5% of the total cropped area. CR cover should not be less than 30% of the total cropped area, measured immediately after planting. There are three categories of distinguished crop cover: 30%–60%, 61%–90%, and >91%. At least three different crops should be grown in rotation including at least one legume (Kassam et al., 2009; Erenstein et al., 2012). The distinction between RCTs and CA is imperative due to the fact that some RCTs if practiced without complementation from other techniques may not be sustainable in the long-term. The ZT, if practiced without CR retention and crop diversification, can be detrimental toward system sustainability and soil health under specific situations (Erenstein et al., 2012; Dey et al., 2016).

In recent times, the CA became globally popular system which is capable of addressing world food need along with improving soil health and agricultural sustainability. The FAO is extending CA to developing countries to combat food shortage and soil degradation. Globally, CA is estimated to be practiced over an area of 180 Mha, corresponding to about 12.5% of cropland. Since 2008, the global area under CA expanded @ 10.5 Mha year^{-1} (Kassam et al., 2019). Relatively temperate countries, namely, Brazil, Argentina, United States of America, Australia etc., adopted CA more rapidly as compared with the tropical countries having more smallholder farmers. In India, CA is practiced approximately in 2.5 Mha (Jat et al., 2020), mostly covering highly productive zones of NW-IGP with higher institutional support. In these rice–wheat (RW) belts of the country, direct-seeded rice under CA prohibited wheat yield loss (amounting up to 1%–1.5% reduction in yield potential per day) due to terminal heat stress, which is very common in otherwise puddled transplanted systems (Erenstein et al., 2012).

7.2 RESOURCE CONSERVATION TECHNOLOGIES AND SOIL HEALTH

7.2.1 TILLAGE MANAGEMENT

In recent years, agricultural scientists and farmers gradually sensed the nonsustainability of conventional intensive tillage, which is extremely detrimental to SOM, aggregation, hydraulic movements, soil biological health, etc. (Chauhan et al., 2012; Dey et al., 2018). Therefore, a reduction in tillage intensity is being promoted worldwide by agricultural researchers (Erenstein et al., 2012; Kassam et al., 2019).

7.2.1.1 CONSERVATION TILLAGE

Conservation tillage can be defined as the sequence of tillage operations aimed at soil and water conservation. The process promotes noninversion of soil with retention of a protective mulch layer (Kassam et al., 2009). The generic term conservation tillage includes: ZT, minimum tillage (MT), RT, strip tillage (ST), etc. Soil is not at all disturbed under ZT, except during sowing and fertilizer application. Chemical herbicides are used as sole weed control measure. For preparation of seedbed, some extent of mechanical disturbances are allowed under MT/RT which helps to break soil crust and loosen the compact soil. Minimum secondary tillages are used as weed control measures along with herbicides (Kassam et al., 2009). The ST is a crop establishment method, where only that portion of soil is disturbed, which contains the seed row. The ST minimizes soil disturbance and retains CR on soil surface. The ST involves a primary tillage that removes residue from seeding zones and placing of soil in ridges, followed by a second operation for placing seed and fertilizer in the seeding zones (Kassam et al., 2009).

Reduced mechanical disturbances under conservation tillage significantly improve stability of soil aggregates as a whole, and especially that of macroaggregates. Abundant fresh CR under conservation tillage acts as aggregation nuclei, around which macroaggregates (250–2000 µm) are being formed, containing coarse intra-aggregate particulate organic matter (iPOM). In the process of macroaggregate formation, fresh SOM binds with microaggregates already present in the soil, thus depicting a shift from proportion of C-poor microaggregates to C-rich macroaggregates, in turn resulting in higher mean weight diameter (MWD), and higher total SOC (Six and Paustian, 2014). Microbial respiration inside macroaggregates breaks coarse iPOM and leads to formation of fine iPOM, around which microaggregates forms within macroaggregates. These "microaggregates within macroaggregates" are very stable in nature, having a larger residence time under ZT (Figure 7.1) (Six et al., 1999; Six et al., 2000). Modak et al. (2019) reported an enhancement of macro- and microaggregate associated C under ZT compared with CT, owing to higher stability of aggregates under ZT (Table 7.1) which led to higher soil bulk density (Dey et al., 2020). As a result, ZT system reports significantly higher C sequestration, compared with conventional tilled plots. Zero-till system has been reported to improve SOC by ~13% in surface soil (Dey et al., 2020) and ~8.3% in 0–60 cm soil layer compared to CT (Liu et al., 2014). The ZT-ZT registers more drastic improvements in the relatively more labile pools of SOC, over CT–CT (Table 7.2).

Tillage alteration has a contrasting effect on soil microbes (Table 7.2). On the one hand, tillage disrupts the continuity of fungal mycelium, and promotes residue mixing throughout the soil profile, negatively affecting build-up of microbial biomass. The pre-requisite of the tillage-promoted microbial growth is the presence of organic matter enriched soil aggregates, which are otherwise protected and gets exposed due to mechanical tillage operations (Spedding et al., 2004; Dey et al., 2018). Both these process determine the effect of tillage on microbial biomass and consequently SOM.

Alterations in tillage management exert a prominent effect on bio-chemical characteristics of SOM, apart from quantity of SOC (Bayer et al., 2002; González-Pérez et al., 2004). Under ZT/RT practices, lack of ploughing does not favor oxidative polymerization of added fresh organic matter. Semiquinone-type free radicals are lesser along with a lower aroma-ticity of compounds which represents a less advanced stage of humification in freshly added organic matter (González-Pérez et al., 2004; Zhang et al., 2011). González-Pérez et al. (2004) reported lowest level semiquinone-type free radical in non-cultivated soil, followed by five years of ZT under maize-pigeon pea cropping system in a Brazilian Oxisol.

Under ZT, lability of SOC increases significantly compared with conven-tionally tilled, reflecting in higher values of lability index or C management index (Jat et al., 2019; Modak et al., 2019; Parihar et al., 2019; Dey et al., 2020). The enhancement in the labile SOC under CA in turn promotes nutrient cycling and improves the nutrient supplying capacity of the soil (Chaudhary et al., 2019). Need-based modifications in fertilizer schedules under CA should be developed for specific agroecological regions. On the other hand, CA enhances non-labile SOC, indicating effective C sequestration (Dey et al., 2020).

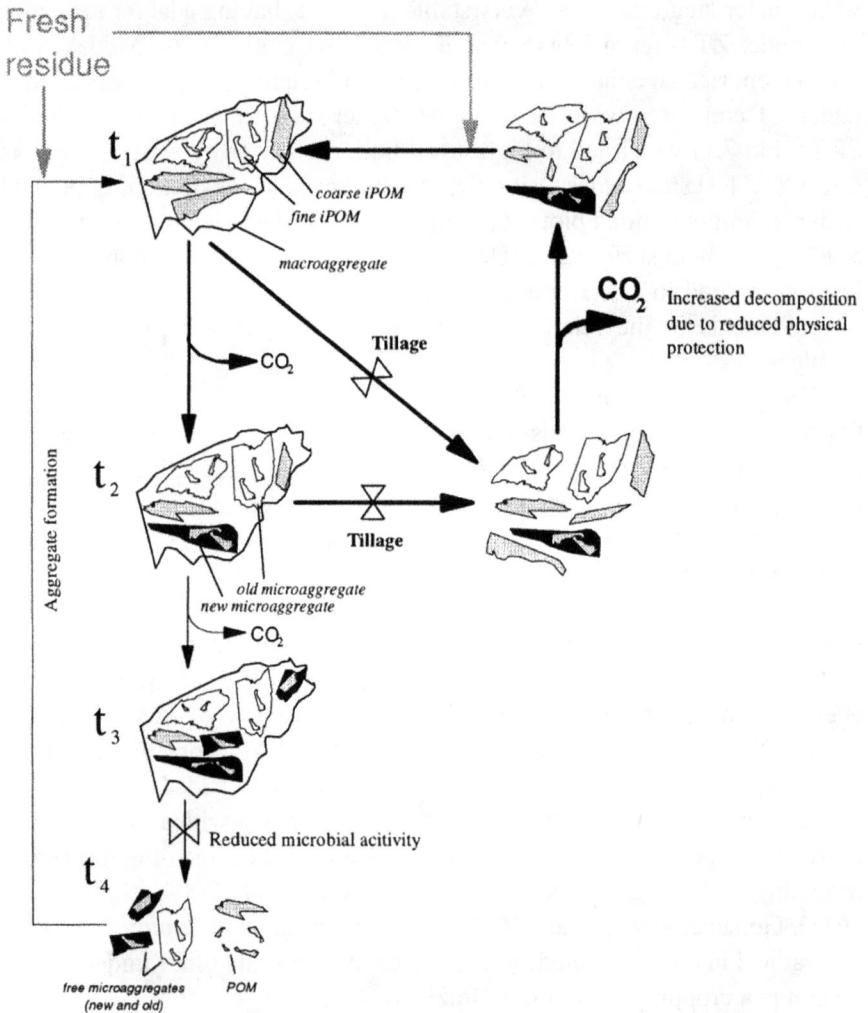

FIGURE 7.1   The cycle of macroaggregate turnover and microaggregate formation.

Source: Reprinted with permission from Six et al., 2000. © Elsevier.

TABLE 7.1 Effect of Tillage and Residue Management on Aggregate Associated C Under a Soybean-Wheat System (Adapted from Modak et al., 2019)

Treatment	Macroaggregate Associated C (g kg⁻¹)		Microaggregate Associated C (g kg⁻¹)	
	0–5 cm	5–15 cm	0–5 cm	5–15 cm
Tillage management				
Zero tillage	10.5a	7.55a	9.43a	7.27a
Conventional tillage	8.34b	6.55b	7.61b	6.14b
Residue management				
No residue	6.66b	5.10d	5.59d	4.69d
Wheat residue	9.78a	6.75c	8.55c	6.30c
Soybean residue	10.5a	7.71b	9.47b	7.58b
Wheat and soybean residue	10.8a	8.64a	10.4a	8.23a

Note: Values followed by a different letter within a column are significantly different ($P < 0.05$).

TABLE 7.2 Total C and Its Different Pools (g kg⁻¹) in Surface Soil Layer Under Conservation Agriculture (Adapted from Dey et al., 2020)

Treatment	Very Labile C	Labile C	Less Labile C	Nonlabile C	Walkley-Black C	Total C	Soil MBC (mg kg⁻¹)
Conventional tillage	2.55b	1.05c	0.76b	18.5b	4.35c	22.9c	203b
Zero tillage	3.50a	2.26ab	1.10a	18.8ab	6.92a	25.7b	222b
Zero tillage + residue	3.84a	2.52a	1.30a	19.9a	7.67a	27.5a	294a
Permanent bed + residue	2.61b	1.84b	1.00ab	20.0a	5.45b	25.4b	258ab

Note: Values followed by a different letter within a column are significantly different ($P < 0.05$).

CRs retained on surface decomposed 1.5 times slower than incorporated residues (Meena et al., 2018). Randall and Iragavarapu (1995) reported ~5% lower NO_3–N losses under ZT compared with conventional tillage. Bhattacharyya et al. (2019) reported ~48% higher N stocks associated with large macroaggregate under ZT compared to conventional plots, along with higher macroaggregate percentage. Conservation tillage promotes conversion of inorganic to organic phosphorus, in turn enhancing soil P availability. Placement of fertilizer P in soil below seed under CA leads to low P-fixation and enhance P availability to plants. CRs placed at soil surfaces favors accumulation of SOM and microbial biomass near surface, which improves the nutrient availability in root zone. Studies on soil K pools under wheat-based systems revealed an improvement or maintenance in the non-exchangeable K (NEK) in CA, whereas a decline in the same was noticed under CT with residue removal. Available K content varied with cropping systems, but the differences due to tillage practices were not as apparent as in case of NEK, suggesting thereby the need for inclusion of NEK (donor pool) in the K fertility evaluation (Meena et al., 2018).

7.2.1.2 PERMANENT RAISED BEDS

In recent years PB based systems have been adopted in both irrigated and dryland farming. Conceptually, PB is a special case of ZT or MT system, where seeds are directly drilled on raised beds, instead of flat surfaces. Permanency of beds means the beds should remain in place at least for several consecutive crops. The PBs should be reshaped after every one or two crop cycles along with light harrowing on the bed surface and seed sowing through raised bed planter. Similar to ZT, PB helps to conserve soil, improve aggregate structure, and reduce soil erosion through minimum mechanical disturbance. Several studies indicated supremacy of PB in C sequestration potential and C stability with lower decay rates over CT (Tables 7.2 and 7.4) (Chaudhary et al., 2019; Parihar et al., 2019; Dey et al., 2020). On the other hand, the PB sequesters lesser amount of C compared to ZT, due to the soil disturbance in reshaping (Table 7.2), which have ~30% more surface area compared to flatbed ZT, leading to higher rate of SOC oxidation and loss to atmosphere. In RW system, raised beds provide a natural opportunity of decreasing compaction, because of limited traffic in the furrow bottoms. The PB system also improves the nutrient availability in soil through nutrient recycling (Table 7.3) (Chaudhry et al., 2019).

7.2.2 RESIDUE MANAGEMENT

Conventional crop establishment methods prescribes removal of CRs prior to sowing, which in spite of providing a good seed establishment environment, is very much detrimental to SOM, both qualitatively and quantitatively, and in turn affects nutrient cycling (Dey et al., 2016; Chaudhury et al., 2019; Ghosh et al., 2019). The RCTs prescribe CR retention on zero-till soil surfaces. Residue retention can be classified in three categories according to their surface coverage *i.e.* 30–60%, 61–90%, and >90% ground cover after planting/sowing. To be considered as CA, surface coverage should be $\geq$30% (Kassam et al., 2009). CR retained on field acts as a source of fresh SOM, which facilitates soil aggregation, and enhances iPOM (Figure 7.1) (Six et al., 2000). A permanent residue cover reduces soil temperature regime, increase water holding and protects the soil from raindrop impact (Chivenge et al., 2007), which in turn promotes slower decay rate of SOM from surface (Dey et al., 2018). Retention of leguminous, cruciferous and cereal CRs improves the SOC pools of variable lability alike (Table 7.4) (Jat et al., 2019). These CRs provides an array of decomposable SOM with different C:N ratios, ranging from resistant lignin to easily degradable carbohydrates. Continuous supply of fresh organic matter often exceeds the metabolic capacity of microbes, which promotes an upsurge in labile SOC. On the other hand, prolonged ZT ensures anaerobic conditions favoring conversion of labile to recalcitrant SOM (e.g., lignin), in turn, enhancing both the labile and non-labile SOM (Tables 7.2 and 7.4) (Dey et al., 2016; Jat et al., 2019). In CA, positive impacts of residue retention on microbial biomass C (Table 7.2) (MBC) is more pronounced compared to tillage reduction (Spedding et al., 2004).

TABLE 7.3 Changes in Surface Soil (05– cm) Fertility (mg kg^{-1}) Due to Long-Term Tillage Modifications Under Different Maize-Based Systems (Adapted from Chaudhary et al., 2019)

Treatment	NO$_3$–N	NH$_4$–N	P	K	S	DTPA-Fe	DTPA-Zn	DTPA-Mn	DTPA-Cu
Tillage practices									
Permanent bed	28.3	15.9	14.3	93.0	17.6	9.46	5.12	3.10	3.09
Zero tillage	27.1	15.1	13.4	87.1	15.4	9.05	5.05	2.97	2.99
Conventional tillage	19.6	10.5	10.1	83.9	9.3	7.67	4.52	2.52	2.64
LSD ($P = 0.05$)	3.5	0.8	0.7	3.1	2.4	0.76	0.4	0.24	0.22
Cropping systems									
MWMb	26.1	14.8	13.5	94.0	16.8	8.37	4.98	2.75	3.00
MCS	28.0	17.2	13.6	86.4	12.8	9.25	4.89	3.03	3.12
MMuMb	24.7	13.4	11.9	82.9	10.8	8.99	5.09	2.94	2.70
Mean	25.0	13.8	12.6	88.0	14.1	8.73	4.90	2.86	2.91
LSD ($P = 0.05$)	2.3	0.9	0.8	3.5	1.0	0.4	0.32	0.29	0.19
Interaction (LSD, $P = 005$)	NS	NS	0.8	NS	1.7	NS	NS	NS	NS

MMuMb: Maize-mustard-mungbean; MWMb: Maize-wheat-mungbean; MCS: maize-chickpea-sesbania; MMS: maize-maize-sesbania.

TABLE 7.4 Soil Organic C Dynamics (0–7.5 cm depth) Under Long-Term CA in an Inceptisol (Adapted from Jat et al., 2019)

Treatment	Total SOC (g kg⁻¹)	Very Labile SOC (g kg⁻¹)	Labile SOC (g kg⁻¹)	Less-labile SOC (g kg⁻¹)	Nonlabile SOC (g kg⁻¹)	C Decay Rate (day⁻¹) at 27 °C	C Decay Rate (day⁻¹) at 37 °C	Temperature Sensitivity (Q_{10})
Cropping systems								
MMuMb	6.36b	1.53a	1.03a	1.99a	1.81a	3.78a	6.77a	1.82a
MWMb	6.72a	1.42a	1.00a	2.25a	2.05a	3.68a	6.61a	1.83a
Tillage and residue management								
PB-R	6.15b	1.37b	0.94b	2.01b	1.82b	3.83a	6.82a	1.82a
PB+R	6.94a	1.58a	1.09a	2.23a	2.04a	3.64b	6.56b	1.83a
N management								
Unfertilized	5.13b	1.11b	0.75b	1.74b	1.54b	3.55a	7.23a	2.08a
N through PU	7.02a	1.60a	1.10a	2.25a	2.06a	3.70a	6.83b	1.87b
N through SCU	6.88a	1.57a	1.08a	2.21a	2.03a	3.84a	6.45c	1.71bc
N through NCU	7.13a	1.63a	1.12a	2.29a	2.09a	3.84a	6.26d	1.64c

Values followed by a different letter within a column are significantly different ($P < 0.05$). MMuMb: Maize-mustard-mungbean; MWMb: Maize-wheat-mungbean.

The ZT do not favor semiquinone-type free radical formation along with lower percentage of aromatic structures of humus, which can be ascribed to continuous supply of labile SOC in terms of fresh CR, often more than microbial metabolic capacity. The humus formed under CA often represents less advanced stage of humification (González-Pérez et al., 2004; Zhang et al., 2011), and benefits nutrient cycling. On the other hand, potential immobilization of soil N may occur under heavy residue load near soil surface in ZT. The substantial SOC build-up, either in the form of microbial or aggregate associated SOM, locks up a sizable amount of N, thus lowering immediate N availability (Du et al., 2010; Dey et al., 2016). Similarly, soluble and loosely bound P increases due to the residue retention while other P pools, namely Al and Fe-bound, reductant soluble, Ca-bound P is negatively affected (Kumawat et al., 2018).

7.2.3 NUTRIENT MANAGEMENT

There are challenges regarding nutrient management in CA when CR is retained on no-tilled surface. A significant amount of fertilizers remains on residue and never come in contact with soil when broadcasted. On the other hand, the nature and amount of retained CRs govern the nutrient cycling. The C/N ratio of retained CR together with its' lignin, polyphenols, and soluble C contents has a major role to play in governing soil N availability (Moretto et al., 2001). Residues with wide C/N ratio favor N immobilization. Good soil aggregation under CA enhances the accumulation and protection of N within the aggregates which may limit immediate N availability (Dey et al., 2018). The tillage management interferes with soil P fixation through a modification in the surface area available for P fixation. Abundant CR retention also acts as a source of organic acids responsible for P solubilization, apart from its role as a P source. Similarly, CR acts as a direct source of available K in soil. The CA practices have been reported to modify different pools, especially the nonexchangeable pools. Therefore, a careful nutrient management protocol can add to potential benefit of tillage and residue management protocols of CA. The 4R nutrient stewardship principle, that is, applying the right source of nutrient, at the right rate, at the right time, and in the right place, should be followed.

7.2.3.1 SITE SPECIFIC NUTRIENT MANAGEMENT

SSNM aims to synchronize fertilizer application schedules with crop's nutrient demands made-to-order to a specific field, environment and crop physiological stages. Indigenous nutrient supplies of the soil can be assessed through nutrient-omission plots. Data on yield targets, native nutrient supply and nutrient use efficiencies is used to develop fertilizer prescriptions. Enhanced rate of annual C-input through SSNM promotes microbial diversity and C use efficiency which helps in improving SOC, mineral N and soil aggregation stability (Tigga et al., 2020), water use efficiency, and radiation conversation efficiency (Parihar et al., 2016) under CA-based systems. The SSNM practices under CA also showed less SOC decomposition (proportional to total SOC), and had a lesser temperature sensitivity of SOC mineralization, in turn enhancing stability of sequestered C (Parihar et al., 2019).

7.2.3.2 SENSOR-BASED TOOLS

Chlorophyll content can be used as an indicator of N content in plant leaves, and leaf greenness sensors can be linked with crop N demand. There are various sensors available in public domain to guide the *in-season* application of N according to crop needs, that is, GS, SPAD, etc. GS measures normalized difference vegetative index (NDVI), calculated from reflectance from plant leaves, which assess plant N status based on greenness (green versus yellow).

$$NDVI = (NIR - Red)/(NIR + Red)$$

Healthy vegetation (chlorophyll) reflects more near-infrared (NIR) and green light. But it absorbs more red and blue light. The GS-based N application after 40–45 days of sowing could curtail 24–48% N fertilizer requirement which enhanced N use efficiencies in maize–wheat systems (Meena et al., 2018).

Soil–Plant Analysis Development (SPAD) meter measures the difference between absorbance of a red (660 nm) and an infrared (940 nm) light through leaf, generating an optical density difference (ODD) value. Using a SPAD value of 44 for N application @ 30 kg N ha^{-1} at maximum tillering of wheat increased yield by 20%. The SPAD-based N application reduced N requirement of rice up to 25%, without any yield loss (Singh et al., 2002).

7.2.3.3 DECISION SUPPORT TOOLS

Decision-support tools such as "Nutrient Manager," NE, and LCC are available in public domain to facilitate site-specific nutrient prescription for different cropping systems (Figure 7.2) (IPNI, 2017). The NE requires some minimum information for developing fertilizer prescription, namely, the current farmers' fertilizer practices and yield of the crops, location/field-specific information, sources of nutrients, application scheduling, irrigation facilities, seed rates, and cost of inputs for profit analysis. Satyanarayana et al. (2013) reported fertilizer saving up to 17%, 56%, and 58% in terms of N, P_2O_5, and K_2O use under NE-based application compared with farmers' practice, in CA- and CT maize.

The NE-based applications promoted split fertilizer application in synchrony to crop demand, ensuring greater nutrient use efficiency.

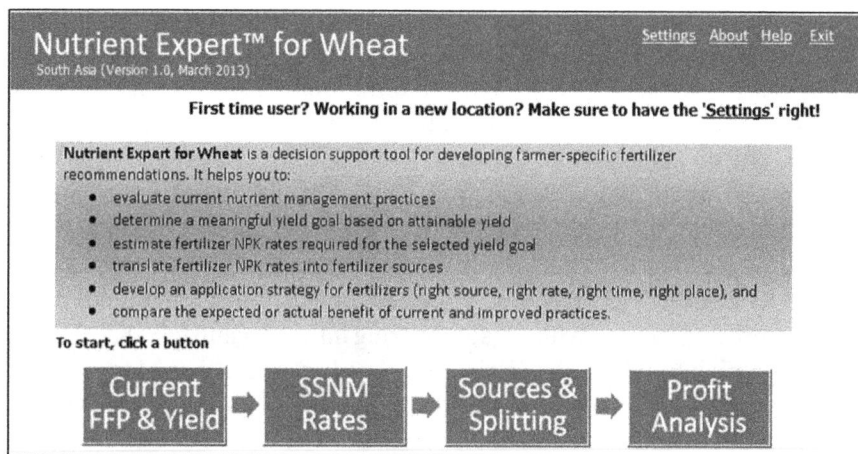

FIGURE 7.2 Working interface of Nutrient Expert for wheat (IPNI, 2017).

LCC is a simple and inexpensive real-time N management tool in cereals developed by IRRI. It recommends the time and dose of N application based on relative leaf greenness.

The LCC is a suitable and good diagnostic tool for detecting N deficiency. Initially IRRI developed four-panel LCC, which was later modified to five-panel LCC for rice by IIRR (for Southern States of India), and six-panel LCC for rice, wheat, and maize by PAU (Figure 7.3) (PAU, 2019).

7.2.3.4 NOVEL FERTILIZER PRODUCTS AND BIO-FERTILIZERS

An array of novel fertilizer products, namely, sulfur-coated urea, zinc-coated urea, neem-coated urea, complex fertilizers, fortified fertilizers, etc., could be used in CA system for enhancing nutrient use efficiency, nutrient availability, and SOC (Table 7.4) (Jat et al. 2019). In the same way, biofertilizers containing living cells of agriculturally beneficial microorganisms also prove beneficial in terms of increasing nutrient use efficiency and enhancing soil health. Most of biofertilizers belong to the categories of N-fixing, P-solubilizing and mobilizing, or plant growth promoting rhizobacteria (PGPR). The N-fixing biofertilizers, that is, *Rhizobium*, *Azospirillum*, *Azotobacter*, blue green algae (BGA), and *Azolla*, fix atmospheric N into plant-usable form. Biofertilizers containing efficient microbial strains when used properly can help curtailing a part of crops' fertilizer N requirement. Legume-*Rhizobium* symbiosis could supply ≥80% of the crop N demand. *Azotobacter*, *Azospirillum*, BGA, and *Azolla* contribute ~20–40 kg N ha^{-1} (Dwivedi et al., 2016). The P solubilizing microbes (PSM) favors the dissolution of sparingly soluble and insoluble soil P. The P solubilizing and K solubilizing microbes are successfully demonstrated in CA experimental fields and proved beneficial for maintenance and improvement of soil health and nutrient use efficiency (Kumawat et al., 2018). Use of PSM (*Bacillus megaterium*) and arbuscular mycorrhizae (AM) could result in high apparent P recovery with values as high as 28.2%. Similarly, use of conventional sources of K like mica, silicate minerals along with K solubilizing microorganisms (KSM) was found beneficial under CA scenario (Dwivedi et al., 2017; Meena et al., 2018). The KSM mainly include bacteria (*Bacillus mucilaginosus*, *Bacillus edaphicus*, etc.) and some fungi (*Aspergillus niger*, *Aspergillus fumigatus*, etc.). Meena et al. (2018) reported that 50% P and K fertilizer savings can be obtained under CA, under application of PSM and AM, and KSB, respectively, along with retention of 4 t crop residue ha^{-1}.

FIGURE 7.3   Six-panel LCC developed by PAU for rice, wheat, and maize (PAU, 2019).

7.2.4   CROP DIVERSIFICATION

Crop diversification in terms of legume inclusion in a cereal–cereal crop-ping systems is an important RCT, and is mandatory practice under CA. Inclusion of legumes improves the productivity of cereal-based systems and regenerates soil fertility. Substitution of rice with pigeon pea in RW systems enhanced nitrogen use efficiency in wheat, and thus helped curtailing fertilizer N requirement of wheat (Dwivedi et al., 2016). The extra-short duration varieties of pigeon pea could be included in cereal systems and where defoliation could be induced at vegetative maturity through foliar spray of 10% urea solution, which could add ~ 1–1.2 t ha^{-1} leaf litter and ~40 kg N ha^{-1} (Meena et al., 2012; Mandal et al., 2013). Inclusion of legumes in maize-based systems enhanced the nutrient availability under CA (Table 7.3) (Chaudhary et al., 2019). Legume inclusion reduces soil compaction, in turn facilitating better root establishment and growth of subsequent crops. Inclusion of legumes may help in minimizing NO_3–N leaching beyond root zone. Minimizing NO_3–N leaching beyond root zone is another advantage of inclusion of legumes. Biochemistry of CR retained on soil surface often controls the rate of decay, lability, and sequestration potential of C under CA (Table 7.4) (Jat et al., 2019). The wheat residue having higher contents of lignin (having ~364.5 days half-life) hinders residue decomposition (Hagin and Amberger, 1974). Inclusion of soybean in a wheat-based system can produce higher amounts of glomalin, (a glycoprotein essential for aggregate stability) under ZT, which might enhance physical protection of aggregate associated SOC leading to higher potential C sequestration (Table 7.1) (Modak et al., 2019). The legume inclusion in the cereal based systems promotes diversity in root exudates, which act as chelating agents. The decomposition of diversified fresh CR provides organic tissue bound micronutrients, as well as natural chelating agents like citric acid, humic acids which in turn enhance the availability of nutrients, especially micronutrients in the soil (Table 7.3) (Shukla et al., 2017; Chaudhary et al., 2019).

7.2.5   WATER MANAGEMENT

PWM strategies and subsurface drip fertigation (SSDF) complements CA through exceptional water saving benefits. The CA enhances water infiltration, water holding capacity, and reduces evaporative losses. Roth et al. (1988) reported complete infiltration of rain water (60-mm rainfall) in a residue retained plot, against only 20% infiltration in a bare plot. Soil cover exponentially reduces run-off. Scopel et al. (2004) reported run-off reduction up to 50% as a result of maintenance of 30% CR coverage on soil. Residue heaps forms succession barriers which facilitates infiltration by giving the rain water excess residing time on soil surface. In the long run, CA impedes crust formation in deeper layers, which in turn drastically improves the infiltration rate, often up to 10 times higher than in conventional tilled plots (Scopel et al., 2004).

The PWM denotes precise and uniform application of irrigation at the right time, right place, and right crop growth stage. The PWM strategies can be easily amalgamated in existing principles of CA. ON the other hand, the SSDF reduces soil evaporation losses, allows better conveyance of water and nutrients to the root zone. The SSDF is instrumental in enhancing nutrient use efficiency along with limiting weed growth and reducing labor cost (Ayars et al., 1999). The SSDF system (with 67.5 cm lateral spacing and 15 cm depth) can save irrigation water by 48–53% in rice and 42–53% in wheat compared to flood irrigation system under CA (Sidhu et al., 2019).

7.3 CONCLUSION

There are a number of RCTs for enhancing the resource use efficiency, but it is often difficult to convince the farmers/stakeholders to adopt RCTs as successful and remunerative crop production technologies. In some cases, it may be difficult to convince the farmers of potential benefits of the RCTs in improving crop productivity and soil health, and reducing production costs. CA is one of the best RCTs though it has several constraints in its adoption. Lack of appropriate seeders, especially for small holders (more than 86% of the country's farmers come under this category), is one of these constraints. Livestock population of India is 535.78 million showing an increase of 4.6% over Livestock Census 2012, which requires more fodder. On the other hand, under rainfed situations, farmers face a scarcity of CRs due to less biomass production of different crops. These CRs have competitive uses as fodder in many areas, with least availability for CR retention essentially required in CA. In other areas, especially in irrigated ecosystem, for timely sowing of the succeeding crop and lack of appropriate CA machinery, farmers often resort to residue burning. This has become a common phenomenon in the cereal–cereal system in Indo-Gangetic Plain, raising serious concerns regarding sustainability of production systems. In view of the ill-effect of CR burning on entire ecosystem and human health, Government initiated several promotional programs including subsidizing CA equipment and microirrigation, especially subsurface drip irrigation. Large scale adoption of RCTs including CA would be possible with increasing awareness and ensuring timely availability of desired machinery and related input.

7.4   WAY FORWARD

RCTs offer a new paradigm for research and development in agriculture with the objective to achieve targeted food production along with resource conservation without deteriorating environmental quality. However, site-specific nutrient and water management protocols for CA under system perspective are to be developed which will promote their adoption at farm level. Availability of machinery/equipment for promotion of RCTs is a prerequisite for sustaining agricultural production and maintaining/restoring soil health. Standardization of metering system for seeding/planting of pulses, oil seeds, rice, and wheat needs to be worked out to have accurate seeding rate. Development of CA-specific machinery/equipment for smallholders should take center stage. Equally important is ensuring timely availability of the same at economical cost or on custom hiring basis. It is important to design long-term CA experiments in experimental fields and farmers' fields to study the impact on soil health, nutrient dynamics, water use efficiency, C sequestration, GHG emission, and ecosystem services. Region-specific interventions must be added in the future for crop diversification through substitution/intensification. Developing complete protocols for CA under prominent cropping systems in different agroecological regions, particularly in rainfed and drylands is the need of hour. Site-specific nutrient prescriptions and precision input management for intensive cropping systems need to be develop to optimize resource use efficiency. Studies on dynamics of weeds, pests, and diseases under CA-based systems would be necessary to formulate strategies for their effective management. Awareness and orientation programs need to be launched for capacity building of stakeholders so as to ascertain wider adoptions of RCTs.

KEYWORDS

- **conservation agriculture**
- **resource conservation**
- **sustainability**

REFERENCES

Ayars, J.E.; Phene, C.J.; Hutmacher, R.B.; Davis, K.R.; Schoneman, R.A.; Vail, S.S.; Mead, R.M. Subsurface drip irrigation of row crops: a review of 15 years of research at the Water Management Research Laboratory. *Agric. Water Manage.* 1999, 42, 1–27.

Bayer, C.; Martin-Neto, L.; Mielniczuk, J.; Saab, S.C.; Milori, D.M.P.; Bagnato, V.S. Tillage and cropping system effects on soil humic acid characteristics as determined by electron spin resonance and fluorescence spectroscopies. *Geoderma* 2002, 105, 81–92.

Bhattacharyya, R.; Das, T.K.; Das, S.; Dey, A.; Patra, A.K.; Agnihotri, R.; Ghosh, A.; Sharma, A.R. Four years of conservation agriculture affects topsoil aggregate-associated [15]nitrogen but not the [15]Nitrogen use efficiency by wheat in a semi-arid climate. *Geoderma* 2019, 337, 333–340.

Chaudhary, A.; Meena, M.C.; Dwivedi, B.S.; Datta, S.P.; Parihar, C.M.; Dey, A.; Sharma, V.K. Effect of conservation agriculture on soil fertility in maize (*Zea mays*)-based systems. Indian *J. Agric. Sci.* 2019, 89, 1654–1659.

Chauhan, B.S.; Mahajan, G.; Sardana, V.; Timsina, J.; Jat, M.L. 2012. Productivity and sustainability of the rice-wheat cropping system in the Indo-Gangetic Plains of the Indian subcontinent: problems, opportunities, and strategies. *Adv. Agron.* 2012, 117, 315–369.

Chivenge, P.P.; Murwira, H.K.; Giller, K.E.; Mapfumo, P.; Six, J. Long-term impact of reduced tillage and residue management on soil carbon stabilization: Implications for conservation agriculture on contrasting soils. *Soil Tillage Res.* 2007, 94, 328–337.

Dey, A.; Dwivedi, B.S.; Bhattacharyya, R.; Datta, S.P.; Meena, M.C.; Jat, R.K.; Gupta, R.K.; Jat, M.L.; Singh, V.K.; Das, D.; Singh, R.G. Effect of conservation agriculture on soil organic and inorganic carbon sequestration, and their lability: A study from a rice-wheat cropping system on a calcareous soil of eastern Indo-Gangetic Plains. *Soil Use Manage.* 2020, 00, 1–10. https://doi.org/10.1111/sum.12577.

Dey, A.; Dwivedi, B.S.; Bhattacharyya, R.; Datta, S.P.; Meena, M.C.; Das, T.K.; Singh, V.K. Conservation agriculture in a rice-wheat cropping system on an alluvial soil of north-western Indo-Gangetic Plains: Effect on soil carbon and nitrogen pools. *J. Indian Soc. Soil Sci.* 2016, 64, 246–254.

Dey, A.; Dwivedi, B.S.; Meena, M.C.; Datta, S.P. Dynamics of soil carbon and nitrogen under conservation agriculture in rice-wheat cropping system. *Indian J. Fert.* 2018, 14 (3), 12–26.

Du, Z.; Ren, T.; Hu, C. Tillage and residue removal effects on soil carbon and nitrogen storage in the North China Plain. *Soil Sci. Soc. America J.* 2010, 74, 196–202.

Dwivedi, B.S.; Singh, V.K.; Meena, M.C.; Dey, A.; Datta, S.P. Integrated nutrient management for enhancing nitrogen use efficiency. *Indian J. Fert.* 2016, 12 (4), 62–71.

Dwivedi, B.S.; Singh, V.K.; Shekhawat, K.; Meena, M.C.; Dey, A. Enhancing use efficiency of phosphorus and potassium under different cropping systems of India. *Indian J. Fert.* 2017, 13 (8), 20–41.

Dwivedi, B.S.; Singh, V.K.; Shukla, A.K.; Meena, M.C. Optimising dry and wet tillage for rice on a Gangetic alluvial soil: effect on soil characteristics, water use efficiency and productivity of the rice-wheat system. *Eur. J. Agron.* 2012, 43, 155–165.

Erenstein, O.; Sayre, K.; Wall, P.; Hellin, J.; Dixon, J. Conservation agriculture in maize-and wheat-based systems in the (sub) tropics: lessons from adaptation initiatives in South Asia, Mexico, and Southern Africa. *J. Sustain. Agr.* 2012, 36, 180–206.

FAO. Voluntary Guidelines for Sustainable Soil Management. Food and Agriculture Organization of the United Nations, Rome, Italy.

Ghosh, A.; Bhattacharyya, R.; Dey, A.; Dwivedi, B.S.; Meena, M.C.; Manna, M.C.; Agnihortri, R. Long-term fertilisation impact on temperature sensitivity of aggregate associated soil organic carbon in a sub-tropical inceptisol. *Soil Tillage Res.* 2019, 195, 104369.

González-Pérez, M.; Martin-Neto, L.; Saab, S.C.; Novotny, E.H.; Milori, D.M.B.P.; Bagnato, V.S.; Colnago, L.A., Melo, W.J.; Knicker, H. 2004. Characterization of humic acids from a Brazilian Oxisol under different tillage systems by EPR, ^{13}C-NMR, FTIR and fluorescence spectroscopy. Geoderma 2004, 118, 181–190.

Hagin, J.; Amberger, A. Contribution of Fertilizers and Manures to a N- and P-Load of Water. A Computer Simulation Model; Final Report to Deutsche Forschungsgemeinschaft Technicon; Israel.

Hobbs, P.R.; Sayre, K.; Gupta, R. The role of conservation agriculture in sustainable agriculture. *Philos. Trans. Royal Soc. B.* 2008, 363, 543–555.

Humphreys, E; Roth, C.H. Permanent beds and rice-residue management for rice wheat systems in the Indo-Gangetic Plain. In ACIAR Proceedings No. 127. Proceedings of a workshop held in Ludhiana, India, 7–9 September 2006.

IPNI. Nutrient Expert® for Wheat—South Asia. 2017. http://software.ipni.net/article/nutrient-expert-for-wheat-south-asia

Jat, M.L.; Chakraborty, D.; Ladha, J.K.; Rana, D.S.; Gathala. M.K.; McDonald, A.; Gerard, B. Conservation agriculture for sustainable intensification in South Asia. *Nat. Sustain.* 2020, 3, 336–343.

Jat, S.L.; Parihar, C.M.; Dey, A.; Nayak, H.S.; Ghosh, A.; Parihar, N.; Goswami, A.K.; Singh, A.K. Dynamics and temperature sensitivity of soil organic carbon mineralization under medium-term conservation agriculture as affected by residue and nitrogen management options. *Soil Tillage Res.* 2019, 190, 175–185.

Kassam, A.; Friedrich, T.; Derpsch, R. Global spread of Conservation Agriculture. *Int. J. Environ. Stud.* 2019, 76, 29–51–.

Kassam, A.; Friedrich, T.; Shaxson, F.; Pretty J. 2009. The spread of conservation agriculture: Justification, sustainability and uptake. *Int. J. Agr. Sustain.* 2009, 7, 292–320.

Kumawat, C.; Sharma, V.K.; Meena, M.C.; Dwivedi, B.S.; Barman, M.; Kumar, S.; Chobhe, K.A.; Dey, A. Effect of crop residue retention and phosphorus fertilization on P use efficiency of maize (*Zea mays*) and biological properties of soil under maize-wheat (*Triticum aestivum*) cropping system in an Inceptisol. *Indian J. Agric. Sci.* 2018, 88, 1184–1190.

Liu, E.; Teclemariam, S.G.; Yan, C.; Yu, J.; Gu, R.; Liu, S.; Liu, Q. Long-term effects of no-tillage management practice on soil organic carbon and its fractions in the northern China. *Geoderma* 2014, 213, 379–384.

Mandal, N.; Dwivedi, B.S.; Datta, S.P.; Meena, M.C.; Tomar, R.K. Soil hydrophysical properties under different nutrient management practices, their relationship with soil organic carbon fractions and crop yield under pigeonpea-wheat sequence. *J. Plant Nutr.* 2019, 42, 384–400.

Mandal, N.; Dwivedi, B.S.; Meena, M.C.; Dhyan-Singh; Datta, S.P.; Tomar, R.K.; Sharma, B.M. Effect of induced defoliation in pigeonpea, farmyard manure and sulphitation pressmud on soil organic carbon fractions, mineral nitrogen and crop yields in a pigeonpea-wheat cropping system. *Field Crops Res.* 2013, 154, 178–187.

Meena, M.C.; Dwivedi, B.S.; Dhyan-Singh; Sharma, B.M.; Krishan-Kumar; Singh, R.V.; Kumar, R.; Rana, D.S. Effect of integrated nutrient management on productivity and soil health in pigeonpea (Cajanus cajan)-wheat (*Triticum aestivum*) cropping system. *Indian J. Agron.* 2012, 57, 333–337.

Meena, M.C.; Dwivedi, B.S.; Mahala, D.; Das, S.; Dey, A. Nutrient dynamics and management under conservation agriculture. 2018. In: System Based Conservation Agriculture (VK Singh et al., eds.), p. 42. Westville Publishing House, New Delhi.

Modak, K.; Ghosh, A.; Bhattacharyya, R.; Biswas, D.R.; Das, T.K.; Das, S.; Geeta-Singh. Response of oxidative stability of aggregate-associated soil organic carbon and deep soil carbon sequestration to zero-tillage in subtropical India. *Soil Tillage Res.* 2019, 195, 104370.

Moretto, A.S.; Distel, R.A.; Didoné, N.G. Decomposition and nutrient dynamic of leaf litter and roots from palatable and unpalatable grasses in a semi-arid grassland. *Appl. Soil Ecol.* 2001, 18, 31–37.

Parihar, C.M.; Singh, A.K.; Jat, S.L.; Ghosh, A.; Dey, A.; Nayak, H.S.; Parihar, M.D.; Mahala, D.M.; Yadav, R.K.; Rai, V.; Satayanaryana, T.; Jat, M.L. Dependence of temperature sensitivity of soil organic carbon decomposition on nutrient management options under conservation agriculture in a sub-tropical Inceptisol. *Soil Tillage Res.* 2019, 190, 50–60.

Parihar, C.M.; Yadav, M.R.; Jat, S.L.; Singh, A.K.; Kumar, B.; Pradhan, S.; Chakraborty, D.; Jat, M.L.; Jat, R.K.; Saharawat, Y.S.; Yadav, O.P. Long term effect of conservation agriculture in maize rotations on total organic carbon, physical and biological properties of a sandy loam soil in north-western Indo-Gangetic plains. *Soil Tillage Res.* 2016, 16, 116–128.

PAU. Package of practices for crops of Punjab, rabi 2019–20, 2019. https://www.pau.edu/content/pf/pp_rabi.pdf

Randall, G.W.; Iragavarapu, T.K. 1995. Impact of long-term tillage systems for continuous corn on nitrate leaching to tile drainage. *J. Environ. Qual.* 1995, 24, 360–366.

Rockström, J; Lannerstad, M.; Falkenmark, M. Assessing the water challenge of a new green revolution in developing countries. *Proc. Natl. Acad. Sci. U. S. A.* 2007, 104, 6253–6260.

Roth, C.H.; Meyer, B.; Frede, H.G.; Derpsch, R. Effect of mulch rates and tillage systems on infiltrability and other soil physical properties of an Oxisol in Parana, Brazil. *Soil Tillage Res.* 1988, 11, 81–91.

Scopel, E.; Da Silva, F.A.M.; Corbeels, M.; Affholder, F.O.; Maraux F. Modelling crop residue mulching effects on water use and production of maize under semi-arid and humid tropical conditions. *Agronomy.* 2004, 24, 383–395.

Shukla, A.K.; Sinha, N.K.; Tiwari, P.K.; Chandra-Prakash; Lenka, N.K; Singh, V.K.; Behera, S.K.; Majumdar, K.; Kumar, A.; Srivastava, P.C.; Pachauri, S.P.; Meena, M.C.; Lakaria, B.L.; Siddqui, S.S.; Singh D. Spatial distribution and management zones forsulfur and micronutrients in Shiwalik Himalayan region of India. *Land Degrad. Dev.* 2017, 28, 959–969.

Sidhu, H.S.; Jat, M.L.,; Singh, Y.; Sidhu, R.K.; Gupta, N.; Singh, P.; Singh, P.; Jat, H.S.; Gerard, B. Sub-surface drip fertigation with conservation agriculture in a rice-wheat system: A breakthrough for addressing water and nitrogen use efficiency. *Agric. Water Manage.* 2019, 216, 273–283.

Singh, B.; Singh, V.; Ladha, J.K.; Bronson, K.F.; Balasubramanian, V.; Singh, J.; Khind, C.S. Chlorophyll Meter- and Leaf Color Chart-Based Nitrogen Management for Rice and Wheat in Northwestern India. *Agron. J.* 2002, 94, 821–829.

Six, J.; Elliott, E.T.; Paustian, K. Aggregate and soil organic matter dynamics under conventional and no-tillage systems. *Soil Sci. Soc. America J.* 1999, 63, 1350–1358–.

Six, J.; Elliott, E.T.; Paustian, K. Soil macroaggregate turnover and microaggregate formation: A mechanism for C sequestration under no-tillage agriculture. *Soil Biol. Biochem.* 2000, 32, 2099–2103.

Six, J.; Paustian, K. Aggregate-associated soil organic matter as an ecosystem property and a measurement tool. *Soil Biol. Biochem.* 2014, 68, 4–9.

Spedding, T.A.; Hamel, C.; Mehuys, G.R.; Madramootoo, C.A. Soil microbial dynamics in maize-growing soil under different tillage and residue management systems. *Soil Biol. Biochem.* 2004, 36, 499–512.

Teixeira, E.I.; George, M.; Herreman, T.; Brown, H.; Fletcher, A.; Chakwizira, E.; DeRuiter, J.; Maley, S.; Alasdair, N. The impact of water and nitrogen limitation on maize biomass and resource-use efficiencies for radiation water and nitrogen. *Field Crops Res.* 2014, 168, 109–118.

Tigga, P.; Meena, M.C.; Dwivedi, B.S.; Datta, S.P.; Dey, A. Effect of conservation agriculture on soil carbon and nitrogen under different fertilizer management practices in maize-wheat cropping system. *Indian J. Agric. Sci.* 2020, 90, 1568–1574.

Unger, P.W. Organic-matter, nutrient, and pH distribution in no-tillage and conventional tillage semiarid soils. *Agron. J.* 1991, 83, 186–189.

Zhang, J.; Hu, F.; Li, H.; Gao, Q.; Song, X.; Ke, X.; Wang, L. Effects of earthworm activity on humus composition and humic acid characteristics of soil in a maize residue amended rice–wheat rotation agroecosystem. *Appl. Soil Ecol.* 2011, 51, 1–8.

CHAPTER 8

Soil Organic Carbon, Its Retention, and Implication to Sustainable Agriculture

ANUPAM DAS[1*], SUBORNA ROY CHOUDHURY[2], and SAMRAT GHOSH[3]

[1]*Department of Soil Science and Agricultural Chemistry, Bihar Agricultural University, Sabour, Bhagalpur, Bihar 813210, India*

[2]*Department of Agronomy, Bihar Agricultural University, Sabour, Bhagalpur, Bihar 813210, India*

[3]*Department of Agricultural Chemistry and Soil Science, Bidhan Chandra Krishi Viswavidyalaya, Mohanpur, Nadia, West Bengal 741 252, India*

Corresponding author. E-mail: anusoil22@gmail.com

ABSTRACT

The "4 per mille" is a historical breakthrough initiative aims toward best management of soil organic carbon (SOC) stocks, which depend on net balance of C outputs and inputs in soil. Soil carbon stabilization comprises both biotic and abiotic processes, which occur concurrently, but understanding of the stabilization mechanisms is still a researchable issue. Soil quality is an important aspect of sustainability, where organic carbon plays a pivotal role. Thus, maintenance of SOC stock is crucial for production sustainability as well as to curb the global warming and enhanced CO_2 concentration in the atmosphere. Net primary productivity is the main substrate for SOC through litter and rhizodeposition, in which microorganisms especially fungi and bacteria and soil physical properties such as mineral fraction, porosity, and structure act on to modify the C stocks. Agricultural practices impinge on SOC stocks by influencing biotic and abiotic processes through modification in land use and management practices such as crop diversification, residue management, application of organic amendments, fertilization, tillage, and so on. Meta-analyses studies and long-term field experiments revealed the interrelation among different processes and factors affecting C stocks. Thus, soil organic matter management is a win–win strategy for improving soil quality, ensuring food security, mitigating climate change, and developing a holistically sustainable production system.

8.1 INTRODUCTION

Carbon is a unique element in the universe known since antiquity. By mass, carbon is the fourth most abundant element in the universe and rated 15th in abundance in Earth's crust (Kring, 1997) constitutes about 0.02% (Lide, 2003). It possesses a unique abundance and diversity in all living and nonliving substances. It is the core element of the organic compound and has an unusual ability to form a vast number and variety of compounds more than any other element, with almost 10 million compounds described to date (Anonymous, 2003). The catenation property enables carbon to form various allotropes including the softest substance graphite and the hardest substance diamond occurring naturally. However, these allotropes have very limited applicability in agriculture. Soil organic carbon (SOC) plays a central role in agricultural science.

SOC comprises about 60% of soil organic matter (SOM) (Paustian et al., 2019). SOM is the decomposing leftover of plants from foliage, crop residues, root exudates, manures, and organic wastes. Farmers have known for millennia that ecosystem services and soil quality mostly depend on the SOM content, even in the earlier 18th century, scientists were mostly interested in studying SOM composition; after that, formal scientific studies have been started (Boussingault, 1841) to determine the impact of various crop and soil management practices on SOM and resultant crop responses and nutrient availability for plant growth. Recently, "4 per mille," a voluntary initiative, come into force throughout the world after the *21st Conference of the United Nations Framework Convention on Climate Change* (UNFCCC) in Paris, where every stakeholders made their efforts toward a better management of soil carbon (C) stocks (Minasny et al., 2017). Some technical advancement in carbon research has been listed in Table 8.1. Thus, SOM management is a win–win strategy for improving soil quality, ensuring food security, mitigating climate change, and developing a holistically sustainable production system.

TABLE 8.1 Chronological Advances in SOM Research

Technical Advances	Reference
^{14}C estimation by liquid scintillation counting	Broser and Kallmann (1947)
Radiocarbon (^{14}C) measurement in soil humus	Tamm and Ostlund (1960)
Pyrolysis of humic and fulvic acids	Nagar (1963)
Detection of ^{14}C by accelerator mass spectrometry	Purser et al. (1977)
Humic materials was characterized by ^{13}C nuclear magnetic resonance spectrometry	Gonzalez-Vila et al. (1976)
Whole soil characterization by solid-state ^{13}C nuclear magnetic resonance spectrometry	Barron et al. (1980)
Stable C isotopes for SOM turnover	Cerri et al. (1985)
Compound-specific *n*-alkyls group detection by the ^{13}C isotopic method	Lichtfouse et al. (1994)
Humic and fulvic acids characterized by Fourier transform ion cyclotron resonance mass spectrometry	Fieve et al. (1997)
Time-of-flight secondary ion mass spectrometry was used for characterization of commercial humic acids	Leis et al. (2000)
Speciation of soil C by near-edge X-ray absorption fine structure	Solomon et al. (2005)
NanoSIMS used for microbial cell detection	Herrmann et al. (2007)
Portable X-ray fluorescence and visible near infrared diffuse reflectance spectrometry used for estimation of soil total carbon and total nitrogen	Wang et al. (2015)
Characterization of dissolved organic carbon was done with Orbitrap	Hawkes et al. (2016)

Modified after Knabner and Rumpel (2018).

8.2 SOIL ORGANIC CARBON

SOC is an easily measureable component of SOM that encompasses 2%–10% of soil's mass (~5% of soil volume) and performing a key role in the soil functions of agricultural soils. SOM contributes to nutrient retention and supply, improvement in soil structure, moisture availability and retention, pollutants degradation, soil resilience, and carbon sequestration. SOC includes animal, plant, and microbial residues at different stages of decomposition. Apart from that, different groups of organic compounds are intimately associated with inorganic soil constituents.

Global carbon cycle encompasses five principal pools, where soil carbon pools is the third largest pool (estimated as 2500 Pg to 1-m depth) followed by oceanic (~38,000 Pg) and geologic pools (~5000 Pg). The SOC pools consist of active humus to inert charcoal carbon. It consists of (1) decomposed residues of animals and plants and (2) substances synthesized from the breakdown products either microbiologically and/or chemically. Depending on the mean residence time (MRT), SOC pools are grouped into three categories: labile pool having the MRT of days to years, intermediate pool with the MRT of years to decades and centuries, and passive pools last for centuries to millennia.

SOC plays a variety of functions such as buffer soil reaction and soil temperature, nutrient bin for major plant nutrient (e.g., N, P, S, Zn, and Mo), influences cation exchange capacity by enhancing surface charge density, enhancing plant available water capacity as well as infiltration capacity, and thereby reducing surface runoff, promotes favorable soil aggregation, uplifts soil biodiversity, and finally enhances input (nutrient, water, etc.) use efficiency. The culmination of these functions has a large-scale impact, for example, it rehabilitates the soil by reducing soil erosion viz-a-viz sediment load in streams and rivers, helps in filtering and biodegradation of soil pollutants and contaminants, and downregulates greenhouse gas (GHG) emissions from the soil.

8.3   STABILITY OF SOC

The Kyoto Protocol, an international treaty extended after UNFCCC in 1992, put forward the central anxiety to reduce GHG emissions, and there was an agenda raised regarding the primary understanding for stabilizing carbon in soils. This consensus arose due to the following facts: (1) soils can act as both sinks and sources of carbon, (2) the global soil carbon (C) pool (to 1-m depth) is about three times of the atmospheric pool and four times that of the biotic pool (Lal, 2004, 2010) and all are in dynamic equilibrium, and (3) past and present agricultural production system has profound influence on global carbon, hydrological, and nutrient cycles (Zomer et al., 2017). Thus, any alteration or deviation in the dynamics and the kinetics of SOC pools potentially influences atmospheric CO_2 concentration vis-a-vis global climate. Carbon storage in soil is principally restrained by two major factors, namely, carbon input through net primary productivity (quality and quantity) and native carbon decomposition rate. However, the carbon stabilization mechanism and its ultimate stabilization potential in soil are still in explanatory stage. Some proposed mechanisms of carbon stabilization are depicted here for clarity of the subject.

8.3.1   CONCEPTUAL MECHANISMS OF SOC STABILIZATION

Central dogma of SOC stabilization is retardation potential of SOM loss by respiration. Net primary productivity is the principal source of carbon in soil, where gaseous carbon dioxide (CO_2) through photosynthesis takes entry into the soil. After that, these carbon are subjected to the various biotic and environment factors, and the resultant are considered to be stabilized SOC (see Figure 8.1). Broadly, three mechanisms of SOC stabilization have been proposed: chemical inertness or chemical stabilization, physical protection of SOC or physical stabilization and organic recalcitrant or biochemical stabilization (Sollins, 1996; Six et al., 2002; Krull et al., 2003; Mayer, 2004). Chemical stabilization is achieved by imposing chemical inaccessibility of SOC to decomposer through the interaction of soil carbon, nitrogen, and soil inorganic constituents. A closed environment or strong chemical bond is formed by the interaction with soil inorganic constituents (e.g., sesquioxide, calcium carbonate, silt, and/or clay) and SOC, where nitrogen acts as a bridging element. Physical stabilization is simply one step ahead of chemical stabilization, where physicochemical stabilization and aggregation accumulate SOC within the aggregate and create a physical barrier between the microbes, enzymes, and their substrates, that is, SOC. Basically, chemical and physical stabilization mechanisms are not a separate entity; rather, they are complementary to one another. Organic recalcitrant is the inherent chemical structure of biomolecules formed through chemical complexing processes such as formation of covalent bond through Maillard reaction, glycation, lipid oxidation, vulcanization, and other reactions, and/or hydrophobicity of labile biomolecules encapsulates them within more recalcitrant ones and makes them resistant to microbial attack. Thus, organic recalcitrance makes SOC biochemically stable (see Figure 8.2).

However, stability of SOC is heterogeneous (Bernoux et al., 1998), and it is vulnerable to agricultural practices (Blanco-Canqui and Lal, 2004). Physically rather than physicochemically stable SOC associated with fine soil particles accounts for at least half (Feng et al. 2016; Di et al., 2017) and at the most 77% of bulk SOC concentration (Wiesmeier et al., 2014) and attributed MRT ranging from years to millennia (Wiesmeier et al., 2014; Cai et al., 2016). Thus, most of the research was concentric to the SOC associated with finer soil particle (<20 μm) (see Table 8.2). It is also a proven fact that highly aggregated soil (stable fraction) did not sequester extra carbon in bulk soil (Chung et al., 2010) under long-term high C inputs scenarios (Chung et al., 2008; Gulde et al., 2008; Di et al., 2014; Du et al., 2014). Hence, SOC has certain maxima for stabilization designated as SOC saturation (Hassink, 1997) reducing the efficiency and rate of stabilization of SOC in soils (Du et al., 2014). Therefore, SOC saturation should be considered before implementing any management practices for carbon sequestration, although its accurate estimation is still dubious (Wiesmeier et al., 2014).

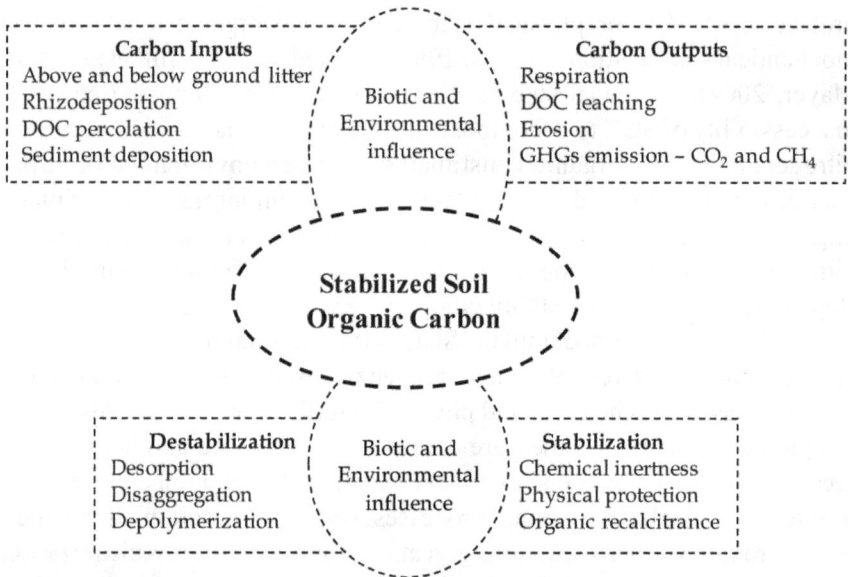

FIGURE 8.1 Conceptual mechanisms of SOC stabilization. Modified after (Sollins, 1996) and (Six et al., 2002).

8.4 SUSTAINABLE AGRICULTURE AND SOC

Sustainable land management designates "the use of land resources, including soil, water, animals and plants, for the production of goods to meet changing human needs, while simultaneously ensuring the long-term productive potential of these resources and the maintenance of their environmental functions (United Nations Earth Summit, 1992)," whereas soil carbon sequestration also ensures sustaining higher agricultural productivity, improvement in input use efficiency, improvement in soil health and biodiversity, and mitigation of adverse environmental impact, that is, climate change. Thus, management practices that maintain and/or improve of SOC status would ensure sustainability in agriculture.

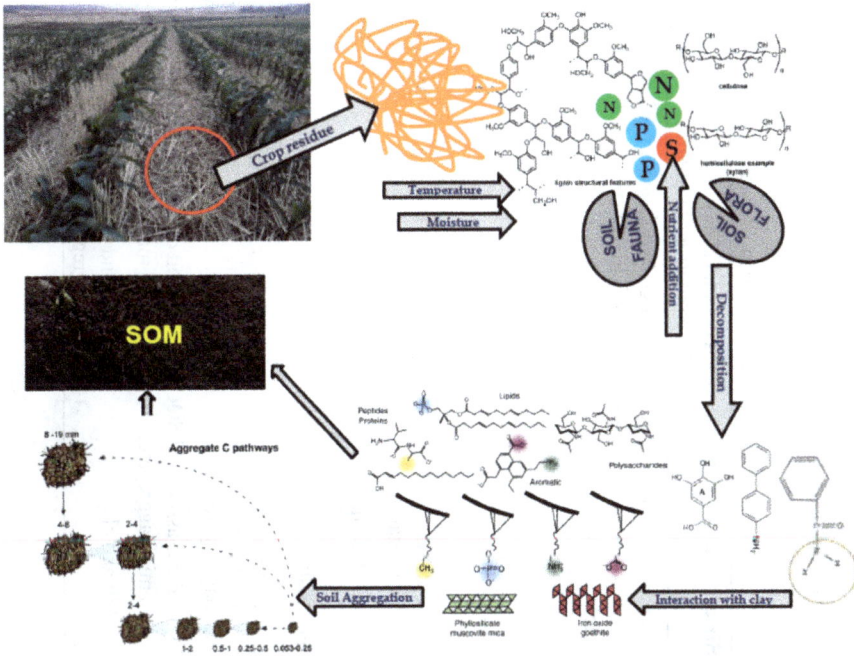

FIGURE 8.2   Schematic diagram of conversion of residue carbon into SOM.

8.4.1   INFLUENCE OF MANAGEMENT PRACTICES ON SOC STOCK

Management practices have profound influence on SOC stocks, which can be moderated either by increasing carbon inputs in soil or by decreasing SOC losses. Net sink for atmospheric carbon can be obtained by adopting best practices such as (1) increasing primary productivity through cover crops, rotations, and agroforestry, (2) improving residue management in soil, (3) integrating nutrient management with organic amendments (manures and composts) and biosolids, (4) avoiding residue burning or fires, (5)

TABLE 8.2 Carbon Fractions as Categorized by Stabilization Mechanism: Physically Protected, Chemically Protected, or Unprotected Organic Matter (OM)

Stabilization Mechanism	Physical Fractionation Method				
	Plaza et al. (2013)	Six et al. (2002) density	Six et al. (2002) Microaggregate isolator	Sohi et al. (2001)	Zimmermann et al. (2007)
Unprotected	(i) Free organic matter (ii) Intra macroaggregate organic matter (iii) Dissolved organic matter	(i) Free Particulate organic matter within large and small macroaggregates	(i) Large Particulate organic matter /Litter (Size - >2000 μm) (ii) Coarse POM in large and small macroaggregates	(i) Free light organic matter	(i) Particulate organic matter (POM) (ii) Dissolved organic carbon
Physical	(i) Intra microaggregate organic matter	i) Free POM within microaggregates ii) Free POM in microaggregates within large and small macroaggregates	(i) Intra microaggregate POM (imPOM) (ii) Intra microaggregate POM within large and small macroaggregates	(i) Intra aggregate light organic matter	(i) Sand and stable aggregates
Chemical	(i) Mineral-associated organic matter	(i) Mineral-associated SOC within large and small macroaggregates (ii) Mineral-associated SOC within microaggregates (iii) Silt and clay	(i) Silt and clay $(s + c)$ (ii) Silt and clay within large and small macroaggregates (iii) Silt and clay within Microaggregates	(i) Mineral associated organic matter	(i) Silt and clay (ii) Resistant soil organic carbon

Modified from (Duddigan et al., 2019)

management of grassland through fertilization and grazing, (6) decreasing heterotrophic respiration (water management and exclusion of tillage), and (7) diminishing erosion rates (see Figure 8.3).

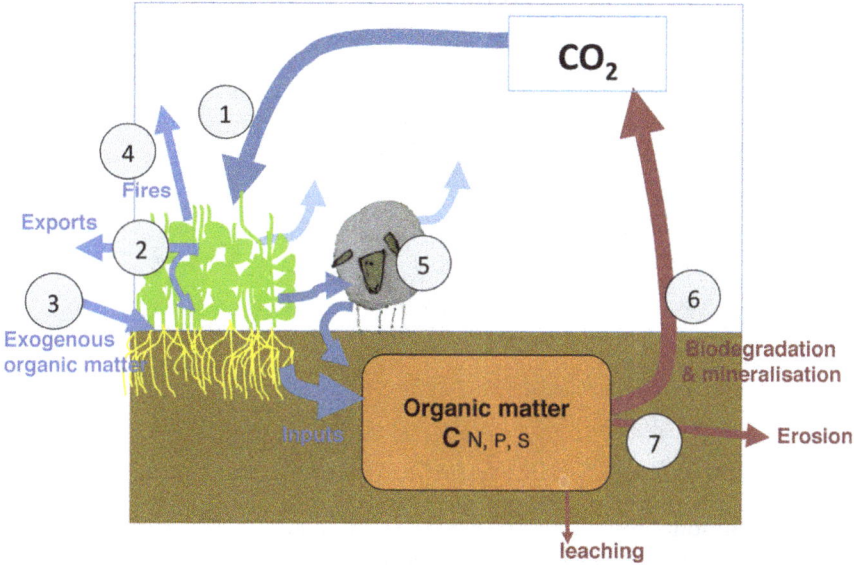

FIGURE 8.3   Management practices to uphold SOC stock [Reprinted with permission from Chenu et al. (2019) © Elsevier].

8.4.2   INCREASING PRIMARY PRODUCTIVITY

Although crop cultivation is known for destructing soil aggregation as well as depleting SOC concentration, cultivated crop species may undoubtedly play role in maintaining SOC and retaining the same for longer period. Crop rotations attribute a profound influence on annual carbon input and subsequently SOC buildup in soil. Mandal et al. (2007) observed that higher C addition through plant residues and root to soils under NPK or NPK+FYM treatment than control was due to the higher yield of the respective treatment over control (see Table 8.3). They also proved that even though the annual C input are similar (R–F–B: 3.17 Mg ha^{-1} yr^{-1}; R–W–F: 3.33 Mg ha^{-1} yr^{-1}), but the annual C build-up rate was significantly varied with the cropping systems. This suggests that both the quantity and chemical composition of the crop residue is vital for SOC buildup and subsequently their stabilization.

TABLE 8.3 Influence of Cropping Systems on Annual Carbon Input and Carbon Buildup in Soil

Cropping System	Annual Crop Residue Carbon Inputs (Mg C ha⁻¹yr⁻¹)			Carbon Buildup Over the Control (%)		Carbon Buildup Over Rate (Mg C ha⁻¹soil⁻¹yr⁻¹)	
	Treatment						
	Control	NPK	NPK+ FYM	NPK	NPK+ FYM	NPK	NPK+ FYM
Rice–Mustard–Sesame	1.88	2.76	3.75	51.8	55.7	1.91	2.05
Rice–Wheat–Fallow	1.82	3.33	3.97	16.8	23.4	0.27	0.37
Rice–Fallow–Berseem	2.45	3.17	4.16	9.3	24.7	0.13	0.36
Rice–Wheat–Jute	2.58	5.08	6.17	14.9	32.3	0.11	0.25
Rice–Fallow–Rice	2.58	3.56	4.30	33.5	54.8	0.28	0.45

FYM: farmyard manure.

Source: Modified from Mandal et al. (2007).

8.4.3 IMPROVED RESIDUE MANAGEMENT IN SOIL

Crop residue sometimes creates a grievous problem to the farmers; somehow, they are compelled to burn the excess crop residue. Several studies revealed that residue retention/incorporation in cereal systems significantly improve the carbon stock (Ghimire et al., 2017). Sometimes, due to lack of proper implementation, farmers fail to retain crop residue in their fields. Under this circumstance, crop residues could be utilized for organic mulching in orchard. Kumari et al. (2020) found that organic mulching with straw (10-cm depth) is beneficial for improving mango fruit yield, soil fertility, and carbon stock in soil (see Figure 8.4).

8.4.4 INTEGRATED NUTRIENT MANAGEMENT WITH ORGANIC AMENDMENTS

Crop fields are often subject to both balanced and imbalanced fertilization. Adoption of such practices may influence C sequestration in soils because of their effects on crop growth. Padbhushan et al. (2016) observed significant variation in SOC content with the application of mineral fertilization (NPK), integrated nutrient management practices (NPK+FYM), as well as sole organic amendment application (see Figure 8.5). Apart from that, different nutrient management practices significantly influence the soil aggregation, which ultimately attributed SOC stabilization. The analysis carried out by Han et al. (2016) confirmed that soil OM and C stocks were improved when organic amendments are used along with conventional fertilizers (see Figure 8.6). Das et al. (2019) revealed an apparent enrichment of SOC stock in soils upon balanced fertilization vis-à-vis imbalanced or no fertilization (see Table 8.4).

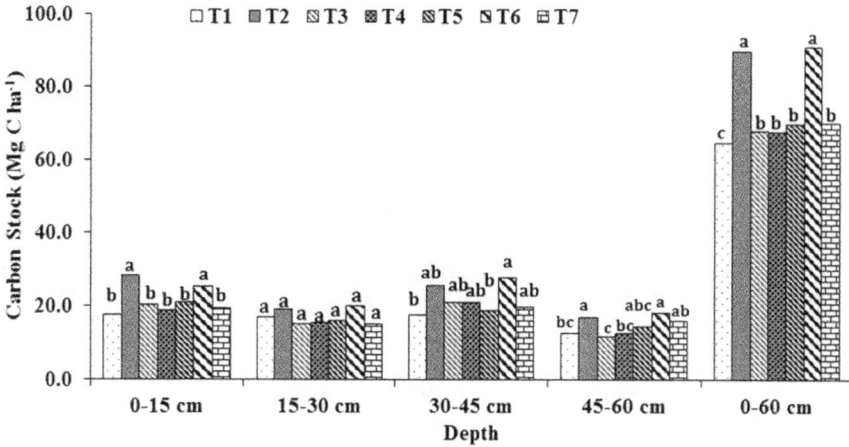

FIGURE 8.4   Effect of INM on carbon stock (Mg C ha⁻¹) of soils in mango orchard. Adapted with permission from Kumari et al. (2020). © John Wiley. T_1: Control (RDF-1000:500:500 g N:P_2O_5:K_2O); T_2: RDF + Organic mulching (10 cm thick); T_3: ½ RDF + 50 kg FYM enriched with *Trichoderma* (250 g); T_4: ½ RDF + 50 kg FYM + *Azospirillium* (250 g); T_5: ½ RDF + 50 kg FYM + *Azotobacter* (250 g); T_6: ½ RDF + 50 kg FYM + Vermicompost (5kg); T_7: ½ RDF + 50 kg FYM + *Pseudomonas fluorescence* (250 g). *Same letters above the bars indicate nonsignificance at* $p = 0.05$.

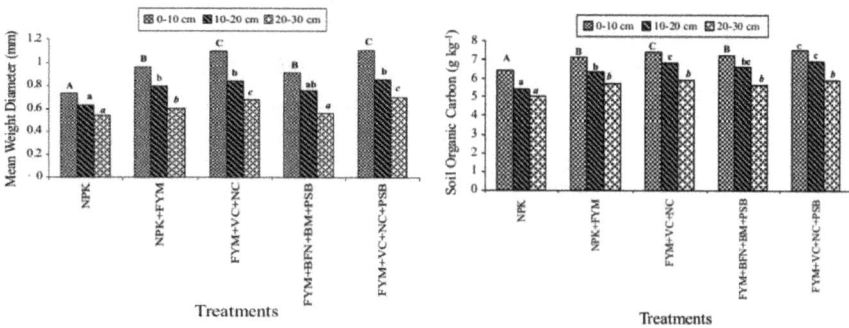

FIGURE 8.5   Impact of INM on mean weight diameter and SOC content in different soil depths. *Same letters above the bars indicate nonsignificance at* $p = 0.05$ Reprinted with permission from Padbhushan et al. (2016)]. © The International Society of Paddy and Water Environment Engineering and Springer Japan 2015

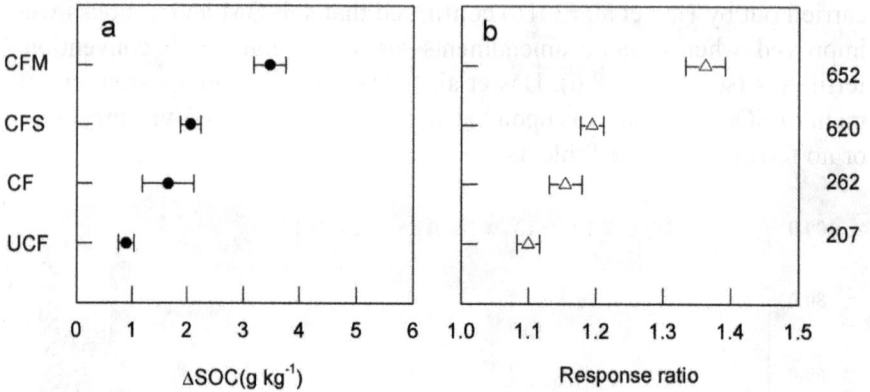

FIGURE 8.6 Difference in SOC (g kg^{-1}) and the relative change under different fertilization schedule. *UCF represents chemical fertilizers, CF represents balanced application of chemical fertilizers, S represents straw application,* and *M represents manure.* Reprinted from Han et al. (2016). http://creativecommons.org/licenses/by/4.0/.

8.5 EFFECT OF TILLAGE IN CARBON STOCK

Tillage can directly influence soil properties and ultimately crop yield. Tillage changes mass–volume relationship, soil water regime, soil temperature, aeration status, aggregation, and, hence, many biogeochemical cycles. Tillage practice mainly influences SOC stock through its aggregate turnover and subsequently heterotrophic respiration. Carbon locked within the aggregate becomes available to the native microbes and evolves CO_2 to the atmosphere that causes global warming. Zero tillage with partial or full residue retention can significantly retain more SOC than conventional tillage and thus possibly reduce CO_2 emission (Das et al., 2013) (see Table 8.5). Recent global meta-analysis also confirmed that no tillage increases SOC stock in surface soil layer (0–15 cm or 0–20 cm); however, low to nonsignificant effect was observed under lower layers (Chenu et al., 2019) (see Table 8.6).

8.6 CONCLUSION

All the above discussion indicates that SOC is crucial for sustainable agriculture. Thorough understanding about the stabilization mechanisms of different SOM pools is utmost important for improving SOC stock by adopting different management practices. Among them, tillage, crop diversity, fertilization, and organic amendments are imperative to SOC stock. However, some

TABLE 8.4 Impact of INM on Bulk Density (BD), SOC Concentration, and C Buildup in Soils (0–15 cm Depth) After 28 Years of Cultivation Under the Rice–Wheat Cropping System

Treatment	BD (Mg m⁻³)	SOC Concentration (g kg⁻¹)	% C Buildup Over Control After 28 Years of Cultivation	% C Buildup Over Initial	C Stock After 28 Year of Cultivation (Mg ha⁻¹ Soil)	C Sequestrated (Mg C ha⁻¹ Soil)	C Buildup Rate (Mg C ha⁻¹ Soil year⁻¹)
Initial	1.46	5.98			13.10		
Control	1.56^a (±0.019)	5.02^f (±0.06)	–	−16.09	11.74^e (±0.85)	−1.36	
50% NPK	1.54^a (±0.034)	5.56^e (±0.06)	10.84	−6.96	12.85^d (±0.66)	−0.25	−0.012
75% NPK	1.53^a (±0.029)	6.54^d (±0.08)	30.26	9.35	15.01^c (±1.24)	1.91	0.023
100% NPK	1.51^a (±0.024)	7.35^c (±0.08)	46.31	22.83	16.64^b (±0.95)	3.54	0.052
50% NPK + FYM*	1.41^b (±0.028)	9.93^a (±0.05)	97.85	66.09	21.01^a (±1.50)	7.91	0.144
50% NPK + Wheat Straw*	1.40^b (±0.036)	9.36^b (±0.07)	86.45	56.52	19.66^a (±0.95)	6.56	0.124
50% NPK + Green manure*	1.43^b (±0.022)	9.58^b (±0.06)	90.86	60.22	20.55^a (±0.62)	7.45	0.131

Values in the parentheses indicates standard deviation; "*" indicates organic amendments supplements 50% N only. Same letters after the values indicate nonsignificance at p = 0.05.

Source: Adapted from Das et al. (2019).

TABLE 8.5 Impact of Tillage and Crop Establishment Methods and Residue Management on Total SOC After Four Years of Cropping

Conservation Agricultural Practices	Total SOC Stock (Mg C ha⁻¹)		
	0–5 cm	5–15 cm	15–30 cm
Tillage and crop establishment			
Conventional tillage–flat sowing	5.91	10.92	17.99 (34.82)
Conventional tillage–raised-bed, sowing (fresh bed)	5.45	10.48	18.30 (34.23)
Zero tillage–flat sowing	6.88	12.12	17.52 (36.51)
Zero tillage–raised bed, sowing (permanent bed)	6.96	11.96	17.70 (36.61)
Residue management			
No residue	6.12	11.03	17.57 (34.73)
Cotton or maize, residue addition	6.17	11.32	17.76 (35.24)
Wheat residue addition	6.36	11.34	18.12 (35.79)
Cotton or maize + wheat residue addition	6.51	11.77	17.91 (36.21)

Data in parentheses indicate total SOC stock (on equivalent depth basis) in the 0–30 cm soil layer.

Source: Adapted from Das et al. (2013).

TABLE 8.6 Effect of Tillage on SOC Stocks Reported by Meta-Analyses

Work	Climate Zone	Number of Sites	Number of Pairs of Plots	Depth (cm)	Duration of Experiment (years)	SOC Stock NT-FIT (kg C m⁻²)	SOC Stock NT-FIT (kg C m⁻² y⁻¹)
West and Post (2002)	Any		93	0–22	≥5	0.62 ± 0.16	0.048
Angers and Eriksen-Hamel (2008)	Any	23	47	0–100	≥5	0.49	0.032
Virto et al. (2012)	Any	37	92	0–30	≥5	0.34	0.022
Haddaway et al. (2017)	Boreo-temperate	29		0–30	≥10	0.46 ± 0.19	
	Boreo-temperate	14		0–150		0.15 ± 0.34	
Meurer et al. (2018)	Boreo-temperate		46	0–30	≥10	0.42 ± 0.18	0.023
			11	0–60		0.15 ± 0.22	<0.01

NT: No tillage; FIT: Full inversion tillage or moldboard ploughing.

Source: Adapted from Chenu et al. (2019).

location-specific management practices such as liming in acid soil, grassland and pasture management, and management of coastal and saline soil are also needed. Hence, soil carbon sequestration/retention is a win–win situation, where soil health will improve, production input use efficiency will improve, and ultimately sustainable agriculture will come into force.

KEYWORDS

- **carbon stock**
- **carbon stabilization mechanism**
- **soil organic carbon**
- **sustainable agriculture**

REFERENCES

Angers, D. A.; Eriksen-Hamel, N. S. Full-inversion tillage and organic carbon distribution in soil profiles: A meta-analysis. *Soil Sci. Soc. Am. J.* 2008, *72*, 1370–1374.

Anonymous. Chemistry Operations. "Carbon". Los Alamos National Laboratory. (December 15, 2003), https://en.wikipedia.org/wiki/Carbon

Barron, P. F.; Wilson, M. A.; Stephens, J. F.; Cornell, B. A.; Tate, K. R. Cross polarization C-13 NMR-spectroscopy of whole soils. *Nature* 1980, *286*, 585–587.

Boussingault, J. B. Ann. Chim. Phys. (III), 1841, 1, 208. (Cited by E. J. Russell. *Soil Conditions and Plant Growth.* 8th ed., Longmans, Green and Co Publisher, 1953).

Broser, I.; Kallmann, H. P. Apha-tielschen, schnelle elektronen und gamma-quanten II. *Z. Naturforsch.* 1947, *2A*, 642–650.

Cerri, C.; Feller, C.; Balesdent, J.; Victoria, R.; Penecassagne, A. Application du tracage isotoique naturel de 13C à l'_etude de la dynamique de la matie`re organique dand les sols. *C. R. Acad. Sci. II* 1985, *9*, 423–428.

Chenu, C.; Angers, D. A.; Barré P.; Derrien, D.; Arrouays, D.; Balesdent, J. Increasing organic stocks in agricultural soils: Knowledge gaps and potential innovations. *Soil Till. Res.* 2019, *188*, 41–52.

Chung, H.; Grove, J. H.; Six, J. Indications for soil carbon saturation in a temperate agro-ecosystem. *Soil Sci. Soc. Am. J.* 2008, *72*, 1132–1139.

Chung, H.; Ngo, K. J., Plante; A.; Six, J. Evidence for carbon saturation in a highly structured and organic-matter-rich soil. *Soil Sci. Soc. Am. J.* 2010, *74*, 130–138.

Das, T. K.; Bhattacharyya, R.; Sharma, A. R.; Das, S.; Saad, A. A.; Pathak, H. Impacts of conservation agriculture on total soil organic carbon retention potential under an irrigated agro-ecosystem of the western Indo-Gangetic Plains. *Eur. J. Agronomy* 2013, *51*, 34– 42.

Das, A.; Rakshit, R.; Padbhushan, R.; Kohli, A.; Sushant, K. S. Effect of annual carbon input on soil carbon sequestration and sustainability under rice-wheat cropping system after the 28th crop cycle. *J. Soil Water Conserv.* 2019, *18*(3), 254–262.

Di, J. Y. et al. Influences of long-term organic and chemical fertilization on soil aggregation and associated organic carbon fractions in a red paddy soil. *Chin. J. Eco-Agr.* 2014, *22*, 1129–1138 (in Chinese with English abstract).

Du, Z. L.; Wu, W. L.; Zhang, Q. Z.; Guo, Y. B.; Meng, F. Q. Long-term manure amendments enhance soil aggregation and carbon saturation of stable pools in North China Plain. *J. Integr. Agr.* 2014, *13*, 2276–2285.

Duddigan, S.; Shaw, L. J.; Alexander, P. D.; Collins, C. D. A comparison of physical soil organic matter fractionation methods for amended soils. *Appl. Environ. Soil Sci.* 2019, *2019*, 3831241. https://doi.org/10.1155/2019/3831241

Fieve, A.; Solouki, T.; Marshall, A. G.; Cooper, W. T. High-resolution Fourier transform ion cyclotron resonance mass spectrometry of humic and fulvic acids by laser desorption/ ionization and electrospray ionization. *Energy Fuel* 1997, *11*, 554–560.

Ghimire, B.; Ghimire, R.; Dawn, V.; Mesbah A. Cover crop residue amount and quality effects on soil organic carbon mineralization. *Sustainability* 2017, *9*, 2316, doi: 10.3390/ su9122316.

Gonzalez-Vila, F. J.; Lentz, H.; L€udemann, H.-D. Fourier transform carbon-[13]NMR spectra of natural humic substances. *Biochem. Biophys. Res. Commun.* 1976, *72*, 1063–1070.

Gulde, S.; Chung, H.; Amelung, W.; Chang, C.; Six, J. Soil carbon saturation controls labile and stable carbon pool dynamics. *Soil Sci. Soc. Am. J.* 2008, *72*, 605–612.

Haddaway, N. R.; Hedlund, K., Jackson; L. E., Kätterer; T., Lugato, E.; Thomsen, I. K.; Jørgensen, H. B.; Isberg, P. E. How does tillage intensity affect soil organic carbon? A systematic review. *Environ. Evid.* 2017, *6*, 2–48.

Han, P; Zhang, W; Wang, G; Sun, W; Huang, Y. Changes in soil organic carbon in croplands subjected to fertilizer management: A global meta-analysis. *Nature Sci. Rep.* 2016, *6*, 27199, doi: 10.1038/ srep27199.

Hawkes, J. A.; Dittmar, T.; Patriarca, C.; Tranvik, L.; Bergquist, J. 2016. Evaluation of the orbitrap mass spectrometer for the molecular fingerprinting analysis of natural dissolved organic matter. *Anal. Chem.* 2016, *88*(15), 7698–7704.

Herrmann, A. M., et al. A novel method for the study of the biophysical interface in soils using nano-scale secondary ion mass spectrometry. *Rapid Commun. Mass Spectrom.* 2007, *21*, 29–34. https://doi.org/10.1002/rcm.2811

Knabner, I. K.; Rumpel C. Advances in molecular approaches for understanding soil organic matter composition, origin, and turnover: A historical overview. *Adv. Agron.* 2018, *149*, 1–48. https://doi.org/10.1016/bs.agron.2018.01.003

Kring, D. Composition of earth's continental crust as inferred from the compositions of impact melt sheets, In: Proc. 28th Conference of Lunar and Planetary Science XXVIII, Houston, TX, USA, March 17–21, 1997.

Krull, E. S.; Baldock, J. A.; Skjemstad, J. O. Importance of mechanisms and processes of the stabilization of soil organic matter for modeling carbon turnover. *Funct. Plant Biol.* 2003, *30*, 207–222.

Kumari, R.; Kundu, M.; Das, A.; Rakshit, R.; Sahay, S.; Sengupta, S.; Ahmad, M. F. Long-term integrated nutrient management improves carbon stock and fruit yield in a subtropical mango (*Mangifera indica* L.) orchard. *J. Soil Sci. Plant Nutrition* 2020, *20*, 725–737. https://doi.org/10.1007/s42729-019-00160-6

Lal, R. Agricultural activities and the global carbon cycle. *Nutrient Cycling Agroecosyst.* 2004, *70*, 103–116.

Lal, R. Managing soils and ecosystems for mitigating anthropogenic carbon emissions and advancing global food security. *Bio Sci.* 2010, *60*, 708–721.

Leis, A.; Lamb, R. N.; Gong, B.; Schneider, R. P. Evidence for the contribution of humic substances to conditioning films from natural waters. *Biofouling* 2000, *15*, 207–220.

Lichtfouse, E.; Elbisser, B.; Balesdent, J.; Mariotti, A.; Bardoux, G. Isotope and molecular evidence for direct input of maize leaf wax n-alkanes into crop soils. *Org. Geochem.* 1994, *22*, 349–351.

Lide David R. Abundance of elements in the earth's crust and in the sea, In: *CRC Handbook of Chemistry and Physics*, 97th ed. Boca Raton, FL, USA: CRC Press, 2003, pp. 14–17.

Mandal, B.; Majumder, B.; Bandyopadhyay, P. K.; Hazra, G. C.; Gangopadhyay A.; Samantaray R. N.; Mishra A. K.; Chaudhury, J.; Saha, M. N.; Kundu, S. The potential of cropping systems and soil amendments for carbon sequestration in soils under long-term experiments in subtropical India. *Global Change Biol.* 2007, *13*, 357–369.

Mayer, L. M. The inertness of being organic. *Marine Chem.* 2004, *92*, 135–140.

Meurer, K. H. E.; Haddaway, N. R.; Bolinder, M. A.; Kätterer, T. Tillage intensity affects total SOC stocks in boreo-temperate regions only in the topsoil—a systematic review using an ESM approach. *Earth Sci. Rev.* 2018, *177*, 613–622.

Minasny, B.; Malone, B. P.; McBratney, A. B.; Angers D. A.; Arrouays D.; Chambers A. Soil carbon 4 per mille. *Geoderma* 2017, *292*, 59–86.

Nagar, B. R. Examination of the structure of soil humic acids by pyrolysis—gas chromatography. *Nature* 1963, *199*, 1213–1214.

Padbhushan, R.; Rakshit, R.; Das, A.; Sharma, R. P. Effects of various organic amendments on organic carbon pools and water stable aggregates under a scented rice–potato–onion cropping system. *Paddy Water Environ.* 2016, *14*, 481–489.

Paustian, K.; Collier, S.; Baldock, J.; Burgess, R; Creque, J. Quantifying carbon for agricultural soil management: From the current status toward a global soil information system, *Carbon Manage.* 2019, *10*(6), 567–587, doi: 10.1080/17583004.2019.1633231.

Plaza, C.; Courtier-Murias, D.; Fern'andez, J. M., Polo, A.; Simpson, A. J. Physical, chemical, and biochemical mechanisms of soil organic matter stabilization under conservation tillage systems: A central role for microbes and microbial byproducts in C sequestration. *Soil Biol. Biochem.* 2013, *57*, 124–134.

Six, J.; Conant, R. T.; Paul, E. A.; Paustian, K. Stabilization mechanisms of soil organic matter: implications for C-saturation of soils, *Plant Soil* 2002, *241*(2), 155–176.

Sohi, S. P.; Mahieu, N.; Arah, J. R. M.; Powlson, D. S.; Madari, B.; Gaunt, J. L. A procedure for isolating soil organic matter fractions suitable for modeling. *Soil. Sci. Soc. Am. J.* 2001, *65*(4), 1121–1128.

Sollins, P.; Homann, P.; Caldwell, B. A. Stabilization and destabilization of soil organic matter: Mechanisms and controls. *Geoderma* 1996, *74*, 65–105.

Solomon, D. J.; Lehmann, J.; Kinyangi, J.; Liang, B.; Sch€afer, T. Carbon K-edge NEXAFS and FTIR-ATR spectroscopic investigation of organic carbon speciation in soils. *Soil Sci. Soc. Am. J.* 2005, *69*, 107–119.

Tamm, C.O.; Östlund, H.G. Radiocarbon dating of soil humus. *Nature* 1960, *185*, 706–707.

Virto, I.; Barré, P.; Burlot, A.; Chenu, C. Carbon input differences as the main factor explaining the variability in soil organic C storage in no-tilled compared to inversion tilled agrosystems. *Biogeochemistry* 2012, *108*, 17–26.

Wang, D.; Chakraborty, S.; Weindorf, D. C.; Li, B.; Sharma, A.; Paul, S.; Ali, M. N. Synthesized use of Vis NIR DRS and PXRF for soil characterization: Total carbon and total nitrogen. *Geoderma* 2015, *243–244*,157–167.

West, T. O.; Post, W. M. Soil organic carbon sequestration rates by tillage and crop rotation: A global data analysis. *Soil Sci. Soc. Am. J.* 2002, *66*, 1930–1946.

Wiesmeier, M.; Rico, H.; Rene, D.; Harald, M.; Peter, S. Estimation of past and recent carbon input by crops into agricultural soils of southeast Germany. *Eur. J. Agron.* 2014, *61*, 10–23.

Zimmermann, M.; Leifeld, J.; Schmidt, M. W. I.; Smith, P.; Fuhrer, J. Measured soil organic matter fractions can be related to pools in the RothC model. *Eur. J. Soil Sci.* 2007, *58*(3), 658–667.

Zomer, R. J.; Bossio, D. A.; Sommer, R.; Verchot, L. V. Global sequestration potential of increased organic carbon in cropland soils. *Sci. Rep.* 2017, *7*, 15554, doi: 10.1038/s41598-017-15794-8.

CHAPTER 9

Hydrothermal Sensitivity of Soil Organic Carbon Under Imminent Moisture and Temperature Stress

AVIJIT GHOSH[1*], ABIR DEY[2], RANJAN BHATTACHARYYA[2], M. C. MANNA[3], and S. K. CHAUDHARY[4]

[1]ICAR-Indian Grassland and Fodder Research Institute, Jhansi, Uttar Pradesh 284003, India

[2]ICAR-Indian Agricultural Research Institute, New Delhi, Delhi 110012, India

[3]ICAR-Indian Institute of Soil Science, Bhopal, Madhya Pradesh 462038, India

[4]Indian Council of Agricultural Research, New Delhi, Delhi 110012, India

[*]Correspondence author. E-mail: avijitghosh19892@gmail.com

ABSTRACT

Past findings found that the temperature response of the substratum of low quality is greater than that of the substratum of high quality. Since soils contain vast volumes of low-quality carbon, knowing their reaction to increasing temperatures can help determine how atmospheric CO_2 can react to climate change. Empirical tests, though, do not offer definitive data for collateral assessing the effect of certain influences, such as variability in moisture. In addition, various methods widely used to evaluate the temperature sensitivity of specific substrates will produce seemingly different and conflicting findings, even though they are focused on the same basic principles. As temperature change in future will be connected with modifications in the moisture content of soil that will also impact the decomposition process, an appropriate framework involving the absolute changes in respiration rates due to changes of the different factors would be very helpful. To tackle the problem, a systematic theoretical method for researching the vulnerability of respiration levels with respect to shifts in several decomposition drivers is being suggested.

9.1 INTRODUCTION

Soil organic carbon (SOC) sequestration is the new silver bullet to combat loss of soil fertility and soil erosion and improve sustainability of farming systems on one hand and help overcome challenges posed by climate change, global warming, and greenhouse gas emission on the other hand. Worldwide, scientists, researchers, and policymakers are realizing the importance of SOC sequestration and promoting the technologies that enhance C input, increase SOC stability, and minimize the losses of SOC. According to the fifth assessment report of the Intergovernmental Panel on Climate Change, the global mean temperature increased by 0.85 °C over the period 1880–2012 and is likely to increase by another ~0.3 °C by the end of 2035 (IPCC, 2014). Likewise, moisture stress is rampant nowadays, due to excessive industrial, agricultural, and domestic use leading to lowering of groundwater levels. Different agricultural crops grow in different soil moisture conditions. Therefore, the stability of sequestered C in soils should be studied under different hydrothermal conditions. The temperature and moisture regimes are the two most important conditions, which regulate growth and activity of soil microorganisms responsible for soil organic matter (SOM) mineralization. An increased ambient temperature within the acceptable range increases the rate of SOM mineralization, whereas optimum moisture is beneficial for these biochemical reactions. The changes in the C mineralization rate with different hydrothermal conditions are known as hydrothermal sensitivity. This chapter deals with the hydrothermal sensitivity of the soil C mineralization process and how it affects the stability of soil C under imminent temperature and moisture stress.

9.2 TEMPERATURE RESPONSE OF SOC MINERALIZATION AND CARBON SEQUESTRATION

Approximately, SOC consists up to 81% of total C, which are actively involved in the global C cycle. The effect of temperature on the global C cycle depends on two opposite processes, that is, vegetation growth and soil respiration, as modified by changes in the thermal regime. Higher temperature promotes both the processes, but studies suggest that C output from soil caused by higher soil respiration is greater than C input to soil mediated by higher vegetative growth (von Lützow and Kögel-Knabner, 2009). The SOM turnover models predict that greater temperature sensitivity of SOC might lead to considerable losses of soil C in the impending global warming scenario. Different pools of SOM respond differently to the changes in the thermal regime. The understanding of discrepancy in the nature of temperature sensitivity of different C pools has major consequences for the C turnover models and, thus, is a matter of highly topical debate nowadays. If labile pools are more sensitive to temperature rise, the feedback between climate change and soil C would be short-lived, as these labile SOC pools consist of very small fraction of global SOC reserve. The present understanding of the subject states that the nonlabile or stable SOC pools have higher values of temperature sensitivity. This scenario leads to a much dangerous ramification, where global warming can lead to a much higher C loss from the soil. The concept of temperature sensitivity of SOM decomposition emerged from two basic theories: the Arrhenius equation and the Michaelis–Menten kinetics (Davidson and Janssens, 2006). While the Arrhenius equation deals with the fact that stabilized SOM has higher temperature sensitivity due to higher activation energy, the Michaelis–Menten kinetics describes the stabilization of SOM through spatial inaccessibility for microbes and microbial enzymes.

9.3   RELATIVE AND ABSOLUTE TEMPERATURE SENSITIVITY

The activation energy is inversely proportional to its quality for the process of decomposition of organic matter; the higher the amount of energy needed to break down a substratum, the lower its quality (Ghosh et al., 2020; Wankhede et al., 2020). Based on this, temperature sensitivity can be treated as absolute and relative sensitivity. The absolute sensitivity ($\delta X/\delta T$) expresses the absolute change of the measure X for a given unit change in temperature, while the relative sensitivity [$(1/X)\, \delta X/\delta T$] expresses this change relative to the actual value of the measured X. Reaction rates of SOC decomposition (k) follow the Arrhenius equation as

$$k = A\exp\left(-\frac{E}{RT}\right) \qquad (9.1)$$

where A is the pre-exponential factor and E is the activation energy. Both are assumed independent of the temperature. R is the universal gas constant and T is the absolute temperature (K). From Equation (9.1), it can be assumed that a low-quality substratum needs large volumes of energy (E) to be deteriorated; thus, its decomposition rate (k) is sluggish. Activation energy is an indicator of substratum efficiency; the greater the activation energy, the lesser the substratum efficiency. As E increases, $k \to 0$.

In Equation (9.2), a formal description for measuring the temperature sensitivity of decomposition is provided as the partial derivative of the rate of decomposition in relation to temperature, indicating that as the temperature rises, the rate of decomposition will also rise for constant values of the activation energy

$$\frac{\partial k}{\partial T} = \frac{EA}{RT^2} \exp\left(\frac{-E}{RT}\right) = K\frac{E}{RT^2} \tag{9.2}$$

On the other hand, temperature sensitivity of decomposition rates is expressed in relative terms or logarithmic form as

$$\frac{\partial lnK}{\partial T} = \frac{EA}{RT^2} = \frac{1}{K}\frac{\partial K}{\partial T} \tag{9.3}$$

Absolute [see Equation (9.2)] and relative [see Equation (9.3)] sensitivities of decomposition rates produce different and apparently contradictory results. Comparing the limiting behavior of absolute and relative sensitivities as the quality of the substrate decreases, Equations (9.4) and (9.5) can be obtained as

$$\lim_{E \to \infty} \frac{\partial k}{\partial T} = 0 \tag{9.4}$$

$$\lim_{E \to \infty} \frac{1}{k}\frac{\partial k}{\partial T} = \infty \tag{9.5}$$

Such limits demonstrate that the absolute and relative sensitivities of the levels of decomposition act in opposite ways, with the absolute sensitivity falling steadily to zero and the relative sensitivity increasing linearly to infinity, while the substratum consistency decreases (Elev).

Another commonly used relative measure of temperature sensitivity is Q_{10} and is denoted as

$$Q_{10} = \frac{K_{T+10}}{K_T} \tag{9.6}$$

9.4   FACTORS GOVERNING TEMPERATURE SENSITIVITY

The Q_{10} mineralization is governed by both intrinsic and extrinsic factors. The quality of SOM is the most important intrinsic factor for controlling the activation energy, decay rate, and the temperature sensitivity (von Lützow and Kögel-Knabner, 2009; Dash et al., 2019). The enzyme kinetics and the Arrhenius equation suggest that the SOM components with higher activation energies mainly contribute to the temperature-mediated increase in SOM decay. Degrading biogeochemically recalcitrant organic matter (those needing higher activation energy to degrade) will usually be more prone to temperature variations than decomposing more labile SOM. The polyphenol content of SOM is positively linked with activation energy and vis-à-vis temperature sensitivity. The SOM having a slower decay rate is generally more sensitive to increased temperature. The semi-quinone types of functional groups, which are common in highly polymerized aromatic structures of humic acids, are often highly temperature sensitive. These chemically recalcitrant pools of SOC are the main contributors to the temperature sensitivity.

On the other hand, the microbial biomass carbon (MBC)/SOC ratio is often negatively correlated with Q_{10} (Ghosh et al., 2020). Chemical recalcitrance, physical defense, and biological accessibility are the determinants of C consistency (Ghosh et al., 2019). Activation energy of SOM decomposition is often higher in subsurface layers compared with surface layers, due to (1) chemical recalcitrance, (2) substrate inaccessibility, (3) limited microbial population, (4) restricted C supply, (5) confined air and water supplies, and (6) higher proportion of recalcitrant C and clay content (Yan et al., 2017). Therefore, the subsurface C storage faces great threat of loss in the impending global warming scenario, which are otherwise considered to be a stable pool of sequestered C. Soil pH is an important factor in influencing the MBC/SOC ratio and, thus, has a negative effect on Q_{10} (von Lützow and Kögel-Knabner, 2009). The largest global C stocks are found in these colder regions, which can topple the global C balance in the impending global warming scenario.

The environmental extremities play a crucial role in SOM decomposition. Deposition of water-repellent molecules in drought-prone areas often restricts the diffusion of enzyme or substrates through water films. Flooding slows oxygen diffusion to decomposition reaction sites, often allowing only anaerobic decomposition, which is comparatively slower enzymatic pathways, compared with aerobic decomposition processes. Freezing of extracellular soil water slows down the diffusion process of substrates and extracellular enzymes. In the near future, the plausible climate change could drastically affect these extremities and, in turn, create an imbalance in the current global C cycle. In the anaerobic peat soils, lower activity of phenol oxidase and hydrolase enzymes slows down SOM decomposition, which can be quickly reversed upon aeration of peatlands. Studies estimate that 400–500 Pg of C in peatlands all over the globe is in the danger of emission, if the anaerobic condition fails to prevail in future (Davidson and Janssens, 2006). Due to summertime drying of peatlands, there are already reports of faster C loss from peat soils and bogs of England and Wales. The desiccated peats are often prone to manmade or natural fires, which will further likely to add to C emission. On the other hand, drying of peatland also decreases methane emission, which may balance the excess CO_2 emission. The permafrost soils across the globe hold a large amount of deep soil C, along with C stored in surface layers, protected from the mineralization process. In a case of a global-warming-induced thawing process in the near future, permafrost area could be reduced by 25%, thus rendering $\sim$ 100 Pg C vulnerable to decay (Davidson and Janssens, 2006). The thawing of permafrost often creates mosaic of flooded areas interspersed within higher dry areas; flooded area facilitates methane emission, while dried uplands contribute to CO_2 emission. The stabilization of C in peatlands or permafrost is often different from stabilization in mineral soils through physical protection, chemical bonding, or biochemical recalcitrance.

9.5   HYDROTHERMAL SENSITIVITY—AN ADVANCEMENT TOWARD ENVIRONMENTAL SENSITIVITIES

Future temperature change will be connected with changes in soil moisture that will also impact the decomposition process. To attempt this, an appropriate framework involving the absolute changes in respiration rates due to changes of the different factors would be very helpful. The gradient of soil respiration (ΔR_s) is controlled by multiple factors and is defined as

$$R_S = f\left(X_1, X_2, X_3, \ldots, X_n\right) \tag{9.7}$$

$$\nabla R_S = \left[\frac{\partial R}{\partial X_1}, \frac{\partial R}{\partial X_2}, \frac{\partial R}{\partial X_3}, \ldots \frac{\partial R}{\partial X_n}\right] \tag{9.8}$$

For calculating the sensitivity of respiration considering simultaneous changes of multiple factors, the direction P is given by

$$\nabla R_S . \vec{P} = \left[\frac{\partial R}{\partial X_1} P_1 +, \frac{\partial R}{\partial X_2} p_2 +, \cdots \right]$$

$$= \sum_{i=1}^{n} \frac{\partial R}{\partial X_1} P_i \qquad (9.9)$$

where P is a unit vector.

If respiration is controlled by three variables, that is, activation energy E, temperature T, and water content W, then we may write

$$R_S = A \exp\left(-\frac{E}{RT} \right)\left(\frac{W}{w+E} \right) \qquad (9.10)$$

9.6 MANAGEMENT PRACTICES TO OPTIMIZE HYDROTHERMAL SENSITIVITY

The soil and crop management practices that promote stabilization of SOC often promote greater values of hydrothermal sensitivity. The micro-aggregate-associated C registers lower activation energy compared with macroaggregate-associated C by virtue of weaker clay–organic matter linkages (Ghosh et al., 2016). The SOM associated with microaggregates inside macroaggregates are the most stable fraction of physically protected C. On the other hand, more humified SOM tends to be more stabilized fractions (Wang et al., 2016). Greater contents of humic acids, abundance of complex aromatic structure, and oxidized functional groups, that is, quinone structures, render biochemical recalcitrance to SOM. Clay–mineral linkages often play crucial role in C stability. Noncrystalline Fe or Al oxides form stable clay–humus complexes by virtue of their huge surface area. Soil or crop management practices that promote stabilization of SOM are often related to higher hydrothermal sensitivity of SOM decomposition (Dash et al., 2019).

Nutrient management exerts great effect on hydrothermal sensitivity of SOM decomposition. Site-specific nutrient management (SSNM) and integrated nutrient management (INM) often promote greater crop biomass production as well as better soil health. Application of organic manures and soil amendments promote better aggregation and ensure greater supply of fresh organic matter, which forms intraparticulate organic matter (iPOM) and is physically protected from water, air, and microbial attack. Application of organic manure often results in solubilization of crystalized sesquioxide minerals and formation of noncrystalline minerals with higher binding capacity with SOM, by virtue of greater proportion of organic acids in soil. Contrarily, sole application of inorganic nitrogen, phosphorus, and potassium fertilizes does not promote Fe, Al, and Si oxides and causes a lower C stability. These C stabilization mechanisms might impart higher hydrothermal sensitivity to the SOM under INM by means of (1) faster desorption of C from the adsorption sites, (2) increased thermal sensitivity of recalcitrant carbon at higher temperatures, and (3) efficacy of biological decomposition processes at these temperatures (Ghosh et al., 2016). On the contrary, greater supply of organic matter under INM enhances the labile portion of SOC, by virtue of the continuous supply, apart from enhancing stability. The freshly formed C-rich macroaggregates contribute largely to labile soil C (Ghosh et al., 2019). Inclusion of organics also boosts the population and activity of soil microbes, facilitating soil respiration. Under organics application, substrate limitation is not there, but often quantity of SOM exceeds the capacity of microbes to digest. This leads to an enhancement of SOC lability, which, in turn, results into lesser humification of SOM, containing more aliphatic structures in humus. These contradictory processes control hydrothermal sensitivity of SOM under different nutrient management options. For an effective and sustainable C management protocol, enhancement of both labile and nonlabile pools is equally effective and, in turn, controls hydrothermal sensitivity of their decomposition. The SSNM protocol along with inclusion of organics seems to be the way to achieve optimum hydrothermal sensitivity of SOM decomposition.

The zero tillage (ZT), reduced tillage (RT), or permanent bed planting (PB) practices, which work on the principle of minimum mechanical disturbances, significantly improve stability of soil aggregates, especially that of macroaggregates. The mean residence time of macroaggregates increases, depicting a shift from proportion of C-poor microaggregates to C-rich macroaggregates under ZT. Microbial respiration inside macroaggregates breaks coarse iPOM and leads to formation of fine iPOM, around which stable microaggregates form within macroaggregates, imparting greater physical protection to SOC (Six et al., 2004). The physical protection of SOC in aggregates, along with smaller pore sizes under ZT, restricts substrate availability to microbes, thus canceling the temperature-mediated increment of SOM decomposition (according to the Michaelis–Menten kinetics). The conservation tillage practices register higher activation energy of SOM. Higher values of the Michaelis–Menten constant (K_m) under ZT/RT/PB conditions combined with lower substrate availability make these conservation tillage practices essentially an in situ temperature-insensitive process that alters the release of easily decomposable substrates and thereby reduces the decomposition rate (Sandeep et al., 2016). Substrate limitations in the form of diminishing marginal return (i.e., a smaller increase in catalysis of SOC by extracellular enzymes than the previous increment) under conservation tillage can create breaks in the positive feedback of microbial-mediated depolymerization, thereby changing a forward Michaelis–Menten model to a reverse mode. On the other hand, ZT/RT practices do not favor oxidative polymerization of added fresh organic matter, in turn inhibiting formation of lower semiquinone-type free radical concentration and lower percentage of aromatic C, which represents a less advanced stage of humification (Zhang et al., 2011). The semi-quinone-type free radical, which might have contributed to temperature-sensitive less labile pool of SOC, is comparatively less in proportion under ZT compared with the conventionally tilled plots (Jat et al., 2019). Therefore, ZT practices might be beneficial for enhancing the stability of SOC through physical protection, not chemical recalcitrance, which grants lower temperature sensitivity. This ensures the stability of sequestered C in the impending global warming scenario (Parihar et al., 2019).

The type of organic matter that enters the soil system guides the recalcitrance or quality of SOC. Especially, under conservation agriculture, the biochemical make-up of the crop residue determines the decomposition rate and, in turn, the lability/stability of the SOM. Addition of cereal residues with high C:N ratio and high lignin contents often contributes to chemical recalcitrance of SOM. The lignin-type biomolecules impart recalcitrance by virtue of an extremely long half-life (364.5 days). Thus, SOM rich in lignin is stable in the current temperature scenario but susceptible to loss under elevated temperatures (Jat et al., 2019). On the other hand, green manuring crops contribute to residues (both in-season leaf fall and after harvest residues) with low C:N ratio, which mainly contributes to more labile pools of SOM. Farmyard manure, by virtue of its higher lignin and phenol content, is more recalcitrant in nature compared with green manures/crop residues. Decomposition of crop residues with high C:N ratio proceeds through a two-step process: (1) decomposition of water-soluble forms during the first phase and (2) decomposition of recalcitrant and structural carbon components during the second phase. These crop residues could release sufficient carbon from its recalcitrant/structural components and stimulate higher microbial activity at higher temperatures, in turn registering a higher Q_{10} value than manure-treated plots (Sandeep et al., 2016).

9.7 CONCLUSION

Temperature and humidity are the two most influential variables that control SOC conditions, independent of land uses, management methods, landscape, and forms of soil. Both temperature and humidity are closely associated with SOC stocks, percentages, and CO_2 efflux. Q_{10} mainly depends on the initial substrates factors, such as supplies of C, chemical structure, microbial culture, nutrient management practice, tillage operation, etc. INM can regulate Q_{10} to minimize C loss; similarly, conservation agriculture reduces temperature sensitivity of C mineralization. However, soils of lower layers are more temperature sensitive than that of the top layer. Q_{10} values have large variations among different ecosystems. Thus, to reliably forecast the reaction of SOC cycling to global warming through climate–carbon cycle models, variations in ecosystem types should be addressed. The more sensitive response to heating of recalcitrant SOC pools than the reactive SOC pools supports the CQT hypothesis and shows that a minor change in Q_{10} of reactivating SOC values has profound implications. Future research should combine moisture as a variable along with temperature to predict C loss pattern more accurately. The equations proposed here need to be validated and checked.

KEYWORDS

- decomposition
- low-quality CO_2
- temperature
- sensitivity
- vulnerability

REFERENCES

Dash, P.K.; Bhattacharyya, P.; Roy, K.S.; Neogi, S.; Nayak, A.K. Environmental constraints' sensitivity of soil organic carbon decomposition to temperature, management practices and climate change. *Ecol. Indic.* 2019, *107*, 105644.

Davidson, E.A.; Janssens, I.A. Temperature sensitivity of soil carbon decomposition and feedbacks to climate change. *Nat. Rev.* 2006, *440*, 165–173.

Ghosh, A.; Bhattacharyya, R.; Dey, A.; Dwivedi, B.S.; Meena, M.C.; Manna, M.C.; Agnihortri, R. Long-term fertilisation impact on temperature sensitivity of aggregate associated soil organic carbon in a sub-tropical inceptisol. *Soil Tillage Res.* 2019, *195*, 104369.

Ghosh, A.; Bhattacharyya, R.; Dwivedi, B.S.; Meena, M.C.; Agarwal, B.K.; Mahapatra, P.; Shahi, D.K.; Salwani, R.; Agnihorti, R. Temperature sensitivity of soil organic carbon decomposition as affected by long-term fertilization under a soybean based cropping system in a sub-tropical Alfisol. *Agr. Ecosyst. Environ.* 2016, *233*, 202–213.

Ghosh, A.; Das, A.; Das, D.; Ray, P.; Bhattacharyya, R.; Biswas, D.R.; Biswas, S.S. Contrasting land use systems and soil organic matter quality and temperature sensitivity in North Eastern India. *Soil Tillage Res.* 2020, *199*, 104573.

IPCC. Climate Change: Synthesis Report. Contribution of Working Groups I, II and III to the Fifth Assessment Report of the Intergovernmental Panel on Climate Change [Core Writing Team, R.K. Pachauri and L.A. Meyer (eds.)]. IPCC, Geneva, Switzerland, 2014.

Jat, S.L.; Parihar, C.M.; Dey, A.; Nayak, H.S.; Ghosh, A.; Parihar, N.; Goswami, A.K.; Singh, A.K. Dynamics and temperature sensitivity of soil organic carbon mineralization under medium-term conservation agriculture as affected by residue and nitrogen management options. *Soil Tillage Res.* 2019, *190*, 175–185.

Parihar, C.M.; Singh, A.K.; Jat, S.L.; Ghosh, A.; Dey, A.; Nayak, H.S.; Parihar, M.D.; Mahala, D.M.; Yadav, R.K.; Rai, V.; Satayanaryana, T.; Jat, M.L. Dependence of temperature sensitivity of soil organic carbon decomposition on nutrient management options under conservation agriculture in a sub-tropical Inceptisol. *Soil Tillage Res.* 2019, *190*, 50–60.

Sandeep, S.; Manjaiah, K.M.; Mayadevi, M.R.; Singh, A.K. Monitoring temperature sensitivity of soil organic carbon decomposition under maize–wheat cropping systems in semi-arid India. *Environ. Monit. Assess.* 2016, *188*, 1–15.

Six, J.; Bossuyt, H.; Degryze, S.; Denef, K. A history of research on the link between (micro) aggregates, soil biota, and soil organic matter dynamics? *Soil Tillage Res.* 2004, *79*, 7–31.

von Lützow, M.; Kögel-Knabner, I. Temperature sensitivity of soil organic matter decomposition—what do we know? *Biol. Fertil. Soils* 2009, *46*, 1–15.

Wang, Y.; Gao, S.; Li, C., Zhang, J.; Wang, L. Effects of temperature on soil organic carbon fractions contents, aggregate stability and structural characteristics of humic substances in a Mollisol. *J. Soils Sediments* 2016, *16*, 1849–1857.

Wankhede, M., Ghosh, A., Manna, M.C., Misra, S., Sirothia, P., Rahman, M.M., Bhattacharyya, P., Singh, M., Bhattacharyya, R. and Patra, A.K., Does soil organic carbon quality or quantity govern relative temperature sensitivity in soil aggregates? *Biogeochemistry* 2020, *148*, 191–206. https://doi.org/10.1007/s10533-020-00653-y

Yan, D.; Li, J.; Pei, J.; Cui, J.; Nie, M.; Fang, C. The temperature sensitivity of soil organic carbon decomposition is greater in subsoil than in topsoil during laboratory incubation. *Sci. Rep.* 2017, *7*, 5181.

Zhang, J.; Hu, F.; Li, H.; Gao, Q.; Song, X.; Ke, X.; Wang, L. Effects of earthworm activity on humus composition and humic acid characteristics of soil in a maize residue amended rice–wheat rotation agroecosystem. *Appl. Soil Ecol.* 2011, *51*, 1–8.

PART IV
Smart Tools for Monitoring Soils

CHAPTER 10

Application of Remote Sensing Technology for Estimation of Soil Moisture

KOUSHIK BANERJEE[1] and BAPPA DAS[2*]

[1]ICAR-Mahatma Gandhi Integrated Farming Research Institute, East Chamaparan, Bihar 845401, India

[2]ICAR-Central Coastal Agricultural Research Institute, Old Goa, Goa 403402, India

*Corresponding author. E-mail: bappa.iari.1989@gmail.com; bappa.das@icar.gov.in

ABSTRACT

Surface soil moisture is one of the integral parts in hydrological cycle, which affects the conversation of water and energy fluxes at the land–atmosphere boundary. Precise estimation of soil moisture along with assessing the spatial and temporal variations is crucial for several ecological studies. Modern scientific developments in a spaceborne remote sensing platform have revealed that soil moisture can be estimated by multiple remote sensing platforms such as ground-based, unmanned-aerial-vehicle-based, and space-based, from higher altitude. This study presents a comprehensive understanding of theoretical principles of different methods available for measuring soil moisture using optical, infrared/thermal infrared, and micro-wave remote sensing. Additionally, different satellite missions such as Soil Moisture Active Passive, Soil Moisture and Ocean Salinity, and Advanced Microwave Scanning Radiometer, which incorporate both land-based and meteorological data, are discussed along with their different application prospects.

10.1 INTRODUCTION

Near-surface soil moisture is highly variable spatiotemporally and is a chief variable for hydrological and climatological research and is playing a fundamental role in various physical processes, which take place at the land–atmosphere boundary, such as, infiltration of water, surface runoff, soil evaporation, exchange of heat and gas, and soil erosion (Amani et al., 2017). Soil hydraulic properties (SHPs) are essential for studying soil water content and movement in agricultural fields. Direct measurements of SHPs at large scales are expensive, tedious, destructive, and estimation technique dependent (Demattê et al., 2010; Xu and Wang, 2015). So, pedotransfer functions (PTFs), which are relations between SHPs for easily measurable soil properties (e.g., soil texture, bulk density, and organic matter content), were generally used to estimate SHPs (Vereecken et al., 2010; Babaeian et al., 2015a). Regardless of the fact that PTFs utilize easy-to-measure soil properties, these predictors are often not available at required spatial resolution or their uncertainty may be too large to give precise predictions of SHPs. On the contrary, the same can be achieved at different regional and national scales through remote sensing techniques, which are noninvasive as well as time and labor proficient (Viscarra Rossel et al., 2006; Demattê et al., 2010). Broadly, soil moisture estimation through the remote sensing technique is classified into three broad groups: optical region (based on the reflectance value), thermal region (based on temperature and emissivity), and microwave (MW) region (based on emission and scattering value) (Ghahremanloo et al., 2019). Due to improvements in both spectral and spatial resolutions, remote sensing techniques offer an imperative means for mapping SHPs. The fundamental notion for quantitative determination of SHPs via the remote sensing platform is the interlinkage among soil hydraulic parameters and soil surface reflectance at visible- and near-infrared (VNIR), mid-infrared (MIR), thermal-infrared (TIR), and MW region.

10.2 GROUND-BASED SOIL MOISTURE RETRIEVAL

The main principle for soil moisture estimation from remote sensing is to develop relationship between soil water content and surface reflectance or emittance or absorption. The relationship can be established using different empirical statistical or process-based models. The empirical statistical models are known as spectral transfer functions (STFs) or spectral pedotransfer functions (SPTFs). STFs only contain remote-sensing-based spectral data, while SPTFs involve other easily measurable soil properties such as soil electrical conductivity (EC), pH, soil organic carbon, soil texture, and soil aggregation, which affect soil moisture content (SMC) with spectral data. Surface spectral reflectance information in the 350–2500-nm wavelength region along with simulated Landsat-ETM+ wavelengths is used for estimating Genuchten–Mualem SHPs (α, n, and K_s). By following this method, parameter n is estimated efficiently, but difficulty is seen for other SHPs such as α and K_s. However, SHPs are estimated from an inadequate amount of soil water content data, which pose the main drawback of the above method (Santra et al., 2009). Similarly, Atterberg limits are also considered good parameters for estimating soil water. These Atterberg limits were assessed effectively using partial least squares regression with feature selection ($PLSR_{FS}$) with VNIR and MIR data delivered the best outcome with R^2 of 0.77 (Gupta et al., 2016). Sarathjith et al. (2014) used diffuse reflectance spectroscopy with data mining algorithms to estimate aggregate size distribution. Santra et al. (2015) assessed soil properties of hot arid western Rajasthan, India using reflectance spectroscopy in VNIR and short-wave infrared (SWIR) regions (400–2500 nm). They reported that sand, clay, and organic carbon content can satisfactorily be estimated but not soil pH and EC. Srivastava et al. (2016) tried to characterize salt-affected soil in the Indo-Gangetic plains of Haryana, India using VNIR spectroscopy. Their hyperspectral-based models were capable to capture >75% of the variability of salinity-related parameters. Mitran et al. (2015) used both Hyperion data and ground reflectance data for retrieval of physicochemical properties of salt-affected soils. Their study revealed that EC, exchangeable sodium percentage, cation exchange capacity, and Mg^{++} can be estimated accurately using partial least squares regression (PLSR) models with feature selection. Sarathjith et al. (2016) used five data mining techniques, namely, PLSR, support vector regression (SVR), discrete wavelet transformation (DWT), DWT-PLSR, and DWT-SVR, with diffuse reflectance spectroscopy for prediction of soil nutrient contents. They have recommended DWT-SVR as the best technique for estimating soil nutrients. Soil moisture can also be measured using sensor techniques as designed by Yin et al. (2013). Their sensor works based on light-emitting diode reflectance data at two wavelengths, namely, 1940 nm (strong water absorption band) and 1800 nm (weak water absorption band). Their study revealed good performance of the developed sensor both for laboratory and field moisture measurements. Similarly, an inter-relationship exists between surface SMC and soil surface evaporation rate. Using this property and band depth analysis at water absorption features, it was seen that 1.44 and 1.93 μm were more sensitive than 0.97 and 1.16 μm for detecting and estimating soil water content change (Tian and Philpot 2015). Soil water content can also be measured using normalized difference soil moisture index, which takes the spectral information at 3.5–2.5 μm region (Haubrock et al., 2008). Oltra-Carrió et al. (2015) tried to retrieve SMC using four local and global spectral indices in VIS to SWIR region (0.4–2.5 μm) both for laboratory and field experiments. Results showed very good to excellent performance of the indices with R^2 greater than 0.75 for laboratory experiments, while it good to excellent for field experiments with R^2 greater than 0.65. Babaeian et al. (2015b) used PTFs, STFs, and SPTFs to predict SMC and van Genuchten and Brooks–Corey hydraulic parameters. Their study revealed accurate performance of point STFs and SPTFs for low to intermediate soil water contents. For water content close to saturation, point PTFs performed better compared to STFs and SPTFs, while the performance of parametric PTFs, STFs, and SPTFs was comparable in estimating the soil–water retention curve. Xu and Wang (2015) tried to retrieve SMC of saline soils using TIR spectra and PLSR. They recommended the PLSR model using the first-order derivative spectrum as it has provided the best result for SMC estimation. However, measuring SMC from the ground-based station is mostly meager and frequently not adequate to show the spatial unevenness and changeability of soil water at bigger spatial scales, compulsory for above applications.

10.3   SATELLITE-BASED SOIL MOISTURE RETRIEVAL

10.3.1   MICROWAVE

Soil moisture estimation and observations over a larger spatial extent can be done using the MW remote sensing technique from the spaceborne platform (Wagner et al., 2003). Different low-frequency MW sensors (both passive and active) using C band, X band, Ku band, and L band have been in use since past few decades in retrieving near-surface soil moisture (Owe et al., 2008; Kerr et al., 2010; Kumar et al., 2018). Soil moisture measurement using MW sensors is based on the principles that the emission from the land surface is a function of land surface temperature (LST), soil surface roughness, vegetation/crop coverage extent, and SMC. Within the MW range, the sensitivity to soil moisture is most significant at higher wavelength, that is, low frequencies (~10–1 Ghz), and is, thus, considered in typical range of satellite remote sensing. MW remote sensors are not unaffected by the cloud shield and nighttime darkness; thus, MW observations are time independent (all-weather capability in remote sensing) (Jackson et al., 1996). Further-more, in the longer MW region (L band; 1–2 GHz), which decreases the effect of vegetation in attenuating MW signal, soil moisture estimation in a much deeper layer (beyond the surface) can be done (Jackson et al., 1982). The active MW sensors (RADAR) give measurements at higher spatial reso-lutions although influenced by native geography, surface unevenness, and crop/vegetation coverage than the passive sensors (Lakshmi, 2013; Kumar et al., 2018). On the other hand, due to poor spatial resolution (~25–50km), passive MW observations are more affected by spatial diversities and scaling effects. Along with this, the existence of snow cover, ice-covered soil, and occurrence of precipitation also restricts the ability of the soil moisture measurements (Parinussa et al., 2011). However, satellite missions, such as soil moisture active passive (SMAP) and soil moisture and ocean salinity (SMOS), with L band passive MW frequency, have revealed encouraging result for mapping close-surface (0–5 cm) soil water content worldwide with a spatial and temporal resolution of 2500–4000 m and two to three days, respectively (Mohanty et al., 2017). Moreover, the close-surface soil water is stretched to the crop root zone depth (up to 100 cm) utilizing process-based methodologies and data integration techniques. Owing to the variations in the spatiotemporal coverage of active and passive MW sensors and because of scarce obtainability of consistent observed ground measure-ments, reliable assessment of the soil water dataset from spaceborne remote sensing is problematic. To address these issues, numerous studies have been conducted, such as land surface model climatology (Kumar et al., 2015), triple collocation approaches (Dorigo et al., 2010), and spectral fitting (Su et al., 2014), to evaluate the comparative superiority of global soil moisture measurements. However, all these methods are based on assumptions of linearity (observed soil moisture versus true soil moisture), signal error, and error orthogonality in the component dataset (Gruber et al., 2016). To enumerate these errors, theoretic information and autoregressive investiga-tion of the time-series dataset from a huge number of current space-based soil moisture missions were used and evaluated the soil moisture reclama-tion products from AMSRE, ASCAT, SMOS, AMSR-2, and SMAP missions using information-theory-based metrics (Kumar et al., 2018). However, SMAP-derived soil moisture retrieval showed much lower errors (similar to in situ observations) along with high information content, higher levels complexity, and lower entropy mainly over moderately grown vegetation areas than AMSRE, ASCAT, SMOS, and AMSR2. Recently, the European Space Agency's (ESA) operational satellite mission has established a new prototype in spaceborne remote sensing applications. ESA's Sentinel-1 radar image data have successfully retrieved surface SMC with high spatiotemporal resolution. Combination of Sentinel-1 and Sentinel-2 SAR image data (100 m resolution) can also be used in retrieving soil moisture from the spaceborne platform. For achieving this, two different algorithms based on the understanding of Sentinel-1 data obtained in the vertical-to-vertical (VV) polarization have been used (Gao et al., 2017).

To understand the influence of vegetation in estimating soil moisture, these polarized data are further united with optical data obtained from Sentinel-2/Landsat-8. In the first algorithm, coarse spatial resolution scatterometer data are used, which are based on change detection. The second algorithm, which is built on the change among backscattered Sentinel-1 radar data taken on two successive days, is defined as a function of normalized difference vegetation index (NDVI). This provided the volumetric moisture estimation to a root-mean-square error (RMSE) of nearly 0.087 and 0.059 m^3/m^3 for the first and second algorithms, respectively. Addition of RADAR data from Sentinel-1 satellite in the SMAP mission can improve surface soil moisture (SSM) approximation more precisely than the root zone depth both spatially and temporally (Lievens et al., 2017). Incorporation of Sentinel-1 observations in the SMAP mission has the harmonizing impact of radar and radiometer observations. Another recently developed methodological approach for retrieving SSM is modified water cloud model (WCM) based on the synergy of Sentinel-1 (SAR) and Landsat-8 (OLI) data in a partially vegetated covered area (Bao et al., 2018). To eradicate the influence of vegetation coverage on SSM estimation, a model was built to quantify vegetation water content using Landsat-OLI-based spectral index. Later, the quantified vegetation water content model was subtracted from the original WCM to develop the modified WCM with a spectral index. In addition, an SSM assessment prototype is constructed based on the improved WCM. The method was verified at two study areas (UK and Spain), where soil moisture data were obtained from observational networks at the ground level (see Figure 10.1).

10.3.2   *SOIL MOISTURE ESTIMATION USING OPTICAL AND THERMAL BANDS*

In thermal methods (TIR band—3.5–14 μm), soil moisture can be estimated using the principle that the LST is much responsive to soil surface moisture content (Mobasheri and Amani, 2016). Henceforth, a relationship may be

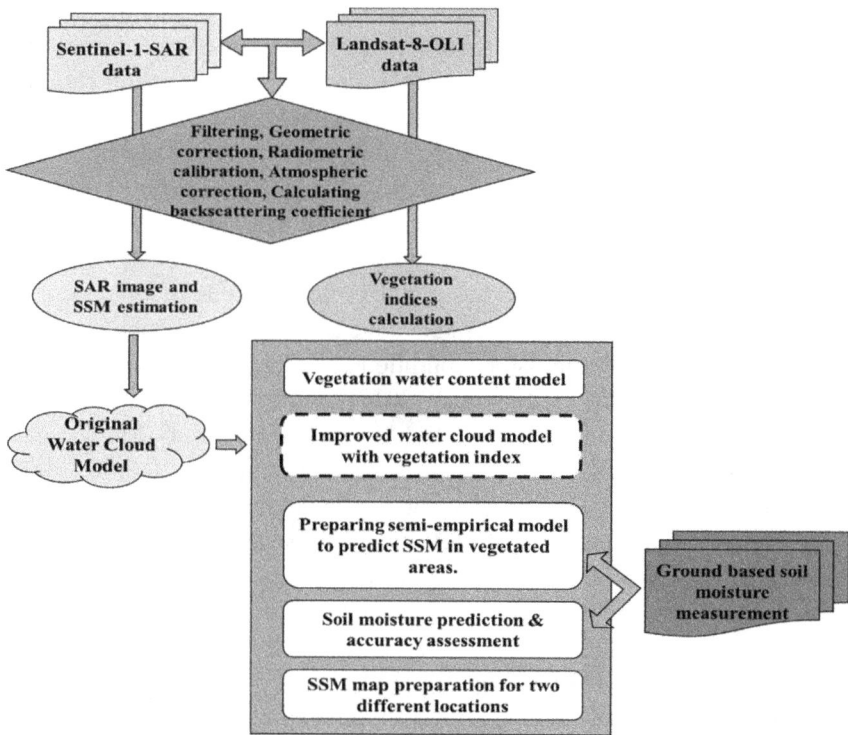

FIGURE 10.1 Flowchart for satellite-based soil moisture estimation [Modified and adapted from Bao et al. (2018)].

developed between soil moisture value and amplitude of soil temperature change values. Using LST product from Meteosat-SEVRI (Spinning Enhanced Visible and Infrared Imager) satellite, a new soil moisture index, called the temperature rising rate vegetation dryness index (TRRVDI), was developed by Zhang et al. (2014). It was found that the ambiguity in soil moisture estimation has been decreased using the TRRVDI index, which considers the temporal LST variations. However, the availability of supplementary data for calculating the theoretical dry edge was the reported limitation of using TRRVDI for estimating soil moisture. To characterize the impact of soil water stress response on wheat crop, Banerjee et al. (2019) has developed a new index called normalized sunlit and shaded index (NSSI) using thermal and visible images under the field condition. Different crop biophysical parameters measured in the study gave significant and higher relationship with the thermal image-based NSSI than the visible image-based NSSI to signify the effect of imposed soil water stress on the wheat growth and development. An elliptical association among diurnal LST cycles and net surface shortwave radiation was found for modeling bare soil surface moisture content (Leng et al., 2014). Additionally, an amalgamation of optical and thermal infrared data has been done to propose an algorithm for estimating soil moisture (Leng et al., 2016). This method was assumed more proficient to estimate the SSM than the commonly available methods, in which the model coefficients parameters are acquired by substantial replicated datasets. The study showed support vector machine method was the best for estimating LAI for the thermal image as compare to visible image of wheat crop grown under moisture conditions. To delineate the soil moisture stress or drought conditions, Son et al. (2019) used a multitemporal Landsat-MODIS data fusion technique in Central America. In this study, MODIS data products, namely, MOD09A1 (500 m, wavelength ranging from visible to SWIR region) and MOD11A2 (1-km eight-day composite product) were first reconstructed using random forest technique. These reconstructed LST data were then fused with the Landsat-8 (optical, SWIR, and thermal bands) data using STARFM tool (ledaps.nascom.nasa.gov) to produce original datasets with two wavelengths (red and NIR) and LST data from both Landsat (30-m resolution) and MODIS (eight-day temporal resolution) satellites. From this, an index was developed called temperature vegetation dryness index (TVDI) to delineate drought conditions. TVDI result obtained by this was validated with the ground measured data, which gave good R^2 values (0.75–0.85). Calibration of radar data either empirically or physically for retrieving and monitoring SSM over the large geological area under different conditions is still in ambiguity mainly at high spatial resolution. An advanced synergistic method for calibrating soil moisture was developed by merging Sentinel-1 (S1) MW and Landsat-7/8 (L7/8) thermal band at high resolution (Amazirh et al., 2018). The method consists of initial normalization of the Radar-S1 backscatter coefficient values (for 2015–2016 season); subsequently, the normalized coefficient parameter value was standardized from reference points with the thermal band-based soil moisture proxy, that is, soil evaporative efficiency (SEE). The maximum and minimum values of SEE were reached under water-saturated and dry soil conditions, respectively. The projected methodology was assessed with the in situ soil moisture measurement values obtained from three bare fields (semi-arid region, Morocco), and they evaluated it with a conventional method using radar data only. The result showed that the VV polarization was better correlated with the soil water data than the vertical to horizontal (VH) with R^2 values of 0.47 and

0.28, correspondingly. However, after combining Landsat 7/8 thermal data with the Radar-S1 (VV) backscattering data, root-mean-square difference between satellite and ground-based soil moisture data was reduced to 0.03 $m^3 m^{-3}$ compared to 0.16 $m^3 m^{-3}$ using S1-VV only.

10.3.3 SOIL MOISTURE ESTIMATION USING SATELLITE AND METEOROLOGICAL DATA

SSM can partition turbulent energy into sensible and latent heat at the interface between land and atmosphere. Similarly, soil moisture also assigns precipitation into infiltration that enters into the soil and surface runoff, which is considered important in maintaining climate and water cycle in terrestrial ecosystems (Wang et al., 2016).

During 2010, the ESA has started estimating the soil moisture through climate change initiative (CCI) project, which has been committed to generating the comprehensive and most reliable soil moisture data product globally using MW sensors including active and passive (Dorigo et al., 2010). The current form of CCI products (v03.2) covering a time span of 37 years (1978–2015) can considerably assist in drought/soil moisture stress monitoring, precipitation, and evapotranspiration (ET) study over an extensive area or most likely at the comprehensive scale (Ghulam et al., 2007).

Though MW emissions have the capability to penetrate clouds layers and raindrops and deliver incessant soil moisture data products, there are few concerns tangled in the presently existing MW derived soil moisture data products for producing all-weather soil moisture data products. Zhang et al. (2017) reported that the MW signals probably interfered by radio frequency interference, particularly in China. These impure image pixels will be mislaid for producing soil moisture data products. Xiao et al. (2016) reported the limitations of using algorithms during soil moisture products preparation in a number of definite circumstances such as the nonintegration of repetitive algorithms and circumstances, where the crop or vegetation moisture amount is beyond assured value. Data gaps will affect significantly in retrieving soil moisture and consequently converting it problematic for producing all-weather soil moisture products from MW remote sensing. For obtaining spatially comprehensive satellite-derived soil moisture products, numerous gap-filling approaches have been established using remotely sensed variables such as NDVI and LAI with more or less clear physical implication (Liu et al., 2017). Soil moisture estimation by combining both meteorological and satellite data was initially done by Moran et al. (1994). They suggested a new theory called vegetation index/temperature (VIT) trapezoid scheme, which is a modification of formerly developed concepts of crop water stress index, to get SMC for a moderately vegetated region. Using the VIT trapezoid scheme, available soil moisture and the ET fraction for any given pixel, composed of LST and fractional vegetation cover (FVC), can be calculated at a better spatial resolution in optical or thermal infrared-based satellite images along with meteorological data (Sun, 2016) (see Figure 10.2).

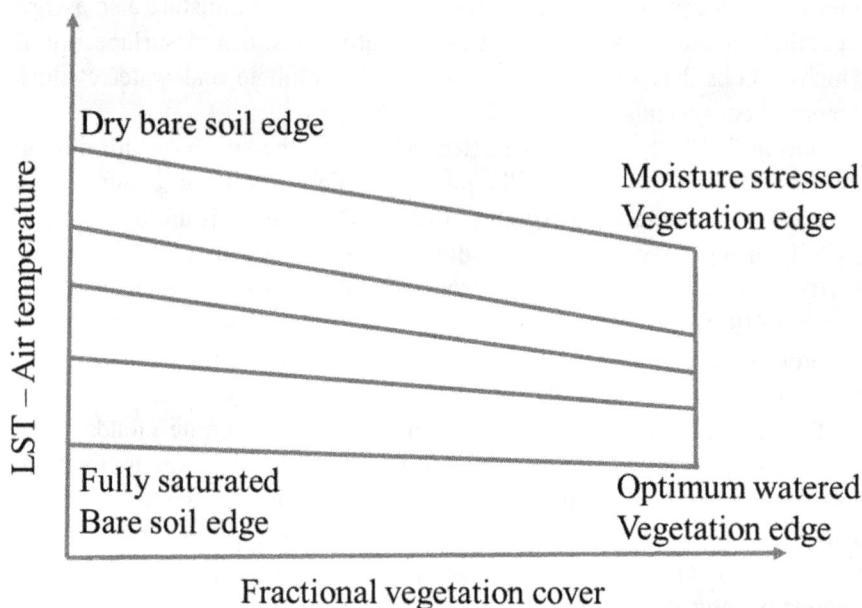

FIGURE 10.2   Conceptual framework of trapezoidal feature space using LST and FVC.

High-resolution meteorological data of necessary weather elements (e.g., bright sunshine hours, wind speed, ambient temperature, and specific humidity) are required for defining the hypothetical restriction boundaries using empirical radiation balance and energy budget equations (Moran et al., 1994). Furthermore, some key parameters such as aerodynamic and external surface resistance can be obtained from the gridwise meteorological datasets, which are spatially complete. This supplementary meteorological information gives another way for getting of soil moisture over the gap or unknown pixels, which is advantageous in developing all-weather soil moisture data products for a regional scale. The VIT approach was used for deriving all-weather soil moisture product (both under clear sky and cloudy conditions) at finer spatial resolution by satellite images and gridwise meteorological data (Moran et al., 1994).

Additionally, meteorological information is applied to compute the necessary parameters for filling missing values, where the trapezoidal model is unavailable. The proposed VIT trapezoid scheme developed was tested in an agricultural field, which gave the most rational estimation of all-weather soil moisture with a satisfactory accurateness between the estimated and observed soil moistures at different depths (RMSE: 0.067–0.079 m^3 m^{-3}). Hence, the developed scheme discloses substantial prospective to give all-weather soil moisture data product by combining presently available spaceborne imageries and weather data at a regional or global scale.

10.4 SATELLITES INVOLVED IN MICROWAVE BAND FOR SOIL MOISTURE REMOTE SENSING

From the earlier two decades, MW remote sensing has been playing an effective role in estimating soil moisture through assessing dielectric characteristics of the soil. Numerous low frequencies (X, C, and L bands) have classically been used to perceive bare or vegetated soil surface moisture content (Calvet et al., 2011).

Various longer wavelength MW satellites using C and X band sensors (such as AMSR-E, ASCAT, RADARSAT, and WindSAT) onboard have revealed the potential in quantifying surface wetness globally. A number of satellite with L band radiometers and radars comprising SMOS (ESA-2009, 1.4 GHz), AQUARIUS (NASA-2011, 1.413 GHz and 1.26 GHz), and SMAP (NASA-2015, 1.41 GHz and 1.26 GHz) have been positioned in trajectory during the preceding number of years for assessing and observing worldwide close-surface (0–5 cm) soil moisture data and ocean salinity content information (Mohanty et al., 2017). Additionally, the Sentinel-1 (from 2016 June) data are being assessed along with SMAP data products due to their identical orbital path and visiting time, which may offer gap-filling (Leng et al. 2017) information and permits data merging at greater resolutions to support wide-area coverage at global scale (see Figure 10.3). Similarly, finer spatial resolution near-surface soil moisture data can be obtained from ALOS-2 PALSAR (14 days revisit time) platform on a regional scale. Upcoming L band space missions NISAR (NASA–ISRO Synthetic Aperture Radar, SAR in L- and S band, the Launch year 2022) and the German Tandem-L mission

(Mohanty et al., 2017) will be capable to retrieve soil moisture at a better spatial resolution and will show the new area of application. Along with this, ASCAT-SG (Advanced Scatterometer) will be the advanced generation scatterometer with several earth science applications, including soil moisture retrieval (see Table 10.1).

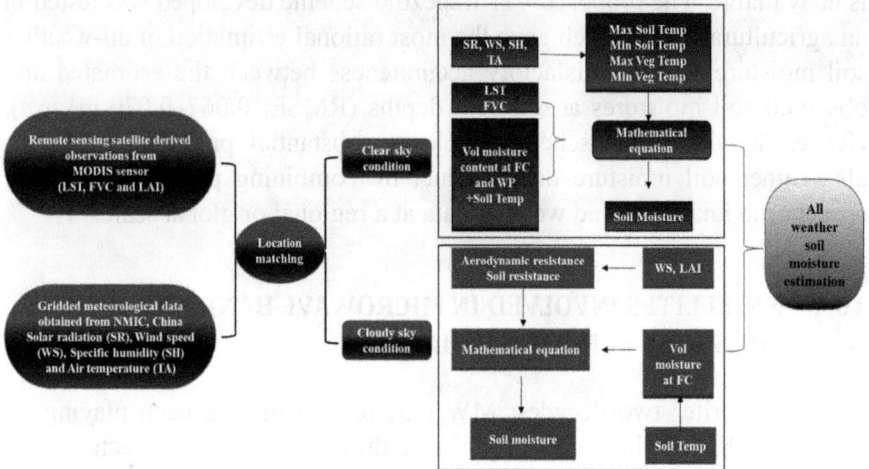

FIGURE 10.3   Flowchart for soil moisture estimation combining satellite and meteorological data [Modified and adapted from Lievens et al. (2017)].

10.5   SATELLITE SOIL MOISTURE VALIDATION

To make satellite-based soil moisture monitoring more useful for global earth resource observation, various in situ invasive or noninvasive soil measurement techniques have been in use around the globe. Mohanty et al. (2017) has reviewed current progress of several nondestructive soil moisture quantifying methods and observing networks. Some promising validation strategies such as cosmic-ray neutron probes using cosmic-ray soil moisture interaction code along with global-navigation-satellite-based reflection data have exposed encouraging performances to validate estimated soil moisture information covering an extent of 100–1000 m². Additionally, a number of soil moisture test centers including various in situ soil moisture sensors with various precision and accuracy have been established, such as TERENO, MOISST, and TxSON, to assess the s-derived soil moisture data products. Besides, several temporary soil moisture evaluation network systems such as NOAA's US Climate Reference Network and the USDA's Soil Climate Analysis Network, along with short-term field-level operations (e.g., SMAPVEX) using aerial and ground-based sampling have been utilized for authenticating the space-based soil moisture data products.

TABLE 10.1 Different Soil Moisture Estimation Satellites

Band Type	Satellite (Instrument)	Frequency (GHz)	Spatial Resolution	Temporal Resolution (Days)
C	ENVISAT (ASAR, Active)	1.41	30–1000 m	5
	MetOP (ASCAT, Active)	5.33	25–50 km	2
	Aqua (AMSR-E, Passive)	6.9–89	25–50 km	2
	RADARSAT-1 & 2 (Active)	1.27	10 m	24
	Sentinel-1A & 1B (Active)		5–20 m	6
	Coriolis 6.8–37 (WindSAF, Passive)	19.35	8–71 km	8
X	Aqua (AMSR-E, Passive)	6.9–89	25–50 km	2
	GCOM-W1 (AMSR-2, Passive)	6.9–89	25–50 km	2
	Coriolis 6.8–37 (WindSAF, Passive)	19.35	8–71 km	8
L	Aquaries (Aquaries, Active /Passive)	1.26	76–156 km	7
	SMOS (MIRAS, Active)	5.25	35–60 km	3
	NISAR (NISAR, Active)		0.1–50 km	12–60
	ALOS (PALSAR, Active)	1.27	10–100 m	46
	Tandem-L (Tandem-L, Active)	5.4	3–20 m	8
	SMAP (SMAP, Active/ Passive)	1.41	3 km, 40 km	2–3
	Coriolis 6.8–37 (WindSAF, Passive)	19.35	8–71 km	8
K	SSM/I (SSM/I, Passive)	1.26	13–69 km	0.5
	Coriolis 6.8–37 (WindSAF, Passive)	19.35	8–71 km	8
S	GCOM-W1 (AMSR-2, Passive)	6.9–89	25–50 km	2
	NISAR (NISAR, Active)		0.1–50 km	12–60

Source: Modified from Mohanty et al. (2017).

The SMAP mission was the earliest real-time validation of satellite-derived soil moisture products with ground-measured soil moisture data (Entekhabi et al., 2010), while AMSR-E is considered the first large-scale soil moisture satellite evaluation program, where Aqua satellite was used to inaugurate a number of moderate-resolution watershed networks (Jackson et al., 2010). In the SMOS mission, a continuous observation was piloted to evaluate these similar watershed networks. Current satellite missions that are launched or are on the program, for instance, the Global Change Observation Mission—Water

(GCOM-W) or the Sentinel series satellite operations, will help in the better work efficiency of in situ resources. Almost all the network systems exist in the International Soil Moisture Network, which has been utilized for authentication of satellite-derived soil moisture products (Wu et al., 2016).

10.6 DIFFERENT SOIL MOISTURE MEASUREMENT MISSION

10.6.1 SMAP MISSION

It is one of the Earth observation satellites (EOS) (launched on January 31, 2015) produced by NASA for National Research Council's Decadal Survey. This satellite mission aims at providing quantification of superficial soil moisture and freezing-thawing state of moisture with a repetitive time of two to three days (Schalie et al., 2016). Additionally, to get the soil moisture at the root zone, SMAP surface measurements are combined with different hydrologic models. These measurements permit applications in the following areas.

1. Apprehend the path which connects the global water, energy, and carbon cycles.
2. Assess water and energy fluxes at the terrestrial surface universally.
3. Measure net carbon flux in boreal sites.
4. Improve weather and climate change prediction proficiency.
5. Prepare better flood forecast and drought observing proficiency.

The SMAP mission was aimed to continue for a time period of three years after introduction. Simultaneously, complete validation of the observation, scientific intervention, and applications programs are executed, and all the image data and satellite observations are accessible openly via NASA collection centers (see Figure 10.4).

10.6.1.1 MEASUREMENT CONCEPT

The SMAP mission comprises a committed spaceship and devices located in a near-polar Sun-synchronous trajectory. The whole measurement organization entails both passive (Radiometer) and active devices (Synthetic Aperture Radar-SAR) working in L band range of MW. This combination of Radiometer and SAR instrument takes benefit of the finer spatial resolution of the SAR and the identifying accurateness of the radiometer. The scientific payloads of these two instruments are given in Table 10.2.

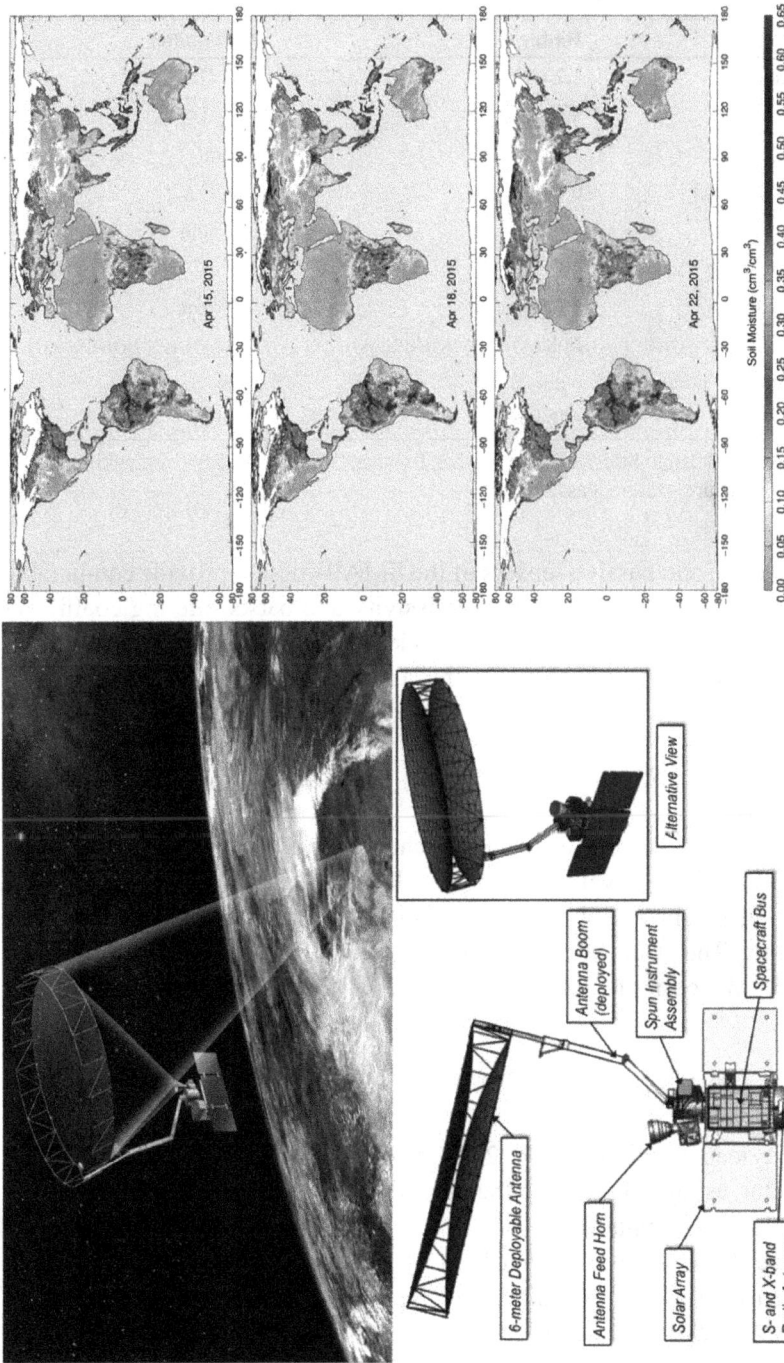

FIGURE 10.4   (a) The SMAP mission model containing L band radar and radiometer. (b) SMAP instrument labels. (c) Universal volumetric soil water content for top 2-in soil depth, created through SMAP mission. (https://smap.jpl.nasa.gov/).

TABLE 10.2   Specification of SMAP Mission Instrument

Features	Radar	Radiometer
Frequency	1.2 GHz	1.41 GHz
Polarizations	VV, HH, HV	V, H, U
Resolution	1–3 km	40 km
Antenna diameter	6 m	6 m
Rotation rate	14.6 rpm	14.6 rpm
Incidence angle	40°	40°
Swath width	1000 km	1000 km
Orbit	Polar, Sun-synchronous	Polar, Sun-synchronous
Local time asc. node	6 am	6 am
Altitude	670 km	670 km

Source: Adapted from Soil Moisture Active Passive. Wikipedia. https://en.wikipedia.org/wiki/Soil_Moisture_Active_Passive.

The active and passive sensors of the SMAP mission provide comparable measurements of the land surface emissivity and backscattering coefficient from the upper 5 cm of soil with moderate vegetation cover to produce estimated soil moisture map or data products.

10.6.2   SMOS MISSION

SMOS is ESA's satellite mission launched on November 2, 2009 with the envision to deliver novel intuitions into Earth's water cycle and climate for monitoring of snow and ice accumulation and to give enhanced weather forecasting. The prime interest of this project is to give global interpretations of land SSM, ocean salinity and to characterize ice and snow-covered surface.

10.6.2.1   PAYLOAD

In its payload, a passive MW 2D radiometer called microwave imaging radiometer using aperture synthesis (MIRAS) device working in L band (1.4 GHz) was installed. MIRAS contains a fundamental construction and three deployable arms that hold 69 antenna elements that are evenly dispersed over the central structure. Besides this, some other features of the SMOS mission are given in Table 10.3. In general, in the MW region, L band has very high sensitivity to soil moisture as well as low sensitivity to atmospheric effects, and surface coarseness is negligible (Wigneron et al., 2000). It gives volumetric soil moisture content (VSMC) with an accuracy of 4% with a spatial resolution of 35–50 km and temporal resolution of three days.

TABLE 10.3   SMOS Mission Features

Category	LEO, Sun-synchronous, polar
Elevation	Min 761.3/Max 788.4 km
Revisit time	149 days
Trajectories for every day	14
Inclination angle	98.42°
Elevation control	Three-axis stabilization with local standard and terrain adjustment

Source: Adapted from earth.esa.int/web

10.6.2.2   MISSION STATUS

On October 31, 2019, SMOS has completed its decadal voyage in orbit. This extraordinary satellite mission has not only surpassed its strategic life span in space orbit, but also additionally outshined its aimed systematic objectives. It was planned to provide SMOS data products, which are considered fundamental constituents of global water cycle. By regularly representing these variables, SMOS is simultaneously improving knowledge of the global moisture cycle and the conversation procedures amid Earth's land surface and the atmosphere; additionally, it is serving to advance weather predictions skill and helping in global weather research. Seeing the excellency of SMOS mission in technical and scientific performances, the program Board for Earth Observations has approved the extension of the SMOS mission in February 2019 (in its 177th Advisory Committee for Earth Observation meeting) (see Figure 10.5, SMOS Newsletter April 2019).

10.6.3   ADVANCED MICROWAVE SCANNING RADIOMETER (AMSR) MISSION

AMSR instrument has worked on three different satellites namely: (1) AMSR-JAXA's ADIOS spacecraft (from 2002 to 2003), failed due to solar

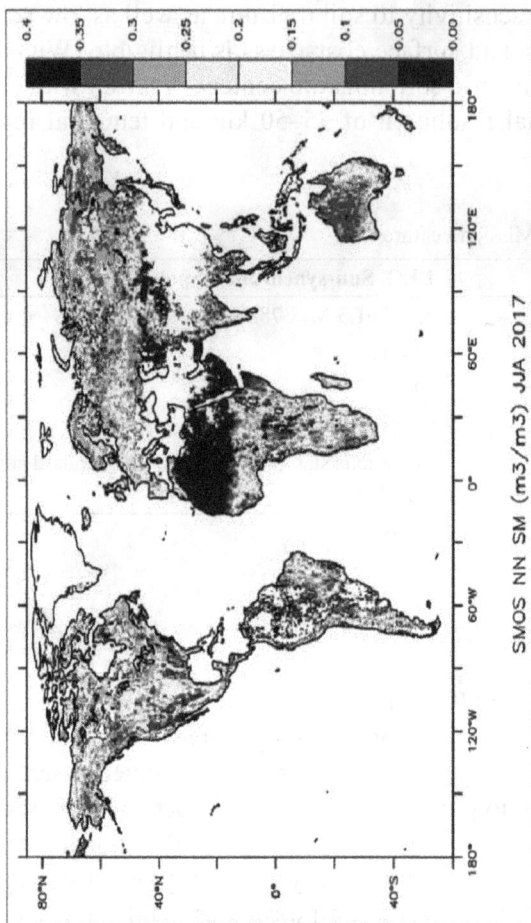

FIGURE 10.5 Global average SMC (for June, July and August 2017). Calculated from swath-based soil-moisture produced using "neural network" in near real time at ECMWF. (Image credit: ECMWF, SMOS.https://earth.esa.int/web/guest/missions/esa-operational-eo-missions/smos/multimedia-book)

panel damage, (2) AMSR-E—NASA's EOS spacecraft called Aqua (from 2002 to 2011), and (3) AMSR-2 JAXA's—GCOM-W1spacecraft (from 2012 to present). Both the AMSR-E and AMSR-2 have estimated soil moisture at global level. Apart from this, AMSR-E was also dedicated for measuring several critical parameters for understanding the Earth's climate and to observe global climate change scenarios such as observation of global precipitation, oceanic water vapor content, measuring cloud water content, measuring wind speed close to earth surface, and determining the change in sea surface temperature, snow cover, and sea ice parameters. Similarly, AMSR-2 has been measuring weak MW emission from the upper surface of the atmosphere and the Earth. It has been placed about 700-km height from the Earth surface and brings the estimation of MW emission and scattering very much precisely. It takes only 1.5 s for AMSR-2 antenna to rotate and attains the data covering an area of 1450 km (swath), and this permits the satellite to cover >99% of the earth surface both during daytime and night-time. Im et al. (2016) used three different machine learning algorithms such as random forest regression, boosted regression trees, and cubist regression to downscale AMSR-E-derived soil moisture data product (25×25 km) under different climatic conditions (South Korea and Australia) using MODIS products (1 km), containing land surface albedo, LST, NDVI, enhanced vegetation index, LAI, and ET. Their study showed that random forest regression approach performed superiorly than other regression models with the highest R^2 value (0.71–0.84) and lowest cross-validation RMSE value (0.049–0.057) to estimate AMSR-E-based soil moisture. Additionally, they also showed that LST, albedo, and ET were the most influential parameters for estimating soil moisture from the satellite. In another study, Wu et al. (2016) evaluated AMSR-2 soil moisture products over the contiguous state of USA with International Soil Moisture Network. They mainly evaluated the JAXA algorithm of measuring AMSR-2 soil moisture product over 598 stations and concluded that the nighttime soil moisture is normally lesser than the daytime, and estimation efficiency was found much better over the Great plain than the temperate forest mountain areas.

10.7 SOIL MOISTURE ESTIMATION FROM AN UNMANNED AERIAL VEHICLE (UAV) PLATFORM

Use of UAV in the civilian and military application has widened extensively from the past few decades (Mesas-Carrascosa et al., 2014). Moreover, rapid scientific improvements have empowered a comprehensive array of responsibilities to be accomplished by UAVs, with tasks starting from entertainment to framing pictures, providing conveyance assets, and perceiving topographically unreachable areas, which cannot be attained by human eyes. Sensing SSM from the long-distance satellite remote sensing platform in recent times become a substitute that can offer an incessant and wide-scale observation of the SSM. Although it is only capable of quantifying SSM data, yet it is considered a worthy basis of information for hydrological balance and agronomic study (Gao et al., 2006; Escorihuela et al., 2010). Thus, for observing the spatial variations of soil moisture from regional to global scale, a rising need has inspired the development of airborne sensors (Famiglietti et al., 2008). In water-related divisions, investigators and scientist have constructed various UAV platforms to confirm the preferred spatial, spectral, radiometric, and temporal resolution of the UAV sensed remote areas where topographical circumstance prevents exhaustive field observations or the poor spatiotemporal resolution of satellite images is not well-matched with the perspective (Mesas-Carrascosa et al., 2014). Several studies were undertaken to estimate the soil moisture using the UAV platform. Hsu and Chang (2019) used UAV areal images (thermal inertia images and multispectral images) to estimate soil moisture and compared with the in situ measured soil moisture. For this, several vegetation indices, that is, TVDI, NDVI, LST, and apparent thermal inertia, were calculated and used as an input in empirical equation for estimating soil moisture. The result showed that the empirical soil moisture estimation using TVDI was more accurate when matched with the in situ ground-based soil moisture data.

To measure the soil moisture in a highly varied landscape is still a challenge. Karst landforms, which account for 10% of the Earth's land surface, is a typical example of Karst topography and performs an imperative part in water supply for the human community (Hartmann et al., 2014). From the report of Liu et al. (2016), this area is in front of drastic environmental provocation caused due to severe climatic hazards and deprivation. Hence, Luo et al. (2019) have undertaken a study for understanding the soil moisture discrepancy and its subsequent estimation in Karst areas, which are necessary for environmental safeguard and managing water resources. Their study has developed a detailed model for the karst mountainous catchment regions after inspecting ecological factors (soil condition, vegetation type, and topographical nature) and relating to different remote sensing platform derived data (Landsat-8, Radarsat-2, ASTER Global Digital Elevation Model (DEM) V002 (ASTGTM2) and UAV). Additionally, this study recognized important governing factors responsible for the spatial distribution of soil moisture for highly heterogeneous landscapes. DEM (high-resolution,

0.12 m) was generated with the images obtained from the UAV platform (DJI Phantom 3) from an altitude of 500 m above the surface. NDVI was obtained from Landsat-8 (30 m) satellites that deliver nine individual bands from 433 to 2300 nm. MW backscatter data of that region were obtained from C band Radarsat-2 (10 m) satellites in four polarization patterns (HH, HV, VH, and VV). They obtained a list of different parameters (including the percentage of SM variability explained by them) that influence soil moisture estimation of that heterogeneous landscape. According to the result, vegetation category (35.7%), aspect ratio (7.7%), height index (4.2%), soil BD (3.3%), soil total N content (3.1%), aspect effect with vegetation type (3.4%), and soil total phosphorus (1.3%) altogether elucidated 58.8% of the soil moisture variability. Along with the above, the study also revealed that visible and IR bands from Landsat-8 and topographic byproducts from UAV photogrammetry-based DEM were intensely associated with soil moisture (as obtained after PLSR analysis) than other datasets.

In another study, Chang and Hsu (2018) reported the impact of land cover change and soil moisture variation using thermal imagers from a UAV platform. Soil-based linear equations were assimilated into thermal image-derived change analysis to SMC data. The aim of the study was to use a payload equipped with a thermal infrared sensor for detecting LST and to gather instantaneous SMC through a mathematical model under various land cover type. TVDI was calculated from the aerial imagery captured with the UAV. In situ soil moisture measurement was done following a conventional method. Empirical equations were calculated to estimate the surface SMC using thermal image-based soil moisture data and the TVDI of the experimental area.

The estimated SSM was then regressed with the in situ measured value, and the result showed that linear empirical relationship was more comparable to the actual measured soil moisture data. Similarly, Hassan-Esfahani et al. (2014) used the Aggie-Air UAV platform that supports multispectral cameras (including VIS, NIR, and thermal camera) and capture aerial imagery with greater spatial and temporal coverage for estimating topsoil moisture. The focus of the study was to assess the efficiency of the Aggie-Air machinery in assisting precision agriculture supervision by giving high resolution (15 cm for VIS-NIR and 60 cm for TIR) topsoil moisture estimates of alfalfa and oats crop fields under sprinkler irrigation system in Utah. Ground-based soil moisture measurement was done on the same date as the UAV mission. In situ soil moisture data and Aggie-Air imagery were subsequently used for training the nonlinear machine learning regression model (RVM—relevance vector machines). Subsequently, after the calibration, the model was applied to entire aerial imagery obtained from the Aggie-Air to get the topsoil moisture map for the given area. The nonlinear RVM model showed good accuracy in estimating topsoil moisture with an RMSE of 3.04% when compared with the observed data. Additionally, the RVM model also produced high-resolution images along with the map of SSM for different dates of flights, which showed the spatial variation of soil moisture status of the sprinkler irrigation system. This geospatial map could be used in irrigation water management and scheduling. The decision-support-based efficient irrigation management strategy provides the quantity and timing of irrigation at given spatial scale. For this, volumetric water content in the root zone depth at an adequate spatial scale is needed. Hassan-Esfahani et al. (2017) presented a Bayesian data mining methodology that considers a known field environment along with remote sensing data for providing likelihood approximations of root zone soil water at three different soil depths (see Figure 10.6). The data mining algorithms were examined and standardized to integrate the UAV sensed spatial data with field observations. A trapezoidal integration scheme was used in estimating VSMC at the three different root zone depth using the modeling approach with good accuracy.

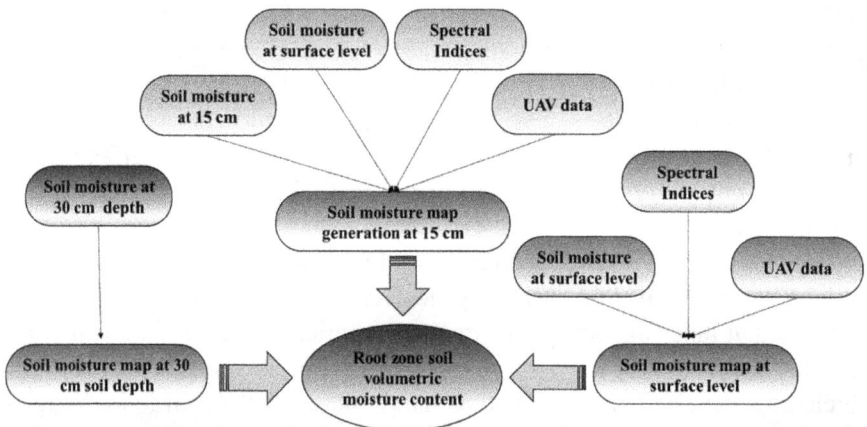

FIGURE 10.6   Flowchart of UAV-based soil moisture estimation [Modified and adapted from Hassan-Esfahani et al. (2017)].

10.8   CONCLUSION

Soil moisture plays an important role in the exchange of moisture and heat energy between the land and atmosphere and thus controls the geohydrological cycle. This chapter provides a detailed description of spatiotemporal estimation of soil moisture at ground-level UAV-based platform and from the satellite-based imaging techniques. Application of different machine learning models, such as PTFs, in estimating soil moisture has been covered, where sensitive electromagnetic bands and/or basic soil properties are used as inputs. The use of optical, thermal, and MW bands in satellite-based soil moisture estimation along with their ground-based validation has been covered in this chapter. Merging of sensitive electromagnetic bands with meteorological data has tremendous potentiality for better soil moisture estimation. Simultaneously, different ongoing satellite-based soil moisture estimation missions such as SMAP, SMOS, and AMSR have been discussed along with their implication in characterizing soil moisture over a larger region. In the end, soil moisture information is mandatory for water reservoir controlling, timely drought warning, field irrigation scheduling, and forecasting crop yield. The use of remote sensing tools in soil moisture estimation will help to provide more insights in vadose zone hydrology in the upcoming periods.

KEYWORDS

- **remote sensing**
- **retrieval algorithms**
- **sensing platforms**
- **soil water**

REFERENCES

Amani, M., Salehi, B., Mahdavi, S., Masjedi, A., Dehnavi, S., 2017. Temperature-vegetation-soil moisture dryness index (TVMDI). *Remote Sens. Environ. 197*, 1–14. https://doi.org/10.1016/j.rse.2017.05.026

Amazirh, A., Merlin, O., Er-Raki, S., Gao, Q., Rivalland, V., Malbeteau, Y., Khabba, S., Escorihuela, M.J., 2018. Retrieving surface soil moisture at high spatio-temporal resolution from a synergy between Sentinel-1 radar and Landsat thermal data: A study case over bare soil. *Remote Sens. Environ.* 211, 321–337. https://doi.org/10.1016/j.rse.2018.04.013

Babaeian, E., Homaee, M., Montzka, C., Vereecken, H., Norouzi, A.A., 2015a. Towards Retrieving Soil Hydraulic Properties by Hyperspectral Remote Sensing. *Vadose Zone J. 14*, 1–17. https://doi.org/10.2136/vzj2014.07.0080

Babaeian, E., Homaee, M., Vereecken, H., Montzka, C., Norouzi, A.A., van Genuchten, M.T., 2015b. A comparative study of multiple approaches for predicting the soil–water retention curve: Hyperspectral information vs. basic soil properties. *Soil Sci. Soc. Am. J. 79*, 1043–1058. https://doi.org/10.2136/sssaj2014.09.0355

Banerjee, K., Krishnan, P., 2020. Normalized sunlit shaded index (NSSI) for characterizing the moisture stress in wheat crop using classified thermal and visible images. *Ecol. Indic.* *110*, 105947. https://doi.org/10.1016/j.ecolind.2019.105947

Bao, Y., Lin, L., Wu, S., Kwal Deng, K.A., Petropoulos, G.P., 2018. Surface soil moisture retrievals over partially vegetated areas from the synergy of Sentinel-1 and Landsat 8 data using a modified water-cloud model. *Int. J. Appl. Earth Obs. Geoinf. 72*, 76–85. https://doi.org/10.1016/j.jag.2018.05.026

Calvet, J.-C., Wigneron, J.-P., Walker, J., Karbou, F., Chanzy, A., Albergel, C., 2011. Sensitivity of passive microwave observations to soil moisture and vegetation water content: L-band to w-band. *IEEE Trans. Geosci. Remote Sens. 49*, 1190–1199. https://doi.org/10.1109/TGRS.2010.2050488

Chang, K.-T., Hsu, W.-L., 2018. Estimating soil moisture content using unmanned aerial vehicles equipped with thermal infrared sensors, in: 2018 IEEE International Conference on Applied System Invention, pp. 168–171. https://doi.org/10.1109/ICASI.2018.8394559

Demattê, J.A.M., Fiorio, P.R., Araújo, S.R., 2010. Variation of routine soil analysis when compared with hyperspectral narrow band sensing method. *Remote Sens. 2*, 1998–2016. https://doi.org/10.3390/rs2081998

Dorigo, W.A., Scipal, K., Parinussa, R.M., Liu, Y.Y., Wagner, W., de Jeu, R.A.M., Naeimi, V., 2010. Error characterisation of global active and passive microwave soil moisture datasets. *Hydrol. Earth Syst. Sci. 14*, 2605–2616. https://doi.org/10.5194/hess-14-2605-2010

Entekhabi, D., Njoku, E.G., O'Neill, P.E., Kellogg, K.H., Crow, W.T., Edelstein, W.N., Entin, J.K., Goodman, S.D., Jackson, T.J., Johnson, J., Kimball, J., Piepmeier, J.R., Koster, R.D., Martin, N., McDonald, K.C., Moghaddam, M., Moran, S., Reichle, R., Shi, J.C., Spencer, M.W., Thurman, S.W., Tsang, L., Van Zyl, J., 2010. The soil moisture active passive (SMAP) mission. *Proc. IEEE 98*, 704–716. https://doi.org/10.1109/JPROC.2010.2043918

Escorihuela, M.J., Chanzy, A., Wigneron, J.P., Kerr, Y.H., 2010. Effective soil moisture sampling depth of L-band radiometry: A case study. *Remote Sens. Environ. 114*, 995–1001. https://doi.org/10.1016/j.rse.2009.12.011

Famiglietti, J.S., Ryu, D., Berg, A.A., Rodell, M., Jackson, T.J., 2008. Field observations of soil moisture variability across scales. *Water Resour. Res. 44*, W01423. https://doi.org/10.1029/2006WR005804

Gao, H., Wood, E.F., Jackson, T.J., Drusch, M., Bindlish, R., 2006. Using TRMM/TMI to retrieve surface soil moisture over the Southern United States from 1998 to 2002. *J. Hydrometeorol. 7*, 23–38. https://doi.org/10.1175/JHM473.1

Gao, Q., Zribi, M., Escorihuela, M., Baghdadi, N., 2017. Synergetic use of sentinel-1 and sentinel-2 data for soil moisture mapping at 100 m resolution. *Sensors 17*, 1966. https://doi.org/10.3390/s17091966

Ghahremanloo, M., Mobasheri, M.R., Amani, M., 2019. Soil moisture estimation using land surface temperature and soil temperature at 5 cm depth. *Int. J. Remote Sens. 40*, 104–117. https://doi.org/10.1080/01431161.2018.1501167

Ghulam, A., Qin, Q., Teyip, T., Li, Z.-L., 2007. Modified perpendicular drought index (MPDI): A real-time drought monitoring method. *ISPRS J. Photogramm. Remote Sens. 62*, 150–164. https://doi.org/10.1016/j.isprsjprs.2007.03.002

Gruber, A., Su, C.-H., Zwieback, S., Crow, W., Dorigo, W., Wagner, W., 2016. Recent advances in (soil moisture) triple collocation analysis. *Int. J. Appl. Earth Obs. Geoinf. 45*, 200–211. https://doi.org/10.1016/j.jag.2015.09.002

Gupta, A., Das, B.S., Kumar, A., Chakraborty, P., Mohanty, B., 2016. Rapid and noninvasive assessment of atterberg limits using diffuse reflectance spectroscopy. *Soil Sci. Soc. Am. J.* 80, 1283–1295. https://doi.org/10.2136/sssaj2015.11.0402

Hartmann, A., Goldscheider, N., Wagener, T., Lange, J., Weiler, M., 2014. Karst water resources in a changing world: Review of hydrological modeling approaches. *Rev. Geophys. 52*, 218–242. https://doi.org/10.1002/2013RG000443

Hassan-Esfahani, L., Torres-Rua, A., Jensen, A., Mckee, M., 2017. Spatial root zone soil water content estimation in agricultural lands using bayesian-based artificial neural networks and high- resolution visual, nir, and thermal imagery. *Irrig. Drain. 66*, 273–288. https://doi.org/10.1002/ird.2098

Hassan-Esfahani, L., Torres-Rua, A., Ticlavilca, A.M., Jensen, A., McKee, M., 2014. Topsoil moisture estimation for precision agriculture using unmmaned aerial vehicle multispectral imagery, In: 2014 IEEE Geoscience and Remote Sensing Symposium, pp. 3263–3266. https://doi.org/10.1109/IGARSS.2014.6947175

Haubrock, S.-N., Chabrillat, S., Lemmnitz, C., Kaufmann, H., 2008. Surface soil moisture quantification models from reflectance data under field conditions. *Int. J. Remote Sens. 29*, 3–29. https://doi.org/10.1080/01431160701294695

Hsu, W.-L., Chang, K.-T., 2019. Cross-estimation of soil moisture using thermal infrared images with different resolutions. *Sens. Mater. 31*, 387. https://doi.org/10.18494/SAM.2019.2090

Im, J., Park, S., Rhee, J., Baik, J., Choi, M., 2016. Downscaling of AMSR-E soil moisture with MODIS products using machine learning approaches. *Environ. Earth Sci. 75*, 1120. https://doi.org/10.1007/s12665-016-5917-6

Jackson, T.J., Cosh, M.H., Bindlish, R., Starks, P.J., Bosch, D.D., Seyfried, M., Goodrich, D.C., Moran, M.S., Du, J., 2010. Validation of advanced microwave scanning radiometer soil moisture products. *IEEE Trans. Geosci. Remote Sens. 48*, 4256–4272. https://doi.org/10.1109/TGRS.2010.2051035

Jackson, T.J., Schmugge, J., Engman, E.T., 1996. Remote sensing applications to hydrology: soil moisture. *Hydrol. Sci. J. 41*, 517–530. https://doi.org/10.1080/02626669609491523

Jackson, T.J., Schmugge, T.J., Wang, J.R., 1982. Passive microwave sensing of soil moisture under vegetation canopies. *Water Resour. Res. 18*, 1137–1142. https://doi.org/10.1029/WR018i004p01137

Kerr, Y.H., Waldteufel, P., Wigneron, J.-P., Delwart, S., Cabot, F., Boutin, J., Escorihuela, M.-J., Font, J., Reul, N., Gruhier, C., Juglea, S.E., Drinkwater, M.R., Hahne, A., Martín-Neira, M., Mecklenburg, S., 2010. The SMOS mission: New tool for monitoring key elements ofthe global water cycle. *Proc. IEEE 98*, 666–687. https://doi.org/10.1109/JPROC.2010.2043032

Kumar, S. V., Dirmeyer, P.A., Peters-Lidard, C.D., Bindlish, R., Bolten, J., 2018. Information theoretic evaluation of satellite soil moisture retrievals. *Remote Sens. Environ. 204*, 392–400. https://doi.org/10.1016/j.rse.2017.10.016

Kumar, S. V., Peters-Lidard, C.D., Santanello, J.A., Reichle, R.H., Draper, C.S., Koster, R.D., Nearing, G., Jasinski, M.F., 2015. Evaluating the utility of satellite soil moisture retrievals over irrigated areas and the ability of land data assimilation methods to correct for unmodeled processes. *Hydrol. Earth Syst. Sci. 19*, 4463–4478. https://doi.org/10.5194/hess-19-4463-2015

Lakshmi, V., 2013. Remote sensing of soil moisture. *ISRN Soil Sci. 2013*, 1–33. https://doi.org/10.1155/2013/424178

Leng, P., Li, Z.-L., Duan, S.-B., Gao, M.-F., Huo, H.-Y., 2017. A practical approach for deriving all-weather soil moisture content using combined satellite and meteorological data. *ISPRS J. Photogramm. Remote Sens. 131*, 40–51. https://doi.org/10.1016/j.isprsjprs.2017.07.013

Leng, P., Song, X., Duan, S.-B., Li, Z.-L., 2016. A practical algorithm for estimating surface soil moisture using combined optical and thermal infrared data. *Int. J. Appl. Earth Obs. Geoinf. 52*, 338–348. https://doi.org/10.1016/j.jag.2016.07.004

Leng, P., Song, X., Li, Z.-L., Ma, J., Zhou, F., Li, S., 2014. Bare surface soil moisture retrieval from the synergistic use of optical and thermal infrared data. *Int. J. Remote Sens. 35*, 988–1003. https://doi.org/10.1080/01431161.2013.875237

Lievens, H., Reichle, R.H., Liu, Q., De Lannoy, G.J.M., Dunbar, R.S., Kim, S.B., Das, N.N., Cosh, M., Walker, J.P., Wagner, W., 2017. Joint sentinel-1 and SMAP data assimilation to improve soil moisture estimates. *Geophys. Res. Lett. 44*, 6145–6153. https://doi.org/10.1002/2017GL073904

Liu, M., Xu, X., Wang, D., Sun, A.Y., Wang, K., 2016. Karst catchments exhibited higher degradation stress from climate change than the non-karst catchments in southwest China: An ecohydrological perspective. *J. Hydrol. 535*, 173–180. https://doi.org/10.1016/j.jhydrol.2016.01.033

Liu, R., Shang, R., Liu, Y., Lu, X., 2017. Global evaluation of gap-filling approaches for seasonal NDVI with considering vegetation growth trajectory, protection of key point, noise resistance and curve stability. *Remote Sens. Environ. 189*, 164–179. https://doi.org/10.1016/j.rse.2016.11.023

Luo, W., Xu, X., Liu, W., Liu, M., Li, Z., Peng, T., Xu, C., Zhang, Y., Zhang, R., 2019. UAV based soil moisture remote sensing in a karst mountainous catchment. *CATENA 174*, 478–489. https://doi.org/10.1016/j.catena.2018.11.017

Mesas-Carrascosa, F.J., Notario-García, M.D., Meroño de Larriva, J.E., Sánchez de la Orden, M., García-Ferrer Porras, A., 2014. Validation of measurements of land plot area using UAV imagery. *Int. J. Appl. Earth Obs. Geoinf. 33*, 270–279. https://doi.org/10.1016/j.jag.2014.06.009

Mitran, T., Ravisankar, T., Fyzee, M.A., Suresh, J.R., Sujatha, G., Sreenivas, K., 2015. Retrieval of soil physicochemical properties towards assessing salt-affected soils using Hyperspectral Data. *Geocarto Int. 30*, 701–721. https://doi.org/10.1080/10106049.2014.985745

Mobasheri, M.R., Amani, M., 2016. Soil moisture content assessment based on Landsat 8 red, near-infrared, and thermal channels. *J. Appl. Remote Sens. 10*, 026011. https://doi.org/10.1117/1.JRS.10.026011

Mohanty, B.P., Cosh, M.H., Lakshmi, V., Montzka, C., 2017. Soil Moisture Remote Sensing: State-of-the-Science. Vadose Zo. J. 16, vzj2016.10.0105. https://doi.org/10.2136/vzj2016.10.0105

Moran, M.S., Clarke, T.R., Inoue, Y., Vidal, A., 1994. Estimating crop water deficit using the relation between surface-air temperature and spectral vegetation index. Remote Sens. Environ. 49, 246–263. https://doi.org/10.1016/0034-4257(94)90020-5

Nawar, S., Buddenbaum, H., Hill, J., Kozak, J., Mouazen, A.M., 2016. Estimating the soil clay content and organic matter by means of different calibration methods of vis-NIR diffuse reflectance spectroscopy. *Soil Tillage Res. 155*, 510–522. https://doi.org/10.1016/j.still.2015.07.021

Oltra-Carrió, R., Baup, F., Fabre, S., Fieuzal, R., Briottet, X., 2015. Improvement of soil moisture retrieval from hyperspectral VNIR-SWIR data using clay content information: From laboratory to field experiments. *Remote Sens. 7*, 3184–3205. https://doi.org/10.3390/rs70303184

Owe, M., de Jeu, R., Holmes, T., 2008. Multisensor historical climatology of satellite-derived global land surface moisture. *J. Geophys. Res. 113*, F01002. https://doi.org/10.1029/2007J F000769

Parinussa, R.M., Meesters, A.G.C.A., Liu, Y.Y., Dorigo, W., Wagner, W., de Jeu, R.A.M., 2011. Error estimates for near-real-time satellite soil moisture as derived from the land parameter retrieval model. *IEEE Geosci. Remote Sens. Lett. 8*, 779–783. https://doi.org/10.1109/ LGRS.2011.2114872

Santra, P., Sahoo, R.N., Das, B.S., Samal, R.N., Pattanaik, A.K., Gupta, V.K., 2009. Estimation of soil hydraulic properties using proximal spectral reflectance in visible, near-infrared, and shortwave-infrared (VIS-NIR-SWIR) region. *Geoderma 152*, 338–349. https://doi. org/10.1016/j.geoderma.2009.07.001

Santra, P., Singh, R., Sarathjith, M.C., Panwar, N.R., Varghese, P., Das, B.S., 2015. Reflectance spectroscopic approach for estimation of soil properties in hot arid western Rajasthan, India. *Environ. Earth Sci. 74*, 4233–4245. https://doi.org/10.1007/s12665-015-4383-x

Sarathjith, M.C., Das, B.S., Vasava, H.B., Mohanty, B., Sahadevan, A.S., 2014. Diffuse reflectance spectroscopic approach for the characterization of soil aggregate size distribution. *Soil Sci. Soc. Am. J. 78*, 369–376. https://doi.org/10.2136/sssaj2013.08.0377

Sarathjith, M.C., Das, B.S., Wani, S.P., Sahrawat, K.L., Gupta, A., 2016. Comparison of data mining approaches for estimating soil nutrient contents using diffuse reflectance spectroscopy. *Curr. Sci. 110*, 1031–1037. https://doi.org/10.18520/cs/v110/i6/1031-1037

Schalie, R. va. der, Kerr, Y.H., Wigneron, J.P., Rodríguez-Fernández, N.J., Al-Yaari, A., Jeu, R.A.M. d., 2016. Global SMOS soil moisture retrievals from the land parameter retrieval model. *Int. J. Appl. Earth Obs. Geoinf. 45*, 125–134. https://doi.org/10.1016/j.jag.2015. 08.005

Seneviratne, S.I., Corti, T., Davin, E.L., Hirschi, M., Jaeger, E.B., Lehner, I., Orlowsky, B., Teuling, A.J., 2010. Investigating soil moisture–climate interactions in a changing climate: A review. *Earth-Sci. Rev. 99*, 125–161. https://doi.org/10.1016/j.earscirev.2010.02.004

Son, N.T., Chen, C.F., Chen, C.R., Molina Masferrer, M.G., Recinos, L.E.M., 2019. Multitemporal landsat-MODIS fusion for cropland drought monitoring in El Salvador. *Geocarto Int. 34*, 1363–1383. https://doi.org/10.1080/10106049.2018.1489421

Srivastava, R., Sethi, M., Yadav, R.K., Bundela, D.S., Singh, M., Chattaraj, S., Singh, S.K., Nasre, R.A., Bishnoi, S.R., Dhale, S., Mohekar, D.S., Barthwal, A.K., 2017. Visible-near infrared reflectance spectroscopy for rapid characterization of salt-affected soil in the indo-gangetic plains of Haryana, India. *J. Indian Soc. Remote Sens. 45*, 307–315. https:// doi.org/10.1007/s12524-016-0587-0

Su, C.-H., Ryu, D., Crow, W.T., Western, A.W., 2014. Stand-alone error characterisation of microwave satellite soil moisture using a Fourier method. *Remote Sens. Environ. 154*, 115–126. https://doi.org/10.1016/j.rse.2014.08.014

Sun, H., 2016. Two-Stage Trapezoid: A new interpretation of the land surface temperature and fractional vegetation coverage space. *IEEE J. Sel. Top. Appl. Earth Obs. Remote Sens. 9*, 336–346. https://doi.org/10.1109/JSTARS.2015.2500605

Tian, J., Philpot, W.D., 2015. Relationship between surface soil water content, evaporation rate, and water absorption band depths in SWIR reflectance spectra. *Remote Sens. Environ. 169*, 280–289. https://doi.org/10.1016/j.rse.2015.08.007

Vereecken, H., Weynants, M., Javaux, M., Pachepsky, Y., Schaap, M.G., Genuchten, M.T. Van, 2010. Using pedotransfer functions to estimate the van genuchten–mualem soil hydraulic properties: A review. *Vadose Zone J. 9*, 795. https://doi.org/10.2136/vzj2010.0045

Viscarra Rossel, R.A., Walvoort, D.J.J., McBratney, A.B., Janik, L.J., Skjemstad, J.O., 2006. Visible, near infrared, mid infrared or combined diffuse reflectance spectroscopy for simultaneous assessment of various soil properties. *Geoderma 131*, 59–75. https://doi. org/10.1016/j.geoderma.2005.03.007

Wagner, W., Scipal, K., Pathe, C., Gerten, D., Lucht, W., Rudolf, B., 2003. Evaluation of the agreement between the first global remotely sensed soil moisture data with model and precipitation data. *J. Geophys. Res. 108*, 4611. https://doi.org/10.1029/2003JD003663

Wang, J., Ling, Z., Wang, Y., Zeng, H., 2016. Improving spatial representation of soil moisture by integration of microwave observations and the temperature–vegetation–drought index derived from MODIS products. *ISPRS J. Photogramm. Remote Sens. 113*, 144–154. https://doi.org/10.1016/j.isprsjprs.2016.01.009

Wigneron, J.-P., Waldteufel, P., Chanzy, A., Calvet, J.-C., Kerr, Y., 2000. Two-dimensional microwave interferometer retrieval capabilities over land surfaces (SMOS Mission). *Remote Sens. Environ. 73*, 270–282. https://doi.org/10.1016/S0034-4257(00)00103-6

Wu, Q., Liu, H., Wang, L., Deng, C., 2016. Evaluation of AMSR2 soil moisture products over the contiguous United States using in situ data from the international soil moisture network. *Int. J. Appl. Earth Obs. Geoinf. 45*, 187–199. https://doi.org/10.1016/j.jag.2015.10.011

Xiao, Z., Jiang, L., Zhu, Z., Wang, J., Du, J., 2016. Spatially and temporally complete satellite soil moisture data based on a data assimilation method. *Remote Sens. 8*, 49. https://doi. org/10.3390/rs8010049

Xu, L., Wang, Q., 2015. Retrieval of soil water content in saline soils from emitted thermal infrared spectra using partial linear squares regression. *Remote Sens. 7*, 14646–14662. https://doi.org/10.3390/rs71114646

Yang, Z., Ouyang, H., Zhang, X., Xu, X., Zhou, C., Yang, W., 2011. Spatial variability of soil moisture at typical alpine meadow and steppe sites in the Qinghai-Tibetan Plateau permafrost region. *Environ. Earth Sci. 63*, 477–488.

Yin, Z., Lei, T., Yan, Q., Chen, Z., Dong, Y., 2013. A near-infrared reflectance sensor for soil surface moisture measurement. *Comput. Electron. Agric. 99*, 101–107. https://doi. org/10.1016/j.compag.2013.08.029

Zhang, D., Tang, R., Zhao, W., Tang, B., Wu, H., Shao, K., Li, Z.-L., 2014. Surface soil water content estimation from thermal remote sensing based on the temporal variation of land surface temperature. *Remote Sens. 6*, 3170–3187. https://doi.org/10.3390/rs6043170

Zhang, Z., Lan, H., Zhao, T., 2017. Detection and mitigation of radiometers radio-frequency interference by using the local outlier factor. *Remote Sens. Lett. 8*, 311–319. https://doi. org/10.1080/2150704X.2016.1266408.

CHAPTER 11

Spectral and Smartphone-Based Tools to Monitor Plant and Soil Nitrogen Status for Site-Specific Nitrogen Management in Crop Plants

GOPAL RAMDAS MAHAJAN[1*], BAPPA DAS[1], BHASKAR GAIKWAD[2], DAYESH MURGAOKAR[1], KIRAN PATEL[1], and RAHUL M. KULKARNI[1]

[1]*ICAR-Central Coastal Agricultural Research Institute, Goa, India*

[2]*ICAR-National Institute of Abiotic Stress Management, Baramati, Pune, India*

Corresponding author. E-mail: gopal.mahajan@icar.gov.in

ABSTRACT

Minimizing the losses of fertilizer nitrogen and improving the nitrogen use efficiency (NUE) in different crops has been one of the important research areas in agricultural research. Though technological advancements to achieve this have happened significantly, it still seems to be difficult due to several constraints for its wide and large-scale adoption by the farmers and crop growers. The chapter covers interesting information compiled for the spectral and smartphone-based tools for foliar nitrogen monitoring in different crops. Use of these tools as a part of package of practice for site-specific nitrogen management will help to enhance the NUE besides improving the crop productivity. The use of digital tools with high mobility to match with the nature of farming, highly accessible cost of the device, and their computing power allows a wide-scale adoption at field level not only for fertilizer nitrogen management but also for several other purposes.

11.1   INTRODUCTION

Currently, the issue of fertilizer nitrogen (N) management is drawing attention among farmers, researchers, and policymakers especially under changing climatic conditions. Nitrogen (N) is a most important essential plant nutrient for the plants as it is required in larger amounts compared to other nutrients. In the year 2017, the total fertilizer N consumption of the world was 107.66 Mt (IFASTAT, 2020). Of this, 51.38 Mt was urea and it accounted for 47.7% of fertilizer N consumption. The fertilizer N consumption was about 55.86% of the total $N+P_2O_5+K_2O$ consumption of 192.72 Mt. Among different fertilizer nutrients, the fertilizer N is the most widely used in agricultura across the work; however, its its use efficiency by crops is low ranging from 25% to 50%. There are several reasons of the low N use efficiency (NUE) in various crops. The applied fertilizer N is lost through various processes. De Datta et al. (1988) reported a loss up to 37.0% of applied nitrogen by several processes. Reliance of the farmers to use the nonfixed inputs like fertilizer compared to the fixed to maximize the cereal production further leads to lowering down the NUE (Bijay-Singh and Ali, 2020; Wang et al., 2017)smallholder farmers in developing countries apply fertilizer nitrogen (N. The major reason for low NUE is that the farmers' fertilizer practice of imbalanced fertilizer application. Poor advisories to the farmers about the right amount of fertilizers, right time, place, and method of application and unawareness among farmers is also a probable reason. Low NUE and excessive fertilizer N loss may result in critical situations affecting our environment, economy, and resources and hence approaches are required for improving NUE. To achieve higher NUE and reduced N losses, increasing crop productivity through efficient N management with environment-friendly conditions is crucial. Crop demand-based fertilizer applications can reduce losses of applied N fertilizer. For efficient management, N application should be matched to crop demand so that the crops do not encounter deficiency throughout. Conventional methods of application include broadcasting, placement, band placement, pellet application, foliar application, etc. Major focus of the past research with respect to fertilizer N management was related to the timings of fertilizer N application and came out with split N application strategy. Dobermann Achim and Fairhurst Thomas (2000) reported three-split N application as 1/3rd at final land preparation, 1/3rd at the mid-tillering, and the rest at 5–7 days prior to panicle initiation growth stage of rice crop as most efficient to improve the grain yield and NUE. Similarly, in case of long duration variety (>150 days), based on the availability of water, the basal N fertilizer dose should be postponed to 15–20 DAT. Nitrogen is an essential nutrient and known to boost the yield of crops however it also has some adverse effects due to its overuse. Excessive N application subsequently causes low dissolved oxygen (hypoxia and anoxia) in water bodies or reservoirs that harm the aquatic organisms and vegetation, accelerate surface water eutrophication, promote certain harmful algal blooms producing harmful toxins; and other undesirable changes that affect food webs and fisheries and it contaminates the groundwater. A significant portion of fertilizer N applied is lost as ammonia and in turn contributes to acidification and eutrophication of the reservoirs when it is received in the form of rain. Nitrous oxide (N_2O) is a greenhouse gas and nitric oxide (NO) is involved in the tropospheric ozone associated processes. Owing to the adverse effect of the losses to the environment, plant demand driven fertilizer N supply is critical to achieve high grain yield levels and improved NUE. The problem associated with low NUE can be solved by efficient N management using tools and technologies available. One of the pragmatic strategies for efficient N management is site-specific nutrient management (SSNM). One of the key features of SSNM is that it is an adjustment made to the fertilizer N prescriptions using tools like leaf color charts (LCC) and chlorophyll meters (CM) (Bijay-Singh, 2012) or other suitable tools. In this chapter, we review and discuss different tools basically working on spectral information and information technology to monitor the plant and soil N status and to subsequently make the fertilizer N prescriptions to improve the yield and NUE.

11.2 DIAGNOSTIC TOOLS TO MONITOR PLANT N STATUS AND FERTILIZER N MANAGEMENT TO IMPROVE YIELDS AND USE EFFICIENCY

11.2.1 LEAF COLOR CHART (LCC)

An LCC is a strip made of plastic with shades of green color differing from light yellowish green to dark green. Initially, it was developed by Japan Furuya (1987), subsequently improved versions with more variable color shades code were also developed like six-panel LCC (IRRI-LCC, six-panel) (IRRI, 1996), LCC (ZAU-LCC, 8 shades—3, 4, 5, 5.5, 6, 6.5, 7, and 8) calibrated for Indica, Japonica and hybrid rice (Yang et al., 2003), 8-panel LCC with green color scale (1–8) with percent leaf N, and their corresponding number values (Boyd, 2020). As far as size is concerned, it varied from 8 cm × 3 cm (Singh et al., 2006) to 13–20 cm × 7 cm (Houshmandfar and Kimaro, 2011). Recently, a refined version of IRRI-LCC with four green color shades (2–5) matching to spectral reflectance of plant leaves was popular and used since the year 2003 (Fairhurst et al., 2007). These have been specifically studied for high yielding rice varieties in Asia (Witt et al., 2005).

It is a useful and rapid technique for the accurate diagnosis of N deficiency in the plant (Singh, 2008). Although more commonly used on rice crop throughout the world, a common problem like yellowing due to improper N supply generally occurs in most of the crops and thus LCC finds its application for various other crops belonging to monocot class like wheat (Varinderpal-Singh et al., 2012) and maize (Varinderpal-Singh et al., 2011) with narrow leaf structure hence gaining importance (Ravi et al., 2007). The LCC is helpful to ensure an optimum fertilizer N supply within the required time interval and need (Witt et al., 2005). Optimum and judicious supply of N to crops through LCC analysis at a crucial stage provides a higher crop yield with the best quality (Sharif, 1992). The LCC is very simple, farmer friendly, and quick to use (Hussain et al., 2000; Singh et al., 2010).

Owing to its easy-to-use nature and accessibility to farmers and stake-holders, the use of LCC to monitor plant N status and its application has been reported widely by the researcher. The strategies for LCC-based N management in rice and maize have been already established (Varinderpal-Singh et al., 2011, 2007). Leghari et al. (2016)the leaf colour chart is being successfully used worldwide for the proper rate of nitrogen application and thus boosting the greatest productivity. Study was begun in 2013 using of diverse literature available on leaf colour chart from various resources. The mirror, paper, painting colours, aluminium and plastic tape, glue, plastic shopper, pencil, geometrical instruments and camel brushes of different sizes were used as material. All processes step by step completed. As result, a new modern leaf colour chart (LCC reported rice yield improvement by 19.9% to 46.2% with N application depending on LCC color. Nachimuthu et al. (2007) found saving of 50% fertilizer N by LCC-based N management in rice. Different studies have been reported to guide fertilizer N application in wheat indicating a single constant threshold LCC value of <4 (Shukla et al., 2004), 4.5 at maximum tillering (Alam et al., 2006), and 5 (Maiti and Das, 2006). According to Varinderpal-Singh et al. (2012), 25 kg N ha^{-1} at planting, 45 kg N ha^{-1} at first irrigation, and then a prescribed dose of 30 or 45 N ha^{-1} at second irrigation stage depending on LCC value of 4 reported showed improved yield and NUE. Researchers have also used the threshold values of

4 and 5 for N monitoring in crops (Gupta et al., 2011; Mathukia et al., 2014; Srinivasagam and Stephan, 2013). Singh (2006) observed an improvement in NUE with use of a single threshold value at all the growth stages of wheat and found that N application after the maximum tillering does not improve NUE. Mahajan et al. (2014) developed a critical LCC score of 4.4 at tillering and panicle initiation and 4.5 at flowering stage of hybrid rice and prescribed that corrective measures should be taken up as and when the observed values of the LCC score fall below these values at the respective growth stage. Growth stage-specific and optimum values of the LCC score were developed by Mahajan et al. (2019) for a wheat crop where they identified an optimum value of 4.5 LCC score at maximum tillering growth stage is required to be maintained to attain the economic optimum grain yield. Researchers have also used LCC as an effective tool to optimize N application and reduction in N losses as a component of climate-smart agricultural practice. Khatri-Chhetri et al. (2017) identified LCC-based N management as a nutrient smart technology suitable as a climate-smart agricultural technology.

11.2.2   CHLOROPHYLL METERS (CM)

The recent modern approach besides LCC for N monitoring is a CM. Minolta SPAD-502, most widely used CM, instantly provides an estimate of leaf N status through indirect measurement of chlorophyll. The circuit consists of two LEDs emitting wavelength of 650 nm (red) and 940 nm (infrared) and a silicon photodiode detector. These both types of radiations pass through the leaf. The transmitted light passed through the leaf is received and converted by the detector into an electrical signal. The amount of light passing through leaf is inversely related to the chlorophyll content of the clipped leaf sample. The values displayed in terms of arbitrary units (0–99.9, unitless). The CM needs to be calibrated with chlorophyll or N content. FieldScout CM 1000 CM is based on a point and shoot technique for rapid chlorophyll measurement. Its sensor adjusts with different light conditions. It can calculate and display average of multiple readings and records each sample in the built-in data logger. This CM has also been used for N management in a few crops and turfgrass.

Unlike LCC, CM provides exact values of the chlorophyll measurements (chlorophyll and N content). Thus, it is a helpful tool for efficient management of N fertilizer in different crops. It could be used as a guide to the farmers to adjusts time of N application (Vetsch and Randall, 2004).

The use of SPAD-based N management resulted in reduced N requirement (12.5%–25%) without losing the paddy yield (Singh et al., 2002).

Several researchers were attracted to undertake research on the CM for N monitoring in various crops and it has also been found very useful. Bijay-Singh and Ali (2020) reported SPAD meter correlating significantly with leaf N content and used to make the N fertilizer applications. Previously, use of SPAD is studied in crops like canola, wheat, barley, potato, and corn (Blackmer and Schepers, 1994; Follett et al., 1992; Peng et al., 1993; Piekkielek and Fox, 1992; Turner and Jund, 1991; Zhu et al., 2012). There are basically two approaches to suggest the application of fertilizer N using SPAD reading and these are absolute or threshold reference values and relative sufficiency values.

11.2.2.1   THRESHOLD VALUE APPROACH

Balasubramanian et al. (1999) reported the SPAD value 35 (dry season) as the for and 32 (wet season) to achieve high NUE and yield levels. The SPAD for fertilizer N management to irrigated rice in the south and northwest India were reported to be 37 and 37.5, respectively. SPAD threshold values in rice were reported by Hussain et al. (2003) and Ali et al. (2015) as most appropriate for supplying fertilizer N. Kyaw (2003) found improvement in boro rice yield with reduction inf the fertilizer N use (3%–12%) than local N recommendation when a strategy of SPAD 35 threshold value based N application was followed. Ali et al. (2015) identified applying N 30 kg N ha^{-1} if SPAD value is less than 37 as an effective N management strategy. SPAD value (44) based application of 20 kg N ha^{-1} was effective to improve the yield of wheat (Kyaw, 2003). Besides the grain yield advantage, Takebe et al. (2006) recommended application of 30 kg N ha^{-1} to maintain a SPAD threshold values between 50 and 52 at full heading to obtain high wheat grain protein content at maturity. They suggested 60 kg N ha^{-1} application when SPAD values fall in between 45 and 50 to get 12% or more grain protein content. Singh et al. (2002) reported a yield increase by 20% with an application of 30 kg N ha^{-1} using a SPAD threshold value of 42 at maximum tillering in wheat. Mahajan et al. (2014) developed a critical SPAD and FieldScout CM 1000 CM values 42.3 (~42) and 285 at tillering stage and 43 and 276 at panicle initiation stage and 41.7 (~42) and 270 at flowering stage of hybrid rice and prescribed that corrective measures should be taken up as and when the observed values of the LCC score fall below these values at a given crop growth stage. Growth stage-specific and optimum values of the

LCC score were developed by Mahajan et al. (2019). In wheat, an optimum SPAD and FieldScout CM 1000 values of 45 and 295 at maximum tillering should be maintained to achieve the economic optimum grain yield.

11.2.2.1   RELATIVE SUFFICIENCY VALUE APPROACH

Nitrogen management based on the relative sufficiency approach has also been reported by several authors (Bijay-Singh, 2012; Hussain et al., 2000; Saudy, 2014; Singh, 2008; Singh et al., 2006; Yu et al., 2012). Saudy (2014) found higher fertilizer NUE (about 40% less fertilizer N requirement) in wheat over the blanket recommendation upon application fertilizer N when sufficiency index value was <95% without compromising the yield. Yu et al. (2012) derived 97% as a threshold index for maize for prescriptions of fertilizer N. Following an N management practice in rice using sufficiency index limit of <90% reduced the fertilizer N amounts by 30 kg ha−1 without compensating the yield (Hussain et al., 2000). Guiding N application to rice based on sufficiency index of 90% saved 50 kg N ha−1 of the prescribed dose 120 kg N ha^{-1} (Singh et al., 2006). The potential use of the CM to monitor rice leaf N status and time of top dressing has already been established (Gholizadeh et al., 2009; Huang et al., 2008).

11.2.3   GREENSEEKER

The GreenSeeker is a recent development and a tool to monitor crop growth and biomass by recording normalized difference vegetation index (NDVI). The NDVI is measured by the red (660 ± 10 nm) and near-infrared (NIR) (780 ± 15 nm) wavelengths sensors. The NDVI is calculated using the formula: $NDVI = \rho NIR - pred\ \rho NIR + pred$ where ρNIR represents the portion of NIR radiation emitted returned from sensed area and pred is fraction of red radiation returned from the sensing area. The measurement area of GreenSeeker is 1 cm × 60 cm on the top of the canopy of crop. This sensor collects more than 10 readings per second and this information is stored in an on-board IPAQ control unit. Researchers used this tool in crops like wheat, rice, and maize to monitor the vegetation condition, biomass and canopy N.

Previous studies concerning NDVI sensor guided approach fertilizer N management was carried out mostly in commercial crops like rice, wheat and maize. Bijay-Singh et al. (2015) reported that the fertilizer N application reduced to 75–97 kg N ha^{-1} using GreenSeeker than previous application rates of 92 to 180 kg N ha^{-1} in rice at 6 out of 19 on-farm locations without compensating the rice grain yield. While investigating dry direct-seeded rice in India, Ali et al. (2015) reported grain yield using the sensor-based strategy (60–90 kg N ha^{-1} in 2–3 splits) was similar to general recommendation. Investigations by Yao et al. (2012) using GreenSeeker-based N application in rice showed an increase in the partial factor productivity by 48% (application of 90–110 kg N ha^{-1} of which 45% N as basal and 20% N at tillering). Xue et al. (2014) observed an improvement in yield with optimal N doses and higher NUE and profits in rice with sensor-based model guided fertilizer management compared to the fertilizer application practices of the farmers. Bijay-Singh et al. (2011) reported wheat grain yield with higher recovery efficiency by 6.7%–16.2% and agronomic efficiency by 4.7–9.4 kg grain kg N applied^{-1} after applying the sensor-guided dose although observed grain yield was at par with that of the standard recommendation. Cao et al. (2016) conclusively proved for wheat and maize, the sensor-based N management reduced fertilizer N dose by 62% and 36%, increased NUE by 68%–123% and 20%–61%, decreased N loss by 81% and 57% and lowered N$_2$O and greenhouse gas emission and reactive N losses by 54%–68% and 20%–42%, respectively. Likewise, Li et al. (2009) reported 61% fertilizer NUE in winter wheat GreenSeeker sensor-based N application. In one of the study carried out in maize grown in two different types of soil (black and aeolian sandy soils) revealed that the sensor-based N application reduces N use by 65% and 62% with improvement in NUE up to 40% and 65%, respectively. Canopy reflectance sensors-based N monitoring requires algorithms to convert sensor readings into N content and further N requirement.

11.3 SMARTPHONE-BASED N MONITORING IN CROPS AND ITS USE FOR N MANAGEMENT

The optimum content of plant tissue N is crucial to maintain productivity and the quality of crops. New advanced tools or models besides direct laboratory measurement techniques (e.g., Kjeldahl digestion method) to determine plant tissue N status could be explored as the conventional methods are expensive and time-consuming when large numbers of samples are needed to be analyzed. Nowadays, indirect methods based on spectral reflectance or transmittance data from the leaves or canopy of the crops plants plants (Erdle et al., 2011; Meyer et al., 1992; Muñoz-Huerta et al., 2013; Thomas and Oerther, 1972) and soils and visual inspection of the color via standards are available. Dedicated devices to measure the spectral reflectance or transmittance include CM (SPAD-502, Konica Minolta, Ramsay, NJ, USA), multispectral radiometer (MSR 16R, CropScan Inc., Rochester, MN, USA), GreenSeeker (Handheld crop sensor, Trimble Ag. Div., Westminster, CO, USA), and Yara N-Sensor (ALS2, Yara UK Ltd., NE Lincolnshire, UK) (Muñoz-Huerta et al., 2013). The principle of the tools relies on the relationship of chlorophyll content with tissue N concentration. However, most of them except LCC are costly, more technical, and used mostly for academic and research only and rural farmers might not be aware or may not have access to these products due to insufficient funds to procure them.

However, some of the tasks that are performed by the devices or tools could be replicated on the smartphones; hence, it is a useful tool to assess various agricultural tasks due to its high mobility to match with the nature of farming, highly accessible cost of the device, and their computing power allowing its wide application at field level for different purposes. Overall, a user-friendly alternative for farmers who already own a smartphone to easily downloadable applications is required. Recently, advancements in smartphones with various types of built-in physical sensors not just limit them to be used as measurement tools alone but also make them a promising tool to carry out various computations and analysis without any additional attachment. Some of the available smartphone-based N monitoring and management tools have been reviewed and discussed in this section.

11.3.1 A SMARTPHONE CONTROLLED SENSOR-BASED IMAGE ANALYZER (FLORICULTURE CROPS)

It consists of a sensor built using Raspberry Pis, camera and filters, a smartphone connected to a sensor (Adhikari et al., 2020) and remotely to a local computer via a web interface with app (MATLAB Mobile, The MathWorks Inc.). MATLAB software (R2018a, MathWorks Inc., MA, USA) on computer accesses image files on the cloud storage and processes it. It was developed with a purpose to instantaneously measure plant N status nondestructively with low-cost sensors in floriculture crops to make decisions on fertilizer N prescription. It demonstrated a linear and positive relationship between a reflectance ratio and laboratory-measured plant N and chlorophyll concentration. Further, it indicated the minimal influence of other plant tissue nutrient during accurate prediction of N content. It enables growers to make real-time decisions about the application of accurate amounts of N fertilizer for improving productivity and quality of crop and reduce environmental pollution due to N fertilizer overuse. Though such a tool is developed for the floriculture crops, it also suggests some possibility for use in field crops after due refinements and systematic research.

11.3.2 SMARTPHONES AS A PORTABLE SOIL NUTRIENT ANALYZER

This approach considers smartphone as a portable reflectometer which utilizes android application (Akvo Caddisfly) available on Google Play Store (https://play.google.com/store/apps/details?id=org.akvo.caddisfly&hl=en). It is combined Quantofix test strips, to test field soil samples, corelate the reaction of color intensity of the strips to the soil available nutrient status. The specific objective is to provide fertilizer recommendations by precise prediction at field level and smartphone-mediated soil analysis to study changes in soil nutrients (Golicz et al., 2019). Unbiased color detection, storing and geotagging results for future use and cost-effectiveness are certain advantages of its use. The app could be employed for plant-available N estimation using the nitrate-N test strips and smartphone. The results have successfully demonstrated the estimations of nitrate-N. Using the smartphone technology along with the agronomic knowledge could be utilized to manage large-scale field fertility data and its translation to the fertilizer recommendations to improve productivity and economic returns.

11.3.3 SMARTPHONE-BASED ANDROID APPLICATION—SMARTSPAD

It is an app to analyze corn leaf chlorophyll content through contact imaging method (Vesali et al., 2015). A CCD sensor-based camera of phone is used to capture the direct light passing through the leaf. To lessen or omit the noise by ambient light, the camera lens of the smartphone is held in close contact with the leaf surface. Unlike, standard image capturing, contact imaging proposes several benefits like no need of segmenting the background and elimination of variability on account of the distance between object (leaf) and sensor. Also, the difference between a focused and unfocused image is deduced. Additionally, the effect of camera-to-camera variation is also minimized. Their purpose was to estimate corn leaf SPAD values. The agreement between these at real-field conditions with the R^2 and RMSE value for validation data was close to 0.74 and 6.2, respectively. The corresponding values for linear model and neural network were 0.82 and 5.10, respectively. These models were implemented on SmartSPAD app of smartphone and evaluated again using an independent dataset. The predictions using the SmartSPAD for SPAD meter values had a prediction accuracy of $R^2 = 0.88$ and 0.72, and RMSE = 4.03 and 5.96 for neural network and linear model, respectively. It can store a large number of chlorophyll content values, or potentially logging the GPS data at the same time.

11.3.4 A MOBILE DEVICE-BASED RICE LEAF COLOR ANALYZER (BAIKHAO)

It is a smartphone-based app working as a leaf color analyzer in rice crop for indirect estimation of chlorophyll and N (Intaravanne and Sumriddetch-kajorn, 2015). The measurements are carried out under natural illumination (sunlight) using a camera-equipped mobile device with an installed application program (BaiKhao) and by keeping a simple white paper as a white reference. The idea is to capture and analyze colored two-dimensional (2D) images simultaneously from the rice leaf and its surrounding reference to omit and reduce the noise. The outcome of this approach was convincing ($R^2 = 99.8\%$) when correlated with the results of the LCC color levels. The field trial determined 1, 2, 3, and 4 level of the leaf color with prediction accuracy of 92%, 85%, 93%, and 93%, respectively. The application has program to compute amounts of the N fertilizer based on the LCC values recorded digitally.

11.3.5 NITROGEN INDEX APP IS A MOBILE APPLICATION COMPATIBLE WITH AN ANDROID SYSTEM

This smartphone or tablet application is available for download at Google Play™ website (https://play.google.com/store/apps/details?id=gov.usda.ars.spnr.driver&hl=en). This technology enables the use of smartphones for on-site evaluation and visit of the farmer at a given site for nitrogen loss risk (Delgado et al., 2013). The app-based assessment correlated ($R^2 = 0.88$) with the observed nitrate values. It provides a virtual simulation for N leaching to be conducted and simultaneously saves the data in the XML-based file format (.NIN), which can be accessed via a Wi-Fi network or by connecting the computer or laptop.

11.3.6   YARA SMARTPHONE N SENSOR AND N TESTER CHIP (YARAIRIX) (HTTPS://WWW.YARA.COM/NEWS-AND-MEDIA/ NEWS/ARCHIVE/2019/YARA-TRANSFORMS-SMARTPHONES-INTO-NITROGEN-SENSORS/)

It is available for Android and iOS users who use the mobile phone camera to predict N demands of crops at earlier stage of growth. It is supplemented by two hardware options (a smartphone clip and a Bluetooth N-Tester) to record the observation at later stages of the growth of a crop. The core function is to carry out the measurement of chlorophyll levels. This enables the farmers to assess his crop N needs and to get prescriptions to N fertilizer amounts.

11.4   VARIABLE RATE FERTILIZER NITROGEN MANAGEMENT

Owing to the losses due to suboptimal application and adverse effects due to overuse, economical and smart methods and tools to compute and apply required amount of N fertilizer are needed. The smart variable rate N delivery systems seem to be the key for nutrient management that can satisfy the minimum of placing right rate of nutrient at the right place. Variable Rate Technology (VRT), an important part of such smart N delivery systems, is used to apply nitrogenous fertilizers to specific levels appropriate to the management zone.

11.4.1 VARIABLE RATE N APPLICATION TECHNOLOGY (VRNAT)

Generally, a Variable rate N application technology (VRNAT) system includes fertilizing rate decision and variable-rate implementation subsystem. The method used for site-specific variable rate implementation defines the hardware and software components of the subsystem. The VRNAT emphasizes on predicting site-specific fertilizer rates and the implementation of N management schemes which takes into account the variations in crop N demand at a specific growth stage or time of growth.

11.4.2   SITE-SPECIFIC N TOPDRESSING RECOMMENDATION APPROACHES

11.4.2.1   LEAF AREA INDEX (LAI) APPROACH

In this approach time of each N application, LAI is computed using the vegetation indices of canopy spectral data. The N fertilizer dose required adjustments (increased or decreased) as per the value of the LAI. The quantities of the fertilizer N are calculated based on the LAI and canopy N required for production of unit LAI. This approach was first given by Wood et al. (2003) for wheat, and later modified by Xue and Yang (2008) for high-yielding rice varieties.

11.4.2.2 N NUTRITION INDEX (NNI) APPROACH

The NNI was estimated using NDVI of the N fertilizer plot ($NDVI_{fert}$) divided by that of optimally N fertilized plot ($NDVI_{ref}$). If NNI is less than 1, indicating an N deficit, higher amounts of N fertilizer is to be applied. If NNI greater than 1, showing N excess, a reduced dose of N fertilizer has to be applied. The N fertilizer to be applied was computed using the data on $NDVI_{fert}$, $NDVI_{ref}$ and the standard dose of N (Denuit et al., 2002).

11.4.2.3 N FERTILIZER OPTIMIZATION ALGORITHM (NFOA) APPROACH

Raun et al (2002) used this approach to calculate an in-season estimate of grain yield (INSEY) with the help of DAS (or DAT) and NDVI. Using the INSEYS, the predicted potential grain yield (kg ha^{-1}) can be estimated. The predicted early-season plant N uptake (PNU, kg ha^{-1}) is computed using NDVI. The critical NUE of an in-season N application for the high-yield crop was set between 0.0 and 1.0. The in-season top-dress fertilizer N requirement (kg ha^{-1}) is determined as = ((Grain N uptake – predicted early season N uptake)/NUE).

11.4.3 VARIABLE-RATE IMPLEMENTATION SUBSYSTEM

The methods of implementing site-specific variable rate implementation within a field include map-based, on-the-go sensor-based, and a combination of sensor-based systems with map overlay (Swinton and Lowenberg-Deboer, 2001; Zhang et al., 2002). The fertilizing rates are obtained from the fertilizing rate decision subsystem at real-time (using on the go sensors) or pre-fed values (map-based).

11.4.3.1 MAP-BASED APPROACH

In map-based approach, the information is collected using the sensors and the collected information could be used by the stakeholders over a period or later for decision-making on inputs. The prescription maps are prepared using these maps by using the GIS tools. This prescription map is then placed in a VR controller on an implement or prime mover to guide VRA of the inputs (Khanna, 2001; Chandel et al., 2016). This approach can save the cost and time involved in the sampling and testing of the soil nutrients.

11.4.3.2 ON-THE-GO SENSOR-BASED APPROACH

In this approach, the needed deterministic parameters are predicted by the sensors from the soil, crop, or environment based on calibration and validation with set points. Further data on these parameters in real-field conditions could be obtained by mounting sensors on the front of implements. Sensors work in real-time, and gather information on soil and other parameters, computes the needs, and deliver a customized amount of fertilizer on site-specific basis at the rear. Decisions on both to record observations and application of input is done simultaneously using the sensor-based VRT (Swinton, 2005). Sensor-based VRT methods have been developed for the application of N (Raun et al., 2002) GreenSeeker, RT-200, N-Yara (Gaikwad, 2016).

11.4.3.3 REAL-TIME WITH MAP OVERLAY

This approach combines the benefits of sensor and map-based VRFA systems mentioned above. The controller acquired the data from GPS, digital maps, and sensor for further processing and possible actions.

11.4.4 COMPONENTS OF VARIABLE RATE N FERTILIZATION SYSTEM

The variable rate N fertilization system comprises of sensors, electronically controllable fertilizer metering devices, and dispensing mechanisms for rate control and placement at the desired location. Its metering device consists of actuator motor (variable speed type or constant speed type coupled with stepless variable transmission) for controlling the amount of granular N fertilizer dispensed from the designed discharger at a site in the field. A proportional electromagnetic solenoid flows control valve with associated control systems is used for liquid and gaseous forms of N fertilizer. The associated control systems have either electro-mechanical, electro-hydraulic, electro-pneumatic hardware components or its combinations, and associated software and firmware mean to achieve three major functions: initializing sensors with an appropriate scanning mode, streaming of measurement data and its transmission. An information management system integrates all parts of the variable-rate N application system. This system should be able to store the collected data for further operations. It should be able to manage different data formats exchange with services that gives computation for fertilizer application.

The design of the components of the variable-rate N fertilization system discussed above depends on the method of fertilizer application. Though manufactured by many, the basic application methods depend on its form, that is, solid, liquid, or gaseous. The availability of fertilizer type at geolocation is dependent on the economic considerations, regulations governing the geolocation, and the target crop. The developed countries of the European Union and the USA have availability and access to all forms of nitrogenous fertilizers, whereas it is mostly restricted to granular fertilizers in the developing. In India, solid inorganic fertilizers are predominantly used because of their easy availability at subsidized rates. The adoption of the suitable VRNAT at geolocation is therefore governed by the availability of fertilizer form, the suitability of crops grown in the area, and availability of crop-specific calibration factors of algorithms governing the fertilizing rate decision subsystem.

11.4.5   PROFITABILITY AND EFFICIENCY

Impact of VRNAT on profitability and efficiency of uniform rate and variable rate application in different crops has been studied by a number of researchers and is summarized in Table 11.1.

11.4.6   SCOPE AND STATUS OF VRNAT AND SSNM IN SOME DEVELOPING COUNTRIES

The VRNAT developed in advanced countries is a challenge for practical adoption in developing countries due to mismatching of the VRNAT to suit small size of the farm holdings and economic constraints. However, developing suitable-sized VRNAT for developing countries has been under consideration due to rapid socio-economical changes in the developing countries. The level of adoption the technologies associated with the precision agriculture and VRNAT varies region to region. Appropriately sized and affordable equipment to suit the farm need for the VRNAT needs to be developed. Like other parts of the world, the rate of adoption is slow and uneven in the developing countries due to various reasons. Some of the practical and affordable technologies for VRNAT and SSNM are summarized for small-sized farms.

TABLE 11.1   Variable Rate Application and Its Effect on Yield, NUE, and N Saving

Treatment	Authors	Crop	Nutrients	Response
Variable rate	Yang et al, 2001	Sorghum	N & P	369 kg ha^{-1} yield increase
Variable rate	Raun et al., 2002	Cereal grain	N	15% increase in NUE
RTNM	Singh et al., 2002	Rice	N	20–30% less fertilizer
SSNM	Dobermann et al 2002	Rice	N & P	13 and 21% nutrient uptake increase
SSNM	Peng Xian-Longl et al 2006	Rice	N	Decreased the average N rate by 33.8%
RTNM	Gaikwad, 2016	Wheat and rice	N	8%–15% urea savings

11.4.6.1   SOFTWARE FOR COMPUTING TOP DRESSING DOSE OF N

A multilingual android app (version 1.02) and an online web app were created and launched by ICAR-Central Institute of Agricultural Engineering (CIAE), Bhopal, India, to calculate the N top dress amounts by calculating the NDVI values recorded by GreenSeeker sensor for the wheat and rice crops (Gaikwad, 2014).

The rate calculation is based on a modified N fertilizer optimization algorithm (Modified NFOA) and customized equations for limited varieties of rice and wheat have been modeled to compute top-dress N dose.

11.4.6.2   VARIABLE-RATE SYSTEM FOR UREA APPLICATION IN RICE AND WHEAT CROPS FOR TOP N DRESSING

An on-the-go variable rate urea application system based on the NDVI sensor (GreenSeeker) was developed for mid-season top dressing urea on rice and

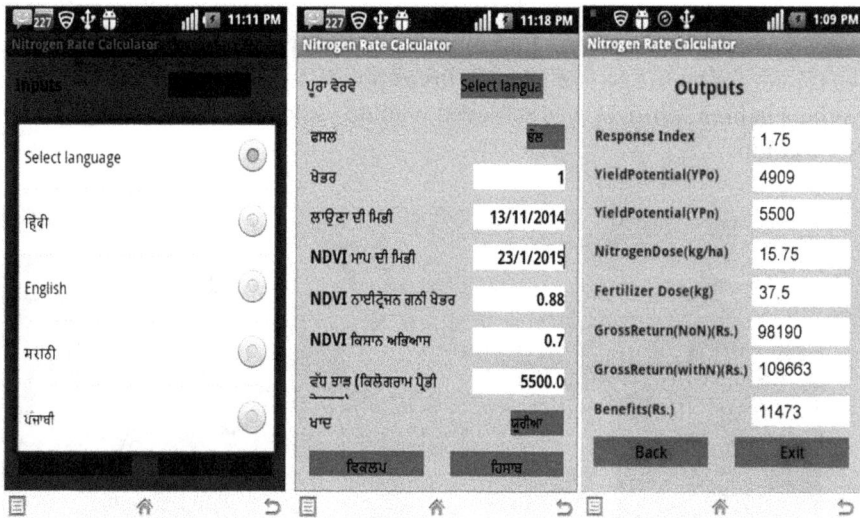

FIGURE 11.1   User interface of the nitrogen rate calculator.

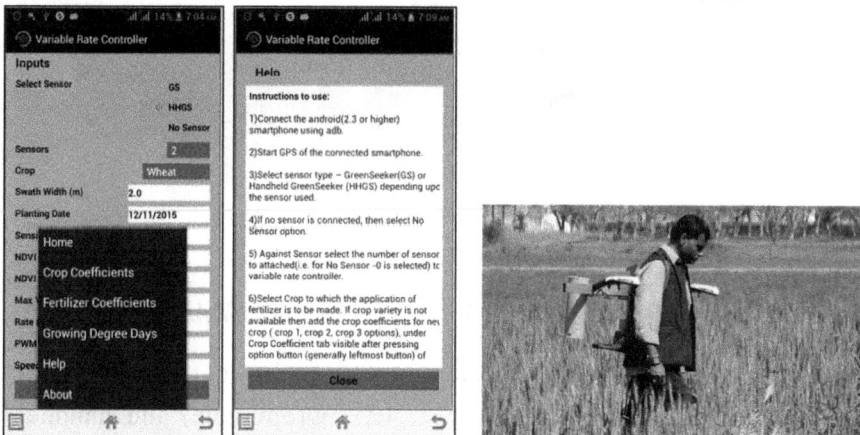

FIGURE 11.2   User interface of the variable rate controller and variable rate urea application system in wheat.

wheat crops at ICAR-CIAE, Bhopal. The applicator can be mounted on the back of the operator and covers a swath width of 4 m. The applicator consists of two GreenSeeker sensors which sense the crop NDVI values. A "Variable Rate Controller" android app installed on an android phone controls the applicator for metering 8.5–30 kg/ha N (18.5–65 kg of urea/ha) at 2 km/h forward speed with 25 mm wide fluted roller. The controller of the applicator is designed for scaling up to four modular units for customized mounting on the tractor for on the go site-specific urea application. An estimated 8%–15% savings in urea fertilizer was achieved with no yield loss in areas with spatial N variation.

FIGURE 11.3   Low-cost CIAE SPAD meter.

11.4.6.3   *LOW-COST CIAE SPAD METER (GAIKWAD ET AL., 2017)*

The unaffordability of commercially available hand-held canopy reflectance sensors (GreenSeeker, OptRx, CropCircle) and handheld CMs has limited the adoption of SSNM in developing countries. A prototype low-cost handheld SPAD meter developed at ICAR-CIAE, Bhopal, India will likely encourage the adoption of SSNM due to its low cost (<\$100) and affordability to small-scale farmers when used on community basis custom hiring. The CIAE-SPAD meter is plugged to OTG-enabled android smartphone for display and data logging of modified SPAD values that can correlate well to the chlorophyll content of the crop and subsequent assessment of the N requirement of the crop using a critical SPAD value approach.

11.5   WAY FORWARD

A plethora of information is available about the plant and soil N management using different diagnostic tools. However, there are still issues that need to be considered while planning future research and development programs. At present, promising technologies are available suitable for diverse systems and has merits and demerits. Suggestions for future research and development in the form of a way forward have been compiled here based on the review made for this chapter.

- Developing a robust technology that could be adopted easily seems to be a needed research area at present. Efforts should be focused to reduce the complexity and cost and improving ease of handling and accessibility of the technology. The diagnostic tools developed should be brought on to a platform easily available to farmers, for instance, smartphones. Also, in this regard, LCC shows some promise being low-cost, easy-to-use and inexpensive. The other tools are suitable for undertaking the research rather than its wide-scale field use and adoption due to cost and complexity in handling.
- Smartphone-based technologies show a promise to act as a platform to host the N diagnostic tools but need rigorous research and calibrations. Though such techniques have been developed, they have not yet been widely adopted by the farmers and stakeholders.
- Focused research on reducing the cost and complexity of the N diagnostic tools will help for its wide-scale adoption.
- Currently, the use of the diagnostic tools is mostly tested in isolation and most specifically in an environment where variability with respect to N is created. Using these tools in a complex environment to develop the climate-smart and energy-smart agricultural practices could add to their utility.

KEYWORDS

- **chlorophyll meter**
- **digital tools for nitrogen management**
- **nitrogen losses**
- **nitrogen use efficiency**

REFERENCES

Adhikari, R., Li, C., Kalbaugh, K., Nemali, K., 2020. A low-cost smartphone controlled sensor based on image analysis for estimating whole-plant tissue nitrogen (N) content in floriculture crops. Comput. Electron. Agric. 169, 105173.

Alam, M.M., Ladha, J.K., Foyjunnessa, Rahman, Z., Khan, S.R., Harun-ur-Rashid, Khan, A.H., Buresh, R.J., 2006. Nutrient management for increased productivity of rice–wheat cropping system in Bangladesh. F. Crop. Res. 96, 374–386. https://doi.org/10.1016/j.fcr.2005.08.010

Ali, A.M., Thind, H.S., Sharma, S., Singh, Y., 2015. Site-specific nitrogen management in dry direct-seeded rice using chlorophyll meter and leaf colour chart. Pedosphere 25, 72–81. https://doi.org/10.1016/S1002-0160(14)60077-1

Balasubramanian, V., Morales, A.C., Cruz, R.T., Abdulrachman, S., 1999. On-farm adaptation of knowledge-intensive nitrogen management technologies for rice systems, in: Resource Management in Rice Systems: Nutrients. Springer, pp. 79–93.

Bijay-Singh, 2012. Plant-Need Based Nitrogen Management in Rice and Wheat. New Delhi, India.

Bijay-Singh, Ali, A.M., 2020. Using hand-held chlorophyll meters and canopy reflectance sensors for fertilizer nitrogen management in cereals in small farms in developing countries. Sensors 20, 1127. https://doi.org/10.3390/s20041127

Bijay-Singh, Sharma, R.K., Jaspreet-Kaur, Jat, M.L., Martin, K.L., Yadvinder-Singh, Varinderpal-Singh, Chandna, P., Choudhary, O.P., Gupta, R.K., Thind, H.S., Jagmohan-Singh, Uppal, H.S., Khurana, H.S., Ajay-Kumar, Uppal, R.K., Vashistha, M., Raun, W.R., Gupta, R., 2011. Assessment of the nitrogen management strategy using an optical sensor for irrigated wheat. Agron. Sustain. Dev. 31, 589–603. https://doi.org/10.1007/s13593-011-0005-5

Bijay-Singh, Varinderpal-Singh, Purba, J., Sharma, R.K., Jat, M.L., Yadvinder-Singh, Thind, H.S., Gupta, R.K., Chaudhary, O.P., Chandna, P., Khurana, H.S., Kumar, A., Jagmohan-Singh, Uppal, H.S., Uppal, R.K., Vashistha, M., Gupta, R., 2015. Site-specific fertilizer nitrogen management in irrigated transplanted rice (Oryza sativa) using an optical sensor. Precis. Agric. 16, 455–475. https://doi.org/10.1007/s11119-015-9389-6

Blackmer, T.M., Schepers, J.S., 1994. Techniques for monitoring crop nitrogen status in corn. Commun. Soil Sci. Plant Anal. 25, 1791–1800. https://doi.org/10.1080/00103629409369153

Boyd, V., 2020. Easy-to-use Leaf Color Chart helps growers gauge crop's mid-season nitrogen needs. [WWW Document]. URL https://www.ricefarming.com/departments/feature/the-color-green/ (accessed 4.11.20).

Cao, Q., Miao, Y., Shen, J., Yu, W., Yuan, F., Cheng, S., Huang, S., Wang, H., Yang, W., Liu, F., 2016. Improving in-season estimation of rice yield potential and responsiveness to topdressing nitrogen application with crop circle active crop canopy sensor. Precis. Agric. 17, 136–154. https://doi.org/10.1007/s11119-015-9412-y

Chandel, N.S., Mehta, C.R., Tewari, V.K., Nare, B., 2016. Digital map-based site-specific granular fertilizer application system. Curr. Sci. 1208–1213.

De Datta, S.K., Samson, M.I., Kai-Rong, W., Buresh, R.J., 1988. NitrogeNUE and nitrogen-15 balances in broadcast-seeded flooded and transplanted rice. Soil Sci. Soc. Am. J. 52, 849–855. https://doi.org/10.2136/sssaj1988.03615995005200030045x

Delgado, J.A., Kowalski, K., Tebbe, C., 2013. The first Nitrogen Index app for mobile devices: Using portable technology for smart agricultural management. Comput. Electron. Agric. 91, 121–123. https://doi.org/10.1016/j.compag.2012.12.008

Denuit, J.P., Olivier, M., Goffaux, M.J., Herman, J.L., Goffart, J.P., Destain, J.P., Frankinet, M., 2002. Management of N fertilization of winter wheat and potato crops using the chlorophyll meter for crop N status assessment. Agronomie. 22, 847–853.

Dobermann, A., Fairhurst, T., 2000. Rice: Nutrient disorders & nutrient management, 1st ed. Philippines: International Rice Research Institute (IRRI).

Dobermann, A., Witt, C., Abdulrachman, S., Gines, H. C., Nagarajan, R., Son, T.T., Tan, P. S.,Wang, G.H., Chien, N.V., Thoa, V.T.K., Phung, C.V., Stalin, P., Muthukrishnan, P., Ravi, V., Babu, M., Simbahan, G.C., Adviento, M.A.A., 2002. Site-specific nutrient management for intensive rice cropping systems in Asia. Field Crops Res. 74, 37–66.

Erdle, K., Mistele, B., Schmidhalter, U., 2011. Comparison of active and passive spectral sensors in discriminating biomass parameters and nitrogen status in wheat cultivars. F. Crop. Res. 124, 74–84.

Fairhurst, T., Witt, C., Buresh, R., Dobermann, A., Fairhurst, T., 2007. Rice: A practical guide to nutrient management. Philippines: International Rice Research Institute (IRRI).

Follett, R.H., Follett, R.F., Halvorson, A.D., 1992. Use of a chlorophyll meter to evaluate the nitrogen status of dryland winter wheat. Commun. Soil Sci. Plant Anal. 23, 687–697.

Furuya, S., 1987. Growth diagnosis of rice plants by means of leaf color. Jpn Agric Res Q 20, 147–153.

Gaikwad, B.B., 2014. Software for estimation of top dressing dose of nitrogen, CIAE Newsletter 23(3–4), 2.

Gaikwad, B.B., 2016. Spectral reflectance based prototype of variable rate urea application system for top dressing in rice and wheat crops, CIAE Newsletter 26(2), 4.

Gaikwad, B.B., 2017. Low cost SPAD meter, CIAE Newsletter 27(1), 1–2.

Gholizadeh, A., Amin, M.S.M., Anuar, A.R., Aimrun, W., 2009. Evaluation of leaf total nitrogen content for nitrogen management in a Malaysian paddy field by using soil plant analysis development chlorophyll meter. Am. J. Agric. Biol. Sci. 4, 278–282.

Golicz, K., Hallett, S.H., Sakrabani, R., Pan, G., 2019. The potential for using smartphones as portable soil nutrient analyzers on suburban farms in central East China. Sci. Rep. 9, 16421. https://doi.org/10.1038/s41598-019-52702-8

Gupta, R.K., Singh, V., Singh, Y., Singh, B., Thind, H.S., Kumar, A., Vashistha, M., 2011. Need-based fertilizer nitrogen management using leaf colour chart in hybrid rice (Oryza sativa). Indian J. Agric. Sci. 81, 1153.

Houshmandfar, A., Kimaro, A., 2011. Calibrating the leaf color chart for rice nitrogen management in Northern Iran. African J. Agric. Res. 6, 2627–2633.

Huang, J., He, F., Cui, K., Buresh, R.J., Xu, B., Gong, W., Peng, S., 2008. Determination of optimal nitrogen rate for rice varieties using a chlorophyll meter. F. Crop. Res. 105, 70–80.

Hussain, F., Bronson, K.F., Peng, S., 2000. Use of chlorophyll meter sufficiency indices for nitrogen management of irrigated rice in Asia. Agron. J. 92, 875–879. https://doi. org/10.2134/agronj2000.925875x

Hussain, F., Zia, M.S., Akhtar, M.E., Yasin, M., 2003. Nitrogen management and use efficiency with chlorophyll meter and leafcolour chart. Pakistan: National AgriculturalResearch Centre.

IFASTAT [WWW Document], 2020. URL https://www.ifastat.org/databases/plant-nutrition (accessed 4.11.20).

Intaravanne, Y., Sumriddetchkajorn, S., 2015. Android-based rice leaf color analyzer for estimating the needed amount of nitrogen fertilizer. Comput. Electron. Agric. 116, 228–233. https://doi.org/10.1016/j.compag.2015.07.005

IRRI, 1996. Use of leaf colour chart (LCC) for N management in rice, in: Crop Resour. Manage. Network Technol. Brief 2. IRRI: Manila, Philippines.

Khanna, M. 2001. Sequential adoption of site-specific technologies and its implications for nitrogen productivity: a double-selectivity model. Amer. J. Agr. Econ. 83, 35–51.

Khatri-Chhetri, A., Aggarwal, P.K., Joshi, P.K., Vyas, S., 2017. Farmers' prioritization of climate-smart agriculture (CSA) technologies. Agric. Syst. 151, 184–191. https://doi.org/10.1016/j.agsy.2016.10.005

Kyaw K.K, 2003. Plot-specific N fertilizer management for improved N-use efficiency in rice-based systems of Bangladesh. Cuvillier, Göttingen, Germany.

Leghari, S.J., Leghari, U.A., Buriro, M., Laghari, G.M., Soomro, F.A., Khaskheli, M.A., Hussain, S.S., 2016. Modern leaf colour chart successfully prepared and used in crop production of Sindh, Pakistan. Eur. Acad. Res. 4, 900–916.

Li, F., Miao, Y., Zhang, F., Cui, Z., Li, R., Chen, X., Zhang, H., Schroder, J., Raun, W.R., Jia, L., 2009. In-season optical sensing improves nitrogen-use efficiency for winter wheat. Soil Sci. Soc. Am. J. 73, 1566–1574. https://doi.org/10.2136/sssaj2008.0150

Mahajan, G.R., Pandey, R.N., Datta, S.C., Kumar, D., Sahoo, R.N., 2019. Monitoring wheat (Triticum aestivum L.) leaf nitrogen using diagnostic tools for fertilizer nitrogen management. J. Indian Soc. Soil Sci. 67, 329. https://doi.org/10.5958/0974-0228.2019.00036.7

Mahajan, G.R., Pandey, R.N., Kumar, D., Datta, S.C., Sahoo, R.N., Parsad, R., 2014. Development of critical values for the leaf color chart, SPAD and fieldscout CM 1000 for fixed time adjustable nitrogen management in aromatic hybrid rice (Oryza sativa L.). Commun. Soil Sci. Plant Anal. 45, 1877–1893. https://doi.org/10.1080/00103624.2014.909832

Maiti, D., Das, D.K., 2006. Management of nitrogen through the use of leaf colour chart (LCC) and soil plant analysis development (SPAD) in wheat under irrigated ecosystem. Arch. Agron. Soil Sci. 52, 105–112. https://doi.org/10.1080/03650340500460875

Mathukia, R.K., Gajera, K.D., Mathukia, P.R., 2014. Validation of leaf colour chart for real time nitrogen management in wheat. J. Dyn. Agric. Res. 1, 1–4.

Meyer, G.E., Troyer, W.W., Fitzgerald, J.B., Paparozzi, E.T., 1992. Leaf nitrogen analysis of poinsettia (Euphorbia Pulcherrima Will D.) using spectral properties in natural and controlled lighting. Appl. Eng. Agric. 8, 715–722.

Muñoz-Huerta, R.F., Guevara-Gonzalez, R.G., Contreras-Medina, L.M., Torres-Pacheco, I., Prado-Olivarez, J., Ocampo-Velazquez, R. V, 2013. A review of methods for sensing the nitrogen status in plants: advantages, disadvantages and recent advances. Sensors 13, 10823–10843.

Nachimuthu, G., Velu, V., Malarvizhi, P., Ramasamy, S., Gurusamy, L., 2007. Home journals contact. J. Agron. 6, 338–343. https://doi.org/10.3923/ja.2007.338.343

PENG Xian-long, LIU Yuan-Ying, LUO Sheng-Guo, FAN Li-Chun, SONG Tian-Xing and GUO Yan-wen. 2007. Effects of site-specific nitrogen management on yield and dry matter accumulation of rice from cold areas of northeastern China. Agric. Sci. China. 6(6), 715–723.

Peng, S., García, F. V, Laza, R.C., Cassman, K.G., 1993. Adjustment for specific leaf weight improves chlorophyll meter's estimate of rice leaf nitrogen concentration. Agron. J. 85, 987–990. https://doi.org/10.2134/agronj1993.00021962008500050005x

Piekkielek, W.P., Fox, R.H., 1992. Use of a chlorophyll meter to predict sidedress nitrogen requirements for maize. Agron. J. 84, 59–65.

Raun, W.R., J.B. Solie, G.V. Johnson, M.L. Stone, R.W. Mullen, K.W. Freeman, W.E. Thomason, and E.V. Lukina. 2002. Improving nitrogeNUE in cereal grain production with optical sensing and variable rate application. Agron. J. 94, 815–820.

Ravi, S., Ramesh, S., Chandrasekaran, B., 2007. Exploitation of hybrid vigour in rice hybrid (Oryza sativa L.) through green manure and leaf colour chart (LCC) based N application. Asian J. Plant Sci. 6, 282–287. https://doi.org/10.3923/ajps.2007.282.287

Saudy, H.S., 2014. Chlorophyll meter as a tool for forecasting wheat nitrogen requirements after application of herbicides. Arch. Agron. Soil Sci. 60, 1077–1090.

Sharif, Z., 1992. Nitrogen fertilizer use efficiency in flooded rice soils, in: In: Proceeding 4th National Congress Soil Science. Islamabad, pp. 141–147.

Shukla, A.K., Ladha, J.K., Singh, V.K., Dwivedi, B.S., Balasubramanian, V., Gupta, R.K., Sharma, S.K., Singh, Y., Pathak, H., Pandey, P.S., 2004. Calibrating the leaf color chart for nitrogen management in different genotypes of rice and wheat in a systems perspective. Agron. J. 96, 1606–1621.

Singh, B., 2008. Crop demand-driven site-specific nitrogen applications in rice (Oryza sativa) and wheat (Triticum aestivum): some recent advances. Indian J. Agron. 53, 157–166.

Singh, B., Gupta, R.K., Singh, Y., Gupta, S.K., Singh, J., Bains, J.S., Vashishta, M., 2006. Need-based nitrogen management using leaf color chart in wet direct-seeded rice in northwestern India. J. New Seeds 8, 35–47.

Singh, B., Singh, Y., Ladha, J.K., Bronson, K.F., Balasubramanian, V., Singh, J., Khind, C.S., 2002. Chlorophyll meter and leaf color chart-based nitrogen management for rice and wheat in Northwestern India. Agron. J. 94, 821–829.

Singh, G., 2006. Nitrogen management in wheat using chlorophyll meter and leaf colour chart. M.Sc. Thesis. Punjab Agricultural University, Ludhiana, India.

Singh, H., Sharma, K.N., Dhillon, G.S., Singh, T.A., Singh, V., Kumar, D., Singh, H., 2010. On-farm evaluation of real-time nitrogen management in rice. Better Crop 94, 26–28.

Srinivasagam, K., Stephan, H., 2013. Integrated nutrient management and LCC based nitrogen management on soil fertility and yield of rice (Oryza sativa L.). Sci. Res. Essays 8, 2059–2067. https://doi.org/10.5897/SRE2013.5643

Swinton, S.M., Lowenberg-Deboer, J. 2001. Global adoption of precision agriculture technologies: who, when, why? In: Proceedings of the Third European Conference on Precision Agriculture ENSA-Montpellier, pp. 557–562.

Swinton, S.M., 2005. Economics of site-specific weed management. Weed Sci. 53, 259–263.

Takebe, M., Okazaki, K., Karasawa, T., Watanabe, J., Ohshita, Y., Tsuji, H., 2006. Leaf color diagnosis and nitrogen management for winter wheat "Kitanokaori" in Hokkaido. Soil Sci. Plant Nutr. 52, 577.

Thomas, J.R., Oerther, G.F., 1972. Estimating nitrogen content of sweet pepper leaves by reflectance measurements. Agron. J. 64, 11–13.

Turner, F.T., Jund, M.F., 1991. Chlorophyll meter to predict nitrogen topdress requirement for semidwarf rice. Agron. J. 83, 926–928.

Varinderpal-Singh, Bijay-Singh, Yadvinder-Singh, Thind, H.S., Gobinder-Singh, Satwinderjit-Kaur, Kumar, A., Vashistha, M., 2012. Establishment of threshold leaf colour greenness for need-based fertilizer nitrogen management in irrigated wheat (Triticum aestivum L.) using leaf colour chart. F. Crop. Res. 130, 109–119. https://doi.org/10.1016/j.fcr.2012.02.005

Varinderpal-Singh, D., Yadvinder-Singh, Bijay-Singh, Baldev-Singh, Gupta, R.K., Jagmohan-Singh, Ladha, J.K., Balasubramanian, V., 2007. Performance of site-specific nitrogen management for irrigated transplanted rice in northwestern India. Arch. Agron. Soil Sci. 53, 567–579.

Varinderpal-Singh, Yadvinder-Singh, Bijay-Singh, Thind, H.S., Kumar, A., Vashistha, M., 2011. Calibrating the leaf colour chart for need based fertilizer nitrogen management in different maize (*Zea mays* L.) genotypes. F. Crop. Res. 120, 276–282. https://doi. org/10.1016/j.fcr.2010.10.014

Vesali, F., Omid, M., Kaleita, A., Mobli, H., 2015. Development of an android app to estimate chlorophyll content of corn leaves based on contact imaging. Comput. Electron. Agric. 116, 211–220. https://doi.org/10.1016/j.compag.2015.06.012

Vetsch, J.A., Randall, G.W., 2004. Corn production as affected by nitrogen application timing and tillage. Agron. J. 96, 502–509.

Wang, X., Chen, Y., Sui, P., Yan, P., Yang, X., Gao, W., 2017. Preliminary analysis on economic and environmental consequences of grain production on different farm sizes in North China Plain. Agric. Syst. 153, 181–189. https://doi.org/10.1016/j.agsy.2017.02.005

Witt, C., Pasuquin, J., Mutters, R., Buresh, R.J., 2005. New leaf color chart for effective nitrogen management in rice. Better Crop. 89, 36–39.

Wood, G., Welsh, J., Godwin, R., Taylor, J., Earl, R. and Knight, S., 2003. Realtime measures of canopy size as a basis for spatially varying nitrogen applications to winter wheat sown at different seed rates. Biosyst. Eng. 84, 513–531.

Xue, L. and Yang, L.Z., 2008. Recommendations for nitrogen fertiliser topdressing rates in rice using canopy reflectance spectra. Biosyst. Eng. 100, 524–534.

Xue, L., Li, G., Qin, X., Yang, L., Zhang, H., 2014. Topdressing nitrogen recommendation for early rice with an active sensor in south China. Precis. Agric. 15, 95–110.

Yang, J., Everitt, H., Bradford, J.M., 2001. Comparison of uniform and variable rate nitrogen and phosphorus fertilizer application for grain sorghum. Trans. ASAE 44(2), 201–209.

Yang, W.-H., Peng, S., Huang, J., Sanico, A.L., Buresh, R.J., Witt, C., 2003. Using leaf color charts to estimate leaf nitrogen status of rice. Agron. J. 95, 212–217.

Yao, Y., Miao, Y., Huang, S., Gao, L., Ma, X., Zhao, G., Jiang, R., Chen, X., Zhang, F., Yu, K., 2012. Active canopy sensor-based precision N management strategy for rice. Agron. Sustain. Dev. 32, 925–933.

Yu, W., Miao, Y., Feng, G., Yue, S., Liu, B., 2012. Evaluating different methods of using chlorophyll meter for diagnosing nitrogen status of summer maize, in: 2012 First International Conference on Agro-Geoinformatics (Agro-Geoinformatics). IEEE, pp. 1–4.

Zhang, N., Wang, M., Wang, N. 2002. Precision agriculture—a worldwide overview. Comput. Electron. Agr. 36, 113–132.

Zhu, J., Tremblay, N., Liang, Y., 2012. Comparing SPAD and at LEAF values for chlorophyll assessment in crop species. Can. J. Soil Sci. 92, 645–648.

CHAPTER 12

Innovative Extension Approaches for Diffusion of Nutrient Management Technologies

SUJIT SARKAR[1*], G. S. MAHRA[2], V. LENIN[2], R. N. PADARIA[2], and R. R. BURMAN[2]

[1]Indian Agricultural Research Institute (IARI), Regional Station, Kalimpong, West Bengal, India

[2]Division of Agricultural Extension, IARI, New Delhi, India

*Corresponding author. E-mail: sujitgovt@gmail.com

ABSTRACT

Nutrient management has evolved to a new concept with the progress of science and technology. The contemporary nutrient management technologies not only deals with production concerns but also deals with issues like protecting water from nutrient losses, protecting environmental quality, and maintaining the safe livelihood. However, the adoption rate of these new technologies is very poor. Hence, different innovative extension approaches were devised and tested. This chapter will discuss those innovative approaches and analyses their impact on soil as well as social subsystem. The major extension approaches on soil health management are—general extension approach which includes ministry-based program like soil health card, soil health management program, etc. Commodity-based extension approach, namely, Field friend, KSHEMAM, RubSis, etc. Participatory extension approach like Technology Transfer and Service Center of Thailand, Network-based approach of North Dakota, T&V system extension approach through KVK and ATMA, ICT-based extension approaches like KCC, mKrishi , Micro-mitra, Soil Nutrient Manager, community radio stations (CRS), *Nutrient expert tool,* IoT/Smartphone-Based Sensors apps like *BaiKhao, SOCiT, PocketLAI,* etc. Project-based extension approach, namely, operational research project, institution-based extension approach, namely, extension approaches of Bharatia Agro-Industries Foundation (BAIF), Maharashtra State Grapes Grower Association, etc. University-based extension approach of the University of Delaware, University of Wisconsin, IARI (Pusa), private extension approach like Mahindra Krishi Vihar, IFFCO Kisan, *Mahadhan Saarrthie Centres, Kisan Suvidha Kendras (KSK),* etc. The knowledge of these approaches will help the policy makers to decide which approach to follow in given situation for maximum impact.

12.1 INTRODUCTION

Agriculture has always relied on the human management of soil, plant nutrients, and other natural resources. Nature has its own balance approach to maintain the equilibrium among these human, soil, plant nutrients, and other natural resources. However, intensive agriculture practiced since green revolution has caused multiple problems of nutrient imbalance, namely, depleting soil fertility, deficiency of secondary and micronutrients, declining water table level and its quality, decreasing organic carbon content, increased soil erosion, and degradation and thus deteriorated the overall soil health.

Soil health, as an attribute of physical, chemical, and biological processes is constantly declining and is considered as one of the reasons for declining crop yields. To ensure sustainable crop production, it becomes essential to understand the soil ecosystem that supports the life and growth of different crops, plants, microorganism, etc., and the functional behavior of different nutrient management technologies on this ecosystem. Nutrient management is the science and practice of connecting soil, crop, weather, and hydrologic factors with cultural, irrigation, soil, and water conservation activities to obtain maximum nutrient use efficiency, crop yields, crop quality, and economic returns. It also involves the principle of reducing off-site leakage of nutrients (fertilizer) that may affect the ecosystem. It is a process of finding an optimum combination of specific soil, climate, and crop management conditions to rate, source, timing, and place (i.e. 4R nutrient stewardship) of nutrient application. A significant number of past studies (Bates and Arbuckle, 2017) indicated that widespread adoption of diverse agro-ecologically important nutrient management and conservation practices could help in reduction of nutrient loss from soil (Castellano and Helmers, 2015; Drinkwater and Snapp, 2007; ISU, 2012; McLellan et al., 2015). The adoption of these nutrient management technologies and practices is a *sine qua non* for sustaining agricultural production system in Indian context keeping the population growth rate in mind.

India's population is expected to touch 1.4 billion by 2025. This will trigger the demand for food grains to about 300 million tons. This surge in demand has to be fulfill from the current available arable land, that is, 141 million ha, out of which 120 million ha is already estimated to be suffering from different forms of degradation (Indian Council of Food and Agriculture Report, 2017). With this, the added concern is the current practice or tendency of applying only nitrogen–phosphorus–potassium (NPK) fertilizers by majority farmers due to less awareness of secondary and micronutrient deficiencies. Each year, India is losing 5,334 million tons soil due to excessive soil erosion and indiscriminant application of chemical-based inputs (Newspaper report of The Hindu dated 26th Nov. 2010, www.thehindu.com). In India, the current consumption of NPK ratio is 6.7:2.4:1, which is extremely biased toward nitrogen as against the recommended ratio of 4:2:1 (Reddy, 2018).

Under this scenario, extension services become essential to update the farmers with latest knowledge and skill about modern nutrient management technologies and practices. However, the adoption of nutrient management technologies in India is minimal and it can be enhanced by following suitable extension approaches with focused targeting, appropriate policy formulation, need-based institutional setup, devising proper incentive system, and facilitating easy adoption process. In an effort to address these challenges related to promotion of nutrient management technologies, the government of India and different agencies has implemented wide-ranging nutrient management program and policies. In-spite of multiple attempts by diverse agencies, even now a significant portion of farming communities are not still assured about the utility of the nutrient management technologies and its proper application for enhanced fertility of the soils.

The major reason for this issue is traced back to the fact of lack of knowledge and information by end users. Therefore, increasing demand for specialized extension approach on promotion of nutrient management technologies with long-term engagement has been witnessed in recent past to bridge this gap. There is also a need for comprehending the different extension approaches from farmer's perspective and reorients the activities in such a way that fulfills the existing gap of awareness, knowledge, and skill of farmers on nutrient management technologies, and facilitates the adoption and application of those technologies for the betterment of their life, livelihood, and environment. In this context, the present chapter will assess the different innovative extension approaches adopted for promoting nutrient management technologies in Indian as well as global context.

12.2   NEED OF FOCUSED EXTENSION SERVICE ON NUTRIENT MANAGEMENT

Extension has a rich tradition of organizing educational programs for farmers on a wide array of issues, but mainly focusing on promotion of improved varieties to enhance the yield. During the green revolution phase, much emphasis was placed on increasing agricultural production through adoption of high-yielding varieties along with use of chemical fertilizers and pesticides. But chemical fertilizers were used indiscriminately for raising the crop yield due to lack of awareness and education about its ill effects on soil health, environment, and human system. Hence, we need to educate the farmers on judicious use of fertilizer and chemical for maintaining soil health.

Access to timely and accurate information by farmers is highly important in dissemination of modern soil fertility management practices. But it is observed that the adoption of different nutrient management technologies has generally trail behind the latest scientific developments, thereby minimizing the possible effect of the technologies. The findings of Vanlauwe et al. (2015), Hu et al. (2007), Pan (2014), Arbuckle (2017), Farouque and Takeya (2009), Ahmed, Karablieh, and AlKadi (2004) also supported the fact that the localized extension services, farmer engagement in organizational social networks, timely delivery of science-based fertilizer recommendations, extension contact wields a definite and statistically significant effect on their attitude toward nutrient management technologies and practices. Hence, we need to formulate or adopt a specialized focused extension approaches for promoting complex nutrient management technologies which generally demand scientific knowledge and sufficient technical skill on user's part for its successful application in field.

However, our extension agents lagged behind in terms of skills and knowledge in this aspect. It has been accepted by all the policymakers that the extension agents require latest technical knowledge and skill to publicize the complex nutrient management technologies and practices among the farming communities. The mere information on crop variety, its agronomic practice of growing and fertilizer dose is not sufficient for promoting nutrient management technologies. Extension workers now need to be trained to use hand held computers for calculating recommended N, P, and K fertilizer dosage.

They need to learn the technique of running and analyzing the sample through on field soil test kit for rapid NPK measurement in the field. They have to learn the operation of DSS, mobile-based apps, data entry and analysis for providing on-field site specific recommendation to the farmers. These demanded a specialized extension team with sufficient knowledge of nutrient management and information technology (IT), and to achieve this we need to have a separate extension approach focusing on nutrient management.

12.3 LACUNAE OF TRADITIONAL EXTENSION SYSTEM IN PROMOTING NUTRIENT MANAGEMENT TECHNOLOGIES

Agricultural extension has historically been focused on promotion of new crop varieties and seeds for a better yield. The information on fertilizer application was provided as a subsidiary information package for growing the crop. Oyinbo et al. (2019) reported that agricultural extension traditionally provided general recommendations to improve soil fertility for a vast cultivable land, namely, a district or region or province (Tittonell and Giller, 2013; Kihara et al., 2016; Shehu et al., 2018). These results in low marginal returns as fertilizer usage are different for each individual farmer and field. The prevailing agricultural extension approaches fail to take into consideration the diverse and composite biological, physical, social, and economic conditions of small land holder especially in developing country (MacCarthy et al., 2018; Kihara et al., 2016). To promote nutrient management technologies, we need site-specific agricultural extension for providing advisories that are customized to the need of each farmer and field. Such advisory services would be more useful to achieve higher yield and productivity targets than traditional extension approaches (Ragasa and Mazunda, 2018). In this regard, Kragt and Llewellyn (2014), Vanlauwe et al. (2015), and Vanlauwe et al. (2017) were of the view that information and communication technology (ICT) offer great potential to improve the capacity of agricultural extension officials to give site-specific extension advices to the farmers. However, very few extension approaches of present days utilized or incorporated the ICT to its potential in promotion of SHM technologies. The absence of precise soil test data and recommended fertilizer dosage information further indicates the grassroot level information vacuum that extension agents confront when suggesting farmers on nutrient management. A blank space exists in the previous studies about the factors that might be correlated with the adoption of diverse nutrient management practices, what should be the extension approaches for diffusing nutrient management technologies in different social system, and how to replicate them in comparable social system for successful adoption of those technologies.

12.4   DIFFERENT EXTENSION APPROACHES ON NUTRIENT MANAGEMENT TECHNOLOGIES

According to Axinn (1998), approach is the style of action within a system and it embodies the philosophy of a system. An extension approach may be defined as the way extension system is organized (its structure, leadership position, resource person, equipment, and facilities used), planning out the program with its mission, sets the vision and specific objectives to be achieved, selecting the methods and implementation strategies with suitable resource person in a cost effective way, finalizing the linkage mechanism with its clientele, and identifying its success measurement technique. It consists of a series of procedures for planning, organizing and managing the extension institutions, as well as for implementing practical extension work using the necessary and appropriately adapted means.

Extension approaches on nutrient management in India since its inception continued to evolve keeping the changing scenario of globalization, privatization, farm mechanization, climate change, and complex nature of agriculture in mind. The journey started with scheme for dry farming development in 1938 to operational research project of 1975 to National Watershed Development Programme of 1986 to Integrated Wasteland Development Programme of 1989 to Balanced and Integrated Use of Fertilizers in 1991–1992 to National Project on Organic Farming of 2004–2005 to National Project on Management of Soil Health and Fertility during 2008–2009 to SHM and soil health card scheme of 2015. It is unfortunate to note that these innovative reforms initiative in nutrient management were not able to address the problem low and unequal access to extension service for a large portion of population. It was found that more than 0% farmers had not used any modern technology for getting relevant information with respect to their farming.

Therefore, several questions from different policy corner were asked, "What extension approach should be followed to reach maximum farmers on nutrient management technologies?," "What nutrient management approach should be suggested to solve the problem of a country in a particular circumstances?," etc. Thus, it becomes highly essential to re-examine the current nutrient management extension approaches in India to understand the nature of information gap. The present book chapter is aimed to enlighten the reader with critical discussion on different nutrient management extension approaches so that they become wise enough to choose the most suitable and appropriate approaches to be followed in a specific context.

Different authors classified the extension approaches in different way. However, in this chapter we classified different approaches in the following categories based on existing nutrient management practices followed by the diverse organization, its operational mechanism, institutional type, type of field personnel involves, resources mobilized, and their broad objectives or mission to be achieved in relation to nutrient management:

1. General agriculture extension approaches
2. Private extension approach
3. Commodity-based extension approaches
4. ICT-based extension approach
5. Educational institute approach
6. Project approach of extension
7. Voluntary organization (VO) or NGO-based extension approach
8. Training and Visit (T&V) system of extension
9. Participatory extension approaches
10. Farming system research and extension (FSR/E)

In the following section, the author will discuss the different extension approaches initiated over time-to time in detail with successful cased both from national as well as international context so that the reader can analyze each approach with their applicability and relevance in different situation and become wise enough to formulate proper policy for diffusion of nutrient management technologies.

12.5 GENERAL EXTENSION APPROACH/PUBLIC EXTENSION APPROACH ON NUTRIENT MANAGEMENT

Soil health and quality has always been a matter of great concern for the Government of India. In the last 60 years, government made huge investment in arresting soil degradation and soil health. For this purpose, several developmental schemes have been implemented through its extension functionaries and system. This is called as general extension system or public extension system.

It is regarded as the most classical and commonly used extension approach on nutrient management. This is a centralized and government controlled approach. Most of the ministry-based extension programs and initiative comes under this approach. This approach is based on the assumption that appropriate technology and knowledge on nutrient management is available but remains mostly unutilized. It is believed that if the proper communication of knowledge with respect to these nutrient management technologies could be done, then surely farm practices would be improved. The major schemes or programs under this approach are as below.

12.5.1   SOIL HEALTH CARD (SHC)

Government of India launched "Soil Health Card" scheme in February 2015 with an aim to provide SHCs for each farmers of India in every three years. The aim of the scheme was to address nutrient deficiencies in fertilization practices adopted by the farmers. SHC provides the estimate of soil fertility in 12 indicators, namely—nitrogen, phosphorus, potassium (macronutrients), sulfur (secondary nutrients), zinc, iron, copper, manganese, boron (micronutrients), pH, electrical conductivity, and OC (physical parameters). It also provides crop wise advisory services on fertilizer application to promote the judicious use of fertilizer.

Under this scheme, SHCs are being distributed to farmers, training is being imparted to professionals for soil analysis, and financial assistance is being given for package of nutrient recommendations. Capacity building, regular monitoring, evaluation, and mission management are other important components of this scheme. The uniqueness of this scheme is the site-specific nutrient management recommendation based on SHC and direct involvement of Panchayati Raj Institutions (PRIs) while implementing the scheme.

While assessing the performance of SHC-based nutrient management approach, Reddy (2018) reported that under cycle 1, so far 2.36 crore sample were tested out of 2.54 crore collected sample. Total 9.62 crore SHC were printed under the scheme but only 9.33 crore SHC were distributed among the beneficiaries. The study further reported that due to SHC interventions there was a drastic reduction in usage of Urea and DAP by 20%–30% in paddy and cotton, which helped to bring down the overall cost of cultivation. The drop in cost of cultivation was ranged between Rs. 1000 and Rs. 4000 per acre. Also, the use of micronutrients mainly gypsum was increased due to SHC scheme. It was reported that the yield of farmers was increased significantly who followed SHC-based nutrient management practices. Thus, with reduction in the cost of cultivation and increase in overall yields, net income of farmers raised by 30%–40% after adopting the recommended nutrient management practices under SHC scheme.

Similarly, National Productivity Council (NPC) reports (2017) indicated a wider acceptance of the SHCs by Indian farmers. It is reported that a decrease of 8%–10% in use of chemical fertilizer and increase of 5%–6% in yield was observed on account of adoption of recommended nutrient management practices as per the SHCs.

However, the findings of different regional study indicated that the impact and efficacy of SHC are not impressive across the region. For example, Grover et al. (2017) in his study on soil health in Punjab reported that the recommendation in SHC was given only for macronutrients in paddy, basmati and maize crops. The farmers generally preferred higher amount of urea and lower quantity of DAP and MOP in contrast to SHC instructions. It was reported that mere 34%–77% farmers were familiar about different character of soil testing and SHC program. Though farmers applied only FYM as organic fertilizer but the study indicated minor drop in usage of inorganic fertilizer, notably N and P in paddy and maize cultivation. Major issues raised by farmers regarding SHC are conducting very few awareness camps on soil testing, difficulties in comprehension of SHC reports and slow distribution of SHC reports. Similarly, Bordoloi and Das (2017) in their study on soil health in Assam reported that the farmers were unaware about integrated nutrient management (INM) technologies and lacked in information about the ongoing programs on Soil Health Mission in the study area. Further, no farmers even from SHC villages attended any training program on SHM. The study found that the farmers applied chemical fertilizers on the basis of their own experience and in consultation with the co-farmers instead of SHC recommendation. Bhayal et al. (2019) reported that the majority of beneficiaries (42.14%) had medium level of awareness about various components of SHC and only 25.71% farmers had high level of awareness on SHC.

Reddy (2018) were of the view that the progress of the SHC scheme in terms of coverage was satisfactory but there were concerns regarding soil sample collection, its analysis and delivery of SHCs to farmers in time. According to this study, only 66% farmers could comprehend the content of SHC. In this regard, just less than half of the respondents (44%) mentioned that the hardly any extension officers explained the content and its implication to their farming. Though 57% farmers were of the view that the suggestions provided in SHC were useful to them but only 53% of them could follow the suggestions. The study suggested that awareness campaign regarding distribution of SHC should be organized before the growing season of crops for easy follow up. Several farmers felt that in addition to current information, SHC should also provide information on important physical and microbiological indicators (like soil texture, water holding potential, and water quality and bacterial content). Farmers highlighted the need of timely distribution of SHC card, raising awareness by conducting more number of camps, preparation of GIS-based soil fertility maps and its wider publicity using mass media like wall posters and display boards at village panchayats level to make the SHC program more effective. Beside this, the Government should make necessary arrangement for easy availability of recommended fertilizers and biofertilizers at the village level at fair prices. Overall, among all the states like Karnataka, Tamil Nadu, Chhattisgarh, Uttar Pradesh, Maharashtra, Telangana, and Andhra Pradesh performed better than other states.

12.5.2   *NATIONAL MISSION FOR SUSTAINABLE AGRICULTURE (NMSA)*

The NMSA program was started by Government of India in April 2014. The aim of NMSA was to make agriculture more productive, sustainable, and climate resilient. Reclamation of problematic soils, namely, alkaline, saline and sodic soil is one of the specific components of this program. The major focus is to raise agricultural productivity mainly in rainfed areas by promoting integrated farming, SHM, and reenergizing resource conservation. All the elements of NMSA, namely, rainfed agriculture, soil health maintenance, organic agriculture, etc. have contributed significantly in realizing the goal of sustainable development. NMSA has some important schemes related to nutrient management and are discussed in following sections.

12.5.2.1   *SOIL HEALTH MANAGEMENT (SHM)*

SHM, an inherent part of National Mission for Sustainable Agriculture (NMSA), was launched in 2004–2005. Under this program, government is encouraging adoption of INM practices, that is, soil test-based judicious use of fertilizers in conjunction with biofertilizers and organic manures like FYM, vermicompost, other compost, and green manure to preserve the soil health and its productivity. The scheme also has the mandate to establish new static as well as mobile soil testing laboratories (STLs) besides strengthening existing laboratories to enable them to undertake micro-nutrient testing.

Objectives:

The major objectives of SHM program are:

1. To promote soil-test-based judicious use of fertilizer and using organic source of plant supplements and biofertilizer.
2. To make agriculture more productive, sustainable, and climate resilient.
3. To promote comprehensive SHM practices.

Components:

The program on SHM includes the following components:

1. Formation of databank on site specific judicious use of fertilizers.
2. Preparing district-wise digital soil fertility maps.
3. Providing easy-to-carry soil testing kits to field officials.
4. Advocacy and distribution of micronutrients.

12.5.2.2 INM AND ORGANIC FARMING

This was another component under NMSA which was launched in 2014–2015 for dissemination of sustainable nutrient management technologies. The major components under this were:

1. Mandate of financial help for establishing mechanized fruit/vegetable market waste/agro waste compost production units.
2. Modern liquid/carrier-based biofertilizer/biopesticide production units.
3. Biofertilizer and organic fertilizer analyzing laboratories and intensifying the existing ones under the Fertilizer Control Order (FCO).
4. Promotion of organic farming in cluster approach under Participatory Guarantee System certification.
5. Support to PGS system for on-line data maintenance and residue test
6. Adopting villages under organic farming for manure-management and biological nitrogen harvesting.

7. Conducting training programs and demonstration on organic farming.
8. Research support for development of new package of practices on organic farming for local farming system of the respective states.
9. Setting up a distinct organic agriculture research and teaching institute.

12.5.2.3 PARAMPARAGAT KRISHI VIKAS YOJANA (PKVY)

PMKVY is a detailed and extended element of SHM program under National Mission of Sustainable Agriculture (NMSA). Under this yojana, cluster approach and PGS certification is being used to promote organic farming in adopted organic villages. The main aim is to enhance soil fertility and production of safe food through organic farming. It wants to empower farmers through institutional development in cluster approach in managing farm practices, input production, quality assurance, value addition, and direct marketing. Under this program, a cluster is formed by making a group of 50 or more farmers having land of 50 acres for organic farming. Every farmer is being assisted with Rs. 20000 per acre in three years' time for supporting in activities ranging from seed sowing to crop harvesting to transporting the produce to market. Participatory guarantee system and PGS-India are the key approaches for quality assurance under PKVY.

The scheme envisages:

- Production of commercial organic products through certified organic farming.
- The preparation of pesticide residue free produce and enhancing the health of consumer.
- Raising farmer's income and creating suitable market for merchants.
- Inspiring the farmers for local resource mobilization for input preparation.

Reddy (2017) in his study on impact of PKVY reported that over 95% clusters adopted organic input production unit and above 92% clusters adopted different biological nitrogen harvest plants like Gliricidia, Sesbania, etc. About 65% were using botanical extract production unit and 18.1% phosphate-enriched organic manure. In backward state, adoption of green leaf manure, compost, and organic seed was than progressive state.

The study found that 19.6% clusters were preparing compost followed by green manure (15.4%) and organic seed (13.1%) which indicates the relative success of the scheme. However, only 7.7% of clusters prepared indigenous inputs like *Panchamruth*, 13.8% prepared *Panchagavya*, and 14.3% prepared *Beejamruth*. The use of green manure increased by 50%. The study further shows that the mean cost per hectare in paddy was less in organic farming by 15.1%, while overall returns declined by 7.3%. The combined net effect of higher decline in costs with slightly decreased gross returns was an increase in net return by 36.7%.

12.5.3   NATIONAL PROJECT ON MANAGEMENT OF SOIL HEALTH AND FERTILITY

The Indian government has taken initiative of "National Project on Management of Soil Health and Fertility" in 2008–2009 to encourage soil-test-based balanced application of fertilizers for enhanced soil health and productivity. The major objectives of this initiative were to encourage and promote INM technologies through balanced use of inorganic fertilizers, micronutrients, organic manures, and biofertilizers for enhancing soil fertility and productivity. For achieving these objectives, government has taken many steps like strengthening existing soil-testing set-up and updating the skill and knowledge of STL staff, extension staff, and farmers.

12.5.4   NATIONAL PROJECT ON ORGANIC FARMING (NPOF)

NPOF was launched in 2004–2005 by National Centre of Organic Farming (NCOF). The mandate of NPOF was to impart training to stakeholders on organic farming and to perform quality analysis of biofertilizers and organic fertilizers as per the provision of FCO, 1985. It also has the provision of assistance for establishing organic input units under capital investment backended subsidy scheme in partnership with the National Bank for Agriculture and Rural Development (NABARD).

12.5.5   RAINFED AREA DEVELOPMENT (RAD) PROGRAM

Integrated Farming System (IFS) is a special focus area of Rainfed Area Development Program and mainly intended to enhance crop productivity and minimizing climatic risks. Under this scheme, the crops/farming system is matched against the farmers' activities like horticulture, fishery, livestock, agro-forestry, apiculture, etc., so that they can maximize farm income on the one hand, and on the other to overcome the sever impact of extreme climatic events (drought, flood, uneven rainfall, etc.).

12.5.6 WATERSHED DEVELOPMENT PROGRAM (WDP)

Watershed Development Program was launched mainly to address soil degradation issues: Under the program, the training is being given to rural communities for raising their capacity on soil moisture conservation. National Watershed Development Programme in Rain-fed Areas (NWDPRA) of Union Agriculture Ministry and WDP of NABARD has initiated and implemented numerous watershed activities on a large scale. Integrated Watershed Management Program (IWMP) of Department of Land Resources (DOLR), Ministry of Rural Development was started in 2009–2010 as a follow-up of Drought Prone Areas Programme, Desert Development Programme, and Integrated Wastelands Development Programme. The program components like capacity building, institution building, and natural resource management activities played an important role in soil conservation and health management.

12.5.7 PRADHAN MANTRI KRISHI SINCHAI YOJANA (PMKSY)

The Pradhan Mantri Krishi Sinchayee Yojana (PMKSY) was inaugurated by Government of India on 1st July, 2015 with the goal of "Har Khet Ko Paani" so that end-to-end solutions in irrigation supply chain (water sources, distribution network, and farm-level applications) can be provided at farmers field. The program mainly focuses on creating new ways of assured irrigation, harnessing rain water at farm level through its innovative components of "Jal Sanchay" and "Jal Sinchan." It further aims to promote micro-irrigation so that the components of "Per drop-More crop" become highly successful. The components of "Per Drop More Crop" primarily focus on increasing water use efficiency at field level through promotion of precision/micro irrigation (drip and sprinkler irrigation). This will not only optimize the use of available water resources but also support other water saving interventions like microlevel water storage or water conservation/management activities to complement and supplement the source creation.

From 2015 to 2016, the IWMP is brought under "Watershed Development Component" of "Pradhan Mantri Krishi Sinchayee Yojana." The basic aim of the program was to harness, conserve and develop degraded natural resources (soil, vegetative cover, and ground water table); arresting soil run-off, rain water harvesting and refilling the ground water table; enhancing crop productivity; promoting multiple cropping; adopting different agro-based activities; supporting sustainable livelihoods and raising the farm income.

12.5.8   BALANCED AND INTEGRATED USE OF FERTILIZERS

This is one of the oldest extensive nutrient management program in India, and launched in 1991–1992 with the primary mandate of information dissemination on judicious use of chemical fertilizer including major nutrient like N, P, K; secondary nutrients like S, Ca, Mg; and micronutrients like Zn, Fe, Cu, B, Mo, and Mn in combination with organic nutrients and biofertilizers. The major elements of the scheme are-setting up compost plants to process biodegradable solid waste into compost, enhancing soil testing facilities, arranging seminar and workshops on adoption of soil test-based fertilizer recommendation. The scheme was operational in subsequent plan period and since 2000 it was merged with macro management plan of agriculture scheme.

12.6   PRIVATE EXTENSION APPROACH

Extension services have traditionally been funded, administered and delivered by the public sector since the inception of organized extension service. However, increasing financial difficulties have made many countries to think of alternative ways for supplementing financial and manpower support to public extension. Privatization of extensions service is considered as the most viable alternatives to reduce the burden on the budget starved public extension system. Chile was the first country to test a privatized extension service through its general economic liberalization program of 1978. The private sector in the form of innovative farmers, companies, crop science industry, primary cooperative societies, and businessman is arranging extension advisory services on nutrient management and mainly covers large farmers and farmers growing cash crops who can pay. Few successful cases of private extension on nutrient management are presented in the following sections.

12.6.1 FARMERS' ASSOCIATIONS

It is widely acknowledged by the different policymakers that the contact intensity among the farmers is highest when the organized themselves in some form of farmers' associations. For example, Maharashtra Organic Farmers Federation (MOFF) is a confederation of 120 CSOs and 142000 farmers which forms organic groups of 20 farmers each. It regularly organizes training program on nutrient management practices (green manuring, composting, biofertilizers, liquid organic manures, etc.) to be followed for organic farming. Similarly, Organic Farming Association of India is advocating SHM technologies and methods in Madhya Pradesh.

12.6.2 COMMERCIAL COMPANIES

The commercial companies also provide extension service to farmers or farmer-based associations on nutrient management. The extension approach of Mahindra Krishi Vihar, a one-stop farm solution center run by the Mahindra and Mahindra Ltd. tractor and utility vehicle company, Tata Kisan Sansar of Tata Chemicals Ltd, Hariyali Kisaan Bazaar, run by DCM Shriram Consolidated Ltd., Godrej Agrovet model, and Jain micro-irrigation are perfect examples of successful private extension service for promoting soil health in India.

Fertilizer companies such as IFFCO devised an innovative ICT-based extension approach named as IFFCO Kisan, a mobile app in collaboration with Airtel and IKSL to provide advisory services to its farmers. Another fertilizer company, that is, Deepak Fertilizer and Petrochemical Corporation Limited (DFPCL) has taken noble initiative to give holistic fertility management services to the farmers through its "Mahadhan Saarrthie Centres." DFPCL has also established a well-equipped Agri-Lab to deliver service likes soil testing, plant tissue, water and organic fertilizer samples analysis. National Fertilizer Limited (NFL) has established "Kisan Suvidha Kendras (KSK)"to provide diagnostic service of major and micro nutrients of soil sample and educating the farmers for adoption of soil-test-based fertilizer application. To bring service under one roof along with soil testing and advisory services, 100 "Kisan Suvidha Kendras (KSK)" were already established by NFL in its marketing territory. Beside these, RCFL, Krishak Bharati Cooperative Limited (KRIBHCO) and Coromandel are taking up extension activities on nutrient management and are independently doing soil-testing activities.

12.6.3   INDEPENDENT ENTREPRENEURS/AGENCIES

Progressive and big farmers are interested to pay for diverse range of extension services that enable them to access to quality inputs, credit procurement and field-based suggestions on technology use. Few agri-business entrepreneurs have already established STLs for delivering diagnostic as well as advisory services. For example, "India Mart" started offering soil-testing services at Mumbai, Thane and Nagpur of Maharashtra through an e-transaction portal. Similarly, "Vision Mark Biotech" provides advisory services and supplies biofertilizers in Maharashtra. However, the evidence of private extension initiatives indicates that the farmers are more willing to pay for a package of services instead of only soil-testing services. Further, they prefer to pay for cash crops more rather than for food crops. Therefore, the private enterprises should accordingly arrange their extension services for wider coverage and impact among the customers.

12.7   COMMODITY-BASED EXTENSION APPROACH

There are many crops of great value that play a crucial role in income generation, poverty mitigation, employment generation, and overall liveli-hood security of farming communities, but they are not considered under existing popular extension schemes. The extension approach being used to promote these high value crops has been named as commodity approach. This approach is generally found for major export crops and commercially important cash crops like tea, coffee, sugar, spices, etc. The organizations like private sector firm or specialized public bodies (tea board, coffee board, spices board, tobacco board, coconut development board) and on few occa-sion profit-oriented small companies/farmers' organization (cooperative, producer organization) or individual entrepreneur are promoting commodity-based extension advisory system for promoting efficient nutrient manage-ment technologies.

The major strengths of commodity-based extension approach are recom-mended technology tends to "fit" to the local agro-ecological condition, adoption rate of technology is very high, strong coordination and network with input supplier as well as marketing agents, timely delivery of need-based message, closer management and supervision, ready availability of market for the produce, easy availability of supplementary extension services like credit, training, inputs, etc.

The few innovative extension approaches promoting nutrient management technologies by different commodity boards are presented in the following sections.

12.7.1   TOBACCO BOARD

The Tobacco Board of India promoted different INM technologies for tobacco growers in India. Already an area of 17122 hectares was covered under green manure crops to improve the soil fertility status (Annual report 2016–2017). Another innovative approach of Tobacco Board is that the board procured various fertilizers at very low prices in bulk quantities directly from manufacturing companies and then supplied to the farmers. This lowers the cost of production for farmers and motivates them to apply micro nutrients. Another unique extension approach of Tobacco board is "Field Friend."

12.7.1.1   FIELD FRIENDS

The Tobacco Board has conceptualized "Field Friends" approach by comprising a team of scientists from CTRI, Field executives of M/s. ITC Ltd and M/s. GPI Ltd and Board's technical officers. The "Field Friend" team visited the tobacco farmers area and provided necessary advisory services on different nutrient management technologies, scientific package of practices on crop production and protection, bringing down the usage of pesticides, topping, de-suckering, harvesting, curing, etc.

12.7.2   COFFEE BOARD

The coffee board has launched an innovative information system for soil health monitoring and management known as KSHEMAM.

12.7.2.1   KAAPI SOIL HEALTH MONITORING AND MANAGEMENT (KSHEMAM)

This is a soil-based nutrient management information system for coffee growers. This is ICT enabled, GIS-based decision support system (DSS) and provides location specific fertilizer recommendation. The system is dependent on the data given by the soil analysis laboratory and also depends on the selection of intercrop between pepper and orange. The system accordingly reproduces recommended fertilizer dosage advisory for coffee and specified intercrop by using the latest software engineering and calculation techniques.

12.7.3   RUBBER BOARD

Rubber Production (RP) department renders extension and advisory services to all stakeholders aiming at Sustainable Natural Rubber through its well-knit extension network focused on participatory group approach. Extension network of rubber board comprises of 155 Field stations, 3 Development Offices located in rubber growing region of the country, monitored by 45 Regional Offices and 3 Zonal Offices, coordinated by the Head Office at Kottayam.

The major innovative approaches of Rubber board on nutrient management are as below.

12.7.3.1   FERTILIZER ADVISORY GROUP (FAG)

These groups provide advisory services on fertilizer recommendation and site specific nutrient management in rubber plantations on the basis of soil testing and leaf analysis. This facility is available at Rubber Research Institute of India (RRII), headquarters and at seven regional STLs functioning in Kerala.

12.7.3.2   RUBBER SOIL INFORMATION SYSTEM (RUBSIS)

This information system helped us to know the current soil fertility condition and information on fertilizer recommendation at the fingertip of rubber growers. RRII, Rubber Board in partnership with National Bureau of Soil Survey and Land Use Planning (NBSS and LUP), ICAR conducted a detailed survey of soils in rubber growing region with an aim to provide soil-test-based fertilizer recommendation in whole rubber growing region. The area where rubber was cultivated for more than three years in each panchayat was mapped with medium resolution satellite image at a scale of 1:50,000. Beside this, one composite soil (0–30 cm) sample was taken from each 50 ha area under rubber plantation and linked with GPS-based coordinates of a specific location, and then the sample was tested in laboratory. Finally, web-based fertilizer recommendation was prepared based on interpolated soil fertility data in association with Indian Institute of Information Technology and Management—Kerala (IIITM-K) overlapping diverse soil fertility parameters and soil depth as per the guidelines of discriminatory fertilizer recommendation.

12.8   ICT/IOT-BASED EXTENSION APPROACH ON NUTRIENT MANAGEMENT

As per the 10th agricultural census, In India, there are 125.86 million small landholdings (Krishnan, B.V., 2018) and the number is gradually increasing with passage of time. D Kumar in 2012 reported that a public extension agent hardly able to spend on an average 40 min time for a farmer in a year. With this kind of scenario of increasing number of small landholders on one hand and skewed extension contact intensity on the other side, an alternative extension approach was needed to fill this gap of low extension to farmer ration. In the changing world, mobile phone has become an integral part of human life and this tool has been utilized by many development agencies to promote their product and service. Indian customers represent approximately 30% of the world feature phone market and has become the second biggest mobile market in the world. In 2015, India had 720 million mobile phone users, and among them 320 million were village or rural mobile phone customers. This number of mobile phone users also includes 50 million smartphone customers having internet connectivity (https://claroenergy.in/tag/farmers/#:~:text=In%202015%2C%20India%20had%20720,users%20with%20access%20to%20internet). Now, the report of techARC, a market research firm, shows that the number of smartphone users jumped to 502.2 million by December, 2019 (www.news18.com, January 30, 2020). Hence, the mobile phone can be used to raise awareness on nutrient management technologies of farming communities who is living in remote corners of country with customized text, voice, or video messages in his own languages. This will accelerate the adoption of different nutrient management technologies across the locations and farming categories. Digital India launched in 2015 by Indian Government for promoting digital literacy and establishing digital infrastructure paved the way for this ICT-based extension services.

The term ICT-based extension service refers to dissemination of information using different electronic media, namely, mobile-based extension, web-based extension, newspaper-based extension, TV/video-based extension, information system, DSS, and IoT-based extension service. Out of these, with the progress and development of android system, smart mobile phone has emerged as the most attractive ICT media for delivering digital extension service on SHM technologies. The major ICT-based platform and cases of successful digital extension services on nutrient management are presented in the following sections.

12.8.1 WEBSITE/WEB PORTAL-BASED DIGITAL EXTENSION SERVICES ON NUTRIENT MANAGEMENT

12.8.1.1 FARMERS' PORTAL

This is a new initiative of Indian government for providing farm information on STL, soil testing reports, fertilizer stock availability, wholesaler, and retailer network details. This also links the farmers with website and portal of related department and organizations on a common platform. The portal has a link to another SHC web portal in which one can register and apply for soil testing service from approved lab. The portal acts as a database on soil health for future reference by providing information on different crops.

12.8.2 TELEPHONIC EXTENSION SERVICE ON NUTRIENT MANAGEMENT

12.8.2.1 KISAN CALL CENTERS (KCC)

The service of Kisan Call Center was started in 2005 by DAC to provide extension services over telephone in 22 local languages in whole country. The service includes instant response to the queries voiced by farmers over land phone or mobile phone by dialing a toll free common telephone number 1800-180-1551. The service is accessible from any mobile networks from 6 am to 10 pm in all the seven days of a week including holidays. The KCC services are absolutely free and managed in two levels. The maximum queries were answered by the experts sitting at Level 1. If any questions remain unanswered, then it goes to Level 2 experts. So, level 2 experts are mainly responsible for replying to difficult questions within a stipulated time period. The questions may range from any aspect of soil nutrient management,, namely, fertilizer rate, organic agriculture, biofertilizer, and other aspect of agriculture.

12.8.2.2 MKISAN

This is a mobile-based agro-advisory service for the farming communities through public organizations functioning up to the block level. The registered farmers are grouped as per the block from where they belong, crop they cultivate and activities done by them for delivering specific SMS message which is relevant to them.

12.8.3   MOBILE APPS (APPLICATION)-BASED DIGITAL EXTENSION SERVICE ON NUTRIENT MANAGEMENT

With the advent of Android application, mobile-based apps have become the most useful and user friendly ICT tools to access the agricultural information. Within the past few years, numerous agencies came up with their own app to disseminate agriculture information. The most widely used apps in promotion of SHM are as below.

12.8.3.1   SOIL HEALTH CARD (SHC) MOBILE APP

SHC Scheme was launched in 2015 to issue soil health card to each farmer in every three years. NIC has designed and launched an Android application to simplify data entry job for sample registration. The apps capture the latitude and longitude data of sample area if "location" is ON in the mobile. The details of farmers, land, crops, and fertilizer usage practices can be entered using this app.

12.8.3.2   MKRISHI

mKrishi is a free mobile application of Tata Consultancy Services (TCS). The app combines diverse technologies to provide specific information on fertilizer rate based on soil status and water status of farmer even from low-cost mobile handset.

12.8.3.3 MICROMITRA

This mobile app is developed by ICAR-National Bureau of Agriculturally Important Microorganisms, Mau, Uttar Pradesh in 2017. The app provides formation on various microorganisms, biofertilizers, biocontrol agents, and decomposers. The farmers can also get information about liquid formulation of NPK providing bacteria for various crops, advantages, application, dosages, precautions, and cost through this app.

12.8.3.4 FARM CALCULATORS

This is a very useful mobile app developed by the University of Agricultural Sciences, Bengaluru, Karnataka in 2015. The app helps to calculate exact quantity of seeds and fertilizers needed to grow crops in offline mode.

12.8.3.5 OIL PALM NUTRIENT MANAGEMENT

This mobile app is developed by ICAR-Indian Institute of Oil Palm Research, Andhra Pradesh in 2016 to furnish offline information on queries related to symptoms of nutrient deficiencies of N, P, K, B, Mg, Fe, Cu, Zn, Mn, and other disorders in oil palm along with management practices.

12.8.3.6 PLANT NUTRITION

The app is released by Vasantrao Naik Marathwada Krishi Vidyapeeth, Parbhani, Maharashtra, in 2017. It includes information on plant nutrition and elements present in the soil fertilizers used for the plants. It also provides information regarding major and minor plant nutrients essential for crop growth and symptoms of nutrient deficiency.

12.8.3.7 SOIL NUTRIENT MANAGER

The ICAR-Research Complex for Eastern Region, Patna, Bihar, has developed this app in 2018. It provides fertilizer recommendations based on native soil fertility status and nutritional requirements of the crop to be grown.

12.8.3.8   NUTRIENT DEFICIENCY DIAGNOSER AND MANAGER FOR APPLE (NDDMA)

This mobile app is developed by ICAR-Central Institute of Temperate Horticulture, Srinagar, Jammu and Kashmir. The app helps in management of nutrient deficiency in apple, to optimize yield and reduce yield losses. Farmers can access supplementary information like soil region, physiographic, sub physiographic, landscape, and land management units.

12.8.3.9   URVARA

This mobile app is developed recently by ICAR-Research Complex for Eastern Region Patna, Bihar in 2018. The app facilitates fertilizer recommendation for a crop in view of soil test report/SHC. It provides the user with appropriate dose and cost of the fertilizer for farmers' area.

12.8.3.10   FERTILIZER CALCULATOR—GOA

The ICAR—Central Coastal Agricultural Research Institute, Goa has developed this app in 2015. It provides offline soil-test-based fertilizer recommender (STFR) of crops. It also helps in calculations according to the area of farm or the number of plants/trees.

12.8.3.11   DIGITAL SHC

This mobile app is developed by Krishi Vigyan Kendra, Sambhalpur, Odisha in 2012. It provides information on SHC of the farmer with recommendation for use of fertilizer for different crops. Under this, image-based extension information service is given to the farmers.

12.8.3.12   MKRISHI PAWS—IISWC

This app developed by ICAR-Indian Institute of Soil and Water Conservation, Dehradun, Uttarakhand in collaboration with TCS Limited in 2012. The major features include online mode Personalized Advisory on Water and Soil in remote and hilly region. The app helps to disseminate the agriculture and soil and water conservation (SWC)-related messages/best practices among farmers of north western Himalaya.

12.8.3.13   GYPCAL-SODIC SOIL RECLAMATION

The ICAR-Central Soil Salinity Research Institute, Haryana has developed this app in 2017. The app gives knowledge on chemical reclamation of sodic soil for getting maximum yield in Indo-Gangetic plains by finding out the gypsum requirement in 50 kg bags. It also calculates the exchangeable sodium percentage of the sodic soil. Further, it finds out the total depth of water needed for leaching to flush out salts. The app further predicts the expected harvest of salt tolerant and traditional varieties of rice–wheat after reclaiming the soil chemically.

12.8.4   TV/VIDEO-BASED DIGITAL EXTENSION SERVICES

12.8.4.1   DIGITAL GREEN

Digital Green is a private organization that works to disseminate agricultural information via digital video with an objective to increase the farm productivity through capacity building of small- and marginal-farmers via short instructional videos. Digital Green in collaboration with CSOs like BAIF and PRADAN promoted the necessary knowledge on nutrient management technologies and developed the capacity of rural farmers through short instructional videos.

12.8.4.2   KISAN TV CHANNEL

A separate Kisan TV channel was launched in 2015 for broadcasting different issues on nutrient management and other areas of agriculture under the "Mass Media Support to Agriculture Extension" and "Focused Advertisement Campaign" schemes fund of Doordarshan (national television) and All India Radio (AIR).

12.8.5   COMMUNITY RADIO STATIONS (CRS)-BASED EXTENSION SERVICES

According to the report of Ministry of Information and Broadcasting, GoI, as on March 31, 2020, total 289 operation community radio stations are broadcasting different developmental issues in local language in India (https://

pib.gov.in/PressReleasePage.aspx?PRID=1626170#:~:text=Today%2C%20 India%20has%20290%20operational%20Community%20Radio%20 Stations). They mainly broadcast different popular schemes in the field of agriculture, water, soil, health, local culture, and different welfare schemes. Radio is a powerful communication tool for disseminating nutrient management technologies in local dialect. The power of community radio station in disseminating nutrient management technologies can be understood through a specific case study of "Vernacular radio programs on SWC in N. Ghana". Chapman et al. (2003) reported that Ghana has initiated several innovative actions for establishing FM stations in the country, namely, Upper Region Radio (URA) Station, Radio Savanna for to promote different developmental message in local language. A radio program was designed to promote SWC practices through music and drama in local language using the platform of community radio station. In the radio program, a story on SWC was aired for raising the awareness of backward farmers on this complex technology. In the story, a group of male farmers discuss SWC issues with each other and their wives. One person tries to explain the possible ways to lower down the risk of soil erosion involving SWC technologies and practices like contour ploughing, mulching, preserving and planting trees, organic farming, green manuring low use of chemical inputs, etc. An extension officer and progressive farmer took part in drama and challenged by villages as to why they should change their current practices. The program captured different nutrient management aspects in an entertaining format so that the farmers can easily comprehend the issue and readily act upon it. It has been acknowledge by different policymakers that the impact of community radio station in promoting complex nutrient management technologies among the rural mass was comparatively more than other ICT media. Hence, government should support establishment of community radio stations in more number and encourage broadcasting of nutrient management program in local dialect through this.

12.8.6   DECISION SUPPORT SYSTEMS/EXPERT SYSTEM/ INFORMATION SYSTEM-BASED APPROACH

Decision support system (DSS) is one type of computerized information system that helps in taking right decision in a given situation based on your inputs. It is basically an interactive software-based platform aimed to support individuals in arranging useful information from elementary data, reports, personal knowledge, and different models to take decision for a given problem. Using DSS, a farmer can prioritize the most relevant nutrient management practices to be followed in a given situation for maximum impact. Few major DSSs or expert systems developed for promoting nutrient management technologies are mentioned in the following paragraphs:

12.8.6.1   DECISION SUPPORT SYSTEM FOR AGRO-TECHNOLOGY TRANSFER (DSSAT)

DSSAT is a software package combining impact of soil, crop phenotype, weather, and management options that facilitate an individual to raise "what if" query and simulate the findings within a minute by processing its preloaded data using crop model and application program. The service of DSSAT was used by scientist of more than 100 countries over 15 years. Using DSSAT, farmers can simulate multi-year output for their different crop management practices at any location in the world.

12.8.6.2   NUTRIENT EXPERT TOOL

This is an easy to use and scalable nutrient management information system, known as Nutrient Expert (NE), developed under a project titled as "Taking Maize Agronomy to Scale (TAMASA)." It was developed mainly to provide site-specific fertilizer recommendation to maize growers in Nigeria, Tanzania and Ethiopia. The tool "Nutrient Expert" works on the principle of applying fertilizer as per crop demand, that is, applying right fertilizer source at right rate in right time at right place (4 R' of nutrient use). The information system produces recommendations based on the following inputs: data on a target maize yield, user current crop management operations, environmental features, soil characteristics, and information on existing market prices of inputs and maize.

12.8.6.3   INTELLIGENT ADVISORY SYSTEM FOR FARMERS (IASF)

IASF is an advisory platform for solving the agricultural problems raised by farmers of northeastern states of India. The advisory system covers five important farming activities (insect management, weed management, disease management, rice variety selection, and fertilizer management) which generally need expert's guidance for remedial measures. In this system, farmers' questions along with its appropriate solution are preserved in a database (called CASE). A farmer can raise any question on the aforesaid farming operations. The IASF system automatically processes the data and came up with the most suitable solution from a large database containing information on all possible queries and the most appropriate solution as per experts' opinion. The system guides the farmer in difficult times when contact to an agricultural expert is not possible or when extension workers are not available for advisory services.

12.8.7 EXTENSION APPROACH WITH IOT/SMARTPHONE-BASED SENSORS IN NUTRIENT MANAGEMENT

Pongnumkul et al. (2015) described, in details, how the IoT-based sensors can be used to provide the advisory and diagnostic services on nutrient management to the farmers. For instances, two smartphone-based color estimator apps namely Baikhao and BaikhaoNK helped in evaluating the color level of rice leaf. Based on leaf color estimate, the apps then suggest the recommended dose of N2 fertilizer to be applied in the field without conducting any lab test. Now smartphone sensor technology is put in use for soil sample analysis by analyzing soil color to provide information on different soil fertility parameters. Here, the mobile phones beings used as soil color sensor and soil color information is analyzed from images captured by mobile in-built camera. For example, a smartphone application was developed, namely "Soil Indicator for Scottish Soil (SIFSS)," to study soil and to provide information on soil pH, soil carbon, N, P, and K content in the soil sample of farming communities by simple analyzing the images taken by the mobile camera. Similarly, SOCiT, another sensor-based mobile application was developed to provide information on carbon content of soil based on farmers' geographic locations. The farmers often remain directionless on deciding how much water to be applied for growth of a crop and seek for experts' advice. The application of IoT and sensor-based mobile apps can help farmers in a minute to decide the right dosage and time of water application for bumper production. An application namely PocketLAI was developed for calculating the Leaf Area Index (LAI) which is the main factor to compute the crop water requirements. The apps estimated LAI using an indirect method with the sensors provided in smartphone. The apps used two sensors: the image sensors helped to estimate LAI from the image of leaf canopy and accelerometer was used to get the angle of smartphones. The apps determine the water requirements from LAI and advised the farmers to adjust their water or irrigation plan accordingly. Therefore,

IoT or sensor-based smartphone apps enable the water flow estimation job much easier and cheap as it does not require any permanent equipment for measurement. Luthi et al. (2014) conceptualized and developed an android application to estimate open-channel flow. The application calculates the level of water, surface velocity, and discharge rate by processing a short smartphone-based video of the water flow between two control points with a known distance. The primary findings indicate that the precision of the water level, surface velocity, and runoff data taken by the smartphone application is about 5% of data received from a commercial radar sensor. Hence, it can be said that the IoT-based application and application of sensor-based mobile apps technology will be the future ICT-based major extension approach for providing diagnostic services on nutrient management technologies and practices.

12.9 EDUCATIONAL INSTITUTE (UNIVERSITY/RESEARCH INSTITUTE)-BASED EXTENSION APPROACH ON NUTRIENT MANAGEMENT

Another approach to agricultural extension involves extension activities of agricultural universities, colleges, and research institute. This approach originated from United States (US) cooperative extension service model which is managed by Land Grant universities in every state. India has partially adopted this approach through state agricultural universities which are responsible for agricultural education and research. The approach assumes that faculties have sufficient technical knowledge on nutrient management which is relevant and useful to farm people. The purpose is to help farmers and extension officials to learn about scientific nutrient management technologies. The approach emphasizes on the transfer of technical knowledge to farm people through nonformal education and adult educational activities. This approach conforms that farmers receive precise and updated advisory service from extension agents of public and private sectors. Few successful cases of educational extension approach by different universities or institutes are mentioned in the following paragraphs.

12.9.1 UNIVERSITY OF DELAWARE

The Nutrient Management Program of the University of Delaware provides need-based relevant, unbiased, and scientific information to the people of

Delaware. The nutrient management was planned with an aim to enhance the nutrient use efficiency for raising profitability from agriculture and horticulture, while preserving and enhancing the quality of environment. The Nutrient Management Program functions in close association with the Delaware Nutrient Management Commission for certification and continuing education programs as per 1999 Delaware Nutrient Management Act.

12.9.2   UNIVERSITY OF WISCONSIN

Genskow (2012) reported that Nutrient Management Farmer Education (NMFE) programs of University of Wisconsin were highly effective in helping farmers to understand nutrient management planning and even accounted for an increase in the frequency of soil testing.

12.9.3   UNIVERSITY OF MISSOURI

Nitrogen Watch Program of University of Missouri monitors spring rainfall and pointed out risky areas having problems of nitrogen loss and deficiency.

12.9.4   UCONN'S NUTRIENT MANAGEMENT PLANNING

Simply, a Nutrient Management Plan is a reservoir of the nutrients prepared on the farm or required by the crops to promote the growth of the intended crop, for whole fields on the farm. The aim of the Program is to assist the farmers to target their nutrients as per their fields and need. UConn Extension's Nutrient Management Planning team is applying remote sensing technology to assist farmers for more effective utilization of manure and fertilizer.

12.9.5   INSTITUTE OF FOOD AND AGRICULTURAL SCIENCES OF THE UNIVERSITY OF FLORIDA (UF/IFAS)

Cooperative Extension specialists and agents at the Institute of Food and Agricultural Sciences of the University of Florida (UF/IFAS) are actively engaged with state agencies to supply research-based information on nutrient management. Under the initiative, soil-test-based suggestions from UF/IFAS were provided based on crop nutrient requirements for realizing optimum production. The relevant detailed information on successful crop production besides soil test report is also presented in an assembly of UF/IFAS Extension publications. A primary focus of the UF/IFAS team was to synchronize and compile all desired revisions to the instructional materials for forage crops, especially on nutrient recommendations.

12.9.6 INDIAN AGRICULTURAL RESEARCH INSTITUTE (PUSA), NEW DELHI

Soil testing service in India started in 1955–1956 with IARI STL as the hub to coordinate with other STL in the country. This laboratory, known as Central Laboratory for Soil and Plant Analysis, is renowned among research and extension functionaries, and farming communities for its state-of-art facilities, precise analysis, and satisfactory advisory service. The laboratory extends soil, manure, plant, and irrigation water testing services to the stakeholders. Advanced level training on soil testing, plant analysis, and water quality analysis is conducted each year for the experts of soil science. To promote the soil-test-based fertilizer recommendation, IARI has designed a low cost digital tool, known as Pusa Soil Test and Fertilizer Recommendation (STFR) Meter. It analyzes as many as 14 soil parameters, that is, pH, lime requirement for acidic soil, gypsum requirement for alkaline soil, EC, organic carbon, available N, P, and available K, available S, available B, available Zn, Cu, Fe and Mn. The instrument has proven to be highly useful especially for the region where soil testing setup is still not available.

12.10 PROJECT-BASED EXTENSION APPROACH FOR NUTRIENT MANAGEMENT

The term project generally means pilot testing of a program on a small scale within a specific period making all-out efforts for its success. This approach assumes that the current extension system has failed in making any significant impact on the diffusion of nutrient management technologies, and much can be attained if extension effort is focused in a particular region within a stipulated time period with large infusion of external resources. The purpose of the approach is to demonstrate what can be achieved within a few years.

The achievement of this extension outlook is measured by seeing the immediate changes (e.g., area covered under nutrient management technologies) at the project locale. Few major extension projects having nutrient management technology component are discussed in the following paragraphs.

12.10.1 OPERATIONAL RESEARCH PROJECT

In 1975, Operations Research Project (ORP) was started to demonstrate the usefulness of diverse SWC technologies in controlling soil erosion on farmers' fields with higher benefit cost ratios. But a major lacuna in this project was low rate of farmer participation in project activities. Another important gap in this scheme was putting more focus soil conservation only neglecting other goals.

12.10.2 NATIONAL PROJECT ON MANAGEMENT OF SOIL HEALTH AND FERTILITY (NPMSF)

The NPMSF was initiated in 2008–2009 with primary objective of promoting INM practices through judicious application of chemical fertilizers along with organic manures and biofertilizers. For this national project, the State Governments can withdraw resources from the "Rashtriya Krishi Vikas Yojana" and "Macro Management of Agriculture." Soil nutrient management is intended to promote location as well as crop specific residue management, organic farming by linking soil fertility maps with macro–micro nutrient management, and proper land use as per land type.

12.10.3 SOIL AND WATER CONSERVATION ON WATERSHED BASIS— RIDF (NABARD ASSISTED)

NABARD sponsored various development projects under Rural Infrastructure Development Fund (RIDF) with an aim to create develop necessary infrastructure backup for land development and controlling soil degradation. Watershed-based projects and projects for drainage protection and flood control are being supported under this initiative. The major focus of the scheme is to ensure better and sustainable crop yield in selected watershed area by adopting scientific SWC practices. The project played an important role in raising ground water level, controlling floods, and droughts in the watersheds area under this project.

12.10.4   NATIONAL WATERSHED DEVELOPMENT PROJECT FOR RAINFED AREAS (NWDPRA)

The program namely NWDPRA is being implemented since 8th Five Year Plan by the Government of India for the development of rainfed areas. The major objectives of the program were overall development of watershed area using scientific natural resource management practices through soil conservation measure. Besides promoting natural resource management activities, income generating activities under Farm Production System and Livelihood Support Initiative or landless farmers were also incorporated under the projects.

12.10.5   BHUCHETANA—A MISSION PROJECT OF ANDHRA PRADESH

ICRISAT-led consortium started a project in Andhra Pradesh in 2011–2012 by mapping out the soil nutrient deficiencies and preparing mandal-wise balanced nutrient recommendations with an aim of raising the average productivity of specified crops in the identified districts by 25% in five years. This also emphasizes upon the development of GIS-based soil maps for both micro and macro nutrients.

Besides these national- and state-level projects on nutrient management, the research institutes of ICAR, namely, IARI, New Delhi, IISS-Bhopal, IISWC-Dehradun, CSSRI-Karnal, NBSSLUP, and different agricultural universities, namely, TNAU, BAU, UBKV, PAU, etc., regularly implements different projects on SHM and promotion of nutrient management technologies.

12.11   INSTITUTIONAL/NONGOVERNMENTAL ORGANIZATION (NGO)/FARMERS ORGANIZATION (FO)-BASED EXTENSION APPROACH OF NUTRIENT MANAGEMENT

Since independence, civil-society-based extension proves its potential as a more efficient and responsive extension service provider for rural development. However, its importance in rural development was recognized in more formal and legal way during the last three decades. In a vast country like India, scientists-farmers interaction and visit of farmers to research center is a time taking affair and difficult task. In this context, adoption of this approach can lower down the cost and work pressure of extension functionaries. As a result, numerous farmers' interest group (FIG), cooperative and self-help group (SHG) have been promoted through NGO as an instrument or strategy for initiating and implementing different development program in rural and difficult areas where other mode of extension approaches failed to deliver.

12.11.1   INNOVATIVE EXTENSION APPROACHES OF NGOS

12.11.1.1   EXTENSION APPROACHES OF BHARATIYA AGRO-INDUSTRIES FOUNDATION (BAIF)

The Bharatiya Agro-Industries Foundation (BAIF) is operating in the farming sector in 12 states with a priority on SWC activities on degraded soils, composting, and agroforestry. BAIF promoted creation of farmers' cooperatives and federations of self-help groups: namely Vasundhara Agri Horti Producers Company Ltd for disseminating knowledge on package of practices including SHM for growing crops. The "wadi" program on establishing orchards, supplemented by SWC activities on degraded land in tribal area is one of its much acclaimed programs.

12.11.1.2   SATMILE SATISH CHANDRA CLUB AND PATHAGAR

This NGO was promoting conservation agriculture and SHM technologies in association with state department, university, Farmers Producer Organization (FPO), Banks, Research Institute and Self-Help group (SHG). They are actively promoting soil health promoting conservation agricultural technologies and practices such as mulching, biofertilizer, zero tillage, direct seeded rice, organic farming, fishery, poultry, duckery, goatery, etc., for enhancing the productivity of mother soil. They also offered soil testing service and make available the necessary inputs including biofertilizer and organic fertilizer for farming communities.

12.11.2   FARMERS' PRODUCER ORGANIZATION (FPO)/FARMERS ASSOCIATION-BASED EXTENSION APPROACH

The concept of FPO is gaining popularity in India for strengthening the economic position of farming communities. Mobilizing farmers into specific producer groups or cooperative enhances the effectiveness and efficacy of extension systems in supplying nutrient management technologies. Manaswi et al. (2020) reported that a large section of participating farmers (48%) had technical efficiency of more than 60% as compared to nonmembers (18%). FPOs played in important role in lowering down the transaction cost and number of middleman leading to the attainment of a higher portion of producer's share in consumer's rupee (65%).

The producer organization like Maharashtra State Grapes Grower Association successfully maintaining the relation with research institute and access information and technologies time-to-time on nutrient management of grape and conduct educational programs for its member farmers. Similarly, farmers' associations, namely: Indian Farmers Association; Farmers' Association on Pomegranate; Turmeric Farmers Association of India; Organic Farming Association of India (OFAI), Association of Farmer Companies; cooperative like IFFCO, etc., are successfully providing the advisory services to its members on nutrient management and arranging their fertilizer requirement in addition to credit, inputs, marketing, agro-processing, and farm extension services.

Another new concept, that is, Producer Companies was started since 2002 in India and already more than 500 Producer Companies are enrolled till date. Indian Organic Farmers Producer Company Limited, Shodh Farmer Producer Company, Vanilla India Producer Company Limited, etc., are some of the successful producer companies on promotion of nutrient management technologies so far in India.

12.12   T&V SYSTEM/TRAINING-BASED EXTENSION APPROACH ON NUTRIENT MANAGEMENT

The T&V system of extension was introduced by World Bank in 1974 in India to speed up the dissemination of green revolution technologies and later spread to more than 70 countries in Asia, Africa, and Latin America. It is a system of technical advice to improve agricultural productivity through convincing farmers to adopt a set of recommended farm practice. This fairly centralized approach is rest on a routinely planned calendar of visits to farmers' field with strict schedule of daily and fortnightly activities and training of extension workers. Extension functionaries are only engaged in technology dissemination of recommended practices. Keeping the success of this T&V system, later different organizations especially Krishi Vigyan Kendra (KVK) has adopted training and demonstration activities as extension approach for promoting nutrient management technologies. The purpose of the approach is to motivate farmers to enhance production of specific crops and to enhance the communications between extension staff-researcher and extension agents-farmer.

12.12.1   KRISHI VIGYAN KENDRA

The first KVK was established in 1974 at Puducherry and later spread to each and every district of India with a basic mandate of vocational training, on-farm testing, and front line demonstration of advance agricultural technologies. To demonstrate and promote the INM technology in different agro-ecosystem, KVKs are regularly conducting training, field demonstration, and on-farm testing to assess the suitability the technology at local level. For effective technology transfer, KVKs are regularly organizing vocational training, on-farm trial, field demonstrations, method demonstrations, field days, and exposure visits besides providing soil testing services to district farmers of the district.

12.12.2 TRAINING INSTITUTIONS

In India, the government has established different level training institute to import need-based skill development program at National, State, and District level. For example, national level training program was organized by Directorate of Extension, Ministry of Agriculture with its regional offices at Nilokheri, Hyderabad, Anand, and Jorhat. The most popular program is eight days Model Training Programme in core areas of SHM for extension professional. State Agricultural Management and Extension Training Institutes (SAMETI) is the state level autonomous institute with a mandate to impart training on diverse aspect of agriculture including nutrient management. Beside this, KVK, state agriculture department, and ATMA regularly organizes different training programs on regional SHM issues at district or block level for farming communities and extension functionaries.

12.13   PARTICIPATORY EXTENSION APPROACH ON NUTRIENT MANAGEMENT

In this approach, local people identify and analyses their nutrient management problems and find out the solution through self-mobilization. The participatory nutrient management approach also means that basic agronomic experiment must integrate farmers' perspectives. The promotion of nutrient management technologies through participatory approach holds its significance because of the fact that soil too has broader ecological and socioeconomic roles for living earth. The important ecological roles of soil are balancing hydrologic and biogeochemical cycles, functioning as a natural channel for reprocessing organic substances, mitigating global climatic change, and act as a genetic reservoir of biodiversity. The important socioeconomic roles of soil include arranging a livelihood and supporting infrastructure, natural, and cultural heritage.

Mentioned about the need of participatory governance for diverse soil functions whereby local people can decide or prioritize the different function at a given moment and space.

Farmers' perception and knowledge on soil quality and observing the SHM practices are primary elements of agro-ecosystems, and influences the decision-making regarding conservation of natural resources. Soil is a part of the environmental subsystem, whereas human decision-making and activities are components of the social subsystem. Both the systems, that is, environmental and social subsystem must establish synergistic relationship for sustainable service to the environment. By comparing and validating scientific knowledge with the farmers' understanding on nutrient management technologies, the soil health can be much easily protected. Hence, participatory extension approach supposed to be more successful than other top-down extension approach.

The basic assumption of this approach is that the farming communities have much wisdom regarding proper nutrient management practices in their own land, which can be improved more by learning from outside scientific body of knowledge. It is assumed that the interaction between "indigenous knowledge system" and "scientific knowledge system" will provide much useful insight and solution to the local problem of nutrient management. It is further assumed that the effective extension cannot be achieved without the active participation of the farmer themselves, as well as of research and related services.

12.13.1   EXAMPLE OF SUCCESSFUL PARTICIPATORY NUTRIENT MANAGEMENT APPROACH

12.13.1.1   AGRICULTURAL TECHNOLOGY MANAGEMENT AGENCY (ATMA)

The ATMA concept has been conceptualized under "Innovations in Technology Dissemination" component of "National Agricultural Technology

Project (NATP)" in 2005 with initial pilot testing in seven States of India. Seeing the success of the project, the Indian government launched a new program namely "Support to State Extension Programmes for Extension reforms" and started establishing ATMA in 588 rural districts. The main aim of the scheme was to assess the feasibility of new organizational arrangements and operational procedures to strengthen the present extension system. One major aim of the scheme was to encourage decentralize decision-making at the district level through ATMA. The second goal was to include farmers' input into program planning and raise resource allocation mainly at the block level, and to increase the accountability of stakeholders. The third major aim was to enhance program coordination for better implementation of focus areas in agriculture, namely, SHM, natural resource management, farming system innovations, farmer organization, bridging technology gaps, etc. Simply, ATMA is a registered society of all the pertinent partners for technology diffusion at district level. It involves adopting bottom-up and participatory planning procedure to make the technology diffusion process more farmers driven and farmer accountable. It works in convergence with all the line department, research organization, Krishi Vigyan Kendra, NGO, FPOs, marketing bodies, and other organization working in agriculture. Establishment of FIAC (Farm's Information and Advisory Centre) at block level is the most successful institutional innovation under ATMA model. FIAC generally formulates bottom-up participatory planning and make block action plan for submission to ATMA for funding. FIAC has two elements: block Technology Team (BTT) and Farm Advisory Committee (FAC). The block-level official of agriculture, horticulture and animal husbandry, and farmers' representative present in BTT to identify the priorities of each block. Thus, it facilitated a demand driven participatory extension system by providing feedback and building better accountability in extension service.

The approach facilitated to highlight the local nutrient management or soil health issues beside other agricultural issues in district agricultural plan and supported to adopt recommended nutrient management activities through demonstration or training. The findings of Singh et al. (2013) regarding impact of ATMA in Bihar indicated that the ATMA have mainly advocated different eco-friendly nutrient management technologies like INM, intercropping, organic farming, use of vermicompost, seed treatment, mixed cropping, green manuring, line sowing, summer ploughing, micro irrigation, zero tillage, biofertilizers, etc., and the adoption rate of INM practices was quite significant in this approach.

12.13.1.2   TECHNOLOGY TRANSFER AND SERVICE CENTER (TTC)

The example of TTC in Thailand is one of the most successful cases of participatory extension approach on nutrient management. Under this, a group of body Technology Transfer and Service Center (TTC) involving the target farmers is formed. The knowledge and methods are transferred through this TTC in project areas. One well-trained volunteer farmer from the villagers was selected as "Soil Doctor" and chief of the TTC committee in every village. They first prioritize and choose the innovations which fit to their condition before promoting any nutrient management technologies. If needed, they modify and changed the technology to suite the farmers' conditions. Further, development of portable soil test kit, fertilizer hand book, and a DSS accelerated the adoption of nutrient management technology. The idea of using a DSS and readymade test kit is an innovative approach to disseminate nutrient management technology. This approach calls for whole hearted involvement of farmers and intended to empower the farming communities so that they can take their own decision on nutrient management technologies to adopt keeping their local condition in mind.

12.13.1.3   NETWORK-BASED APPROACHES FOR SOIL HEALTH RESEARCH AND EXTENSION PROGRAMMING IN NORTH DAKOTA, USA

This is another case of successful participatory extension approach on nutrient management. The approach believes that adoption of soil health promoting technology can be increased through knowledge network with program and resource support integrating technical, social, and experiential learning technique. To promote SHM technologies using network-based approach, 32 "Soil Health Café Talks," an informal discussion group was formed during 2014–2012. NodeXL software package was used to develop knowledge network of all the members. The ten most promising members in the network comprises of two scientists, five farmers, and one crop consultant and extension expert. As a result of this, the participants raised the frequency of discussion about cover crops among themselves and also shared farm equipment in more numbers. It was reported that more than 25% participants followed operations using cover crops after attending the café talks. The participants also expanded the utilization of online resources on soil health like Twitter (22%), YouTube (23%), and the web page (21%) as follow-up information to Café Talks. A network-based initiative with participants was proven to be highly fruitful in convincing the farmers for adoption of soil health promoting technologies and practices.

12.14   FARMER FIELD SCHOOLS (FFS)-BASED EXTENSION APPROACH ON NUTRIENT MANAGEMENT

The term has been originated from the Indonesian expression *Sekolah Lapangan* meaning *field school*. The first farmer field school was conceptualized and implemented in Central Java of Indonesia by the UN FAO in 1989 to reduce the dependency of farmer on pesticide usage in the context of devastating outbreaks of insecticide induced brown plant hopper (*Nilaparvata lugens*) surge covering more than 20,000 ha in Java alone. The Farmer Field School is one type of adult education, and it is based on the assumption that farmers learn optimally from the field observation and experimentation.

Basically, the FFS extension approach is a community-based learning model and each FFS comprises of 20–30 neighboring farmers who assemble in morning once in every week for complete crop growing season (typically 14 weeks). It is group-based experiential learning (i.e., learning by-doing) or "discovery learning" that helps farmers informally to discuss their farming activities and then decide which IPM practices should be adopted and evaluated to solve their problem (Manoj, 2013). Though it has been conceptualized for promotion of IPM practices but later it was extensively used for promotion of INM practices as well.

The schedule of the FFS was prepared on the assumption that farmers can only practice INM once they had learnt the skill to apply their own analysis, take their own decisions, and organize their own activities. The focus is on empowerment process, rather than adoption of specific INM technologies is what makes FFS distinguish from other extension approach.

The major advantages of FFS-based extension approach are: it is the most suitable extension model to disseminate natural resource management technologies, confirms life-long learning through self-experience, promote group dynamism and community action, and learn better leadership, communication and management skill, facilitates research-extension-farmer linkage, and finally it empowers the farmers to make their own farming decision.

Bunyatta et al. (2006) in his study on the effectiveness of FFS on dissemination of soil and crop management technologies in Kenya reported there was a significant difference in knowledge acquisition in SandCM technologies by FFS farmers as compared to non-FFS farmers. About 50% of

FFS farmers had gained high to very high level of the knowledge while the majority (>80%) of the NFFS farmers had gained less than 50% of the same knowledge. Almost 45% FFS farmers had followed 50% of the technology components against mere 17% of the non-FFS farmers. The FFS farmers were significantly ($P < 0.05$) better SandCM technology disseminator than the non-FFS farmers. Roya et al. (2015) reported that the maximum proportion (76%) of the FFS participants perceived FFS as moderately effective followed by highly effective (15%). This shows that FFS is the most suitable extension approach for promoting complex nutrient management technologies especially in less-developed areas, among the less-educated farmers with poor resource support.

12.15   CONCLUSION

It can be concluded from the above discussion that no extension approach is best and suitable to all the situations. The selection of any particular extension approach depends upon the objective to be achieved, institutional context, beneficiary characteristics, fund availability, and time limit within which the objectives to be achieved. For example, if the objective is to achieve a desired result within short period over a specific geographic area then project-based extension approach is most suitable than other approach. If the objective is to fulfill some national agenda then public-based general extension approach is more suitable. If the objective is to diffuse some complex nutrient management technologies within some poor socioeconomic populace then participatory approach or VO-based approach works better than others. To impart education and skills among the farming communities on nutrient management technologies, the FFS or T&V system of extension will be more appropriate. If the purpose is to disseminate nutrient management technologies on some cash crops, then commodity-based extension approach will work better. However, if the beneficiary is economically strong and literate, then we can suggest private extension system to reduce the load on government extension system. Thus, the extension policymakers has to decide before implementing any nutrient management program which extension approach will work best in the given situation. For this, the policymakers should have prior knowledge on diverse extension approaches for promoting nutrient management technologies. Further, the chapter showed that the traditional extension approach emphasized only on dissemination of crop varieties ignoring the SHM aspect which led to the current situation of ecological crisis in different agro-ecosystem. Hence, we need separate innovative extension approaches focusing on nutrient management and SHM. Finally, we have to update and capacitate our extension workers with new technologies and innovations about nutrient management technologies and ICTs usage for providing necessary advisory services on nutrient management to the farming communities. This shows that we still need miles to go for diffusing different nutrient management technologies successfully among the farming communities.

KEYWORDS

- **CRS**
- **DSST**
- **Digital Green**
- **IASF Mobile App**

REFERENCES

Ahmed, S. A., Karablieh, E. K. and AlKadi, A. S. 2004. An investigation into the perceived farm management and marketing educational needs of farm operations in Jordan. *Journal of Agricultural Education*, 45(3), 34–43.

Annual Report. 2016–2017, Tobacco Board. www.toabccoboard.com

Arbuckle, J. G. 2017. Understanding predictors of nutrient management practice diversity in Midwestern agriculture. 55(6), Article #v55-6a5. Feature Hanna Bates Program Assistant Iowa Water Center Ames, Iowa hbates@iastate.edu.

Axinn, G.H . 1998. Guide on Alternative Extension Approaches. Prepared under the guidance and sponsorship of Agricultural Education and Extension Service (ESHE), Human Resources Institutions and Agrarian Reform Division.

Bates, H. and Arbuckle J. G. 2017. Understanding predictors of nutrient management practice diversity in Midwestern agriculture. *Journal of Extension*, 55(6), 6FEA5.

Bhayal, V., Wankhede, A., Choudhary, S., Jain, S. K. 2019. Impact and awareness of soil health card with reference to maize production in Dhar District of Madhya Pradesh. *International Journal of Applied Agricultural Research*, 14(2), 79–85.

Bordoloi, J. and Das, A. K. 2017. Impact of soil health card scheme on production, productivity and soil health in Assam. Study No: 148. Agro-Economic Research Centre for North-East India Assam Agricultural University, Jorhat, Assam.

Bunyatta, D. K., Mureithi, J. G., Onyango, C. A. and Ngesa, F. U. 2006. Farmer field school effectiveness for soil and crop management technologies in Kenya. *Journal of International Agricultural and Extension Education*, 13(3). DOI: 10.5191/jiaee.2006.13304.

Castellano, M. and Helmers, M. 2015. How Iowa can improve water quality. *The Des Moines Register*. Retrieved from http://www.desmoineregister.com.

Chapman, R., Blench, R., Berisavljevic, G. K and Zakariah, A. B. T. 2003. Rural radio in agricultural extension: the example of vernacular radio programs on soil and water conservation in N. Ghana Network Paper No. 127. AgREN.

Drinkwater, L. E. and Snapp, S. S. (2007). Nutrients in agroecosystems: rethinking the management paradigm. *Advances in Agronomy*, 92, 163–186.

Farouque, M. G. and Takeya, H. 2009. Adoption of integrated soil fertility and nutrient management approach: farmers' preferences for extension teaching methods in Bangladesh. *International Journal of Agricultural Research*, 4, 29–37.

Genskow, K. D. 2012. Taking stock of voluntary nutrient management: measuring and tracking change. *Journal of Soil and Water Conservation*, 67(1), 51–58.

Grover, D., Singh, J. M., and Kumar, S. 2017. Impact of soil health card scheme on production, productivity and soil health in Punjab impact of soil health card scheme on production, productivity and soil health in Punjab. AERC study no. 41. https://www.researchgate.net/publication/334490902.

https://claroenergy.in/tag/farmers/#:~:text=In%202015%2C%20India%20had%20720,users%20with%20access%20to%20internet

https://pib.gov.in/PressReleasePage.aspx?PRID=1626170#:~:text=Today%2C%20India%20has%20290%20operational%20Community%20Radio%20Stations

https://www.news18.com/news/tech/smartphone-users-in-india-crossed-500-million-in-2019-states-report-2479529.html

https://www.thehindu.com/sci-tech/agriculture/India-losing-5334-million-tonnes-of-soil annually-due-to-erosion-Govt/article15717073.ece

Hu, R., Cao, J., Huang, J., Peng, S., Huang, J., Zhong, X., Zou, Y., Yang, J., and Buresh, R. J. 2007. Farmer participatory testing of standard and modified site-specific nitrogen management for irrigated rice in China. *Agricultural Systems*, 94, 331–340.

Iowa State University. (2012). Iowa nutrient reduction strategy: a science and technology-based framework to assess and reduce nutrients to Iowa waters and the Gulf of Mexico. Retrieved from http://www.nutrientstrategy.iastate.edu/.

Kihara, J., Huising, J., Nziguheba, G., Waswa, B. S., Njoroge, S., Kabambe, V., Iwuafor, E., Kibunja, C., Esilaba, A. O., and Coulibaly, A. 2016. Maize response to macronutrients and potential for profitability in sub-Saharan Africa. *Nutrient Cycling in Agroecosystems*, 105, 171–181.

Kragt, M. E. and Llewellyn, R. S. 2014. Using a choice experiment to improve decision support tool design. *Applied Economic Perspectives and Policy*, 36(2), 351–371.

Krishnan, B. V. 2018. What the agricultural census shows about land holdings in India. The Hindu reports dated October 3, 2018. https://www.thehindu.com/sci-tech/agriculture/indian-farms-getting-smaller/article25113177.ece.

Luthi, B., Philippe, T. and Pena-Haro, S. 2014. Mobile device app for small open-channel flow measurement. *Proceedings of the 7th International Congress on Environmental Modelling and Software (iEMSs '14)*, vol. 1, pp. 283–287.

MacCarthy, D. S., Kihara, J., Masikati, P., and Adiku, S. G. K. 2018. Decision support tools for site-specific fertilizer recommendations and agricultural planning in selected countries in sub-Sahara Africa. *Nutrient Cycling of Agroecosystems*, 110, 343–359.

Manaswi, B. H., Kumar, P., Prakash, P., Anbukkani, P., Kar, A., Jha, G. K., Rao, D. U. M. and Lenin, V. 2020. Impact of farmer producer organization on organic chilli production in Telangana, India *Indian Journal of Traditional Knowledge*, 19(1), 33–43.

Manoj, A. 2013. Impact of farmers' field schools on farmer's knowledge, productivity and environment. PhD thesis. Division of Agricultural Extension, IARI, New Delhi.

McLellan, E., Robertson, D., Schilling, K., Tomer, M., Kostel, J., Smith, D. and King, K. 2015. Reducing nitrogen export from the Corn Belt to the Gulf of Mexico: agricultural strategies for remediating hypoxia. *Journal of the American Water Resources Association,* 51(1), 263–289.

National Productivity Council (NPC) reports (2017). http://164.100.117.97/WriteReadData/userfiles/Study%20on%20Impact%20of%20Soil%20Health%20Card%20Scheme.pdf.

Oyinbo, O., Chamberlin, J., Vanlauwe, B., Vranken, L., Kamara, Y. A., Craufurd, P. and Maertens, M. 2019. Farmers' preferences for high-input agriculture supported by sitespecific extension services: Evidence from a choice experiment in Nigeria. *Agricultural Systems,* 173, 12–26.

Pan, D. 2014. The impact of agricultural extension on farmer nutrient management behavior in Chinese rice production: a household-level analysis. *Sustainability,* 6, 6644–6665. DOI: 10.3390/su6106644.

Pongnumkul, S., Chaovalit, P. and Surasvadi, N. 2015. Applications of smartphone-based sensors in agriculture: a systematic review of research. *Journal of Sensors.* DOI: 10.1155/2015/195308

Ragasa, C. and Mazunda, J. 2018. The impact of agricultural extension services in the context of a heavily subsidized input system: the case of Malawi. *World Development,* 105, 25–47.

Reddy, A. A. 2017. *Impact Study of Paramparagat Krishi Vikas Yojana.* National Institute of Agricultural Extension Management (MANAGE), Hyderabad, Telangana, India, pp. 210.

Reddy, A. A. 2018. *Impact Study of Soil Health Card Scheme.* National Institute of Agricultural Extension Management (MANAGE), Rajendranagar, Hyderabad, Telangana, India, p. 106.

Roya, D., Farouque, M. G., Rahmanc, M. Z. 2015. Effectiveness of farmer field school for soil and crop management. *International Journal of Sciences: Basic and Applied Research (IJSBAR),* 20(2), 1–11.

Shehu, B. M., Merckx, R., Jibrin, J. M., Kamara, A. Y. and Rurinda, J. 2018. Quantifying variability in maize yield response to nutrient applications in the northern Nigerian Savanna. *Agronomy,* 8(2), 1–23.

Singh, K. M., Singh, R. K. P. and Kumar, A. (2013). Agricultural Technology Management Agency (ATMA): A study of its impact in pilot districts in Bihar, India. https://www.researchgate.net/publication/235975667_Agricultural_Technology_Management_Agency_ATMA_A_Study_of_its_Impact_in_Pilot_Districts_in_Bihar_India

Tittonell, P. and Giller, K. E. 2013. When yield gaps are poverty traps: the paradigm of ecological intensification in African smallholder agriculture. *Field Crops Research,* 143, 76–90.

Vanlauwe, B., AbdelGadir, A. H., Adewopo, J., Adjei-Nsiah, S., AmpaduBoakye, T., Asare, R., Baijukya, F., Baars, E., Bekunda, M., Coyne, D., Dianda, M., DontsopNguezet, P. M., Ebanyat, P., Hauser, S., Huising, J., Jalloh, A., Jassogne, L., Kamai, N., Kamara, A., Kanampiu, F., Kehbila, A., Kintche, K., Kreye, C., Larbi, A., Masso, C., Matungulu, P., Mohammed, I., Nabahungu, L., Nielsen, F., Nziguheba, G., Pypers, P., Roobroeck, D., Schut, M., Taulya, G., Thuita, M., Uzokwe, Vanlauwe, B., van Asten, P., Wairegi, L., Yemefack, M., and Mutsaers, H. J. W. 2017. Looking back and moving forward: 50 years of soil and soil fertility management research in sub-Saharan Africa. *International Journal of Agricultural Sustainability,* 15(6), 613–631.

Vanlauwe, B., Descheemaeker, K., Giller, K. E., Huising, J., Merckx, R., Nziguheba, G., Wendt, J. and Zingore, S. 2015. Integrated soil fertility management in sub-Saharan Africa: unravelling local adaptation. *Soil,* 1, 491–508.

PART V
Nutrient Management Approaches

CHAPTER 13

The Panorama of Phosphorus Fertilization in Crop Production

TRISHA ROY[1*], I. RASHMI[2], ABHIJIT SARKAR[3], and JUSTIN GEORGE K.[4]

[1]*ICAR-Indian Institute of Soil and Water Conservation, Dehradun, Uttarakhand*

[2]*ICAR-Indian Institute of Soil and Water Conservation, RC-Kota, Rajasthan*

[3]*ICAR-Indian Institute of Soil Sciences, Bhopal, Madhya Pradesh*

[4]*Indian Institute of Remote Sensing, ISRO, Dehradun, Uttarakhand*

Corresponding author. E-mail: trisha17.24@gmail.com

ABSTRACT

Phosphorus (P) is one of the essential elements for sustenance of all forms of life on the planet. It is irreplaceable in the living system and does not come with any substitute. However, the present rate of exploitation of available P reserves in the globe has led to eutrophication of water bodies on one hand and threatens the occurrence of "peak P" on the other hand where a huge gap between demand and supply is expected. Also, the world soils are depleted in P and need fertilization through external sources, but the use efficiency of P fertilizers still remains a meager 10%–15%. This chapter tries to explore the various new avenues of P fertilizer development along with the historical review of P fertilizers, the scenario of world reserves and resources of P, the hypothesis of "peak P," and how best we can manage the P resources to get maximum output from agriculture to feed the world in the future. Use of nano-rock phosphate, control release fertilizers, use of low-molecular-weight organic acids, genetic modification for better P acquisition, microbial interventions, and use of high end precision agriculture techniques for better agronomic management are few of the strategies, which have to be explored at commercial levels for increasing the P use efficiency and optimize the utilization of the P reserves.

13.1 INTRODUCTION

The discovery of phosphorus (P), as an element, was more of a serendipity in search of the "Philosopher's stone" by alchemist Hennig Brand in the year 1669, while its inclusion in the list of essential plant nutrients occurred in 1839 by C S Sprengel. P along with N and K are the three primary essential nutrients required for crop production and it plays a major role not only in the life cycle of plants but all animals as well. It is a component of the energy molecule adenosine triphosphate and adenosine diphosphate, the energy currency of the cell. The element is vital in converting or harvesting solar energy for growth and reproduction of the plants. The most prime physi-ological process, that is, photosynthesis is heavily dependent on adequate P supply, in absence of which the photosynthetic activity is limited with a direct consequence on crop yield (Rao and Terry, 2000). Thus, P deficiency in soil often leads to yield marginalization and poor crop performance.

Though P is the 11th most abundant element in the earth's crust, its highly reactive nature in the soil system renders it unavailable to plants and globally, 43% soils are reported to suffer from P deficiency (Liu et al., 2012). In lieu of such widespread deficiency of P, supplying cropping systems through external application of the element through fertilizers become imperative. In India, P deficiency is reported across 42% soils (Motsara, 2002) and P fertilizers are unavoidable to obtain optimum crop production. Since the attainment of Green Revolution in India, the fertilizer consumption has witnessed a steady increase and resulted in significant production of food grains. This, in turn, fueled the demand of chemical fertilizer over last few decades.

13.2 P DEFICIENCY IN PLANTS AND ITS IMPACT

The expression of P deficiency in plants does not manifest immediately and is a function of the P reserves within the plant system. The P reserves act as a buffer during the P deficiency and can help tackle short-term fluctuations of P supply to plants. Majority of the P acquired by the higher plants are stored in the form of inorganic P within the cell vacuoles. Under P stress, this inorganic P is translocated within the plant body to meet up the metabolic P requirement (Grant et al., 2001).

A concentration less than 0.2% of P in the plant tissues is considered deficient and visible symptoms start appearing in the plants. P is highly mobile within the plant system and the older leaves show the first symptoms of deficiency. The foliage becomes dark green or purplish in color due to P deficiency. Basic metabolic activities such as respiration and photosynthesis are affected by low P content and under conditions where respiration rates are lower than photosynthesis, the accumulation of carbohydrate occurs in the leaves imparting dark green color. The purplish or red color exhibited during P deficiency is due to accumulation of pigment anthocyanin, which is particularly visible in crops such as maize, guava, apple, and strawberry (Bruulsema, 2016). However, not all P-deficient plants show accumulation of anthocyanin. The common symptoms of P deficiency include stunted plant growth, darker foliage, reduced tillering and biomass production, and overall yield reduction.

As high as 78% and 79% yield reduction in wheat and maize, respectively, are reported due to omission of P from fertilizer schedules (Dai et al., 2013), while application of P and organic manures could increase soybean yield by 93% and wheat yield by 197% in Vertisols (Reddy et al., 2000) indicating the importance of P for obtaining optimal yield.

The P requirement is highest during the initial growth stages and any deficiency during this period may lead to irreversible yield losses. The reason attributed to such losses is mainly because a lack of P during the early growth stages interferes with C metabolism of the plant. The plants exhibit slower rate of leaf appearance and shrunken leaf size, particularly for the lower leaves. This reduces the photosynthetic efficiency and light interception by the leaves and eventually dampens the overall crop growth and development (Grant et al., 2001).

13.3   P CHEMISTRY IN SOIL

The biogeochemical cycle of P is open in nature, unlike N or C cycles which are closed. This indicates that the P, once lost from different sources through erosion, runoff, sedimentation, etc., does not enter back into the cycle. The starting point of P cycle is the weathering of P bearing rocks and minerals, which releases P in the environment. This inorganic P is distributed in soil and water and is taken up by the plants, animals, and aquatic organisms.

These organisms, in turn, contribute toward the organic P pool in soils and sediments. It is estimated that only 0.1% of the total P in the soil is available for uptake by plants (Zhou et al., 1992) and the rest remains fixed in the soil as different chemical compounds driven by the soil pH, which might become available with time.

Typically, the total P concentration in the soil system may vary from anywhere between 200 and 2000 mg P kg^{-1} soil with a mean value of 1000 mg P kg^{-1} soil (Brady and Well, 2002). The inorganic P (Pi) generally accounts for 35%–70% of total P in soil, while the organic P (Po) ranges from 30 to 65% of the total P depending upon the soil organic matter content, climatic condition, vegetation, soil texture, land use, etc. (Harrison, 1987). According to the fractionation scheme presented by Kuo (1996), the inorganic P fractions include: (1) the saloid P (sal–P), (2) calcium P (Ca–P), (3) aluminum P (Al–P), (4) iron P (Fe–P), and (5) reductant-soluble P (ReS–P) as presented in Figure 13.1.

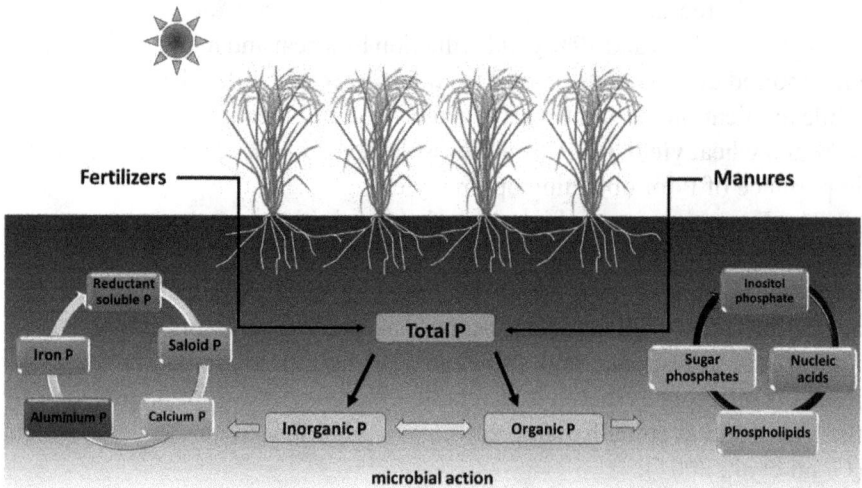

FIGURE 13.1 Dynamic equilibrium of different fractions of organic and inorganic P in the soil system.

Ca-bound P compounds are found in soil having neutral and alkaline pH, while Fe-/Al-bound P minerals are more frequent in acidic soils. Further, the inorganic P compounds can be classified as the primary and secondary minerals, which vary in their solubility and P supplying capacity. Apatites $[(Ca_{10}(PO_4)_6 \cdot H_2O)]$, strengite ($FePO_4$), and variscite ($AlPO_4$) represent the primary P minerals, which are very stable and they release P through their weathering process, which is a sluggish phenomenon. On the contrary, secondary P minerals include Ca, Fe, and Al phosphates having varying rates of dissolution, which are mainly governed by particle size of the minerals and soil pH (Pierzynski et al., 2005). Also, solubility of Fe/Al phosphates is enhanced under neutral–alkaline condition, while acidic pH helps to solubilize the Ca–P (Hinsinger, 2001). Also, there are phosphate compounds, which remain adsorbed on various clays and Al/Fe oxides can be released through desorption reactions. Thus, there exists a complex equilibrium between all these forms of P ranging from very stable, sparingly available P pools, to plant-available P pools such as labile P and solution P (Shen et al., 2011).

The organic P pools are important from plant nutrition point of view as their mineralization can be directed through microbial modifications in the soil and also through plant roots. The soil organic P (Po) mainly represents the stabilized forms of inositol phosphates, phosphonates, orthophosphates, and organic polyphosphates (Turner et al., 2002; Condron et al., 2005). The P nutrition by plants is derived mainly from the inorganic P sources and sometimes from the organic P sources and both these fractions play an important role in P availability of plants.

13.4   HISTORY OF PHOSPHATIC FERTILIZERS

13.4.1   ANCIENT AGRICULTURE AND P FERTILIZATION

Looking back into history when concepts of fertility and organized agrarian societies were uncommon and people practiced hunting and gathering mostly for livelihood, agriculture survived mainly on manures and crop residues which returned back the nutrients removed from soil. As early as 40,000 years ago, the Australian Aboriginals practiced "fire stick" farming where controlled burning of land patches was done to improve productivity. The Australian soils low in P showed greater P availability due to burning (Cordell, 2001), which can be compared with the present day "jhum cultivation" practice.

In the middle Asian countries of China and Japan, the agrarian people used night soil or human excreta as a source of nutrients for their crop (Ashley et al., 2011; Vacarri et al., 2011). Another interesting approach to P fertilization is evidenced in Middle East in Roman and Byzantine era during the time frame of 1st century BC–1st century AD. The people reared pigeons for meat as well as for manure as the pigeon excreta is rich in P (Tepper, 2007). Similar practice of grazing peasants sheep in the fields of the landlords existed in medieval England to enrich the land with sheep droppings rich in P and other nutrients for a bountiful harvest (Driver et al., 1999). Thus, even without the knowledge of elements in agriculture and their functions in the plant and animal system, ancient people knew the importance of giving back to the soil to sustain agriculture production and the society as a whole.

13.4.2   EARLY DEVELOPMENT IN P FERTILIZERS

It was during 1840, Liebig proposed mineral theory that provided a scientific explanation of how nutrients such as N, P, and K were essential elements that circulated continuously between dead and living matter. He also highlighted that it was the chemical form or inorganic form of the elements, which were essential for crop growth rather than the organic form. This transformed the entire agriculture scenario and led way to the modern inorganic-based agriculture system by replacing the age old traditional organic farming. It was during this period, Sir Joseph Henry Gilbert (who had studied under Liebig) and Sir John Bennet Lawes together founded the Rothamsted Research Station in 1840s in England, to undertake long-term trials for studying the effectiveness of mineral and organic fertilizers on crop yields (Rothamsted Research, 2006). On the other hand, in Europe, escalating soil degradation and famines was common, which triggered a search for external sources of fertilizers to boost crop yields (Marald, 1998; Emsley, 2000). Such an instance was reported from England where large volumes of crushed bones (rich in calcium phosphate) were imported from mainland Europe to apply to British farmlands. This was the inception of the development of P fertilizers where the reaction between crushed bones and sulfuric acid produced the first superphosphates and was patented during mid-1800s (Russell and Williams, 1977). Later, in the coming years, these bones were dissolved in sulfuric acid to create a liquid fertilizer (Liu, 2005; Rothamsted Research, 2006). The first superphosphate plant was built in 1854 at Ipswich, England with a sulfuric acid manufacturing facility as well. Simultaneously, during this time, the significance of guano (bird and bat droppings) in P fertilization was unfurled in costal Peru and the US government took initiatives to harvest guano from uninhabited island ecosystems for utilization as P fertilizers.

The demand for P fertilizers slowly gained momentum during the mid-20th century when the world ushered Green Revolution to feed the ever-increasing human population, which was a direct outcome of the rapid industrial growth coupled with urbanization. The Green Revolution introduced several high-yielding varieties, which needed huge chemical inputs in the form of fertilizers to produce the targeted yield. The mining and production of P fertilizers along with nitrogenous fertilizers skyrocketed as a result and the world witnessed six times increase in fertilizer use during 1950–2000 (IFA, 2006).

Rock phosphate (RP) forms the core or base for manufacture of all P fertilizers and naturally occurs in the earth's crust mostly as sedimentary deposits. The first deposit of RPs was discovered during the early 19th century and the mining process was initiated in 1867 in South Carolina, USA (Jasinski, 2013). Subsequently, Florida RP, Tennessee RP, different deposits in Morocco, Tunisia, Algeria, Gafsa, etc., were discovered from which RP mining started extensively. In India, the major RP deposits were discovered in Jhamar Kotra, Rajasthan during the 1960s. Till date, no substitute to RP for the P fertilizer industry has been discovered which makes this a priceless resource for sustaining the agriculture production across the world.

13.4.3   DEVELOPMENT OF P FERTILIZERS IN INDIA

The journey of P fertilizers in India began way back in 1906 when the single superphosphate (SSP) manufacturing plant was established in Ranipet, Tamil Nadu. For a long time, after 1906, there was no further development in the fertilizer manufacturing sector. India in the preindependence era suffered from food insufficiency and under the British rule, agriculture crashed due to lack of modern technologies and unavailability of irrigation facilities. Also, there was only promotion of cash crop which made agriculture more impoverished. The country lurked under severe famine, drought, and food shortage during this period.

Postindependence, the Indian government took serious note of this food shortage and strategic introduction of high-yielding varieties, improvement in irrigation facilities, and increased consumption of fertilizers and other agriculture chemicals that helped in ushering the Green Revolution during the 1960s. Just prior to this, three large public sector fertilizer plants came up in Sindri (Bihar), Nangal (Punjab), and Trombay (Maharashtra). All these plants were then clubbed under the Fertilizer Corporation of India in 1961, which is the main body for controlling the production and quality of fertilizers produced by the public sector industries. Chronology of phosphatic fertilizer development in India has been listed in Table 13.1.

TABLE 13.1   Chronological Development of P Fertilizers in India

Fertilizers	Year of Manufacture
SSP	1906
Ammonium phosphate	1960
Nitrophosphate	1965
Diammonium phosphate (DAP)	1967
Triple superphosphate	1968
Urea ammonium phosphate	1968
NPK complex fertilizers	1968

The consumption of P fertilizers in the pre-Green Revolution era was very meager and only one-fourth of the world consumption. Just before the onset of Green Revolution (1966–1967), the total fertilizer consumption was around 1 million ton which increased to 2.26 million tons during 1970–1971. This increment in fertilizer consumption is often acknowledged to contribute 50% rise in food grain production during Green Revolution (Patra et al., 2016), while some believes that it contributed to only one-third of the increase in grain production (Bumb and Baanante, 1996).

Initially, during the 1950s, the use of P fertilizers in comparison to N fertilizers was much less but increased significantly during the 1960s and 1970s. The share of P in primary nutrient fertilizer consumption has increased from 13.5% in the 1950s to 25.6% during 2000s and declined the share of N fertilizers (Sharma and Thaker, 2011). Over time, the consumption of SSP, which was the starting point of P fertilizers, declined and came to be replaced with DAP and NP complexes due to the higher content of P in these complex fertilizers. In 2016–2017, the production of DAP was 43.65 million tons and complex fertilizers was 79.66 million tons (Annual Report, DoF, 2018). The total consumption of DAP in 2016–2017 was 8963.5 million tons, while in 2017–2018 it increased to 9294.1 million tons (faidelhi.org).

13.5   STATUS OF P RESERVES AND RESOURCES

The RP is the main source of P fertilizer in the globe. The term RP is used widely to include both unprocessed raw material as well as beneficiated ores, which are used by the fertilizer and other manufacturing sectors. The term includes any kind of phosphate rock, which has apatite as some form of mineral component. The RPs are classified mainly into two types based on their source or origin:

1. The sedimentary RPs
2. The igneous RPs

The sedimentary deposits supply almost 80%–90% of the global P requirement and comprise of carbonate apatite $[Ca_{10}(PO_4,CO_3)_6]$. The igneous deposits, which contribute rest 10%–20% of the world production, are mainly the fluoro-, chloro-, and hydroxyapatites in the earth's crust. The sedimentary deposits are mostly of marine origin and easy to mine, while the igneous deposits are formed as intrusive masses or sheets existing in the hydrothermal veins and are more difficult and uneconomic to mine.

The term reserves and resources are often ambiguous and sometimes not distinguished by authors. But United States Geological Survey Bulletin (USGS, 1976) distinguishes between the two, which has been further simplified by Kauwenbergh, 2010.

*   *Reserves*: The phosphate rocks, which can be economically extracted in the present times using the existing technologies.
*   *Resources*: This includes phosphate rock of any grade including the reserves that may be extracted or mined in the future. The reserves are, thus, a part of the existing larger resource base.

The distribution of RP reserves is highly skewed globally. More than 95% of the world RP reserves are distributed in 10 countries. Out of this, Morocco and Western Sahara house 75% of the RP reserves followed by China which has 6%, Algeria and Syria having 3%, United States, Jordan, and Russia having 2% each, and Peru and Saudi Arabia having 1% (USGS Mineral Commodity Summary, 2013).

13.5.1 PHOSPHATE RESERVES/RESOURCES IN INDIA

The USGS Mineral Commodity Summary, 2019 estimates the total amount of Indian RP reserves to be 46,000 tons, which is only 0.065% of the total world reserves. However, the total resources in the country are estimated to be 313.67 million tons (Mines IBO, 2018). The distribution of the resources is presented in Figure 13.2.

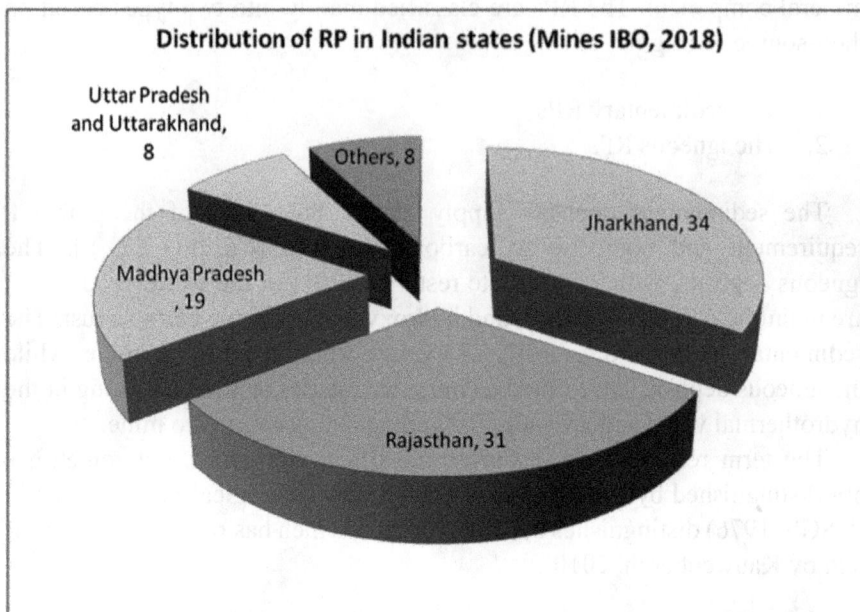

FIGURE 13.2   Distribution of RP resources in India (adapted from Mines IBO, 2018).

On comparing various grades of RPs, IBM (2016) reported that low grade accounted for 37%, followed by beneficiable (29%), blendable (11%), chemical fertilizer and soil reclamation (8% each), and remaining unclassified and not-known grades (about 7%). Though the reserves exist in almost eight states, mining is done commercially only in Rajasthan and Madhya Pradesh. The former accounts for 96% of the total country's production, while only 4% is contributed by the latter. Generally, RPs available are low grade and, therefore, very little amount is used for direct application. According to an estimate, nearly 74,000 tons of RP from Jhamar Kotra, Rajasthan were used for direct application to soil. The RPs reserves from this place were of several grades with 52% of 15–20% P_2O_5 grade followed by 40% of 30–35% P_2O_5 grade and 4% each of 25%–30% P_2O_5 grade and 20%–25% P_2O_5 grade.

13.5.2   PRODUCTION AND CONSUMPTION OF P FERTILIZERS AT GLOBAL SCALE

Of the three primary nutrients N, P, and K produced widely as fertilizers across the globe, the demand and production of N is the highest compared to P and K. However, over time, agriculture practitioners and farmers have realized the significance of other nutrients for sustainable management of land, soil, and crops and concurrently this has led to the increased production and consumption of P and K fertilizers as well. The projected value of total P fertilizer production capacity of the world in 2020 is 64.67 million tons, while for 2019 it is 63.04 million tons. However, when the actual supply of P fertilizer is considered, it is 52.36 million tons for 2019 and predicted to be 53.07 million tons for 2020 (FAO, 2017). The supply capacity is enhanced only when there is a steady state increase in the demand and the demand for different fertilizers has witnessed an increase over time. In the 5-year period during 2015–2020, the projected average annual increase in demand for the three primary nutrient fertilizers is expected to be 1.9%. For P fertilizer alone, the growth rate is higher at 2.2% during the same period. In 2019, the demand for P fertilizers is 45.01 million tons which is likely to increase to 45.8 million tons for 2020 (FAO, 2017).

The different regions of the world, however, show different trends in their demand and supply pattern. According to latest report by FAO (2015), huge demand of P_2O_5 (3,900,000 tons) during the period from 2014 to 2018 was from Asian countries (58%), followed by USA (29%), Europe (9%), and Africa (4%). It is also predicted that among Asian countries, demand of P will be highest from India (27%), followed by China (10%), Indonesia (5%), Pakistan (3%), and Bangladesh (2%). Contrastingly, if we focus on the supply scenario, a negative balance is expected in these regions (FAO, 2017). In the South Asian region, countries such as India, Bangladesh, Bhutan, Nepal, Sri Lanka, and Maldives are all expected to have a negative balance for all the primary nutrient elements. With respect to P fertilizer alone, the negative balance between demand and supply is estimated to be 8.09 million tons which could adversely impact the agricultural scenario.

For India, the obscurity in this regard is more than the prospects. Primarily, India is insufficient in high-grade phosphate (>28% P_2O_5) ores and is entirely dependent on imports for manufacturing commercial phosphatic fertilizers. India has only 15.3 million tons of high-grade reserves, which can be used for manufacturing fertilizers. This can suffice only 35% of the total P demand of the country, while the rest 65% has to be met through imports of high-grade RP (Kumari and Phogat, 2008). For the year 2010–2011, nearly 8.1 Mt of P_2O_5 were consumed for production of 235 Mt of food grain production (SubbaRao et al., 2015). In this, almost 50% of P_2O_5 (4.3 Mt) was directly imported as DAP or NP/nitrogen phosphorus potassium (NPK) fertilizers to meet demand. The remaining amount of

46% was met through import of sulfur and high-grade RPs. Presently, India imports 90% of the phosphatic fertilizers either in the form of raw materials such as sulfur, phosphoric acid, or in terms of finished goods such as DAP and triple super phosphate (Annual Report, GoI, Ministry of Fertilizers and Chemicals, 2017–2018). According to FAI (2016), the projected phosphate fertilizers (P_2O_5) requirement in 2020–2021, 2025–2026, and 2030–2031 would be 7.83, 9.01, and 10.69 million tons, respectively. During August 2018, sharp increase in demand of DAP to the tune of 87.9% was recorded (careratings.com) and the major import of RP were from China (45%), Saudi Arabia (31%), USA (13%), and Jordan (5%). The present worldwide distribution of RP resources indicates that only few countries can sustain P fertilizer production for long period such as Morocco and Algeria where the reserves are estimated to last for 1000 years or more. However, in China, USA, and Brazil, the time is limited to 50 years only. Thus, agricultural sustainability is highly dependent on the tactical and judicious use of available RP by either optimizing its use or improving the use efficiency and reducing losses (SubbaRao et al., 2015). For India, in particular where agriculture survives on import of majority P fertilizers, strong steps should be taken to prevent P loss and enhance efficiency of P fertilizers to improve soil fertility.

13.6   PEAK P AND THE DWINDLING P RESERVES

The nonrenewable, finite nature of P reserves and the open ended structure of inorganic P biogeochemical cycle make it a precious, irreplaceable commodity in agriculture. According to Cordell and White (2011), a scenario referred to as "peak P" is likely to occur during 2033 when the rate of exploitation of the P reserves would surpass the existing good quality P reserves. The unrestricted, nonjudicious usage of the natural P reserves would create a huge gap between demand and supply of P fertilizers. The concept of "peak P" is similar to the concept of "peak oil" quoted by Hubert in 1949. But unlike oil, which can be substituted by renewable energy sources such as solar, wind, or biofuel, there is no substitute to P in the living system. Thus, the "peak P" scenario could be actually grimmer than it appears and paralyze the entire global food production system.

The "peak P" philosophy has been debated widely and many articles quote it to be hoax. Articles such as the "peak P is not happening" (peakoil.com, 2015) or "there is no phosphorus shortage: stop designing foolish systems to recycle it" (forbes.com/sites/timworstall, 2015) tend to downgrade the seriousness of the situation by debating P to be the 11th most abundant element in the earth's crust with a total amount of 4×10^{15} tons present in the globe. This huge amount of P should make us feel secure and safe and ensure that scenarios such as that of "peak P" do not occur. However, Cordell and White (2011) explain in detail why "peak P" is more of a reality to be concerned than just a mere false claim. The cycling of P in the biosphere between the living organisms and the dead organic matter occurs at a temporal scale of days to years, while for cycling of mineral P occurring between lithosphere and hydrosphere the time taken is millions of years. Thus, the concentrated mineral P reserves once lost from the lithosphere become a nonrenewable resource.

The existence of elemental P in the earth's crust is a rare phenomenon and it generally occurs as phosphorus pentoxide (P_2O_5) and the extraction of P for industrial and other purposes is done from the igneous and sedimentary deposits of P resources, which are only 0.007% of the total P concentration of the earth. Again of all the igneous and sedimentary P resources, only 20% is classified as the RP reserves which can be exploited economically with the available extraction techniques for industrial and manufacturing sectors (Cordell and White, 2011; Kauwenbergh, 2010). The peak P does not indicate the absolute exhaustion of the RP reserves; rather, it is indicative of a situation where the production would reach its maxima followed by a sharp decline. This decline would negatively impact the agricultural production, which has to increase many folds (by roughly 56% during 2050) to feed a global population of 10 billion (Ranganathan et al., 2018). Thus, the peak P is a serious issue which has triggered researchers to find solution for better utilization of the existing reserves and improving the P use efficiency for more crop productivity. The longevity of the P reserves has been estimated by various researchers, which vary in time frame. Kauwenbergh (2010) gives the most optimistic predictions where the P reserves are likely to be exhausted in coming 300–400 years and he describes the scenario as "peak phosphate in my view will not be a peak phosphate on the supply side, which is the arguments being raised right now. In my view, it will be a peak phosphate on demand and that will be probably within the next 40 years." The International Fertilizer Development Center estimated phosphate rock reserves to be 60,000 Mt compared to the earlier estimates of 16,000 Mt of phosphate rock by the US Geological Survey (Vaccari, 2009), which again provides hope related to availability of the resources.

13.7   PATHWAYS OF P LOSS FROM THE AGRICULTURAL SYSTEM

There are different pathways through which P is completely lost from the agroecosystems. The fixation of soil P should not be accrued in this regard, since the soil P remains in the system and can be retrieved through different mechanisms. The major loss of P from the soil is by water erosion (8 Mt P/y), which accounts for about 30% of the P lost from food-related human activity (Rustomji et al., 2008). In India, the rate of soil loss is estimated to be 15.35 tons ha^{-1} yr^{-1} (Sharda and Ojasvi, 2016), with an estimated nutrient loss of 5.37–8.4 Mt including N, P, and K. The erosion losses of soil results in maximum P losses from the surface, which gets deposited in the surface water bodies leading to degradation in water quality. Surface application of highly soluble P fertilizers causes huge losses through runoff and the fertilizers are lost from soil following successive runoff events (Vadas et al., 2008). The phenomenon of excessive accumulation of P in the surface water bodies has strong negative impact on aquatic ecosystems and animals and is termed as eutrophication. The cycling of P in the biosphere also causes significant loss of P from the agroecosystem. Of the total P accumulated by the livestock through grazing on arable and nonarable land, only 53% is recycled back, while the rest is lost to the environment (Cornish, 2011). Similarly, the P accumulated by the humans is excreted out from the system without proper management and strategies for its recycling and up to 90% of the P is lost to water bodies as sewage discharges and landfills (Cornish, 2011). Thus, considering the importance of P in living system and the scarce availability in the future, research is being directed in the directions to maximize the utilization of available resources along with reduction in the losses.

13.8   STRATEGIES FOR ENHANCING P FERTILIZER USE EFFICIENCY

Over decades, researchers have tried and tested different methods with varying success rates to improve the P use efficiency by the plants. Some of the strategies include:

1. Modification of the soil properties to increase the P fertilizer use by plants.
2. Adoption of appropriate conservation measures to reduce soil erosion *vis-à-vis* P loss.
3. Development of new fertilizer materials to improve P use efficiency.

4. Use of microorganisms to optimize the use and absorption of P from soil.
5. Agronomic management strategies related to right time, right method, and right place of fertilizer application and precision agriculture techniques.
6. Engineering of the crop rhizosphere and selection of P-efficient cultivars.

All these strategies are discussed in details to understand how the P fertilization can be made more effective in the agricultural field.

13.8.1 MODIFICATION OF THE SOIL PROPERTIES TO INCREASE THE P FERTILIZER USE BY PLANTS

Phosphorus in the soil system is highly reactive element and hardly remains in the available form once applied to the soil system. Immediately after application of the soluble fertilizer granules such as DAP, the concentration of P in the surrounding increases which triggers precipitation of the P. Over time, the P diffuses into the soil and its concentration gradually reduces which initiates the adsorption reaction of P. The soil pH and the ionic composition are the main driving force, which determine the fate of P fertilizer applied to soil. Under acidic conditions, Al- and Fe-phosphates are formed whose solubility increases with the increase in soil pH, while in neutral and alkaline conditions, where Ca is the dominant ionic species, Ca-phosphate is formed. In different pockets of the world, Al toxicity is a major cause for limiting P availability and it affects almost 39% of the land area in South and Central America, 19% in Sub-Saharan Africa, and 13% area in the Asia Pacific (Haefele et al., 2014). Thus, modification of soil pH plays a major role in improving the P use efficiency. The traditional management approaches such as liming for acid soils or application of gypsum and green manuring in alkaline soil help to improve the P availability to plants (Yadesa et al., 2019).

13.8.2 ADOPTION OF SOIL CONSERVATION PRACTICES TO REDUCE SOIL EROSION VIS-À-VIS P LOSS

Loss of P from surface soil due to wind and water erosion is an irreversible loss as the lost P by no means can be extracted back into the system. Also, such losses are of major concern from the viewpoint of pollution of water bodies by eutrophication. Thus, adoption of proper soil conservation practices particularly for the undulating topography such as contour bunding, planting along the contours, maintenance of soil cover through retention of crop residues or mulching to reduce the dispersing action of the raindrops, reducing soil disturbance through no-tillage or reduced tillage practices, intercropping practices, or maximizing canopy cover are some of the practices recommended to reduce runoff losses of P from the soil surface. This can have far-flung effect not only in improving the fertilizer P use efficiency, but have a gross positive impact on the environment as a whole. Also, the critical soil P concentration or the threshold soil P concentration and its relationship with soil P in runoff water are essential for practical approach of P-fertilizer management (Sharpley, 1995).

13.8.3   DEVELOPMENT OF NEW FERTILIZER MATERIALS TO IMPROVE P USE EFFICIENCY

Various approaches to develop fertilizers for improving the phosphorus use efficiency have been attempted by researchers extensively. Sarkar et al. (2018) used different coating agents, namely, polyvinyl alcohol and liquid paraffin to coat commercially available RP (17% P_2O_5) to produce novel control release products. Both the coating agent proved effective than commercial DAP in terms of P recovery and between the agents, the polyvinyl alcohol coated products showed higher recovery compared to liquid paraffin. Use of nano-RP in high phosphate fixing soils such as Andisols can reduce P retention by soils up to 87.2% and increase the phytoavailability of P (Devnita et al., 2018). Use of RP in nanoform for direct field application has been compared with commercially available SSP and the apparent P recovery was at par with SSP. However, the additional benefit of nano-RP is mainly in its residual effect for the next crop and its lower cost which can be exploited for commercial use (Adhikari et al., 2014). Low-molecular-weight organic acids such as citric, oxalic, malic, and acetic present in the rhizosphere are important for P solubilization and enhancing P availability to the crops. Combination of low-grade RP along with citric acid loaded in nanoclay polymer composites served as an effective alternative to commercial DAP in increasing the crop yield in a rice–wheat cropping sequence (Roy et al., 2018a, 2018b).

Graphene oxides with its high surface area are another platform, which can facilitate the retention and slow release of P (Andelkovic et al., 2018). Double-layered hydroxides produced from Mg and Al and intercalated with phosphate anions have been equally effective as commercial mono-ammonium phosphate and triple superphosphate in terms of P uptake and seedling growth of barley, maize, and wheat, particularly in the acid soils (Bernando et al., 2018; Benicio et al., 2016). However, all these new materials are still in their nascent stages and commercial exploitation of the full potential of such products and their performance at wide scale remains unexplored.

13.8.4 USE OF MICROBIAL INOCULANTS/BIOFERTILIZERS TO INCREASE P USE EFFICIENCY

A substantial number of soil microorganisms is capable of solubilizing inorganic P or mineralizing organic P and increase the P availability to plants (Alori et al., 2017). The microbes are commonly referred to as phosphate solubilizing microorganisms (PSMs) and play a critical role in improving the P availability in soils. The name of different microbial strains capable of phosphorus solubilization is presented in Box 13.1.

- **Bacteria:** *Bacillus, Pseudomonas, Azotobacter, Burkholderia, Enterobacter, Erwinia,* and *Paenibacillus*
- **Fungi:** *Penicillium, Alternaria, Aspergillus, Cephalosporium, Cladosporium, Curvularia, Cunninghamella, Rhizoctonia,* and *Trichoderma*
- **Actinomycetes:** *Actinomyces, Micromonospora,* and *Streptomyces*
- **Mycorrhiza:** *Glomus* and *Endogene*

BOX 13.1   Different microorganisms involved in P solubilization.

Duarah et al. (2011) used different strains of phosphate solubilizing bacteria in isolation or as a consortium in a rice–legume cropping system, which resulted in higher germination index, root and shoot length, and higher biomass for both the crops. Several authors have reported enhanced crop growth, increased P uptake, and soil P concentration following the inoculation with PSMs (Xiao et al., 2013; Mamta et al., 2010; Wang et al., 2015; Kumar et al., 2014; David et al., 2014).

The PSMs have different approaches for solubilizing the inorganic and organic phosphorus in soil. For solubilization of inorganic P compounds, the production of different organic acids or siderophores, release of hydrogen or hydroxyl ions, and evolution of carbon dioxide are the various mechanisms (Sharma et al., 2013; Alori et al., 2017). The release of P from organic compounds such as inositol phosphate, phospholipids, and nucleic acids is mainly accomplished with the help of enzymes (Rodriguez and Fraga, 1999). There are two groups of enzymes as follows:

1. The phosphomonoesterase or phosphatases
2. Phytase

Besides the PSMs, the mycorrhizal fungus with its extensive hyphal network plays an important role in mobilizing P from P-deficient soils (Nguyen et al., 2019). The PSMs can be successfully explored for solubilizing P from soils, which have high reserves of P fixed in the soil. This can pave the way for more sustainable and effective use of the existing P resources in agriculture. Also, the mycorrhizal fungus is very effective in extracting P from P-deficient soils and helps in survival of plants under P stress conditions.

13.8.5 AGRONOMIC MANAGEMENT APPROACHES AND PRECISION FARMING

The 4R concept underlines the whole aspect of improving P use efficiency via agronomic practices. The 4R referring to right place, right time, right rate, and right source of fertilizer application that varies with soil type, cropping system, and climatic conditions and is effective in achieving maximum yield per unit of nutrient applied (Roberts and Johnston, 2015).

The right source is often guided by the market availability of different fertilizer materials rather than the suitability of the products for a particular crop or soil. In areas with very high rainfall and susceptible to leaching or surface runoff, P fertilizers with lower solubility should be preferred (Chien et al., 2009). In acid soils, with high P fixing capacity, the direct application of RP (Chien et al., 2009) or partial acidulation products of RP is highly effective (Biswas and Narayanswamy, 2006). However, in India, the easy availability of DAP makes it the most popular choice among farmers with almost half of the consumption demand met through DAP (48.75%) (Figure 13.3), while direct application of RP is only 0.16% though 90 Mha of acid soils is present in India.

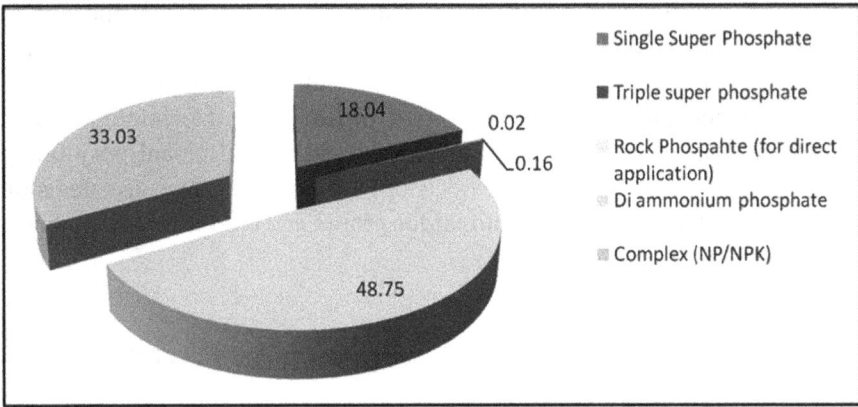

FIGURE 13.3 Percent share of different fertilizer products consumed in India during 2017–2018 (FAI, 2019).

Since most of the P demand is during the initial crop growth stages and highly important for root development, the time of application of fertilizer is very essential. Thus, application of P fertilizer as basal dose is essential to obtain optimum crop yield (Grant et al., 2001; Grant and Flaten, 2019). The rate of application of P fertilizer is driven by the soil testing values (Grant and Flanten, 2019). The rate of P application should be very critically monitored so as not to limit the crop production by under application as well as to minimize the chances of environmental pollution by excessive or over application. The proper placement of P fertilizer is very important for increasing the use efficiency. Since P fertilizer is highly immobile within the soil system, broadcasting generally leads to lesser use efficiency and higher rates of fixation (Fixen, 1992). Band placement or seed placing of fertilizer P has proved to be more efficient in terms of improving P availability to crop plants, reducing its fixation, and increasing crop yield. Also, placement of fertilizer little below the surface gives an advantage to the crop plants in accessing the P as most of the weeds are shallow rooted and placement at the lower layer reduces the competition between crops and weeds (Blackshaw and Molnar, 2009).

In present day agriculture system, precision farming practices follow the 4R approach for nutrient management in soil. The adoption of precision farming at large scale helps in saving of resources and optimum utilization of the inputs without jeopardizing with the environment. Variable rate technology for application of P fertilizer has been effective in reducing the

P fertilizer consumption and improve the water quality (Mallarino, 2006) as this technique allows site-specific nutrient management, which ensures application of exact quantity of fertilizer or any other input taking into account the on-farm variability existing in the field. Use of technologies such as remote sensing, geographic information system, and global positioning system are essential prerequisite to adopt technologies such as precision farming, which can be very beneficial for proper and optimum use of nonrenewable resources such as RP.

13.8.6   ENGINEERING OF CROP RHIZOSPHERE AND SELECTION OF P-EFFICIENT CULTIVARS

The mobility of P is very less within the soil system, which demands close placement of the fertilizer near the root zone. Also, managing the rhizosphere and engineering it according to the needs of the crop can be effective in making the P available to the crop plants. Plant physiological traits can be effectively utilized for detecting P deficiency in standing crop and promoting efficient use of P fertilizers (Frydenvang et al., 2015). The time-resolved chlorophyll fluorescence has been identified as a unique tool to detect the bioactive P in plants and subsequently the P deficiency has been successfully demonstrated in *Hordeum vulgare* (Frydenvang et al., 2015). Plants, as a part of survival strategies, have developed various mechanisms such as modification of the roots for acquiring nutrients such as P (Shenoy and Kalagudi, 2005). Higher plants develop bottlebrush-like proteoid root system to acquire more P (Watt and Evans, 1999), while some have increased number of trichoblast files (Zhnag et al., 2001) as response to P deficiency.

Genes related to the uptake of P and its transportation have been identified successfully and characterized (Duncan and Carrow, 1999). Wang et al. (2009) successfully modified the soybean crop with *APase* gene PAP15 from *Arabidopsis*, which showed significant improvement in dry matter yield and plant P content in sand culture as well as in acid soils. In case of rice crop (*Oryza sativa*), phosphate transporter traffic facilitator 1 (PTTF 1) regulates the exit of phosphate transporters from endoplasmic reticulum to plasma membrane. Modification of PTTF 1 promoter resulted in significant enhancement in tolerance toward low P concentration in soil (Ruan et al., 2015). Application of N fertilizers in the ammoniacal form leads to acidification of the rhizosphere and has enhanced the P availability in crops such as maize (Jing et al., 2010).

13.9 CONCLUSION

Though ambiguity exists in the exact prediction of the longevity of the global P reserves and resources, researchers and agricultural managers across the length and breadth of the world have accepted and acknowledged the importance and irreplaceable nature of the P in sustenance of all life forms. The major challenge in efficient utilization of P fertilizer is restricted by its inherent chemically reactive nature in the soil, which leads to huge losses by fixation or its irreversible loss by surface runoff into water bodies.

With the background of phenomenon such as "peak P" and the environmental issues such as eutrophication, a balance must be struck to achieve a holistic and integrated approach toward P management. The various strategies explored and discussed in this chapter are few among many, which have been successfully implored at laboratory or controlled conditions to improve the P use efficiency. However, research findings need to be strengthened with policy support and reform for global sensitization regarding the issue.

From the Indian perspective, the country has almost equal proportion of dependency on imported P and the domestic available RP (Cordell and White, 2015). This domestic RP reserve needs to be more judiciously managed. Research oriented toward the better utilization of these RP and creation of market suitable new fertilizer materials is the need of the hour and could prove an effective solution to deal with the upcoming P scarcity and increasing expenses of the P fertilizers.

KEYWORDS

- peak P
- P fertilizers
- phosphorus use efficiency
- nonrenewable resource
- new advances in P fertilizers

REFERENCES

Adhikari, T.; Kundu, S.; Meena, V. and Rao, A. S. Utilization of nano rock phosphate by maize (*Zea mays* L.) crop in a Vertisol of central India. *J. Agric. Sci. Tech.* 2014, 4(5A), 384–394.

Alori, E. T.; Glick, B. R. and Babalola, O. O. Microbial phosphorus solubilization and its potential for use in sustainable agriculture. *Front. Microbiol.* 2017, 8, 971.

Andelkovic, I. B.; Kabiri, S.; da Silva, R. C.; Tavakkoli, E.; Kirby, J. K.; Losic, D. and McLaughlin, M. J. Optimization of phosphate loading on graphene oxide–Fe (III) composites—possibilities for engineering slow release fertilizers. *N. J. Chem.* 2019, 43 (22), 8580–8589.

Annual Report, DoF, 2018 http://fert.nic.in/sites/default/files/Annual_Report_2017-2018.pdf (Accessed on 4th January, 2020).

Annual Report, GoI, Ministry of Fertilizers and Chemicals, 2017–2018 http://fert.nic.in/sites/default/files/Annual_Report_2017-2018.pdf (Accessed on 10th January, 2020).

Ashley, K.; Cordell, D. and Mavinic, D. A brief history of phosphorus: from the philosopher's stone to nutrient recovery and reuse. *Chemosphere.* 2011, 84(6), 737–746.

Benício, L. P. F.; Constantino, V. R. L.; Pinto, F. G.; Vergutz, L.; Tronto, J. and Marciano da Costa, L. Layered double hydroxides: new technology in phosphate fertilizers based on nanostructured materials. *ACS Sustain. Chem. Eng.* 2016, 5, 399–409. doi: 10.1021/acssuschemeng.6b01784.

Bernardo, M. P.; Guimaraes, G. G. F.; Majaron, V. F. and Ribeiro, C. Controlled release of phosphate from layered double hydroxide structures: dynamics in soil and application as smart fertilizer. *ACS Sustain. Chem. Eng.* 2018, 6, 5152–5161. doi: 10.1021/acssuschemeng.7b04806.

Biswas, D. R. and Narayanasamy, G. Rock phosphate enriched compost: an approach to improve low-grade Indian rock phosphate. *Biores. Technol.* 2006, 97(18), 2243–2251.

Blackshaw, R.E. and Molnar. L.J. Phosphorus fertilizer application method affects weed growth and competition with wheat. *Weed Sci.* 2009, 57, 311–318. doi: 10.1614/WS-08-173.1.

Brady, N. C. and Weil, R. R. The Nature and Properties of Soils, 13th edition, Pearson Education (Singapore) Indian Branch, New Delhi, 2002, 960.

Bruulsema, T. The colors in phosphorus deficient plants. *Better Crops.* 2016, 100(1), 14–16.

Bumb, B. L. and Baanante, C. A. World trends in fertilizer use and projections to 2020 (No. 567-2016-39006). International Food Policy Research Institute. *2020 Brief 38.* 1996, 4, 675.

http://www.careratings.com/upload/NewsFiles/Studies/Fertilizer%20Q1%20FY%20 2019%20Update.pdf (Accessed on 13th January, 2020).

Chien, S.H.; Prochnow, L. I. and Cantarella, H. Recent developments of fertilizer production and use to improve nutrient efficiency and minimize environmental impacts. *Adv. Agron.* 2009, 102, 267–313.

Condron, L. M.; Turner, B. L. and Cade-Menun, B. J. Chemistry and dynamics of soil organic phosphorus. In Phosphorus: Agriculture and the Environment. Sims, J. T. and Sharpley, A. N. eds., American Society of Agronomy, Crop Science Society of America, Soil Science Society of America, Inc., Madison, WI, 2005, 87–121.

Cordell, D. Improving Carrying Capacity Determination: Material Flux Analysis of Phosphorus through Sustainable Aboriginal Communities. BE (Env) Thesis, University of New South Wales (UNSW), Sydney, 2001.

Cordell, D. and White, S. Peak phosphorus: clarifying the key issues of a vigorous debate about long-term phosphorus security. *Sustainability.* 2011, 3(10), 2027–2049.

Cordell, D. and White, S. Tracking phosphorus security: indicators of phosphorus vulnerability in the global food system. *Food Security.* 2015, 7(2), 337–350.

Cornish, P. S. Peak phosphorus: implications for agriculture. *World Agric.* 2011, 2(2): 21–26.

Dai, X.; Ouyang, Z.; Li, Y. and Wang, H. Variation in yield gap induced by nitrogen, phosphorus and potassium fertilizer in North China plain. *PLoS One.* 2013, 8(12), e82147.

David, P.; Raj, R. S.; Linda, R. and Rhema, S. B. Molecular characterization of phosphate solubilizing bacteria (PSB) and plant growth promoting rhizobacteria (PGPR) from pristine soils. *Int. J. Innov. Sci. Eng. Technol.* 2014, 1, 317–324.

Devnita, R.; Joy, B.; Arifin, M.; Hudaya, R. and Oktaviani, N. Application of nanoparticle of rock phosphate and biofertilizer in increasing some soil chemical characteristics of variable charge soil. In AIP Conference Proceedings, 2018, February, Vol. 1927, No. 1, 030027.

Driver, J.; Lijmbach, D. and Steen, I. Why recover phosphorus for recycling, and how? *Environ. Technol.* 1999, 20, 651–662.

Duarah, I.; Deka, M.; Saikia, N. and Boruah, H. D. Phosphate solubilizers enhance NPK fertilizer use efficiency in rice and legume cultivation. *3 Biotech.* 2011, 1(4), 227–238.

Duncan, R. R. and Carrow, R. W. Turf grass molecular genetic improvement for abiotic/edaphic stress resistance. *Adv. Agron.* 1999, 67, 233–305.

Emsley, J. The 13th Element: The Sordid Tale of Murder, Fire, and Phosphorus. John Wiley & Sons, New York, 2000, 327, ISBN: 0-471-39455-6.

FAI, 2019 https://www.faidelhi.org/general/FAI-AR-18–19.pdf (Accessed on 16th January, 2020). faidelhi.org https://www.faidelhi.org/general/FS%20contents%20-2017.pdf

FAO, 2015 World fertilizer trends and outlook to 2018, summary report http://www.fao.org/3/a-i4324e.pdf (Accessed on January 15, 2020).

FAO, 2017 World fertilizer trends and outlook to 2020, summary report http://www.fao.org/3/a-i6895e.pdf (Accessed on January 15, 2020).

Fixen, P. Optimum fertilizer products and practices for temperate-climate agriculture. In Phosphorus and the Environment. Schultz, J. J. ed., International Fertilizer Development Center, Tampa, FL, 1992, 77–85.

forbes.com/sites/timworstall, 2015 https://www.forbes.com/sites/timworstall/2013/06/27/there-is-no-phosphorus-shortage-stop-designing-damn-fool-systems-to-recycle-it/#1e45c712e463 (Accessed on 16th January, 2020)

Frydenvang, J.; van Maarschalkerweerd, M.; Carstensen, A.; Mundus, S.; Schmidt, S. B.; Pedas, P. R. and Husted, S. Sensitive detection of phosphorus deficiency in plants using chlorophyll a fluorescence. *Plant Physiol.* 2015, 169(1), 353–361.

Grant, C. A. and Flaten, D. N. 4R management of phosphorus fertilizer in the northern great plains. *J. Environ. Qual.* 2019, 48(5), 1356–1369. doi: 10.2134/jeq2019.02.0061.

Grant, C. A.; Flaten, D. N.; Tomasiewicz, D. J. and Sheppard, S. C. The importance of early season phosphorus nutrition. *Can. J. Plant Sci.* 2001, 81(2), 211–224.

Haefele, S. M.; Nelson, A. and Hijmans, R. J. Soil quality and constraints in global rice production. *Geoderma.* 2014, 235, 250–259.

Harrison, A. F. Soil Organic Phosphorus—A Review of World Literature. CAB International, Wallingford, Oxon, UK, 1987, 257.

Hinsinger, P. Bioavailability of soil inorganic P in the rhizosphere as affected by root-induced chemical changes: a review. *Plant Soil.* 2001, 237, 173–195.

https://www.faidelhi.org/general/Prodn-imp-cons-fert.pdf.

https://peakoil.com/geology/peak-phosphorous-is-not-happening (Accessed on 16th January, 2020).

Indian Bureau of Mines (IBM). Apatite and Rock Phosphate Indian Minerals Yearbook 2016 (Part-III: Mineral Reviews), 55th edition, Government of India, Ministry of Mines, Indian Bureau of Mines, 2016.

International Fertilizer Industry Association, 2006 Production and international trade statistics. IFIA, Paris. http://www.fertilizer.org/ifa/statistics/pit_public/ pit_public_statistics.asp (Accessed on 22nd December, 2019).

Jasinski, S. M., 2013 Mineral resource of the month: phosphate rock earth. https://www. earthmagazine.org/article/mineral-resource-month-phosphate-rock (Accessed on 13th January, 2020).

Jing, J.; Rui, Y.; Zhang, F.; Rengel, Z. and Shen, J. Localized application of phosphorus and ammonium improves growth of maize seedlings by stimulating root proliferation and rhizosphere acidification. *Field Crops Res.* 2010, 119, 355–364.

Kauwenbergh, V. S. World Phosphate Reserves and Resources, International Fertilizer Development Center (IFDC), Washington, DC, USA, 2010, 43.

Kumar, S.; Bauddh, K.; Barman, S. C. and Singh, R. P. Amendments of microbial biofertilizers and organic substances reduces requirement of urea and DAP with enhanced nutrient availability and productivity of wheat (*Triticum aestivum* L.). *Ecol. Eng.* 2014, 71, 432–437. doi: 10.1016/j.ecoleng.2014.07.007.

Kumari, K. and Phogat, V. K. Rock phosphate: its availability and solubilization in the soil—a review. *Agric. Rev.* 2008, 29, 108–116.

Kuo, S. Phosphorus. In: Methods of Soil Analysis, Part 3-Chemical Methods. Agronomy Monograph, Sparks, D. L. eds., ASA and SSSA, Madison, WI, 1996, vol. 9, 869–920.

Liu, Y. Phosphorus Flows in China: Physical Profiles and Environmental Regulation. PhD-Thesis, Wageningen University, Wageningen, 2005. ISBN: 90-8504-196-1.

Liu, Y.; Feng, L.; Hu, H.; Jiang, G.; Cai, Z. and Deng, Y. Phosphorus release from low-grade rock phosphates by low molecular weight organic acids. *J. Food Agri. Environ.* 2012, 10, 1001–1007.

Mallarino, A. P. and Borges, R. Phosphorus and potassium distribution in soil following long-term deep-band fertilization in different tillage systems. *Soil Sci. Soc. Am. J.* 2006, 70, 702–707. doi: 10.2136/sssaj2005.0129.

Mamta, R. P.; Pathania, V.; Gulati, A.; Singh, B.; Bhanwra, R. K. and Tewari, R. Stimulatory effect of phosphate-solubilizing bacteria on plant growth, stevioside and rebaudioside-A contents of *Stevia rebaudiana Bertoni*. *Appl. Soil Ecol.* 2010, 46, 222–229. doi: 10.1016/j. apsoil.2010.08.008.

Mårald, E. I. Mötetmellanjordbrukochkemi: agrikulturkeminsframväxtpå Lantbruks akademiensexperimentalfält 1850–1907. Institutionenföridéhistoria, Univ, Umeå, 1998.

Mines, I. B. O. Indian Minerals Yearbook 2016. Government of India, Ministry of Mines, Nagpur, 2018.

Motsara, M. R. Available nitrogen, phosphorus and potassium status of Indian soils as depicted by soil fertility maps. *Fertiliz. News.* 2002, 47(1), 15–21.

Narayanasamy, G.; Arora, B. R.; Biswas, D. R. and Khanna, S. S. Fertilizers, manures and biofertilizers. In Fundamentals of Soil Science. Goswami, N. N.; Millar, C. E. and Turk, L. M. eds., Indian Society of Soil Science, New Delhi, 2009, 579–622.

Nguyen, T. D.; Cavagnaro, T. R. and Watts-Williams, S. J. The effects of soil phosphorus and zinc availability on plant responses to mycorrhizal fungi: a physiological and molecular assessment. *Sci. Rep.* 2019, 9(1), 1–13.

Patra, S.; Mishra, P.; Mahapatra, S. C. and Mithun, S. K. Modeling impacts of chemical fertilizer on agricultural production: a case study on Hooghly district, West Bengal, India. *Model Earth Syst. Environ.* 2016, 2(4), 1–11.

Pierzynski, G. M.; McDowell, R. W. and Sims, J. T. Chemistry, cycling, and potential moment of inorganic phosphorus in soils. In Phosphorus: Agriculture and the Environment. Sims, J. T. and Sharpley, A. N. eds., American Society of Agronomy, Crop Science Society of America, Soil Science Society of America, Inc., Madison, WI, 2005, 53–86.

Ranganathan, J.; Waite, R.; Searcchinger, T. and Hanson, C. How to Sustainably Feed 10 Billion People by 2050, in 21 Charts? World Research Institute, Washington DC, USA, 2018.

Rao, I. M. and Terry, N. Photosynthetic adaptation to nutrient stress. In Probing Photosynthesis: Mechanism, Regulation and Adaptation. Yunus, M.; Pathre, U. and Mohanty, P. eds., London, Taylor & Francis, 2000, 379–397.

Reddy, D. D.; Rao, A. S. and Rupa, T. R. Effects of continuous use of cattle manure and fertilizer phosphorus on crop yields and soil organic phosphorus in a Vertisol. *Biores. Technol.* 2000, 75(2), 113–118.

Roberts, T. L. and Johnston, A. E. Phosphorus use efficiency and management in agriculture. *Resour. Conserv. Recy.* 2015, 105, 275–281.

Rodríguez, H. and Fraga, R. Phosphate solubilizing bacteria and their role in plant growth promotion. *Biotechnol. Adv.* 1999, 17, 319–339. doi: 10.1016/S0734-9750(99)00014-2.

Rothamsted Research. Guide to the Classical and Other Long-term Experiments, Datasets and Sample Archive. Harpenden, UK, 2006.

Roy, T.; Biswas, D. R.; Datta, S. C. and Sarkar, A. Phosphorus release from rock phosphate as influenced by organic acid loaded nanoclay polymer composites in an Alfisol. *P Natl. A Sci. India B.* 2018a, 88(1), 121–132.

Roy, T.; Biswas, D. R.; Datta, S. C.; Sarkar, A. and Biswas, S. S. Citric acid loaded nanoclay polymer composite for solubilization of Indian rock phosphates: a step toward sustainable and phosphorus secure future. *Arch. Agron. Soil Sci.* 2018b, 64(11), 1564–1581.

Ruan, W.; Guo, M.; Cai, L.; Hu, H.; Li, C.; Liu, Y. and Mo, X. Genetic manipulation of a high-affinity PHR1 target cis-element to improve phosphorous uptake in *Oryza sativa* L. *Plant Mol. Biol.* 2015, 87(4–5), 429–440.

Russel, D. A. and Williams, G. G. History of chemical fertilizer development 1. *Soil Sci. Soc. Am. J.* 1977, 41(2), 260–265.

Sarkar, A.; Biswas, D. R.; Datta, S. C.; Roy, T.; Moharana, P. C.; Biswas, S. S. and Ghosh, A. Polymer coated novel controlled release rock phosphate formulations for improving phosphorus use efficiency by wheat in an Inceptisol. *Soil Till. Res.* 2018, 180, 48–62.

Sharda, V. N. and Ojasvi, P. R. A revised soil erosion budget for India: role of reservoir sedimentation and land-use protection measures. *Earth Surf. Process. Landf.* 2016, 41(14), 2007–2023.

Sharma, S. B.; Sayyed, R. Z.; Trivedi, M. H. and Gobi, T. A. Phosphate solubilizing microbes: sustainable approach for managing phosphorus deficiency in agricultural soils. *Springer Plus 2.* 2013, 4, 587–600. doi: 10.1186/2193-1801-2-587.

Sharma, V. P. and Thaker, H. Demand for fertilizers in India: determinants and outlook for 2020. *Indian J. Agric. Econ.* 2011, 66, 1–24.

Sharpley, A. N. Dependence of runoff phosphorus on extractable soil phosphorus. *J. Environ. Qual.* 1995, 24(5), 920–926.

Shen, J.; Lixing, Y.; Zhang, J.; Haigang, L.; Bai, Z.; Chen, X.; Zhang, W. and Zhang, F. Phosphorus dynamics: from soil to plant. *Plant Physiol.* 2011, 156(3), 997–1005.

Shenoy, V. V. and Kalagudi, G. M. Enhancing plant phosphorus use efficiency for sustainable cropping. *Biotechnol. A.* 2005, 23(7–8), 501–513.

SubbaRao, A.; Srivastava, S. and Ganeshamurty, A. N. Phosphorus supply may dictate food security prospects in India. *Curr Sci.* 2015, 108(7), 1253–1261.

Tepper, Y. Soil improvement and agricultural pesticides in antiquity: examples from archeological research in Israel. In Proceeding Middle East Gardens Traditions: Unity and Diversity, Dumbarton Oaks Colloquium on the History of Landscape Architecture, Washington DC, 2007, vol. 31, 41–52.

Turner, B. L.; Paphazy, M. J.; Haygarth, P. M. and McKelvie, I. D. Inositol phosphates in the environment. *Philos. Trans. R Soc. Lond. B Biol. Sci.* 2002, 357, 449–469.

USGS Mineral Commodity Summary, https://pubs.er.usgs.gov/publication/mineral2013 (Accessed on 23rd December, 2019).

USGS Mineral Commodity Summary, https://www.usgs.gov/centers/nmic/mineral-commodity-summaries (Accessed on 15th December, 2019).

USGS, 1976 https://pubs.er.usgs.gov/publication/pp1000 (Accessed on 15th December, 2019).

Vaccari, D. A. Phosphorus: a looming crisis. *Scientif. Am.* 2009, 300(6), 54–59.

Vaccari, D. A. Sustainability and the phosphorus cycle: inputs, outputs, material flow, and engineering. *Environ. Eng.* 2011, 12, 29–38.

Vadas, P. A.; Owens, L. B. and Sharpley, A. N. An empirical model for dissolved phosphorus in runoff from surface-applied fertilizers. *Agric. Ecosyst. Environ.* 2008, 127, 59–65. doi: 10.1016/j.agee.2008.03.001.

Wang, H.; Liu, S.; Zhal, L.; Zhang, J.; Ren, T. and Fan, B. Preparation and utilization of phosphate biofertilizers using agricultural waste. *J. Integr. Agric.* 2015, 14, 158–167. doi: 10.1016/S2095-3119(14)60760-7.

Wang, X.; Wang, Y.; Tian, J.; Lim, B.; Yan, X. and Liao, H. Overexpressing AtPAP15 enhances phosphorus efficiency in soybean. *Plant Physiol.* 2009, 151, 233–240.

Watt, M. and Evans, J. R. Proteoid roots: physiology and development. *Plant Physiol.* 1999, 121, 317–323.

Xiao, C.; Zhang, H.; Fang, Y. and Chi, R. Evaluation for rock phosphate solubilization in fermentation and soil–plant system using a stress-tolerant phosphate-solubilizing *Aspergillus niger* WHAK1. *Appl. Microbiol. Biotechnol.* 2013, 169, 123–133. doi: 10.1007/s12010-012-9967-2.

Yadesa, W.; Tadesse, A.; Kibret, K. and Dechassa, N. Effect of liming and applied phosphorus on growth and P uptake of maize (*Zea mays* subsp.) plant grown in acid soils of west Wollega, Ethiopia. *J. Plant Nutr.* 2019, 42(5), 477–490.

Zhang, M. A.; Walk T. C.; Marcus, A. and Lynch, J. P. Morphological synergism in root hair length, density, initiation and geometry for phosphorous acquisition in *Arabidopsis thaliana*: a modeling approach. *Plant Soil.* 2001, 236, 221–235.

Zhou, K.; Binkley, D. and Doxtader, K. G. A new method for estimating gross phosphorus mineralization and immobilization rates in soils. *Plant Soil.* 1992, 147, 243–250.

CHAPTER 14

Novel Potassium Management Strategies for Improvement of Soil Health

RAJEEV PADBHUSHAN[1*], AMARJEET KUMAR[1], UPENDRA KUMAR[2], ARUN KUMAR[3], RAGINI KUMARI[1], and ANSHUMAN KOHLI[1]

[1]*Department of Soil Science and Agricultural Chemistry, Bihar Agricultural University, Sabour, Bhagalpur, Bihar*

[2]*ICAR-National Rice Research Institute, Bidyadharpur, Cuttack, Odisha*

[3]*Department of Seed Science and Technology, Bihar Agricultural University, Sabour, Bhagalpur, Bihar*

[*]Corresponding author. E-mail: rajpd01@gmail.com

ABSTRACT

The present collection of studies emphasizes the potassium (K) status in soil, water, and atmosphere, forms of K, role in plant and soil, and K management strategies through the use of inorganic sources only or integrating with organics or crop residues or biofertilizers to meet the soil and crop demand in cultivation. Biofertilizers from organic formulations and crop residues of K exhausting crops are the rich source of K (75%–80% of K uptake by plants remains in crop residues) and be part of K management strategies for soil and crop in totality. Also plant-based strategy, field-specific management, and advanced technology such as the decision-support systems for Agrotechnology Transfer (DSSAT) software that was developed to predict the crops productivity and also support in K management through fertilizer with the use of data from soil and weather parameters which is gathered from the sites on daily basis. The use of K-based fertilizer either synthetic alone or integrated with organic sources in the recommended dose of fertilizer has improved the crop yield by 0.04% and 10.6% respectively, over the inorganic sources without K fertilization on the 20-year long-term study. The addition of K, either in form of inorganic fertilizers or crop residue, enhances soil organic carbon, available K apart from decreasing soil bulk density, which directs toward better soil health. Therefore, the addition of K through inorganic or organics/crop residues can harness natural resources appropriately and improve soil productivity. Hence, novel K management strategies are required for proper soil health management.

14.1   INTRODUCTION

Soil health has declined due to overexploitation and therefore its maintenance is the main concern nowadays in the current conditions. The alleviated decomposition rate due to prevailing high temperature and improper management practices under intensive cultivation has resulted decline in nutrient availability and organic matter (OM) content in the soil. Lower OM in the soil has affected the microbial population and nutrient solubility. Indian soils are facing similar challenges of nutrient imbalance and low soil carbon level. Potassium (K) is one of the important chemical parameters which impact several vital activities in the soils, vegetation, animals, and human nourishment (Hasanuzzaman et al., 2018; Islam et al., 2018). The importance of K in the agricultural cultivation is well documented (Manning, 2010). As a primary macronutrient, K plays a major role in the growth and development of plants and the sustenance of global human societies through ecosystem services and sustainable development goals.

K deficiency in the soil may affect the K supply to the plants, responsible for yield deterioration. Crops are more susceptible to several plant diseases when K nutrition is limited, which can affect product quality. The injudicious application of inorganic potassic fertilizer also affects crop quality and soil productivity (Mamatha, 2006) and a complete shift to the organic sources alone cannot meet the crop demand and ensure food security (Sharma et al., 2019). On optimum application in the soil, it improves agronomic productivity and sustainability in different crops (Mengel, 1985). To achieve healthy and productive harvest, application of balanced nutrients is required. Integration of organic and inorganic sources can harness natural resources appropriately and improvement of the living of the population in the region by improving crop production (Timsina, 2018).

The, majority of regions of the cultivable soils of the globe are found to be K-deficient owing to the insufficient application of K in the soils either through organics or inorganic sources of fertilizer (Lu et al., 2017). In India, there was no or negligible response on K application before 1970 (Goswamy and Banerjee, 1978; Randhwa and Tandon, 1982) which was due to high K-supplying capacity of the soil. In the year 1970, introduction of high yielding fertilizer responsive system depleted the K soil resources and crops response to K have increased extensively. Studies on various soils of eastern Indian states have reported low or deficient K level (Sarkar, 2001; Mishra and Mitra, 2001; Mandal and Khanda, 2001; Boakakati et al., 2001). The application rate of potassic fertilizer in India is suboptimal of the recommended dose due to the high cost of the fertilizer (Srinivasarao et al., 2014). The minimal amount of production of potassic fertilizer in India has caused imports at a very large scale from other countries to fulfill the requirement in the country.

K in plants during the initial stage of their growth accumulates at a faster rate to that of the nitrogen (N), and hence it supplies support the utilization of N in the plants and ultimately affecting the growth rate and productivity at maturity. The sustainable K management explains the scientific background of K impacts on the growth and yield of crop by considering the facts of K supplying capacity in terms of crop growth stages and the amount of its application. The K sufficiency in the soil can be partly met by the use of organic K sources that ultimately manages the soil health.

14.2 POTASSIUM IN SOIL, WATER AND ENVIRONMENT

Potassium is found in soil, water, and environment and the K reserve in soil, water and atmosphere on earth has been presented in Table 14.1.

TABLE 14.1 The Potassium Reserves in Soil, Water and Atmosphere on Earth

Sr. No.	Sources of Deposition	Amount of Deposition (value $\times 10^9$ in tonnes)	References
1.	Soil	57.7 (soil solution and exchangeable) 57.7 (nonexchangeable) 3773–7662 (primary minerals)	Soil Survey Staff (1999) and Hoeft et al., (2000)
2.	Oceans	552, 686	Pawlowicz (2013)
3.	Atmosphere	0.066	FAO (2008)
4.	Fertilizer mineral deposits on earth	6.7–14.6	Soil Survey Staff (1999)

14.2.1 POTASSIUM IN SOIL

K in soil exists in different pools available for plant root uptake. Water-soluble K constitutes 0.1%–0.2% of total K in surface soils which is readily available form. The exchangeable form of the nutrient gets adsorbed to the negatively charged site of the soil colloids which constitutes about 1–2% of total K. This form is available to the plants and contributes toward yield and K uptake (Sharma et al., 2013). Non-exchangeable K constitutes 1%–10% of the total K and contributes significantly to maintenance of the exchangeable form of K. Mineral K constitutes 90%–98% of total K and rich source of micas and feldspars. This form is unavailable to the plant. K is present in soil water in dissolved form, adhered to clay particles and OM, and also held inside the crystal structures of primary minerals. The K is present in a small amount in OM in soils. This may be due to since K is not the constituent of biomolecules; and therefore, it is leached effortlessly and rapidly from the foliage due to its quick solubilization. The plant available K that is soil solution K is easily leached by runoff. The total soil K concentrations are higher than total soil N and phosphorus (P) concentrations; however, plant available concentration for K is lower than that of the plant available N and P concentration (Sardans and Peñuelas, 2007; Sardans et al., 2008).

14.2.2 POTASSIUM IN WATER

Water reservoir contains K in the soluble form and the irrigation water adds K in the soil system when used. The seawater contains about 0.04% of K. The K in seawater settles and consequently ends up in sediment mostly. In general, river water contains approximately 2–3 mg L^{-1} of K. The seawater contains more K as compared to river water. This may be due to the presence of oceanic basalts in seawater. The quality of irrigation water is determined by the concentration of K in water when ranges from 1 to 2 mg L^{-1}. The canal water, tube wells, boring wells, and river water are the main sources of irrigation that contains this level of K and add into cultivated soil when irrigated.

14.2.3 POTASSIUM IN ENVIRONMENT

K in the atmosphere is also an important source of K existing in ionic forms. The atmospheric K is considered as "free public good" which is freely available and does not outlay anything to generate, procure, and deal out when compared with market available potassic fertilizer. The atmospheric K is not uniformly spread within the landscape and changes in presence based on the dynamism in nature.

14.3 POTASSIUM IN WORLD

K in the soils of the world varies from region to region (Sheldrick et al., 2002). Globally this variation in content is due to differences in soil taxonomy (Table 14.2), mineralogy, and management practices adapted in the region. The K content was highest under the soil taxonomy Aridisols, Mollisols, and Vertisols and the lowest in the Inceptisols, Andisols, and Spodosols (Sardans and Peñuelas, 2015). The regions dominated with the K-feldspar and mica contains the more K in the soils. Organically, rich soils are rich in K. Several regions of the world are adapting such management practices that are either improving or managing the K content in the soils.

TABLE 14.2 Potassium Status on Soil Taxonomy Basis in the World

Soil Taxonomy	Total Surface Area in the World ($\times 10^6$ km^2)	Mean Available K (mg/g) (soil depth, cm)	Mean BD (g/cm)	Global Total Available K ($\times 10^9$ t)
Alfisol	12.62	0.123 (0–130)	1.73	3.51
Andisol	0.91	0.119 (0–200)	0.80	0.17
Aridisol	15.73	0.620 (0–279)	1.50	40.10
Entisol	21.14	0.040 (0–152)	1.50	1.70
Gelisol	11.30	0.072 (0–110)	1.41	1.57
Histosol	1.53	0.147 (0–100)	1.50	0.34
Inceptisol	12.83	0.005 (0–73)	1.20	0.06
Mollisol	9.00	0.706 (0–165)	1.48	21.8
Oxisol	9.81	0.024 (0–100)	1.32	0.31
Spodosol	3.35	0.125 (0–100)	1.50	0.21
Ultisol	11.05	0.064 (0–280)	1.65	3.25
Vertisol	3.35	0.311 (0–234)	1.95	4.69
World total	130.80			57.70

Source: Soil Survey Staff, 1999.

14.4   POTASSIUM IN INDIAN SOIL

In 1969, Ramamurthy and Bajaj prepared soil available K map and showed one-fifth of the total area deficient in K and four-fifth area sufficient to high in range. After 7 years of investigation, Ghosh and Hasan (1976) summarized more or less same result. They published their findings based on 4.5 million collected soil samples that out of 310 districts, 63 districts are low in K content, 130 districts are medium in range, and 117 districts are high in soil available K. The states like Rajasthan, Madhya Pradesh (present Madhya Pradesh and Chhattisgarh), Haryana, and Gujarat were high in soil K status; Punjab, Bihar, Andhra Pradesh, southern states, Orissa, and West Bengal were medium in status, whereas Uttar Pradesh, Meghalaya, Mizoram, Tripura, Pondicherry, Assam, Himachal Pradesh, and Ladakh and union territory Jammu and Kashmir. were low in status. Later, about 3.65 million soil samples were analyzed between 1997 and 1999 (Motsari, 2002) and based on 11 million soil samples (Hasan and Tiwari, 2002), observed the states of Tripura and Jammu and Kashmir were low in soil K status; Assam, Andhra Pradesh, Haryana, and Himachal Pradesh.

Kerala, Meghalaya, Mizoram, Odisha, Puducherry, Punjab, Uttar Pradesh, and West Bengal were medium status while Andhra Pradesh, Maharashtra, Karnataka and Tamil Nadu, Gujarat, Madhya Pradesh were in high soil K status (Naidu et al., 2010, 2011).

The interpretation of K status in soils of India was recorded medium to high range till 1976 and later on the trend moves from medium to low in the year 2002. The obtained result was coherent with the benchmark soils present in various states of the country (Bansal et al., 2002; Sekhon 1999). The report of Dey et al. (2017) on the K status of India concluded that except in one case, the percent soils per district under low and medium K fertility categories are far higher than the high and very high category districts. The crop yield of the low to medium fertility soils was responsive when K fertilizer was applied whereas no K fertilizer application restricted to achieve the optimum yield. K input–output balance was calculated using the IPNI-NuGIS approach and was observed that K balance was negative in most of the states of India and insufficient K fertilizer application over the K uptake by the crops caused K mining from the native soil (Dutta et al., 2013).

14.5   FUNCTIONS OF POTASSIUM IN PLANT

14.5.1   ENZYME ACTIVATION

K acts a vital role in activation of enzymes. Total 60 diverse enzymes are involved in the growth and development of plants. These enzymes serve as catalysts involved in several chemical processes in the plant, being not consumed during this phenomenon. K alters the substantial form of these enzymes molecules and exposes the suitable chemical site for chemical activity. K reacts with different organic anions and other molecules of various compounds and neutralizes them further within the plant system. This process helps to stabilize the soil reaction to neutral to slightly alkaline required for normal functioning of the enzymes. The concentration of K in the cell determines the extent of enzyme activation and the rates of chemical reactions at which this precede. Thus, the rate of chemical reactions is regulated by the pace at which the K molecules enter into cell (Van Brunt and Sultenfuss, 1998).

14.5.2   STOMATAL ACTIVITY

K controls the activity of stomata in plants. Stomata are actually small pores present in the leaves through which the exchange of gases and water vapor takes place with the atmosphere. When K enters inside the guard cells, the water gets accumulated and causes the stomatal pore opening and allowing carbon dioxide and oxygen to move freely inside and outside, but when supplied water is not sufficient, K pumps out from the guard cells and stomatal pores get closed and minimize the water losses (Thomas and Thomas, 2009). K deficits inhibit the stomatal function and plants are further prone to water strain.

14.5.3   PHOTOSYNTHESIS

K acts a significant role in the photosynthetic process in plants and involves in energy production. K activates the enzyme and produces ATP that regulates the rate of photosynthesis. The electrical charge balance at the site of this energy molecule is maintained by these K ions. When these ions are deficient in plants, the photosynthesis process and energy production get reduced and the activities controlled by ATP are slow down resulted increase in plant respiration which further contributes to minimal plants development (Van Brunt and Sultenfuss, 1998).

14.5.4 TRANSPORTING OF SUGARS

K deficits in the plants result into less energy production and further break down transport system of sugar. Sugar formed in the photosynthesis process needs to be moved to other portions of the plant through the phloem for further consumption and stowage. The breakdown of transporting system causes build-up of photosynthates in the plants and further reduces the speed of photosynthesis. K sufficiency supports the plant processes and transport systems function properly (Van Brunt and Sultenfuss, 1998).

14.5.5 WATER AND NUTRIENTS TRANSPORT

K sufficiency in plants also helps in the transportation of water and minerals in the plant system through xylem. If the K amount in the plant system is reduced, it depresses the translocation of nutrients like nitrates, phosphates, and divalent cations calcium and magnesium (Schwartzkopf, 1972). The transport systems in the plant are efficiently operated when K supply is sufficient which are in conjunctions with the specific enzymes and plant growth hormones (Thomas and Thomas, 2009).

14.5.6 PROTEIN AND STARCH SYNTHESIS

The enzymes responsible to activate the protein and starch are blended by the K. Several major steps of the synthesis process are not possible in the absence of K. The enzyme nitrate reductase catalyzes the formation of protein will not work properly if K is deficient even though abundant N is available (Patil, 2011).

14.5.7 CROP QUALITY

K acts as a vital role in the enhancement of crop quality. The presence of K in sufficiency improves the diseases resistance, physical texture, and shelf-life of the produce and added value to the produce. Fiber quality of the cotton can be increased. It added winter hardiness and reduces the lodging of grains. Therefore, deficiency of K especially during the critical crop growth stage can result into losses in crop yield and quality.

14.6   SOIL POTASSIUM DYNAMICS

The information of K dynamics in soil correlates positively with fertilizer recommendation for various crops. The distribution of K forms controls the function of K availability in growing crops. It occurs in various forms: water-soluble K, exchangeable K, non-exchangeable K (fixed K), and mineral K.

The K availability is affected by K-fixation or retention in the soil. The retention processes of K are restricted to interlayer ions in the area formed by holes and oxygen sheets (Barshad, 1951). The force complexes in the interlayer reaction in clays are the electrostatic attractions between positive-charged and negative-charged ions, and extensive forces due to ion hydration (Kittrick, 1966). The extent of K-fixation in clay and soil depends on the clay-type mineralogy and charge density, the extent of interlayering, the moisture availability, pH of the ambient solution bathing soil, and K ions concentration (Rich, 1968; Sparks and Huang, 1985).

The clay minerals liable for this process are 2:1 or 2:2 types' minerals like montmorillonite, vermiculite, and weathered micas. According to Rich (1968), dioctahedral vermiculite fixes K in acid soils, weathered micas in moist as well as dry condition, and montmorillonite in dry condition. The high charge density fixes added K than those with low charge density. There is importance of interplay hydroxyaluminum, (Al) and hydroxy Fe^{3+} (ferric) material on K-fixation (Rich and Obenshain, 1955). They stated that hydroxy Al and hydroxy Fe^{3+} interlayer groups declined K-fixation. Further Rich and Black (1964) observed the overview of hydroxy Al groups into Libby vermiculite augmented k_G (Gapon selectivity coefficient) from 0.057 to 0.111 L $mol^{-1/2}$. Wetting and drying and freezing and thawing affect the K-fixation in the soil (Hanway and Scott, 1957a, 1957b; Cook and Hutcheson, 1960). The degree of K-fixation or release depends on the types of colloid present and the level of K^+ ions present in the solution. K-fixation by 2:1 clay minerals may be hugely affected by the adsorbed cations or the anions within the system. Soil pH changes affect K-fixation (Volk, 1934; Coleman and Harward, 1953).

The reservoirs of K are mica and feldspar that replenish the K in the labile pool. K releases depend on two processes: (1) Conversion of micas to expansible 2:1-layer silicate minerals by substituting the K^+ with hydrated cations and (2) the dissolution of micas ensuing for forming weathered products. The ions first release by edge weathering, and then second by layer weathering. The K release depends on several factors that include cation exchange reactions, dissolution processes, biological activity, and temperature.

K is a mobile ion in the soil and so substantial amounts can be vanished by the leaching process (Quemener, 1986) that affects the efficiency of the K fertilizers added. The amount of lost K through leaching can be increased when K inputs exceed the soil retention capacity and plant demand, in the presence of drainage (Johnston, 2003).

14.7   POTASSIUM MINING IN INDIAN SOIL

One of the major concerns has been the low use of K in crops. The last five decades in India showed that K contributed to less than 10% of the total fertilizer nutrient consumption (Majumdar et al., 2017). However, recent agronomic studies have documented the significant impact of imbalanced and low K application on crop yield and farm profitability (Sanyal et al., 2014; Majumdar et al., 2016). In fact, historically nutrient management in India is N-driven, followed by P and very less K. This has been identified as one of the key reasons for diminishing fertilizer response and low nutrient use efficiency (NUE) in Indian agriculture (Pathak, 2010). Low use of K in crops could be traced back to the legacy of perception that the Indian soils are rich in K due to the abundance of K-bearing minerals; and reduction in K use will help the national exchequer as India imports its entire potash fertilizer requirement (Sanyal et al., 2014; Majumdar et al., 2016). These ideas have been propagated for a long time with the result that farmers also neglect the use of K fertilizer. Farmers invariably used far lesser amounts of K compared to N and P (Singh, 2015). However, fact remains that not all Indian soils have an abundant supply of K-bearing minerals, and the presence of K-bearing minerals in the soils does not necessarily mean that crops have access to the required amount of K at the right time (Sanyal, 2014; Sanyal et al., 2014; Majumdar et al., 2016). Rather, application of less than the required amount of K to the crops leads to much higher uptake and removal of potash from the soil compared to addition, thereby leading to K mining. Responses to added K can be seen in the range of soil test values found in Vertisols of India if good agronomic practices are followed, and soil test-based K fertilization is practiced for achieving higher yield targets (Dey, 2013). Applying suboptimal K or no K reduces crop productivity, causes a loss in farmers' income, and induces mining of native reserve K, which is detrimental to K fertility and soil health.

14.8 POTASSIUM MANAGEMENT IN SOIL

The plant nutrients from various sources can be categorized under inorganic and organic sources of nutrients. The inorganic sources of fertilizers are high analysis forms that are loaded with very few impurities; however, organic sources of fertilizer are low analysis fertilizers that are rich in a wide range of nutrients and also retain various organic compounds that are in very high impure form. These sources of nutrients are important as far as farming is concerned and it is important to understand the properties of sources before utilization. Therefore, a judicious application of these materials is required; and also required knowledge on soil and environment fate so that its effect on system will be minimal.

K in the plants exists in the ionic form (Chhibba, 2010). The addition of K in soil through organics addition and crop residue management, and the use of inappropriate K-fertilizers is insufficient to supply the K as removed through plants and under such conditions it is required to use K fertilizers to supply soil as well as plants. The present existing knowledge of about nutrient balances explains that there is a gap prevailing between application and supply of K to the crops, and hence suggesting about the proper and timely application of the K-rich fertilizers. Unfortunately, the efforts to assess the fertility status of soil available K, or the status of any other plant nutrient, have been scattered and the bigger picture.

Farmers of India only prefer to add N-based fertilizer, no potassic fertilizer that has resulted low K status of the soils. Improving the crop responses and soil health management of individual nutrient are required. Likewise K management is important for the better crop sustainability and managing soil health.

14.8.1 POTASSIUM MANAGEMENT THROUGH INORGANIC FERTILIZER

The inorganic K fertilizers which are a rich source of K nutrient are applied to the crops in a proper rate and amount that ultimately supports crop growth and development. K applied at basal helps in root proliferation and helps initial plant from diseases and drought. K management through inorganic fertilizer not only affects crop growth and development, but also soil quality by enriching the physical, chemical, and biological nature of soil. Long-term fertilizer studies explained that fertilizer supplied considerably influenced the different forms of K (total, exchangeable, and mineral K) on the surface. Different dosages of inorganic K added at various crop growth stages have been reported that various forms of K augmented with increase in K quantity and further reduced from tillering to harvest growth stages (Thippeswamy et al., 2000). The exchangeable and nonexchangeable forms of K are the main vital forms contributing toward K nutrition in the rice–wheat system in acid Alfisols of Palampur (Sharma and Verma, 2000). Also the exchangeable water soluble and HCl extractable K is the most important K fractions causing significantly increase toward K uptake by the maize crop. Incessant use of synthetic fertilizers enriched all the K fractions in the soil over without K application and the nutrient uptake was greater (Sood, 2005). However, water-soluble K was reduced in long-term fertilizer experiment under synthetic fertilizers (Sood et al., 2008) and changed different K forms like exchangeable, nitrate, reserve, and total (Rajani et al., 2010).

The common K fertilizers used in agriculture are potassium chloride (KCl), potassium sulfate (K_2SO_4), potassium-magnesium sulfate ($K_2SO_4 \cdot 2MgSO_4$), and potassium nitrate (KNO_3). Among all K fertilizers, KCl accounts for 95% of the total potassic fertilizer. It is also called a muriate of potash which is the least expensive source of K. The K content of KCl is 60% and used on most of the cultivated crops. $K_2SO_4 \cdot 2MgSO_4$ contains 22% of K and least commonly used among all K fertilizers. K_2SO_4 is preferred in chloride-sensitive crops such as tobacco. The K content in this fertilizer is 50% and the most K expensive source of fertilizer. KNO_3, also called as saltpeter, is an excellent source of K. The K content is 44% and commonly used in the foliar application of K. Foliar application of K is commonly done in horticultural crops and sometimes to cereal crop when required to apply in the standing crop showing K deficiency symptoms. Since K is an immobile nutrient, and hence required to be applied in the root zone either basal dose or split application preferring the basal dose as one of the split application of K. The split application of K fertilizer improves the fertilizer use efficiency.

14.8.2 INTEGRATED POTASSIUM MANAGEMENT THROUGH INTEGRATION OF INORGANICS AND ORGANICS

It is well known that integrating chemical fertilizers along with organic manures proved to be beneficial in improving crop yields and in sustaining soil health. The process of compost preparation is an ecofriendly process through which organic waste materials are disposed and then applied to the soil system as organic amendments. Table 14.3 depicts the average K content in the bulky manures which are the main sources of organic amendments.

TABLE 14.3 Average Potassium (K_2O) Content of Bulky Manure

Manures	% K_2O content
Animal refusal	0.1–0.3
Fresh cattle dung	0.3–0.4
Fresh horse dung	0.3–1.0
Fresh poultry manure	0.8–0.9
Dried sewage sludge	0.2–0.5
Activated-dried sewage sludge	0.5–0.7
Cattles urine	0.5–1.0
Horses urine	1.3–1.5
Man urine	0.2–0.3
Sheeps urine	1.8–2.0
Coal ashes	0.53
Household ashes	2.3–12.0
Wood ashes	1.5–36.0
Dried rural compost	0.8–1.2
Dried urban compost	1.0–2.0
Dried FYM	0.3–1.9
Filter-press cake	2.0–7.0
Rice hulls	0.3–0.5
Groundnut husks	1.1–1.7
Dried banana	1.00
Cotton	0.66

Source: **Dhama** (1996),

The product obtained after decomposition contains several nutrients in sufficient proportion that supports better crop growth (Zia et al., 2003). The effect of these amendments will be more sustainable when applied on a long-term basis. When these organic materials are integrated with inorganic source of fertilizer it supports the recommended supply of nutrients and measures for improving productivity. Now researchers are promoting the use of these approaches at larger scale. Long-term integrated nutrient management (INM) improved K fractions in soil over control (Sawarkar et al., 2013). Inclusion of plant growth-promoting K solubilizing rhizobacteria (biological K-fertilizers) enhanced the availability of this nutrient in cultivable soils when integrated with the application of waste mica that might minimize the negative effect on the Indian economy (Meena et al., 2014).

Figure 14.1 represented the effects of K application with or without in the treatments supplied with inorganics alone or INM on yield (average yield of wheat and maize, t ha^{-1} from long-term 1991 to 2010). The average yield increased by 259.9% (synthetic fertilizer with K compared to control); 0.04% (inorganics with K over the inorganics without K), and 10.6% (integrated inorganics and organics K compared to synthetic fertilized K). This depicts the importance of K use on long-term basis (Xue-yun et al., 2014).

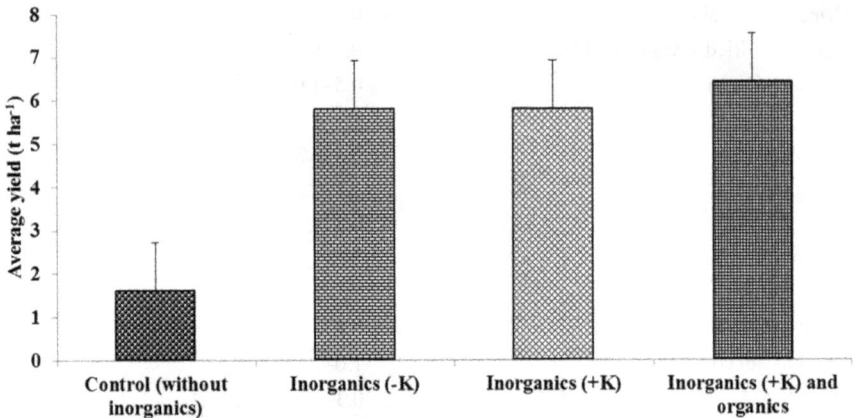

FIGURE 14.1   The effects of K application with or without in the treatments supplied with inorganics alone or INM on yield (average yield of wheat and maize, t ha^{-1} from long-term 1991 to 2010). Graph plotted using the mean + standard error of the values (source: Xue-yun et al., 2014).

14.8.3   USE OF POTASSIUM SOLUBILIZING BACTERIA (KSB) FOR POTASSIUM MANAGING

Biofertilizers are organic formulations containing living microorganisms like bacteria, fungi, and algae that are capable of exhibiting various plant growth promoting functions. There are a wide range of biofertilizers, which are used in the cultivation of crops like formulation containing nitrogen (N) fixing microorganisms (azolla, cyanobacteria, azotobacter, azospirillum, etc.), phosphate, K, zinc-solubilizing microorganisms (pseudomonas, bacillus, aspergillus, arbuscular mycorrhiza, etc.), and sulfur-oxidizing microorganisms. By replacing the 30%–50% of the requisite urea-N, cynobacteria and azolla can supplement the N need of rice crop. Heterotrophic N-fixing microbes can fix nitrogen from 15.6 to 22 kg N ha^{-1} year^{-1} in rice. Phosphate and K-solubilizing bacteria (KSB) can reduce P and K fertilizer application by 15%–30% without any significant reduction in the rice yield. Figure 14.2 shows the effect of KSB on seed yield of mung bean (Meena et al., 2018). There is a 4% yield advantage when KSB is applied alone. However, yield advantages of 7% and 12% were obtained when 100% K was applied through inorganic source and 75% K through inorganic source is integrated along with KSB, respectively, over the no K application.

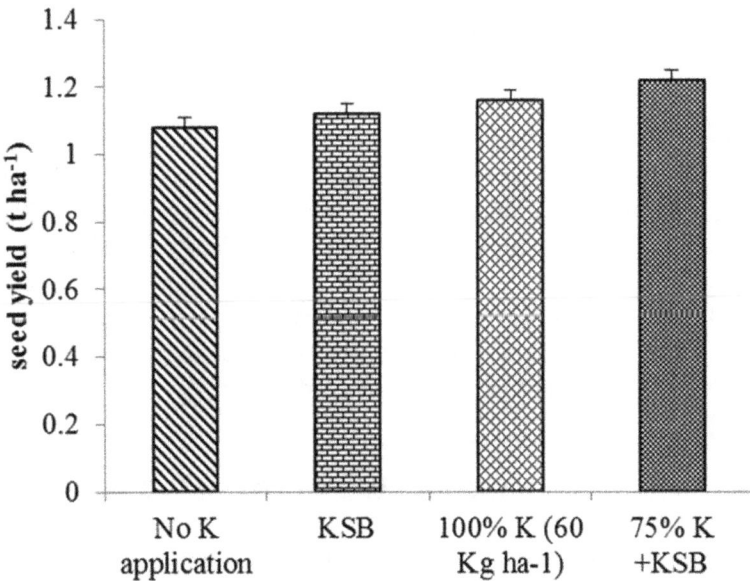

FIGURE 14.2 The impact of potassium solubilizing bacteria (KSB) on grain yield of mung bean (Source: Meena et al. (2018).

Biofertilizers are easier to produce and with low-cost value. Several studies suggest that the use of biofertilizer improves the soil nutrient status and microbial activity. Hence, even economically weaker farmers can easily afford and use in their field for better crop growth. Now, the agriculture department workers are more emphasizing on producing liquid or carrier-based K biofertilizers on large scale for converting the insoluble K of the soil to soluble K and meet the soil and crop demand. Therefore, use of biofertilizers can one of the K management strategies for improving the soil health.

14.8.4   USE OF CROP RESIDUES FOR POTASSIUM MANAGEMENT

Crop residues are the key natural sources that are responsible for retaining stability in the agricultural ecosystem (Singh, 2003). Approximately 75%–80% of K remains in cereal residues and can be part of the K management strategies and worthful for farmers (Chhibba,, 2010). In existing cropping systems mainly cereals based produces a huge amount of crop residues. A major portion of the residues are fed to cattle and leftover is either burnt by the most of the farmers or used as fuel source. If the left over residues are utilized for incorporating in the soil can 75% of K requirement of crop and making them valuable nutrient sources. The declining SOC from the cultivated soil is problematic as it affects the soil health. Figure 14.3 represents the effects of crop residues and FYM on available K in the incubation study and it has been found that more available K when crop residue and FYM is added in the soil as compared to inorganics applied only. However, crop residues not only provide K, but it is also a rich source of SOC. Rice straw retention as mulch on the field under zero tillage markedly improved NH_4OAc-K in surface layer in rice–wheat cropping sequence (Kharia et al., 2016). The incorporation of crop residues can reduce the application of K through synthetic fertilizers. Therefore, crop residue management can be the one best strategy for managing K for soil and crop.

14.8.5   ADVANCE TECHNOLOGY FOR POTASSIUM MANAGEMENT

There is a need to manage K as per the plant-based strategy; field-specific management and advance technology such as the Decision-Support Systems for Agrotechnology Transfer (DSSAT) software that was developed to predict the productivity of the crops and also support in nutrient management by fertilizer with the use of data from soil and weather parameters which are gathered from the sites/location on daily basis. The software has the capacity to predict K-fertilizer recommendations for most crops. Site-specific nutrient management (SSNM) can be applied through the use of approaches soil test-based and plant-based where responses of crop under nutrients limited condition are used indirectly to estimate the nutrients supplying capacity of particular soil. This manner gives a thought of managing nutrients in the spatial and temporal variable way during production (Timsina et al., 2013). At this moment, adoption of the 4R Nutrient Stewardship concept and its implementation via nutrient expert along and various crops and cropping sequences in a holistic way is highly appropriate. SSNM-based recommendations ensure minimal K nutrient depletion and manage the crop productivity, improving NUE, and ultimately goal of sustainable system. The new approaches of fertilizer management can improve the fertilizer use efficiency.

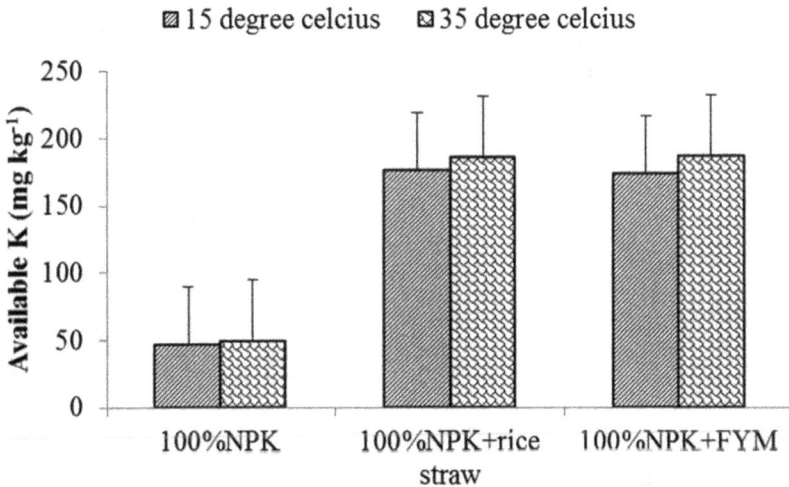

FIGURE 14.3   The effects of rice straw and FYM with inorganic fertilization on two different temperatures on K availability in incubation study (source: Kaur and Benipal, 2006).

14.9   EFFECT OF POTASSIUM MANAGEMENT STRATEGIES ON SOIL HEALTH

For clear findings, some datasets from the different studies (seventy-seven data pairs from the different studies of India) were compiled and the data was interpreted (Figure 14.4). Data collected was further distinguished into four treatments included inorganics without K (−K), inorganics with K (+K), INM without K (−K), and INM with K (+K), without barrier of soil classification, soil type, sources of fertilizers/manures, and other management practices. K management is source dependent that is use of synthetic alone or integration with organics affects the soil health. Long-term studies provide the information in this direction and direct toward soil management for better soil productivity. Soil pH is directly related with nutrient availability and nutrient uptake by root. The presence of OM buffers the soil quality. OM is rich in K and when added to the soil provides K to the soil and further crop. Data from the compiled studies pointed moderation of soil reaction take place on the addition of K for both inorganic and INM treatments. In case of bulk density (BD), it was decreased for both treatments on K addition, whereas soil organic carbon (SOC) and soil available K increased for both treatments on K addition. The decreased in BD on K addition were 7.5% and 1.9%, and SOC increased were 7.0% and 4.5%, for inorganic alone and INM treatments, respectively. The increased soil available K were 60.3% and 4.7%, respectively, for the same treatments. This synthesized data analysis directs toward the improvement of soil health as SOC is increasing on K management.

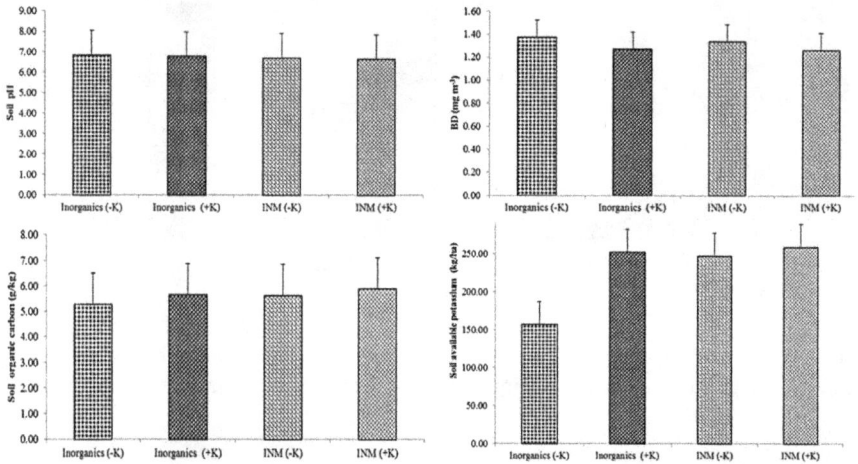

FIGURE 14.4 The effects of K application with or without in the treatments supplied with inorganics alone or INM on soil properties (soil pH, BD, soil organic carbon, and soil available potassium) in the 77 datasets collected from Indian studies. Graph plotted using the mean + standard error of the values.

14.10   CONCLUSION AND FUTURE PROSPECTS

K deficiency in the soil has affected adversely to the crop production system. K is one of the essential and major nutrient that is being neglected in the fertilizer management either due to its cost or lack of awareness among farmers about K uses. Balancing of K either in the soil systems or meeting the crop requirement is prerequisite for better crop growth and soil health. Potassium can be managed either through soil application or foliar application. Soil application through inorganic alone, integrating inorganic sources with organic sources, incorporating crop residue, and use of biofertilizers are the important K management strategies. Among these management practices except the use of inorganic fertilization all others are not only improved the crop productivity but also managed the soil health. During recommendation of K fertilizer all fractions of K should be considered while planning for K management strategy as all K fractions are contributing to crop yield. Split application of K and FSNM based K fertilization also improved K use efficiency.

Several obstacles are prevalent while using the crop residues in the K management strategy as the *in situ* or *ex situ* crop residue management is very difficult in crops and nutrients availability point of view as the crops are not able to receive the nutrients especially K at peak time of nutrients requirement. Therefore, proper crop residue management strategy is required for rapid decomposition and provision for nutrients availability. Knowledge of clay mineralogical formation is essential as this information will provide the standardization of K reserve pools and the extent of mining. This will help planning agencies to formulate an effective K management strategy. Also integration approach of K management in FSNM using biofertilizers/organics needs to be work out at the field level for generating the recommendation for the farmers.

ACKNOWLEDGMENT

The authors are thankful to all the researchers whose contributions have been used for writing and compiling this book chapter.

KEYWORDS

- soil organic carbon
- potassium forms
- plant uptake
- organics

REFERENCES

Bansal, S.K.; Srinivasa, R. Ch.; Pasricha, N.S.; Imas, P. Potassium dynamics in major benchmark soil series of India under long-term cropping. In: 17th World Congress of Soil Science, Bangkok, Thailand, August, 14–21, 2002.

Barshad, I. Cation exchange in soil: I Ammonium fixation and its relation to potassium fixation and to determination of ammonium exchange capacity. *Soil Sci.* 1951, 72, 361–372.

Boakakati, K.; Bhattacharyya, H.C.; Karmakar, R.M. Nutrient mining in agro-climatic zones of Assam. *Fert. News.* 2001, 46 (5), 61–63.

Chhibba, I.M. Rice-wheat production system: soil and water related issues and options. *J. Indian Soc. Soil Sci.* 2010, 58, 53–63.

Coleman, N.T.; Harward, M.E. The heats of neutralization of acid clays and cation-exchange resins. *Am. Chern. Soc.* 1953, 75, 6045–6046.

Cook, M.G.; Hutcheson, T.B. Soil potassium reactions as related to clay mineralogy of selected Kentucky Soils. *Soil Sci. Soc. Am. Proc.* 1960, 24: 252–256.

Dey, P.; Santhi, R.; Maragatham, S; Sellamuthu, K.M. Status of phosphorus and potassium in the Indian soils vis-à-vis world soils. *Indian J. Fert.* 2017, 13 (4): 44–59.

Dharma, A.K. Organic Farming for Sustainable Agriculture. Agro Botanical Publishers (India). 1999, 115–130.

Dutta, S.K.; Majumdar, K.; Khurana, H.S.; Sulewski, G.; Govil, V.; Satyanarayana, T.; Johnston, A. Mapping potassium budgets across different states of India. *Better Crops— South Asia.* 2013, 7, 28–31.

FAO. Current world fertilizer trends and outlook to 2014. 2008. ftp://ftp.fao.org/ag/agp/docs/cwfto14.pdf.

Ghosh, A.B.; Hasan, R. Available potassium status of Indian soils. In: Potassium in soils, crops and fertilizers. Bulletin No.10, 1976, *Indian Soc. Soil Sci.* New Delhi. 1976.

Goswamy, N.N.; Banerjee, N.K. Phosphorous, potassium and other macro elements. Soils and Rice. IRRI Publication. Los Banos, Philippines. 1978. pp. 561–568.

Hanway, J.J.; Scott, A.D. Soil potassium moisture relations. II. Profile distribution of exchangeable K in Iowa soils as influenced by drying and rewetting. *Soil Sci. Soc. Am. Proc.* 1957a, 21: 501–504.

Hanway, J.J.; Scott, A.D.; Stanford, G. Replaceability of ammonium fixed in clay minerals as influenced by ammonium or potassium in the extracting solution. *Soil Sci. Soc. Am. Proc.* 1957b, 21, 29–34.

Hasan, R. Potassium status of soils in India. *Better Crops Int.* 2002, 16 (2), 3–5.

Hasanuzzaman, M.; Bhuyan, M.H.M.; Nahar, K.; Hossain, M.S.; Mahmud, J.A.; Hossen, M.S.; Masud, A.A.C.; Moumita; Fujita, M. Potassium: a vital regulator of plant responses and tolerance to abiotic stresses. *Agronomy.* 2018, 8 (3), 31.

Hoeft, R.G.; Nafziger, E.D.; Johnson, R.R.; Aldrich, S.R. Modern Corn and Soybean Production. MCSP Publications. Champaign, IL, USA, p. 353, 2000

Islam, S.; Timsina, J.; Salim, M.; Majumdar, K.; Gathala, M.K. Potassium supplying capacity of diverse soils and K-use efficiency of maize in South Asia. *Agronomy.* 2018, 8 (7), 121.

Johnston, A.E. *Understanding Potassium and Its Use in Agriculture.* European Fertilizer Manufacturers Association. 2003

Kaur, N. and Benipal, D.S. Effect of crop residue and farmyard manure on K forms on soils of long-term fertility experiment. *Indian J. Crop Sci. 2006, 1(1–2): 161–164.*

Kharia, S.K.; Thind, H.S.; Sharma, S.; Kumar, S. Long term effect of rice cultivation methods, tillage and rice straw management on chemo-enzymatic properties in wheat. *Green Farming.* 2016, 7 (6), 1389–1393.

Kittrick, J.A. Free energy of formation of kaolinite from solubility measurements. *Am. Mineral.* 1966, 51, 929–146.

Lu, D.; Li, C.; Sokolwski, E.; Magen, H.; Chen, X.; Wang, H.; Zhou, J. Crop yield and soil available potassium changes as affected by potassium rate in rice–wheat systems. *Field Crops Res.* 2017, 214, 38–44.

Majumdar, K.; Sanyal, S.K.; Dutta, S.K.; Satyanarayana, T.; Singh, V.K. Nutrient mining: addressing the challenges to soil resources and food security. In *Biofortification of Food Crops*, Springer, India, 2016, 177–198.

Majumdar, K.; Sanyal, S.K.; Singh, V.K.; Dutta, S.K.; Satyanarayana, T.; Dwivedi, B.S. Potassium fertilizers management in Indian agriculture: current trends and future needs. *Indian J. Fert.* 2017, 13 (5), 20–30.

Mamatha HN. Effect of organic and inorganic sources of nitrogen of nitrogen on Yield and Quality of onion (*Allium cepa* L.) and soil properties in Alfisols. MSc (Agriculture) Thesis, University of Agricultural Sciences, Dharwad. 2006.

Mandal, B.K.; Khanda, C.M. Nutrient mining in agro-climatic zones of West Bengal. *Fert. News* 2001, 46 (4), 63–71.

Manning, D.A.C. Mineral sources of potassium for plant nutrition: a review. *Agron. Sustain. Dev.* 2010, 281–294.

Meena, M.D.; Biswas, D.R. Phosphorus and potassium transformations in soil amended with enriched compost and chemical fertilizers in a wheat–soybean cropping system. *Commun. Soil Sci. Plant Anal.* 2014, 45: 624–652.

Meena, H.N., Rana, K.S., Kumar, A.; Shukla, L.; Shekhar, M.; Kumar, A.; Kumar, A.; and Meena, S.K. Influence of crop residue and potassium management on yield, quality, nutrient uptake and nutrient use efficiency by mung bean in maize (*Zea mays* L.)—wheat (*Triticum aestivum* L.)—mung bean cropping system. *Int. J. Curr. Microbiol. App. Sci.* 2018, 7 (4), 3284–3295.

Mengel, K. Potassium movement within plants and its importance in assimilate transport. In: R.D. Munson (ed.). *Potassium in Agriculture.* American Society of Agronomy Madison, WI, USA, 1985, pp. 397–411.

Misra,U.K.; Mitra, G.N. Nutrient mining in agro-climatic zones of Orissa. *Fert. News.* 2001, 46 (4), 73–79.

Motsari, M.R. Available NPK Status of Indian Soils as depicted by soil fertility maps. *Fert. News.* 2002, 47 (8), 15–21.

Naidu, L.G.K.; Ramamurthy, V; Ramesh, S.C.K. Potassium deficiency in soil and crops. *Indian J. Fertil.* 2010, 6 (5), 32–38.

Naidu, L.G.K.; Ramamurthy, V.; Sidhu, G.S.; Sarkar, D. Emerging deficiency of potassium in soils and crops of India. *Karnataka J. Agric. Sci.* 2011, 24 (1), 12–19.

Pathak, H. Trend of fertility status of Indian soils. *Curr. Adv. Agri. Sci.* 2010, 2 (1), 10–12.

Patil, R.B. Role of potassium humate on growth and yield of soybean and black gram. *Inter. J. Pharm. Bio. Sci.* 2011, 2 (1), 242–246.

Pawlowicz, R. Key physical variables in Ocean: temperature, salinity, and density. *Nature Edn Know.* 2013, 4(4), 13.

Quemener, J. Important factor in potassium balance sheets.In International Potash Institute (ed.) Nutrient Balances and the Need for Potassium. International Potash Institute, Madison, WI, USA, 1986, pp. 41–72.

Rajani, A.V.; Butani, B.M.; Shobhana, H.K.; Naria, J.N.; Golakiya, B.A. Dynamics of potassium fractions in a calcareous VerticHaplustepts under AICRP-LTFE soils. *Asian J. Soil Sci.* 2010, 5 (1), 55–59.

Ramamurthy, B.; Bajaj, J.C. Soil Fertility Map of India, Indian Agricultural Research Institute, New Delhi, 1969.

Randhwa N.S.; Tandon, H.L.S. Advances in soil fertility and fertilizer use research in India. *Fertil. News*. 1982, 27, 11–26.

Rich, C.I. Mineralogy of Soil Potassium (abstract).

Rich, C.I.; Black, W.R. Potassium exchange as affected by cation size, pH, and mineral structure. *Soil Sci.* 1968, 97, 384–390.

Rich, C.I.; Obenshain, S.S. Chemical and clay mineral properties of a red-yellow podzolic soil derived from mica schist. *Soil Sci. Soc. Am, Proc.* 1955, 19, 334–339.

Sardans, J.; Peñuelas, J. Drought changes phosphorus and potassium accumulation patterns in an evergreen Mediterranean forest. *Funct. Ecol.* 2007, 21, 191–201.

Sardans, J.; Pe~nuelas, J.; Ogaya, R. Drought-induced changes in C and N stoichiometry in a Quercusilex Mediterranean forest. *Forest Sci.* 2008, 54, 513–522.

Sanyal, S.K. Potassium—the neglected major plant nutrient in soil crop management practices. Journal of the Indian Society of Soil Science 62 (Supplement), S117–S130. Pathak, H. 2010. Trend of fertility status of Indian soils. *Curr. Advan. Agricul. Sci.* 2014, 2 (1), 10–12.

Sanyal, S.K.; Majumdar, K.; Singh, V.K. Nutrient management in Indian agriculture with special reference to nutrient mining—a relook. *J. Indian Soc. Soil Sci.* 2014, 62, 307–325.

Sardans, J.; Peñuelas, J. Potassium: a neglected nutrient in global change. *Global Ecol. Biogeogr.* 2015, 24, 261–275.

Sarkar, A.K. Nutrient mining in agro-climatic zones of Jharkhand. *Fert. News.*2001, 46 (4), 103–107.

Sawarkar, S.D.; Khamparia, N.K.; Thakur, R.; Dewda, M.S.; Singh, M. Effect of long-term application of inorganic fertilizers and organic manure on yield, potassium uptake and profile distribution of potassium fractions in Vertisol under soybean wheat cropping system. *J. Indian Soc. Soil Sci.* 2013, 61, 94–98.

Schwartzkopf C. Potassium, calcium, magnesium—how they relate to plant growth mid-continent agronomist, us green section role of potassium in crop establishment from agronomists of the potash & phosphate institute. 1972.

Sekhon, G.S. Potassium in Indian Soils and Crops, Proceedings of the INSA. 1999, B65 (3 & 4), 83–108.

Sharma, S., Chander G., Verma, T.S., and Verma, S. Soil potassium fractions in rice-wheat cropping system after twelve years of lantana residue incorporation in a northwest Himalayan acid Alfisol. *J. Plant Nutr.* 2013, 36, 1809–1820.

Sharma, R.P.; Verma, T.S. Effect of long term lantana addition on soil potassium fraction, yield and potassium uptake in rice-wheat cropping in acid Alfisols. *J. Potash Res.* 2000, 16, 41–47.

Sharma, S.; Padbhushan, R.; Kumar, U. Integrated nutrient management in rice–wheat cropping system: an evidence on sustainability in the Indian subcontinent through meta-analysis. *Agronomy.* 2019, 9, 71. https://doi.org/10.3390/agronomy9020071

Sheldrick, W.F.; Keith, S.; Lingard, J. A conceptual model for conducting nutrient audits at national, regional, and global scales Nutrient Cycling in Agroecosystems. 2002, 62(1), 61–72.

Singh, Y. Crop residue management in rice-wheat system, RWCCIMMYT: addressing resource conservation issues in rice-wheat systems of south Asia: A Resource Book, Vol. 153, 2003. Rice-Wheat Consortium for the Indo-Gangetic plains—CIMMYT, New Delhi.

Singh, V.K. Inferences drawn from the presentation made at National Seminar on Soil Health Management and Food Security, Role of Soil Science Research and Education, October 8–10, 2015, NBSS&LUP Regional Center, Kolkata, 2015.

Soil Survey Staff. Soil taxonomy: A basic system of soil classification for making and interpreting soil surveys. 2nd edition. Natural Resources Conservation Service. U.S. Department of Agriculture Handbook 1999, p. 436.

Sood, B.; Subehia, S.K.; Sharma, S.P. Potassium fractions in acid soil continuously fertilized with mineral fertilizers and amendments under maize-wheat cropping. *J. Indian Soc. Soil Sci.* 2008, 56 (1), 54–58.

Sparks, D.L.; Huang, P.M. Physical Chemistry of Soil Potassium. In: Munson, R.D., Ed., Potassium in Agriculture, ASA, CSSA, and SSSA, Madison. 1985, 201–265.

Srinivasarao, C.; Kundu, S.; Ramachandrappa, B.K.; Reddy, S.; Lal, R.; Venkateswarlu, B. Potassium release characteristics, potassium balance, and finger millet (*Eleusine coracana* G.) yield sustainability in a 27-year long experiment on an Alfisol in the semi-arid tropical India. *Plant Soil.* 2014, 374 (1,2), 315–330.

Thippeswamy, H.M.; Shiva kumar, B.G.; Balloli, S.S. Potassium transformation studies in lowland rice (*O. sativa* L.) as influenced by levels and time of K application. *J. Potassium Res.* 2000, 16, 7–11.

Thomas, T.C.; Thomas, A.C. Vital role of potassium in the osmotic mechanism of stomata aperture modulation and its link with potassium deficiency. *Plant Signal Behav.* 2009, 4 (3), 240–243.

Timsina, J. Can organic sources of nutrients increase crop yields to meet global food demand? *Agronomy.* 2018, 8 (10), 214. https://doi.org/10.3390/agronomy8100214

Timsina, J.; Singh, V.K.; Majumdar, K. Potassium management in rice–maize systems in South Asia. *J. Plant Nutr. Soil Sci.* 2013176, 317–330.

Van Brunt, J.M.; Sultenfuss, J.H. Better crops with plant food. In Potassium: Functions of Potassium. 1998, 82 (3), 4–5.

Volk, N.J. The fixation of potash in difficultly available form in soils. *Soil Sci.*1934, 37, 267–288. DOI: 10.1097/00010694- 193404000–00003

Xue-yun, Y., Ben-hua, S.; and Shu-lan, Z. Trends of yield and soil fertility in a long-term wheat-maize system. *J. Integrative Agri.* 2014, 13 (2), 402–414.

Zia, M.S.; Khalil, S.; Aslam, M.; Hussain, F. Preparation of compost and its use for crop-production. *Sci. Tech. Dev.* 2003, 22, 32–44.

CHAPTER 15

Micronutrient Management in Soils and the Route to Biofortification

AMIT KUMAR PRADHAN*, KASTURIKASEN BEURA, and
MAHENDRA SINGH

*Department of Soil Science and Agricultural Chemistry,
Bihar Agricultural University, Bhagalpur 813210, Bihar, India*

Corresponding author. E-mail: amyth_bhu88@rediffmail.com

ABSTRACT

Micronutrients and more specifically micronutrient cations form a very important constituent of natural compounds or metabolites indispensable for plant metabolism and human health. Contrary to the fact that micronutrients are required in lower quantities by both plants and humans, their essentiality cannot be compromised. A lack of these micronutrients more importantly iron and zinc in food can consequently lead to deficiency-related disorders or health problems in humans, especially children and pregnant women. Though an increasing attention has been paid toward research emphasizing on the micronutrient enrichment of edible agricultural products which include food grains, ambiguity still remains on the most effective strategy to do so. Micronutrients can be externally added to the processed food through a process called fortification which eventually turns out to be very expensive and obviously unsuitable for the low-income populace. The concentration of micronutrients in grains or the edible plant parts and ultimately the human body can be enhanced naturally or artificially by the process of biofortification which includes manipulation at the gene level, transgenic approaches, and exogenous strategies popularly termed as agronomic biofortification. The long and complicated translocation pathway of micronutrients from soil to the grains affects their allocation to the edible tissues. Other limiting factors like nutrient competition, presence of antinutrients like phytic acid and important soil properties further act as determining aspects for micronutrient bioavailability. Though genetic and transgenic approaches are considered to be the most effective biofortification strategies, agronomic practices, and integrated-micronutrient management can prove to be cost-effective and easy methods to biofortify edible plant parts subsequently used for human consumption.

15.1   INTRODUCTION

Micronutrients are indispensable for optimum plant growth and development owing to their essentiality in the regulation of various metabolic activities. At the same time, they are also integral to human health. The concept and problems associated with hidden hunger with respect to micronutrients have been encountered and well established in the last few years or perhaps decades. Consequently, the importance of micronutrients, their nutrition for plant and humans and accompanying research is being given due attention. The cationic micronutrients, namely, iron (Fe), zinc (Zn), manganese (Mn), and copper (Cu) are particularly important given that these have been proven to be essential micronutrients for all higher organisms. Even moderate micronutrient deficiencies can result in many health problems in humans ranging from gentle to severe like suboptimal metabolism, decreased immunity, and reduced productivity (Tulchinsky, 2010). In humans, Fe and Zn deficiencies have been found to be the most prevalent form of micronutrient deficiencies, reportedly affecting around two billion people and leading to more than 0.8 million annual deaths (World Health Organization, 2016). The problem is higher in countries with low to very low income where the probable risk is high for micronutrient deficiencies, especially for Zn (40%) and Fe (5%) (Joy et al., 2014). Mn and Cu deficiencies in humans are relatively less prevalent when compared to Fe and Zn. These micronutrient deficiencies singly or in combination pose a significant threat to human health.

Human diet to a major extent comprises of plants, but all major food crops lack certain essential micronutrients (Zhu et al., 2007). As evident from Figure 15.1, there exists inadequate micronutrient intake in several parts of the world (Beal et al., 2017). The endosperm of staple cereals like rice, wheat and maize is the most important source of calories to humans, providing around 23%, 17%, and 10% of total global calories, respectively (Rawat et al., 2013). However, the endosperm of grain tissues lack adequate amounts

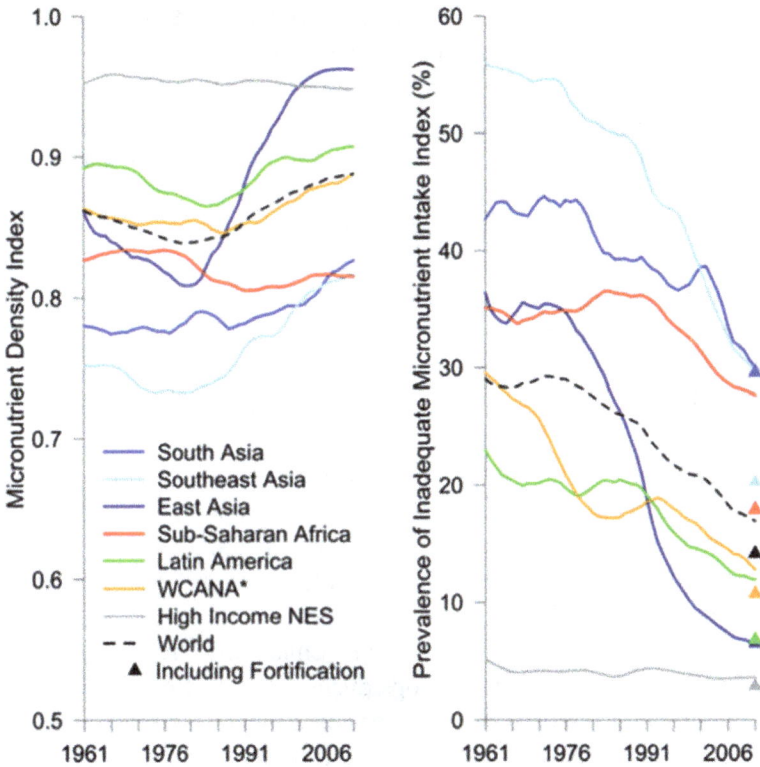

FIGURE 15.1 Micronutrient Density Index and inadequacy of prevalent (estimated) micronutrient intake for the world and seven major regions.
Source: Reprinted from Beal et al. (2017). https://creativecommons.org/licenses/by/4.0/.

of nutrients particularly iron and zinc (Gomez et al, 2010). Plant availability of micronutrients in soil is strongly influenced by the cropping pattern along with the techniques of fertilization (Wei et al., 2006; Li et al., 2007), and any deficiency or even an excess concentration of micronutrient cations in the top soil may lead to a possible yield and quality reduction in crops. Therefore, maintaining soil micronutrients in optimum quantities not only meets the plant requirement, but also prevents any potential buildup. Replenishment rate from soil solid into the soil solution determines the micronutrient availability in soils. In almost all soils, micronutrient replenishment is naturally aided by the mineral reserves of the pedosphere or lithosphere. The variations in pedospheric characteristics further lead to a spatial variation in micronutrient availability in soils (Katyal and Sharma, 1991). Continuous use of inorganic fertilizers, both nitrogenous and phosphatic combined with an intensive cropping system with less use of organic manures results in quick depletion of micronutrients from soils (Dhane and Shukla, 1995). For effective remediation of a nutrient deficiency in the field, it is necessary to get information about the inherent capacity of soil to supply nutrients from its surface as well as the lower soil layers (Setia and Sharma, 2004).

All micronutrients especially micronutrient cations have similar chemistry in soils under proper and uniform management practices. The availability of all micronutrients to plants depends heavily on the soil pH and soil organic matter. The availability of most mineral cations increases with decreasing soil pH and vice-versa. Organic matter on the other hand chelates or forms complexes with the metallic cations due to the presence of negatively charged organic functional groups, consequently increasing or maintaining their availability to plants irrespective of the soil pH.

Hidden hunger can effectively be countered directly by nutrition-specific and indirectly by nutrition-sensitive strategies (Ruel et al., 2013). Nutrition-specific interventions focus on the consumption behavior and include different considerations like dietary diversification and micronutrient fortification or supplementation. On the contrary, nutrition-sensitive interventions address the major determinants and factors leading to malnutrition and also focus on biofortification. Enhancing the phytoavailability of essential mineral nutrients in crops during stages of plant growth either through genetic or agronomic practices is considered under the process of biofortification (Bouis et al., 2011). It is an effective alternative to existing and traditional methods of countering micronutrient deficiencies which include preparation of fortified food which is expensive, and also micronutrient supplementation. These methods though conventional are difficult for the larger proportion of population to afford, owing to their limited resources, and even lower income (Haas et al., 2005). Increasing the concentration of micronutrients in grains by the process of biofortification has immense potential to combat the deficiency of micronutrients and has a great impact on human health. Biofortification through genetic means involves genetic engineering and classical breeding approaches (Saltzman et al., 2013). Agronomic biofortification on the other hand is achieved either through the application of recommended micronutrient fertilizers to the soil and/or direct foliar spray on the crop foliage, or other methods not involving genetic manipulation.

As a matter of fact, any approach which could increase micronutrient uptake by roots and result in a high transfer of micronutrients from the soil to the plant parts can play a crucial role in biofortification. Several alternative exogenous and cost-effective methods of biofortifying crops with micronutrients have been evaluated for their potential role in enhancing the bioavailability of micronutrients if not on a larger scale.

15.2 MICRONUTRIENT FRACTIONS IN SOIL

Environmental conditions readily modify micronutrient bioavailability and remobilization. In soil, physiochemical changes can change the equilibrium significantly between the soil solution and organic or inorganic nutrient fractions (Gismera et al., 2004). Knowledge about chemical reactions of micronutrients within components of soil and their geochemical behavior is fundamental to the scientific prediction of their plant availability. Micronutrient cations have been found to be adsorbed on organic matter or forming carbonate and oxide complexes which can be estimated by sequential fractionation procedures to estimate their distribution in different soil fractions (Rauret et al., 1999). Distribution of the micronutrient fractions and their further dynamics depends on important soil properties like organic matter content, pH, texture, structure, clay mineral types (Spark and Wells, 1995), cation exchange reactions, and sesquioxide content in soil (Guadalix and Pardo, 1995). The total soil micronutrient content does not necessarily account for the actual bioavailability, since the available concentration easily takes part in different chemical reactions, which further varies according to the soil and water characteristics resulting in a short-lived increment in available micronutrients followed by a decrease in phytoavailability (Impa et al., 2012). Therefore, the micronutrient distribution between the fractions in soil must be clearly differentiated.

Organic amendments help in improving the soil's physical, chemical, and biological properties. A positive effect on soil properties also plays an important role in enhancing the micronutrient nutrition of crops due to higher uptake by proliferating roots and also acts as an indirect source of supplementary micronutrients to the soil (Rengel et al., 1999). The binding of metal cations to organic matter to form organometallic complexes leads to the reduction in the concentration of free ions in soil solution and enhances the plant availability of minerals at the root-rhizosphere interface since the concentration of total dissolved ions increases though availability also depends on the dissociation kinetics and mobility of the formed complexes. Chelation of micronutrient cations with organic matters also prevents the formation of insoluble carbonates and oxides in soil (Schulin et al., 2009).

15.3   MICRONUTRIENT MANAGEMENT AND FERTILIZATION

Micronutrient deficiency in agricultural soils can lead to lower plant uptake and a gradual decline in production. A decrease in plant availability can directly be correlated with a lower bioavailability in the economic plant parts used for human consumption. Also, approximately 30% arable soil is reportedly calcareous where the bioavailability of micronutrient cations and crop productions are evidently restricted. In such cases, fertilization is the most common and accepted practice to alleviate micronutrient deficiency through agriculture (Rengel et al., 1999).

Soil application of micronutrients depends on crop species, soil conditions, and crop genotypes. Soil application of micronutrients like Zn can lead to an increase grain yields and grain micronutrient concentrations. Fe quickly tends to get converted into plant-unavailable Fe(III) forms and application of inorganic Fe-containing fertilizers to Fe-deficient soils becomes ineffective, and if we target an increase in grain Fe concentration, this method is likely to be a costly strategy. Even the most effective source of micronutrients in form of fertilizers may vary with need and rhizospheric limitations. Synthetic Fe-chelates have been reported to be more effective for the correction of Fe deficiency (Abadia et al., 2011). Once micronutrient fertilization is exceeded the recommended dose, further fertilization can cause a reduction in crop yield and also decrease the density of micronutrients in grains. This may be attributed to most soil-applied micronutrients getting fixed into plant-unavailable forms and not readily transported down the soil profile. These factors decrease the effect of micronutrient fertilizer application (Rengel et al., 1999).

Alternatively, since micronutrients are required in smaller amounts by the plants, the foliar application can be better than soil application in terms of suitability. Plants can absorb soluble nutrients through leaves and this phenomenon is utilized for delivering plant nutrients directly to the growing foliage avoiding the root–shoot translocation pathway. There should be more emphasis on the effects of foliar application of micronutrients on improving their density in grains and other quality parameters in edible parts of crop. Foliar application of soluble micronutrient fertilizers is an efficient strategy to enhance grain concentration because micronutrients like Fe or Zn have been found to easily retranslocate from foliage to grain. Yin et al., (2016) and Cakmak et al., (2008) showed that foliar application of Zn increased Zn concentration in grains.

Integrated macronutrient management along with applications of organic amendments can resultantly lead to a decrease in soil pH and increase in available Fe and Zn content in soils, ultimately increasing the crop yield and concentration of Fe and Zn in grains (Shahzad et al., 2014). Alternating interaction effect between different nutrient elements should be taken into consideration while formulating proper micronutrient management strategies. The bioavailability of micronutrients like Fe and Zn is affected tremendously by rhizospheric pH. Fertilizers containing NH_4^+-N decrease the pH in the apoplast of root and rhizosphere, thus favoring Zn or Fe activation for uptake. NO_3^--N on the other hand increased pH and decreased the bioavailability of Fe or Zn. On the other hand, nutrients like phosphorus (P) can decrease the plant uptake of Zn due to the formation of insoluble phosphates and diluting grain Zn concentration.

There may be inconsistencies in the micronutrient need by crops and their supply by plant roots at critical or later stages of crop growth like reproductive stages, thus restricting the yields and micronutrient content accumulation in grains. So, micronutrient manipulation at different crop stages can be useful tool to increase the use efficiency of micronutrients, more specifically Zn and Fe by soil application or foliar spray.

15.3.1 ENSURING PROPER MICRONUTRIENT MANAGEMENT IN SOILS

- It must be made sure that a relatively poor crop growth in a field is due to micronutrient deficiency and not the result of other growth factors.
- In a particular crop or specific soil type, deficiency of micronutrient(s) has been encountered in previous cropping seasons or not.
- The affected crop(s) need to be analyzed for specific micronutrient deficiency symptoms.
- Complete chemical analysis of soil and plant samples from micronutrient deficient areas.
- Once micronutrient deficiency is established, and the specific micronutrient(s) can be applied at recommended dosage to the deficient areas.
- Choice of the most adequate micronutrient fertilizer must be based upon its solubility, safety parameters and best application methods (i.e., soil applied or foliage sprayed).
- If mode of application of any micronutrient fertilizer is recommended to be seed-treated, band placement, sprayed onto foliage or from a chelated material, the application rates would be relatively lower than when broadcasted or applied in nonchelated form.

- It may be taken into consideration that certain agrochemicals, lime, or other amendments can supply additional micronutrients or to some extent affect the micronutrients availability in soil.

15.4 MICRONUTRIENT ABSORPTION BY PLANT ROOTS

Plant roots potentially absorb adequate quantities of micronutrients to fulfill their metabolic requirements even from soils containing lower contents of soluble micronutrients. This may be attributed to plants having regulatory mechanisms to manipulate the rhizospheric region so that they can increase the amount of soluble mineral nutrients readily available for their uptake (Chaney, 1987). The possible mechanisms include solubilization and mobilization of micronutrients from soluble solid fractions to root surfaces for uptake; and an enhanced diffusion gradient for soluble micronutrients from the soil solution to the absorption sites across the root-cell membranes (Romheld and Marschner, 1986). Some mechanisms also involve acidification of the root zone by H^+ efflux, increment in root-cell plasma membrane reductase activity which further leads to a direct reduction of existing micronutrient complexes, thus allowing for dissociation of micronutrient metal-complex and higher absorption of the released micronutrient ion (Kochian, 1991).

15.4.1 BIOAVAILABILITY OF MICRONUTRIENTS: UPTAKE AND TRANSLOCATION PATHWAY

The amount of nutrients including micronutrients that accumulate in economic plant parts depends on their availability in the soil and subsequently on the plant's ability to acquire and translocate these nutrients to the respective sink organs. Micronutrient movement in the soil and its transfer from soil into the human body via seeds, shoots, and grains depends upon its relocation in seeds, mechanistic pathways in the plant and subsequent dietary intake (Figure 15.2).

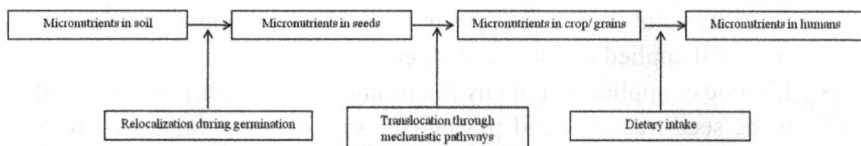

FIGURE 15.2 Micronutrient translocation pathway from soil to the human body and the factors influencing their bioavailability. Modified from Valenca et al. (2017).

15.4.1.1   CONVENTIONAL UPTAKE OF NUTRIENT IONS

Crops take up micronutrients in ionic form and the process involving the transport of cationic micronutrients is complicated owing to the chemical nature of their ionic species which take part in chemical reactions. There is a natural tendency of the cations to form natural complexes which are mainly metal-organic complexes having varying charge and stability (Tiffin, 1972). Rietra et al., (2015) observed that the formation of complexes may lead to competition among the micronutrient cations for their entry into the plant system, and an otherwise requisite process of nutrient uptake surprisingly becomes a limiting factor for the uptake of micronutrients which may be specific to some. Mechanisms involved in uptake of micronutrients consequently affect their accumulation in different parts of the plant including grains. Conversion of micronutrient cations to soluble ionic forms that is Fe^{3+} to Fe^{2+}, Mn^{4+} to Mn^{2+} and Cu^{2+} to Cu^+ at the plasma membrane of the cell is important for their possible uptake. This process is notably performed by nonselective enzymes like ferric reductases (Dimpka et al., 2015a) which also occur in cells of both root and shoot. Plant uptake for Zn occurs in form of a divalent ion while it does not take part in any redox reactions. After the conversion of Fe, Cu, Mn, and Zn to a plant-available form, they are transported into the plant through the roots involving divalent enzymes like iron-regulated transporters present in root and shoot cells (Sinclair and Krämer, 2012). In some plants, the root exudates like phytosiderophores and organic acids stimulate the uptake of nutrient ions like Fe, Cu, Zn, and Mn sometimes by similar compounds having metal-binding characteristics (Keuskamp et al., 2015).

Metal homeostasis acts as a driving force for the micronutrient uptake from the soil. Nutrient transport into the roots from the rhizosphere can be identified by the proteins and mechanisms involved in the nutrient translocation within plants. Acquisition of micronutrients like Fe by plants is reportedly based on process of reduction in non-Poaceae plants and chelation in Poaceae plants (Sperotto et al., 2014). Recent studies suggest that certain monocots like rice take up nutrients by both reduction and chelation (Ricachenevsky and Sperotto, 2014).

15.4.1.2   ROOTS TO CROP AND GRAINS

Several soil factors like pH, organic matter, aeration, moisture, and nutrient interactions directly or indirectly influence the micronutrient bioavailability to the edible portions of the crop. Also, the crops and their respective varieties due to the structure and characteristic functions of their root systems play a major role in micronutrient uptake and further translocation within the plants (Alloway, 2009). Some plants have the ability of rhizospheric regulation by excreting organic acids or H^+ ions which in turn enhance the availability of micronutrients and their uptake (Zhang et al., 2010; Marschner, 2012). Interactions and functional interrelationships between nutrient elements also influence their uptake by roots and bioavailability. For example, P in soil can have contrasting roles in Zn nutrition to plants. It may stimulate root growth and consequently enhance Zn uptake, while application of P fertilizers can also lead to the precipitation of soil Zn which is already low in concentration, thus triggering its deficiency (Zingore et al., 2008). Phosphorus application also induces deficiency of Zn reportedly through dilution effect and its interference with translocation of Zn into the plants from the roots (Singh et al., 1988). Liming material or organic amendments when used for soil management can create variations in important soil properties like pH, thus stimulating the bioavailability of micronutrients and their uptake by crops. Mycorrhizal fungi which form a fungal network and act as extensions of the natural root system and increase the surface area of nutrient absorption can increase uptake of nutrients like P and Zn which show low to moderate solubility in soil (Smith and Read, 1997).

Rice is one such Graminaceous plant which possesses sophisticated and manipulative mechanisms for micronutrient acquisition from soil and their transport from roots into shoots and subsequently to grains by phytosiderophore secretion which chiefly belongs to the mugineic acid family and has the potential to solubilize the precipitated or insoluble Fe, Zn, Cu, and Mn (Treeby et al., 1989).

After the uptake of Zn by the roots, the pathway from the root symplast to the seed includes apoplastic barriers which need to be overcome to complete its journey from source to sink (as depicted in Figure 15.3). Substantial membrane potentials exist across the plasma membrane and membranes of vacuole and chloroplast. Zn transport out of the vacuole into the cytoplasm or chloroplast is favored by the existing energy, thus requiring only passive transporters (depicted by arrows in Figure 15.3). On the contrary, its transport out of the cytosol or chloroplast into the vacuole requires active or secondary active transporters (depicted by round and square symbols, respectively).

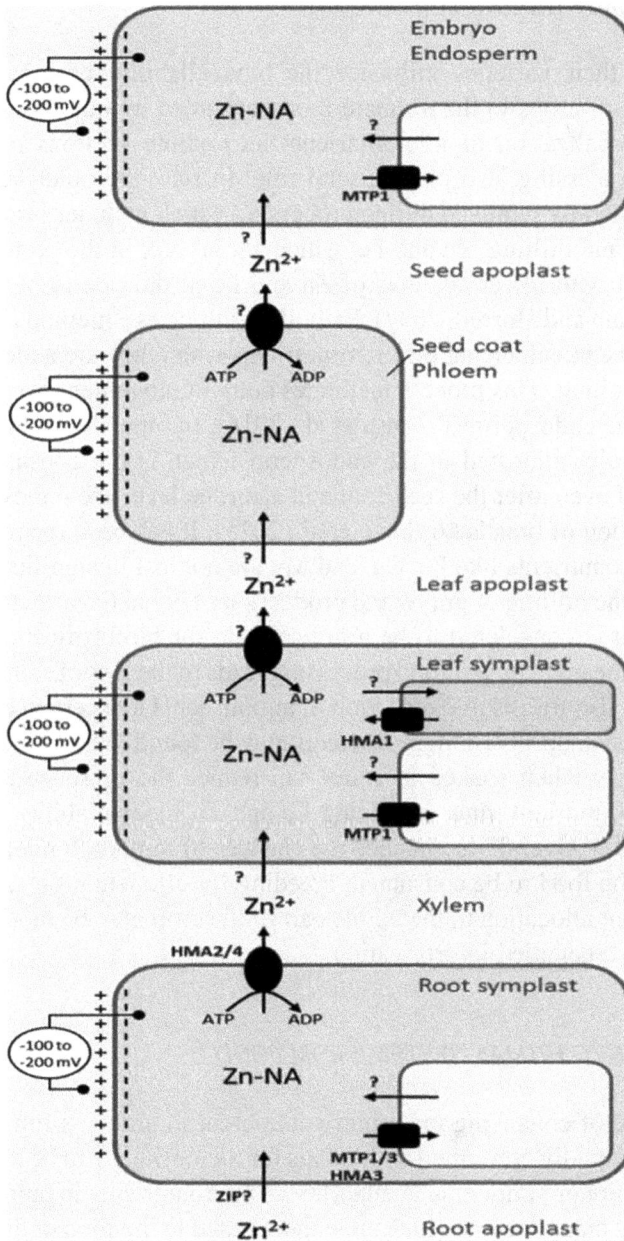

FIGURE 15.3 Zn transport through transport complex formed with nicotianamine (NA) into and out of the apoplast.

Source: Reprinted from Olsen and Palmgren (2014). https://creativecommons.org/licenses/by/4.0/.

15.4.1.3   CROP GRAINS TO FOOD

Crops and their varieties influence the bioavailability of micronutrients from grains of crops to the ultimate food consumed by humans. This justifies the relocalization of micronutrients into edible portions of the crop and food processing also plays a vital role. In rice, the outer layer of the grains is generally removed during processing which includes processes like de-husking and milling. Zn and Fe, which are present in the protein bodies, are thus left with lower contents of Zn and Fe in the rice to be consumed (Zimmermann and Hurrell, 2007). Parboiling in rice as a method is considerably effective in enhancing the micronutrients when they are added to water during parboiling. This process facilitates nutrient movement from the germ layer into the endosperm (Prakash et al., 2016). In other cereals like wheat, Zn is generally allocated in the endosperm which is the consumable part and is intact even after the seed coat and aleurone layer are removed during the preparation of breads (Ajiboye et al., 2015). It has been reported that in wheat, micronutrients like Fe, Cu, and Mn are not lost in significant quantities during the milling of grains and production of bread (Lyons et al., 2005). Thus, wheat is considered to be more suitable for biofortification through agronomic means. Though food processing leads to the loss of some essential nutrients, it also results in a reduction of antinutrients to a certain extent, and thus the bioavailability of micronutrients may be found to have an increasing trend. Cereals when soaked in water can reduce the presence of phytate, a proven antinutrient, thus enhancing Fe and Zn bioavailability (Hotz and Gibson, 2007). Overall, to enhance the chances of important micronutrients to exist in the food to be consumed, breeding for efficient crop varieties for micronutrient allocation to the edible part of the crop may be more effective and is termed genetic biofortification.

15.4.1.4   INTO HUMAN SYSTEM FROM FOOD

The purpose of enhancing micronutrient content in grains is fulfilled when they are eventually consumed by humans for their efficacy to be established. Bioavailability or simply the availability of micronutrients in human food is regulated by many factors which are either related to the food or the host that is humans themselves (Gibson, 2007). One of the essential factors is intake through diet. This is because the bioavailability of micronutrients depends on the nature of diet, chemical forms, amount consumed and interaction with other nutrients and components of food that may regulate their absorption in the gastrointestinal tract (Sandström, 2001). In fruits and vegetables, ascorbic acid acts as a potential enhancer that can increase the bioavailability of micronutrients like Fe, and polyphenols like PA are major inhibitors forming Fe and Zn complexes and limiting their uptake and translocation in the human body (Clemens, 2014). The health, age, sex, genotype, and physiological makeup of an individual also affect the bioavailability of micronutrients from consumed food (Gibson, 2007). The micronutrient status in human body also impacts its absorption; for example, individuals having Fe or Zn deficiency show higher absorption for the nutrients (Hallberg, 2000).

The potential of biofortification, whether agronomic or genetic, is judged by the final enhancement of micronutrient bioavailability at different stages, starting from availability in soil, uptake by plants and leading to their entry into the human food chain. Interactions and metabolisms may lead to variation in relationships among nutrients, soil properties, crop varieties, and human system and also, there are several steps at which micronutrient losses occur. These steps must be considered when we assess the effectiveness of any biofortification strategy and its role in alleviating hidden hunger.

In comparison to macronutrients like N, P, and K, micronutrients are generally remobilized to a lower extent into developing grains (Marschner, 2012). Moreover, the amount of micronutrients retranslocated depends upon their phloem mobility. Notably, Zn is highly mobile in the phloem (Marschner, 2012) while Fe has a comparatively moderate mobility (Kochian, 1991). Grusak et al., (1999) reported that about 70% of the Zn can be retranslocated into the grains from the vegetative portion of wheat.

Phosphorous in seeds of cereals and legumes is in the form of PA (myo-inositol hexakisphosphate) which is its primary storage form accounting for 50%–80% of total seed P. It is formed during the development of seed in the form of Fe and Zn phytate or salts of other divalent cations. PA forms metal complexes with cations from food in the gastrointestinal tract and reduces their intestinal absorption. Because of this mechanistic process, micronutrients like Fe and Zn in consumable food remain underutilized from cereal and legume diets containing a significant amount of PA (Welch, 2002).

15.5 STORAGE OF MICRONUTRIENTS

Biofortification strategies have been devised to develop food or grains with higher contents of essential nutrients for consumption by humans or animals.

Seed loading with essential mineral nutrients is homeostasis-regulated and depends on the activity and specificity of metal transporters. Mechanisms which regulate the translocation and movement of micronutrients especially Fe, Cu, Zn, and Mn in plants are important considerations. The chemical form in which micronutrient cations circulate within cells plays an important role in the regulation of micronutrient movement in plants and their storage in seeds. Ligands and metal–ligand complexes in the sap of xylem and phloem as well as apoplastic fluids result in the absence of micronutrients in significant quantities as free ions in the plant sap. Micronutrients are probably present in less reactive chemical forms which also control their accumulation in seeds and subsequent loading.

The mechanistic pathways of Fe transport from the roots into the seed were studied by Grillet et al., (2014) who identified the chemical forms of Fe during its transport between symplast and apoplast along with phloem loading. They also reported that Fe bioavailability in seeds depends on localization and storage forms of Fe in the plant tissue. The same phenomenon was studied by Vasconcelos et al. (2017) in soybean plants where the iron reductase gene was overexpressed to see changes in Fe accumulation in plant tissues so that the reductase activity can be evaluated for its rate-limiting ability as far as mineral acquisition by seeds is concerned.

15.6   BIOFORTIFICATION STRATEGIES

Biofortification refers to the increase in bioavailability of nutrient elements to the food crops and human body that is achieved using conventional plant breeding, biotechnology, and agronomic practices. Around 792.5 million people across the world are estimated to be malnourished, out of which 780 million people belong to developing countries (Food and Agriculture Organization). Intake of essential micronutrients in low or inadequate amounts has led to around two billion people across the world suffering from their hidden hunger (Ma et al., 2007) despite a visible increase in the production of food grains. Agricultural practices have a marked focus or emphasis on enhancing production and productivity which has resulted micronutrient deficiencies in edible parts of the crops that is grains, thus leading to micronutrient malnutrition among the end-users that is humans. Recent trends suggest a shift toward the production of food grains rich in sufficient quantities of micronutrients. This can prove to be an important strategy to eradicate or counter micronutrient malnutrition which is predominant in developing countries with dietary habits that include micronutrient-poor food grains or other edible agricultural produce (Frossard et al., 2000).

15.6.1 *CONVENTIONAL BREEDING AND TRANSGENIC APPROACHES*

The conventional breeding approach for micronutrient biofortification is reportedly the most accepted and potent strategy for biofortification. Parent lines are screened and selected for the presence of higher nutrient concentration followed by crossing with recipient lines having desirable agronomic characteristics. This leads to the production of plants with desired agronomic traits and nutrient concentrations. Sometimes, breeding strategies rely on the availability of limited genetic variation which can be countered by crossing with lines which are distant relatives, thus transferring the traits into commercial cultivars alternatively achieved by mutagenesis.

The transgenic approach is based on the mechanism of transfer and/or expression of desirable target genes from one species to another. The gene pools do not depend on their evolution or taxonomy. When the presence of any specific micronutrient is not natural in certain crops, transgenic approaches can prove to be an effectively feasible option for enhancing its bioavailability to crop grains (Bouis et al., 2011). The development of transgenic crops depends on the identification and characterization of gene functions, followed by the utilization of those genes to engineer or scientifically manipulate the plant metabolism. Transgenic approaches are also promising techniques to simultaneously incorporate genes responsible for the enhancement of micronutrient bioavailability along with the reduction in the antinutrient concentration which limits the nutrient bioavailability in plants. Genetic modifications can also be used for micronutrient redistribution between tissues, thus boosting the efficiency or even reconstruction of biochemical pathways. The development of transgenic crops undoubtedly involves substantial costs, time, and efforts initially and during research and development, but biotechnologists report it to be a cost-effective and sustainable technique in the long run. There are no taxonomic constraints associated with genetic engineering and even genes synthesis and construction are possible for further use. Transgenic crops with enhanced concentrations of micronutrients can potentially counter micronutrient malnutrition in people of developing countries.

15.6.2   EXOGENOUS BIOFORTIFICATION STRATEGIES

The edible portions of plants contain micronutrients in varying concentrations taken up by plant roots from the soil. Soil micronutrient levels can be improved by the application of micronutrient fertilizers which in turn can contribute to decrease any micronutrient deficiency in humans. Also, targeted application of soluble micronutrient fertilizers to the roots or foliage can prove effective when crops are grown in micronutrient deficient soils or under conditions where immediate unavailability of elements in the soil and/or translocation to the edible plant tissues happens to be a major constraint. Agronomic strategies formulated to increase the concentrations of essential nutrients in edible tissues generally include the application of mineral fertilizers and/or improvement of the solubilization and mobilization of mineral nutrients in the soil.

Agronomic biofortification is mechanistically simple and equally cost-effective, but requires special attention for the source of nutrients, application methods, and their possible effects on the environment. In addition to the chemical micronutrient fertilizers, soil microorganisms with plant growth-promoting characteristics can be used to enhance the mobility of nutrients and their translocation from soil to edible parts of plants.

15.6.3   ALTERNATIVE EXOGENOUS BIOFORTIFICATION STRATEGIES

15.6.3.1   BIOPRIMING USING POTENTIAL NUTRIENT SOLUBILIZING MICROORGANISMS

Plant growth-promoting microorganisms have been widely reported to be able to enhance micronutrient contents in food crops through mechanisms like nutrient solubilization and phytosiderophores production. Integrated inoculation of potential solubilizing microorganisms with mineral fertilizer application can prove to be effective in enhancing the crop growth and yield with a substantial increase in micronutrient uptake. Biofortification of staple food crops with micronutrients by utilizing plant growth-promoting microorganisms is potentially effective in enhancing the uptake of mineral elements, thus increasing the possibilities of increased concentration of micronutrients in edible plant parts. This process is known as biopriming which involves biological seed treatment and is a combination of seed hydration and seed inoculation with any beneficial microorganism.

The solubility of micronutrient cations like Zn is highly pH and moisture dependent, hence leading to zinc-deficient arid and semiarid areas of

India. Nutrient-solubilizing rhizobacteria can make promising contribution to sustainable agriculture. As evident from Figure 15.4, bacteria which can solubilize micronutrients like Zn have various plant growth-promoting properties such as nutrient solubilization phytohormone production including 1-aminocyclopropane-1-carboxylate (ACC) deaminase and siderophores, hydrogen cyanide, and ammonia. These bacteria also secrete organic acids that solubilize the complexed form of nutrients to available form, which simultaneously enhances plant growth promotion, yield, and nutrient uptake. Several rhizobacterial genera belonging to Pseudomonas spp. and Bacillus spp. are reported to solubilize cationic micronutrients from insoluble complexes or compounds in the soil or minerals. These microbes solubilize the metal forms by secreting protons, chelated ligands, and formation of oxidoreductive systems on the cell membranes (Wakatsuki, 1995).

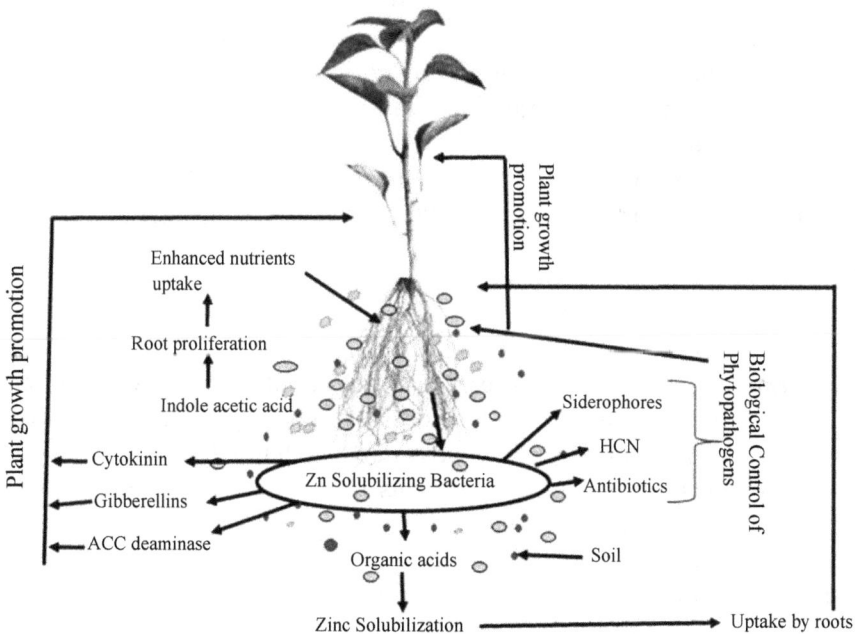

FIGURE 15.4   Mechanism of Zn solubilization by rhizobateria.
Source: Used with permission from Kumar et al. (2019). © Springer.

15.6.3.2   REDUCTION IN PH

Micronutrient availability in soil is directly related to the soil reaction and even a little variation in soil pH may have a huge impact on the solubility of micronutrients in soil. Reportedly, Zn availability decreases 100 times with every unit increase in pH (Havlin et al., 2005). Reducing the pH of an alkaline soil can enhance the bioavailable fraction of Zn to an appreciable level. Certain rhizobacteria like *Pseudomonas fluorescens* have been reported to lower the soil pH by increased extrusion of protons (Wu et al., 2006), which may also occur due to secretion of some low molecular weight organic acids like gluconic acid and 2-ketogluconic acid in the culture during solubilization insoluble micronutrient complexes like Zinc phosphate.

15.6.3.3   CHELATION

Micronutrient cations have a high-affinity interaction with the soil particles due to which their persistence of their plant available fractions in the soil solution is low to very low (Alloway, 2009). However, the bioavailability of these cations can be increased by chelating compounds (Obrador et al., 2003) which are either synthetic or released by potential rhizobacteria or the plant roots into the rhizosphere leading to improved bioavailability to the plants. The chelates are the metabolites of rhizobacteria which form complexes with the cationic forms of Fe, Zn, Cu, and Mn and reduce their reaction with the soil components (Tarkalson et al., 1998).

15.6.3.4   NUTRIPRIMING

Nutripriming is the technique where seeds are partially hydrated with nutrient solutions to allow normal metabolic processes without germination, followed by drying to their initial weight for better handling (Bradford, 1986). In micronutrient nutripriming, micronutrients act as the osmotica (Singh, 2007). Primed or treated seeds show better and synchronized germination (Farooq et al., 2009) because of lower imbibition time combined with the formation of germination-enhancing metabolites (Brocklehurst and Dearman, 2008).

Treatment of seeds with standardized concentrations of micronutrient solutions leads to the increase in micronutrient content in seeds, thus increasing the possibility of direct entry into the plant parts like coleoptiles and radical during germination and not allowing the factors affecting uptake through plant roots come into play. It can be exemplified by the fact that there are significant changes in Fe localization during germination, 36 h after sowing and particularly in the embryo and relocalization into the scutellum, coleoptile, epithelium and also to the leaf primordium and radicle. Zn, unlike

Fe tends to be distributed unevenly in possibly all parts of the seed, and more significantly in the embryo and aleurone layer (Takahashi et al., 2009). A more dynamic flow of Zn is observed during germination which was comparatively more when compared to Fe and Mn.

Priming or treatment of seeds with micronutrients has a potentially positive effect on the biochemical, physiological, and even molecular makeup of the plants. Seed treatment as a mode of application has surprisingly performed better than other methods of micronutrient application. Seed priming can prove to be an easy and cost-effective option for increasing micronutrient bioavailability, esp. to the resource-poor farmers. This approach can effectively enhance micronutrient content in seed and supply micronutrients to seedlings adequately, thus incorporating the micronutrient ions in the natural translocation pathway leading to an improvement in their remobilization to grains.

15.6.3.5 NITROGEN MANAGEMENT

Interactions among nutrients affect the absorption and distribution of individual nutrients (Robson and Pitman, 1983). Experiments involving N management formulated with the objective of N limitations alleviation which in turn can potentially influence the micronutrient concentrations in plants and soils (Wang et al., 2016). Some studies report that the application of N fertilizers increased the concentrations of Zn, Cu, Fe, and Mn in aboveground tissues (Tian et al., 2016). The available soil micronutrient concentrations also tend to exhibit various levels under the application of N. Most studies conducted to study the effect of N management on micronutrient concentrations concluded that N addition results in significant increases in the availabilities of micronutrients like Cu, Mn, and Fe in soils (Tian et al., 2016; Wang et al., 2016).. Thus N management represents a promising agronomic strategy to regulate micronutrient contents in plants esp. grains but large-scale scientific studies at multiple locations and in various crops need to be conducted to validate the approach.

15.7 CONCLUSION

Improved micronutrient bioavailability or biofortification still requires simultaneous enhancement of uptake from the rhizosphere, translocation from roots to shoots, phloem loading, and remobilization. Understanding micronutrient uptake and translocation forms the basis of any biofortification strategy. Any single approach of micronutrient management and uptake enhancement cannot be considered best as of now. Thus an integrated micronutrient management is urgently needed to meet the need of crops and humans alike.

KEYWORDS

- **biofortification**
- **genetic**
- **exogenous**
- **micronutrients**
- **transgenic**

REFERENCES

Abadia, J.; Vazquez, S.; Rellan-Alvarez, R. Towards a knowledge–based correction of iron chlorosis. *Plant Physiol Biochem.* 2011, 49(5), 471–482.

Ajiboye, B.; Cakmak, I.; Paterson, D.; de Jonge, M.D.; Howard, D.L.; Stacey, S.P.; Torun, A.A.; Aydin, N.; McLaughlin, M.J. X-ray fluorescence microscopy of zinc localization in wheat grains biofortified through foliar zinc applications at different growth stages underfield conditions. *Plant Soil.* 2015, 392, 357–370.

Alloway, B.J. Soil factors associated with zinc deficiency in crops and humans. *Environ Geochem Health.* 2009, 31, 537–548.

Beal, T.; Massiot, E.; Arsenault, J.E.; Smith, M.R.; Hijmans, R.J. Global trends in dietary micronutrient supplies and estimated prevalence of inadequate intakes. *PLoS One.* 2017, 12(4), e0175554.

Black, R.E. Zinc deficiency infectious disease and mortality in the developing world. *J Nutr.* 2003, 133(1), 1485S–1489S.

Bouis, H.E.; Hotz, C.; McClafferty, B. Biofortification: a new tool to reduce micronutrient malnutrition. *Food Nutr Bull.* 2011, 32(1), S31–S40.

Bradford, K.J. Manipulation of seed water relations via osmotic priming to improve germination under stress conditions. *Hort Sci.* 1986, 21, 1105–1112.

Brocklehurst, P.A.; Dearman, J. Interaction between seed priming treatments and nine seed lots of carrot, celery and onion. II. Seedling emergence and plant growth. *Ann. Appl. Biol.* 2008, 102, 583–593.

Cakmak, I. Enrichment of cereal grains with zinc: agronomic or genetic biofortification? *Plant Soil.* 2008, 302(1), 1–17.

Clemens, S. Zn and Fe biofortification: the right chemical environment for human bioavailability. *Plant Sci.* 2014, 225, 52–57.

Dhane, S.S.; Shukla, L.M. Distribution of DTPA-extractable Zn, Cu, Mn and Fe in some soil series of Maharashtra and their relationship with some soil properties. *J Indian Soc Soil Sci.*, 1995, 43, 597.

Farooq, M.; Basra, S.M.A.; Wahid, A.; Khaliq, A.; Kobayashi, N. Rice seed invigoration. In: E. Lichtfouse (ed.). Sustainable Agriculture Reviews, Springer: the Netherlands. pp. 137–175. 2009.

Frossard, E.; Bucher, M.; Machler, F. Potential for increasing the content and bioavailability of Fe, Zn and Ca in plants for human nutrition. *J Sci Food Agr.* 2000, 80(7), 861–879.

Gibson, R.S. The role of diet- and host-related factors in nutrient bioavailability and thus in nutrient-based dietary requirement estimates. *Food Nutr.* 2007, 28(1), 77–100.

Gismera, M.J.; Lacal, J.; Silva, P.; Garcia, R.; Sevilla, M.T.; Procópio, J.R. Study of metal fractionation in river sediments: a comparison between kinetic and sequential extraction procedures. *Environ Pollut.* 2004, 227, 175-182.

Gomez-Galera, S.; Rojas, E.; Sudhakar, D.; Zhu, C.; Pelacho, A.M.; Capell, T.; Christou, P. Critical evaluation of strategies for mineral fortification of staple food crops. *Transgenic Res.* 2010, 19, 165–180.

Grillet, L.; Mari, S.; Schmidt, W. Iron in seeds–loading pathways and subcellular localization. *Front. Plant Sci.* 2014, 4, 535.

Grusak, M.A.; Pearson, J.N.; Marentes, E. The physiology of micronutrient homeostasis in field crops. *Field Crop Res.* 1999, 60, 41–56.

Guadalix, M.E.; Pardo, M.T. Zinc sorption by acidic tropical soils as affected by cultivation. *J. Soil Sci.* 1995, 46, 317–322.

Haas, J.D.; Beard, J.L.; Murray-Kolb, L.E.; del Mundo, A.M.A.; Felix, G.B. Iron-biofortified rice improves the iron stores of nonanemic Filipino women. *J Nutr.* 2005, 135, 2823–2830.

Hallberg, L.; Hulthén, L. Prediction of dietary iron absorption: an algorithm for calculating absorption and bioavailability of dietary iron. *Am J Clin Nutr.* 2000, 71(5), 1147–1160.

Havlin, J.L.; Beaton, J.D., Tisdale, S.L.; Nelson, W.L. Soil fertility and fertilizers: an introduction to nutrient management. Prentice Hall, Upper Saddle River, New Jersey. 2007, pp 244–289.

Hotz, C.; Gibson, R.S. Traditional food-processing and preparation practices to enhance the bioavailability of micronutrients in plant-based diets. *J Nutr.* 2007, 136, 1079–1100.

Impa, S.M.; Sarah, E.; Johnson, B. Mitigating zinc deficiency and achieving high grain Zn in rice through integration of soil chemistry and plant physiology research. *Plant Soil.* 2012, 361, 3–41.

Joy, E.L.; Ander, S.D.; Young, C.R.; Black, M.J.; Watts, A.D.C.; Chilimba, B.; Chilima, E.W.P., Siyame. A.A.; Kalimbira, R.; Hurst, S.J.; Fairweather; Tait, A.J.; Stein, R.S.; Gibson, P.J.; White, M.R. Broadley dietary mineral supplies in Africa. Physiol Plant. 2014, 151, 208–229.

Katyal, J.C.; Sharma, B.D. DTPA extractable and total Zinc, Copper, Manganese and Iron in Indian soils and their association with some soil properties. *Geoderma.* 1991, 49, 165–179.

Kochian, L.V. Mechanisms of micronutrient uptake and translocation in plants. In: Micronutrients in Agriculture, Cox, F. R.; Mortvedt, J.J.; Shuman, L.M.; Welch, R.M. Madison (eds), WI: Soil Science Society of America. 1991, pp. 229–296.

Kumar, A.; Dewangan, S.; Lawate, P.; Bahadur, I.; Prajapati, S. Zinc-solubilizing bacteria: a boon for sustainable agriculture. In: Plant Growth Promoting Rhizobacteria for Sustainable Stress Management. Microorganisms for Sustainability. Sayyed R., Arora N., Reddy M. (eds) Springer: Singapore, 2019; vol 12.

Lyons, G.H.; Genc, Y.; Stangoulis, J.; Palmer, L.T.; Graham, R.D. Selenium distribution in wheat grain, and the effect of postharvest processing on wheat selenium content. *Biol. Trace Elem. Res.* 2005, 103, 155–168.

Ma, G.; Jin, Y.; Li, Y.P. Iron and zinc deficiencies in China: what is a feasible and cost-effective strategy? *Public Health Nutr.* 2007, 10, 1017–1023.

Marschner, H. Marschner's Mineral Nutrition of Higher Plants, 3rd edition. Elsevier Ltd., 2012.

Mondal, S.; Bose, B. Impact of micronutrient seed priming on germination, growth, development, nutritional status and yield aspects of plants. *J Plant Nutr.* 2019, 42, 19, 2577–2599.

Obrador, A.; Novillo, J.; Alvarez, J.M. Mobility and availability to plants of two zinc sources applied to a calcareous soil. *Soil Science Soc Am J.* 2003, 67, 564–572.

Olsen, L.; Palmgren, M. Many rivers to cross: The journey of zinc from soil to seed. *Front Plant Sci.* 2014, 5, 30.

Prakash, O.; Ward, R.; Adhikari, B.; Mawson, J.; Adhikari, R.; Wess, T.; Pallas, L.; Spiers, K.; Paterson, D.; Torley, P. Synchrotron X-rayfluorescence microscopy study of the diffusion of iron, manganese, potassium and zinc in parboiled rice kernels. LWT-Food Sci. Technol. 2016, 71, 138–148.

Rauret, G.; López-Sánchez, J.F.; Sahuquillo, A. Improvement of the BCR three step sequential extraction procedure prior to the certification of new sediment and soil reference materials. *J Environ Monitor.* 1999, 1, 57–61.

Rawat, N.; Neelam, K.; Tiwari, V.K.; Dhaliwal, H.S. Biofortification of cereals to overcome hidden hunger. *Plant Breed.* 2013, 132, 437–445.

Rengel, Z.; Batten, G.D.; Crowley, D.E. Agronomic approaches for improving the micronutrient density in edible portions of field crops. *Field Crop Res.* 1999, 60, (1–2), 27–40.

Ricachenevsky, F.K.; Sperotto, R.A. There and back again, or always there? The evolution of rice combined strategy for Fe uptake. *Front. Plant Sci.* 2014, 5, 189.

Rietra. R.P.J.J.; Heinen, M.; Dimpka, C.; Bindraban, P.S. Effects of nutrient antagonism and synergism on fertilizer use efficiency. VFRC Report 2015/5. Virtual Fertilizer Research Centre, Washington, DC. 2015.

Ruel, M.T.; Alderman, H. Nutrition-sensitive interventions and programmes: how can they help to accelerate progress in improving maternal and child nutrition? (Maternal and Child Nutrition Study Group) *Lancet.* 2013, vol. 382; pp. 536–551.

Saltzman, A.; Birol, E.; Bouis, H.E.; Boy, E.; DeMoura, F.F.; Islam, Y.; Pfeiffer, W.H. Biofortification: progress toward a more nourishing future. *Glob Food Secur.* 2013, 2, 9–17.

Sandström, B. Micronutrient interactions: effects on absorption and bioavailability. *Br J Nutr.* 2001, 85 (S2), 181.

Shahzad, Z.; Rouached, H.; Rakha, A. Combating mineral malnutrition through iron and zinc biofortification of cereals. *Comprehen Rev Food Sci Food Safety.* 2014, 13(3), 329–346.

Singh, M.V. Efficiency of seed treatment for ameliorating zinc deficiency in crops. In: Zinc Crops, Improving Crop Production and Human Health, 24–26 May, 2007, Istanbul, Turkey, 2007.

Smith, S.E.; Read, D.J. Mycorrhizal Symbiosis. Academic Press: San Diego. 1997, 607.

Spark, K.M.; Wells, J.D. Characterizing trace metal adsorption on kaolinite. *Eur J Soil Sci.* 1995, 46, 633–640.

Sperotto, R.A. Zn/Fe remobilization from vegetative tissues to rice seeds: should I stay or should I go? Ask Zn/Fe supply! *Front. Plant Sci.* 2013, 4, 464.

Stahl, R.S.; James, B.R. Zinc sorption by B horizon as a function of pH. *Soil Sci Soc Am J.* 1991, 55, 1592–1597.

Stoltzfus, R.J. Iron-deficiency anemia: reexamining the nature and magnitude of the public health problem. Summary: implications for research and programs. *J Nutr.* 2001, 131 (2), 697–700.

Tarkalson, D.D.; Jolley, V.D.; Robbins, C.W.; Terry, R.E. Mycorrhizal colonization and nutrient uptake of dry bean in manure and composted manure treated subsoil and untreated topsoil and subsoil. *J Plant Nutr.* 1998, 21, 1867–1878.

Tian, Q.; Liu, N.; Bai, W.; Li, L.; Chen, J.; Reich, P.B. A novel soil manganese mechanism drives plant species loss with increased nitrogen deposition in a temperate steppe. *Ecology.* 2016, 97, 65–74.

Treeby, M.; Marschner, H.; Romheld, V. Mobilization of iron and other micronutrient cations from a calcareous soil by plant borne, microbial and synthetic metal chelators. *Plant Soil.* 1989, 114, 217–226.

Tulchinsky, T.H. Micronutrient deficiency conditions: global health issues. *Public Health Rev.* 2010, 32, 243–255.

Vasconcelos, M.W.; Gruissem, W.; Bhullar, N.K. Iron biofortification in the 21st century: setting realistic targets, overcoming obstacles, and new strategies for healthy nutrition. *Curr Opin Biotechnol.* 2017, 44, 8–15.

Wakatsuki, T. Metal oxidoreduction by microbial cells. *J Indus Microbiol.* 1995, 14 (2), 169–177.

Wang, R.; Dungait, J.; Buss, H.; Yang, S.; Zhang, Y.; Xu, Z.; Jiang, Y. Base cations and micronutrients in soil aggregates as affected by enhanced nitrogen and water inputs in a semi-arid steppe grassland. *Sci Total Environ.* 2016, 575, 564–572.

Wei, X.; Hao, M.; Shao. M.; Gale. W.J. Changes in soil properties and the availability of soil micronutrients after 18 years of cropping and fertilization. *Soil Till Res.* 2006, 91, 120–130.

Welch, R.M. Breeding strategies for biofortified staple plant foods to reduce micronutrient malnutrition globally. *J Nutr.* 2002, 132, 495–499.

Welch, R.M.; Graham, R.D. Breeding for micronutrients in staple food crops from a human nutrition perspective. *J Exp Bot.* 2004, 55, 353–364.

Yin, H.J.; Gao, X.P.; Stomph, T. Zinc Concentration in Rice (Oryza sativa L.) Grains and allocation in plants as affected by different zinc fertilization strategies. *Commun Soil Sci Plant Anal.* 2016, 47 (6), 761–768.

Zhang, Y.Q.; Deng, Y.; Chen, R.Y. The reduction in zinc concentration of wheat grain upon increased P–fertilization and its mitigation by foliar zinc application. *Plant Soil.* 2012, 361 (1), 143–152.

Zhu, C.; Naqvi, S.; Gomez-Galera, S.; Pelacho, A.M.; Capell, T.; Christou, P. Transgenic strategies for the nutritional enhancement of plants. *Trends Plant Sci.* 2007, 12,548–555.

Zimmermann, M.B.; Hurrell, R.F. Nutritional iron deficiency. *Lancet.* 2007, 370 (9586), 511–520.

Zingore, S.R.; Delve, J.; Nyamangara, J.; Giller, K.E. Multiple benefits of manure: the key to maintenance of soil fertility and restoration of depleted sandy soils on African smallholder farms. *Nutr Cycl Agroecosyst.* 2008, 267–282.

PART VI
Nanotechnological Interventions

CHAPTER 16

Escalating Nutrient Use Efficiency through Nanotechnological Intercessions

AKHILA NAND DUBEY[1], SAMRAT ADHIKARY[2], SATDEV,
NINTU MANDAL[3*], and NILANJAN CHATTOPADHYAY[3]

[1]Department of Soil Science and Agricultural Chemistry,
Institute of Agricultural Sciences, Banaras Hindu University, Varanasi,
Uttar Pradesh 221005, India

[2]Department of Agricultural Chemistry and Soil Science, Bidhan Chandra
Krishi Viswadidyalya, Haringhata, West Bengal 741252, India

[3]Department of Soil Science and Agricultural Chemistry,
Bihar Agricultural University, Sabour, Bhagalpur, Bihar 813210, India

*Corresponding author. E-mail: nintumandal@gmail.com

ABSTRACT

Nanotechnology is simply the control or association of atoms, atoms, or individual subatomic groups in structures to deliver materials and gadgets with completely new and divergent properties. "Nano" alludes to a scale of size somewhere in the range of 1–100 nm, in any event in one measurement. Production of various types of nanoparticles has been reported by several researchers, such as Mg nanoparticle production was best with the treatment of $MgSO_4$ and *Aspergillus fumigatus* as well as by the use both *Pochoniachlamy dosporium* and *Aspergillus fumigates*. Valid characterization techniques for nanoparticle biosynthesis are dynamic light scattering, transmission electron microscopy imaging, and high-resolution transmission electron microscopy. Energy-dispersive X-ray energy spectroscopy and Fourier transform infrared spectroscopy confirmed the presence of a high percentage of an elemental iron signal or the presence of extracellular proteins. ZnO nanoparticles were also significantly enhanced acid phosphatase (51%–108%) and alkaline phosphatase activity (80–209%). The application of ZnO nanoparticles on mung bean (20 mg kg^{-1}) and chickpea (1 mg kg^{-1}) was showed best in terms of seedling growth and development, whereas little more concentration of recommended dose become harmful with the appearance of toxic effect on the plants. Zincated nanoclay polymer composite (ZNCPC) at half dose of $ZnSO_4 \cdot 7H_2O$ increased diethylenetriaminepentaacetic-acid-extractable Zn, field crop characteristics, as well as grain and straw Zn content in rice. Application of ZNCPC at the rate 100% observed highest Zn concentration in grain and apparent zinc recovery. The foliar application of nano-ZnO (40 ppm) would be a viable option in enhancing Zn concentration in grain. Microbe-mediated synthesis of nanofertilizers seems to be promising in days to come. Further studies are to be undertaken for standardization of an economic method for mass production of nanoparticles by microorganisms. Commercialization of such products is the most important future challenge.

16.1 INTRODUCTION

Soil microorganisms assume a basic job in the breakdown of natural substances, the supplement cycle, and the preparation of the soil. Without the pattern of the components, the quest for life on earth would be incomprehensible, since fundamental supplements would be immediately consumed by living beings and secured up a structure that could not be utilized by others. Soil microorganisms are both constituents and producers of organic carbon in the soil, a substance that binds carbon in the soil over long periods of time. The abundance of organic carbon in the soil improves soil fertility and water retention capacity. More and more research supports the hypothesis that soil microorganisms, and in particular fungi, can be used to extract carbon from the atmosphere and bind it in the soil. Soil microorganisms can be an important means of reducing atmospheric greenhouse gases and can help limit the effects of climate change caused by greenhouse gases.

Nanotechnology is a smart field nowadays that is being used practically in all the popular fields of science. Nanofertilizer application in plant development and advancement and for sickness control is an ongoing practice. The nanomaterials were influenced on the dirt microbial system in setting subordinate (nature, kind of nanomaterials, parcel, soil types, sorts of creatures, etc.). Nanomaterials are used logically because of the prohibitive properties (to the extent of the high point; to the extent of the surface volume), making them important for abundant applications in agriculture. Considering the strategy for use, there is each opportunity of its passageway into the earth structure and genuinely or in an indirect manner interfacing with an extent of soil parts. Today, the synthesis and detailing of biodegradable novel particles with a smart transport system are being tried. In any situation, it is not yet clear whether the impacts of nanoparticles (NPs) are gainful or unsafe to plant development. NPs blended from various natural sources can be utilized in agribusiness (Kaur et al., 2018). The reaction of microbial populaces and compounds related to exogenous upgrades (use of nano-Zn and Fe) is a promising region of research around the world. Resistance and resilience of the soil system were influenced by the engineered nanomaterials (Kumar et al., 2019).

The blend of nano-SiO_2 and nano-TiO_2 purportedly increments nitrate reductase action in soybean seeds and consequently expands seed germination and development. It has been discovered that a fitting convergence of nano-TiO_2 improves the development of spinach by advancing photosynthesis and nitrogen digestion. Root expansion fortified via carbon nanotubes nanofunctionalized in onions and cucumbers. Nano-ZnO has been generally utilized in industry for a very long while. Be that as it may, no examination on the conceivable utilization of nano-Zn in agribusiness has been done. The job of NPs in farming is known as nano-agri-business; new innovations are regularly used to improve crop yields (Duhan et al., 2017). The current investigation manages the blend of different NPs by microorganisms and their portrayal and the impact of the suspension of nano-ZnO particles as a micronutrient on the development of mung bean and chickpea plants.

16.2 BIOSYNTHESIS OF VARIOUS NANOPARTICLES BY MICROORGANISMS

16.2.1 BIOSYNTHESIS OF IRON NANOPARTICLES

The *Aspergillus oryzae* growth was segregated from agricultural research farm in a study conducted by Tarafdar and Raliya (2012a). The pure mushroom culture was confined by spreading the inoculum on medium of Martin Rose Bengal Agar (pH 7.2), and samples are collected after serial dilutions. The chloramphenicol anti-infection was autoclaved at a convergence of 10 *μm* g/L added to forestall bacterial sullying. The plates of vaccinated were incubated at 28 °C *for 72 h* in the bio-oxygen demand incubator. Individual fungal colonies were picked and further purified by subculturing on potato dextrose agar (PDA) media (HiMedia, India). The primer distinguishing proof of fungal isolated depended on morphological properties.

16.2.2   BIOSYNTHESIS OF GOLD NPS

Rhizoctonia bataticola TFR-6 growths (NCBI quality bank get to number JQ675307) were developed in a 250-mL Erlenmeyer flagon with 100-mL potato dextrose stock Raliya and Tarafdar, 2013a. The medium pH was modified at 5.8 and afterward stirred continuously the culture media on a turning shaker (150 rpm) at 28 °C for 72 h. After incubation completion, the mycelia parasitic balls were isolated from the way of life stock by separating strategies utilizing Whatman No. 1 filter paper under a biosafety bureau; at that point, the parasitic mycelia were refined multiple times with a sterile water wash. The collected contagious mycelium (Wet weight 20 g) was resuspended in a 250-mL Erlenmeyer flagon with 100 mL of sterile Milli-Q water and again positioned in a turning shaker (150 rpm) at 28 °C for 12 h. Incubation after the acellular filtrate was obtained by isolating the contagious biomass utilizing a 0.45 size layer channel (Whatman, England). Utilizing an acellular filtrate, a saline arrangement of watery $HAuCl_4$ (Merck, Germany) with a last centralization of 0.1 mM in Erlenmeyer flagons was prepared, which end up being the ideal salt fixation for the combination of mono scatter gold NPs. The entire blend was kept on a rotating shaker at 28 °C and 150 rpm. The response was done for 4 h. The biotransformed item was gathered intermittently to describe the molecule size utilizing a particle size analyzer.

16.2.3   BIOSYNTHESIS OF ZNO NANOPARTICLES

The fungi were separated in pure culture from soil tests taken at the Research Institute of the Central Arid Zone of Jodhpur under dry biological systems (Tarafdar et al., 2012b). The fungi were segregated by serial dilution technology, which was washed through the individual spore. The fungi were recognized by the Agharkar Research Institute, Pune, India, and the unadulterated societies were put away on the PDA medium.

The DNA sequences of the fungus was submitted to NCBI and got accession number JF 681300 for *Aspergillus terreus* and JF 681301 for *Aspergillus flavus*. Both the fungi (*A. terreus* and *A. flavus*) have been reported to release huge amount of acid phosphatase, alkaline phosphatase, and phytase. Fungal mycelium from 72 h old culture of each of these fungi was crushed in sterile distilled water to obtain fungal spore suspension of 0.6 OD at 600 nm for inoculation. One mL of fungal suspensions was added in 25-mL potato dextrose broth medium in a 100-mL conical flask. Each fungus was subjected to three treatments, namely, control (only fungal culture), culture with pure ZnO compound (1000 g mL^{-1}), and culture with ZnO NPs (1000 g mL^{-1}). ZnO NPs of 16 nm average size used in this study were prepared by the sol–gel technique. The sterilized distilled water of 1 mL was added in control flask to maintain the same volume. Triplicate flasks of each treatment were incubated at 25±2 °C for 72 h. The method for extraction of extracellular polysaccharide involved removal of fungal mat from the suspension broth. After incubation of 72 h, the mass of fungal was removed from the broth culture by filtration through Whatman No. 1 filter paper to the added 1 mL of 10% tricarboxylic acid cycle (TCA) with each filtrate flask to precipitate the extracellular protein S11 present in the medium secreted by the fungi. The extracellular polysaccharides were quantified spectrophotometrically as total sugar using anthrone method at 620 nm. The polysaccharide content (*μ*g mL^{-1}) was determined by using standard curve of glucose. To determine the content of intracellular polysaccharides, 10 mg of fungal biomass were ground with 10 mL of phosphate buffer (pH 6.5) in a controlled sterilized chamber. Then, the suspension was centrifuged for 30 min at 6000 rpm. Supernatant was treated with 1 mL of 10% TCA to precipitate the proteins. Acid and alkaline phosphatases were tested according to the standard method. Inorganic phosphate (Pi) released was measured as phytase activity.

16.2.4   BIOSYNTHESIS OF SILVER NPS

On the PDA slants, fungal isolate was kept at 28 °C temperature with customary subculture on new media (Jain et al., 2010). The mother culture of (four days old) was inoculated in 100 mL of malt-extractglucose-yeast extract-peptone medium at pH 7.0 in a conical flask with the capacity of 250 mL. The immunized vials of four days incubated at 28 °C on a mechanical shaker (150 rpm). The growth mycelium was isolated from the way of medium life by centrifugation (8000 rpm, 10 min at 4 °C and washed multiple times with sterilized water. Ordinarily, 10 g weight of biomass (new weight) were resuspended in sterile Milli-Q deionized water of 100 mL and again incubated for 72 h in an Erlenmeyer carafe and mixed under conditions like those depicted previously. The biomass was filtered by the Whatman No. 1 channel paper and acellular filtrate was obtained after incubation.

Silver NPs are the amalgamation; a watery arrangement of nitrate of silver at a last grouping of 1.0 mM was mixed to the response vessels, which contained filtrate sans cell, and incubated at 28 °C on a revolving shaker at 150 rpm without presence of light. Acellular filtrate of control condition (without nitrate of silver) in beneficial arrangement and pure nitrate of silver (without acellular filtrate) as controls of negative were additionally completed at the same time with the test carafe in three repetitions (Jain et al., 2010).

16.2.5   BIOSYNTHESIS OF SULFIDE NPS

Despite nano-oxide particles, sulfide NPs have pulled in inconceivable thought as quantum-spot fluorescent biomarkers and optical properties (Yang et al., 2005). Minimal circles crystal of nano was one ordinary kind of sulfide NPs and had been consolidated by microbial. Cunningham and Lundie uncovered that *Clostridium thermoaceticum* could quicken CdS on the phone surface as well as in the medium from $CdCl_2$ in the proximity of cysteine hydrochloride in the improvement medium where cysteine in all probability goes about as the wellspring of sulfide (Cunningham and Lundie., 1993). Nanocompounds of ZnS and PbS were viably consolidated by the regular structures. *Rhodobacter sphaeroides* and *Desulfobacteraceae* had been used to get nano-ZnS intracellularly with 8 nm and the average separation over 2–5 nm, independently (Bai et al., 2006).

16.3   CHITOSAN-GRAFTED ZINC CONTAINING NANOCLAY POLYMER BIOCOMPOSITES (CZNCPBCS)

Acetic acid 1% (v/v) of 100 mL was taken in a reaction vessel arranged on the engaging stirrer with hot plate. Ammonium persulfate (APS) of 0.761 g weight was taken and added as a radical-free initiator followed by mixture of 2-g chitosan (w/w). After heating the mixture at 50–60 °C for 15–20 min, 22-mL acrylic acid was taken in four neck response vessels containing a gas line of N_2, a thermometer, and a condenser over the stirrer. Acrylic acid was mixed (60% of neutralization degree) by addition of 14.4-mL ammonia. Then, 4.6-g acryl amide was added followed by mixing of various rates (w/w) of monomer (8%, 10%, and 12%), clay content (w/w) (bentonite and nanobentonite), and citrate of zinc as a Zn^{2+} transporter. The crosslinker power of N,N-methylenebisacrylamide (0.132 g) (w/w) was blended into the mixture followed by the addition of 0.3208-g APS (w/w). The reaction polymerization occurred under atmosphere of N_2. The temperature of the response vessel was around 60–70 °C. Polymer composites were synthesized and collected in Petri plates; under vacuum oven, they will be dried by reduced pressure (Mandal et al., 2018).

16.4   CHARACTERIZATION OF NANOPARTICLES

Various instruments are utilized for the characterization of NPs, for example, dynamic light scattering (DLS), energy-dispersive X-ray spectroscopy (EDaX), UV–Visible spectroscopy, Fourier transform infrared spectroscopy (FTIR), transmission electron microscope (TEM) imaging, and high-resolution transmission electron microscopy (HR-TEM).

16.4.1   SCANNING ELECTRON MICROSCOPY (SEM)

Morphology characterized of zincated nanoclay polymer composites (ZNCPCs) was researched by the scanning electron microscope (Mandal et al., 2018). Limited quantity of dried powder was set on carbon tape and faltered covered with slender conductive layer of gold-palladium. The SEM image was taken with the model of Zeiss EVOMA 10 instrument electron seeing enhancement at 20.00 kV/EHT and 10 Pa between 13.00 KX to 15.00 KX after 24-nm covering with palladium. Broken morphology with some specific topography was expressed in the SEM image of ZNCPCs. There were irregularities in surface with the increment in the content of clay-specific surface topologies (Mandal et al., 2018).

16.4.2   TRANSMISSION ELECTRON MICROSCOPY AND HIGH-RESOLUTION TRANSMISSION ELECTRON MICROSCOPY

To confirm size and shape of the biosynthesized iron NPs, TEM measurements were performed utilizing a drop covering strategy, in which a solution drop containing NPs was set on the Cu networks secured with carbon and held under vacuum drying for the time being before stacking onto an example holder. TEM and HR-TEM pictures of the example were taken utilizing the TEM model JEM-2100F instrument. The instrument was operated at an accelerating 4 V of 200 kV.. TEM estimations were utilized to examine the morphological affirmation of the Fe NPs. A TEM image (see Figure 16.1) presents a good distribution of the spherical of nano-Fe. This result also observed in the HR-TEM image, which also supports the crystalline nature of biosynthesized Fe NPs.

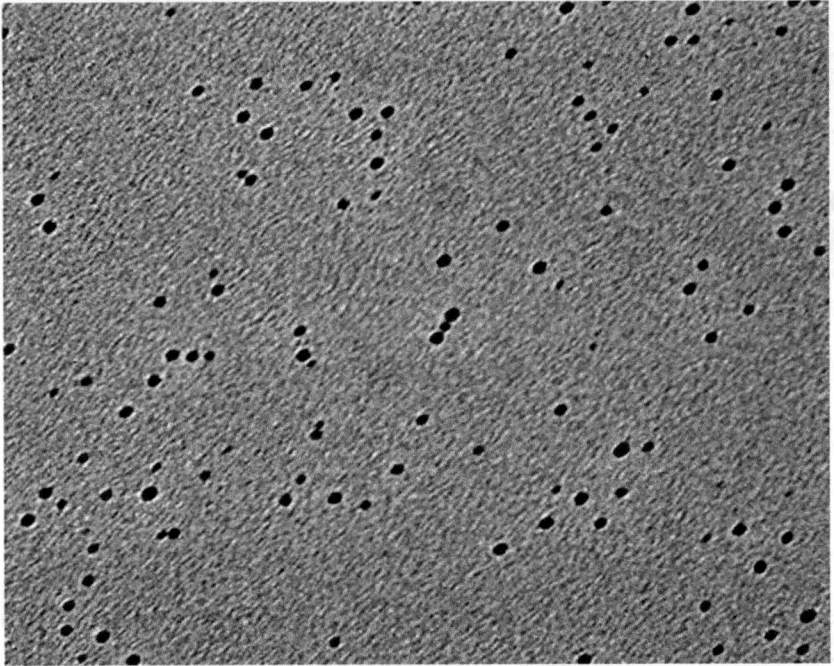

10 nm.

FIGURE 16.1   Micrograph of biologically synthesized iron NPs through TEM.

16.4.3   DYNAMIC LIGHT SCATTERING

The size of particle of iron NPs was examined utilizing DLS estimations, which decided the size of the particles by estimating the rate of change in the intensity of the scattered laser light by the particles during their diffusion by the solvent.

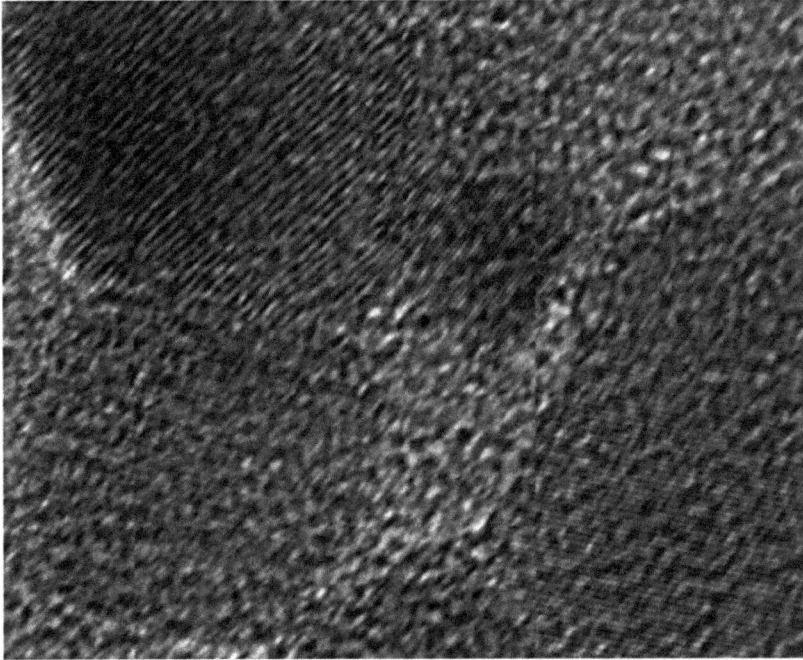

1 nm.

FIGURE 16.2   Micrograph of single iron NPs through HR-TEM.

The analyzer of the particle size (Beckman DelsaNano C, USA) was utilized to gauge the estimate and affirm the molecule size circulation of the NPs. The size of particles of the biosynthesized Fe NPs was dissected with an analyzer of particle size. The histogram plainly expressed the size in the range of the particle between 10 and 24.6 nm. The polydispersity index (PDI) was 0.258, which presents a high monodisperse nature of the particles. The PDI mirrors the consistency of the incorporated NPs and is estimated with DLS.

16.4.4   ENERGY-DISPERSIVE X-RAY SPECTROSCOPY

The elemental composition of the sample was analyzed by EDS from the C-coated Cu grids and dried in vacuum for 3 h before being loaded onto a holder of sample. Elemental composition of individual particles and purity of the sample in atomic percentage of metal was analyzed using Thermo-Noran EDS accessory equipped with TEM (JEM-2100F). EDaX analysis of particles showed a strong single for the 6.4-keV Fe NPs, which is a characteristic of Fe metal.

The quantitative measurement results obtained from the analysis of EDaX reflect that 84% of the atomic particles were made of metal of ferrous, which prove the ferrous metal purity in the biotransformed product. The other peaks of chlorine (12.4% atom) and oxygen (3.6% atom) are owing to used precursor salt for biosynthesis of iron nanoparticles.

16.4.5   UV–VISIBLE SPECTROSCOPY

The synthesis of the NPs was characterized by checking with by UV–visible spectroscopy techniques by sampling aliquots (1 mL) at various time intervals. The spectra absorption was measured with a model of Jasco V-630 UV–visible spectrophotometer (Jasco Corporation, Tokyo, Japan), which was operated in the range up to 200–900 nm with a 1-nm resolution. UV–visible spectroscopy was used to checked the synthesized of nano-ZnO. A subsequently increase in absorption to around 375 nm was seen compared to the reaction time, which represents the synthesis of ZnO NPs. Synthesis was relatively slow at the initial time intervals (0, 6, 12, and 24 h), but significant changes in the size of absorption were observed at subsequent time intervals (36, 48, and 72 h). The reaction start (0 h) and end (72 h) were observed at the pH value of 7.4 and 7.8, respectively.

This lower modification of the pH makes the present protocol "environmentally friendly," which is very advantageous compared to the chemical synthesis protocols, which have large fluctuations in pH. The NP stability in the solution, which was stored for more than three months at room temperature (25 °C) under dark chamber condition after the end of the reaction, was measurements by UV–visible spectroscopy.

16.4.6   FOURIER TRANSFORM INFRARED SPECTROSCOPY

For FTIR measurements, the biotransformed products, which were present in the filtrate without cell of fungal, were lyophilized, and potassium bromide was used for diluted with the ratio of 1:100. The spectra of FTIR of samples were observed on a FimShimazdu IR Prestige-21 spectrometer (Shimadzu, Nakagyo-ku, Japan) with a fixation in diffuse reflection mode (DRS-8000) (Shimadzu Corporation, Nakagyo-ku, Japan). All measurements were carried out in the range of wavenumbers 400–4000 cm^{-1} at a resolution of 4 cm^{-1}. FTIR analysis of freeze-dried samples showed an intense band in the vicinity of wavenumbers 400–600 cm^{-1} that centered around wavenumber 430 cm^{-1} and is attributed to ZnO vibrations (Figure 16.5b). The presence of bands at wavenumbers 1625 and 1550 cm^{-1}corresponds to the bending vibrations of the amide I and amide II of proteins, respectively. The band at wavenumber 3450 cm^{-1} has been reported to occur due to stretching vibrations of amide I superimposed on the side of hydroxyl group band. Moreover, the bands at wavenumbers 2995 and 2350 cm^{-1} are due to the stretching vibrations of amide II and the presence of atmospheric CO_2, respectively. The name near the wavelength numbers 400–600 cm^{-1} showed lyophilized sample, centered on the spectra 430 cm^{-1} and the ZnO vibrations attributed. The bending vibrations of amide I and amide II of proteins were observed on spectra band at wave numbers 1625 and 1550 cm^{-1}. The band has been reported to occur at 3450 cm^{-1} of wavelength due to the stretching vibrations of amide I superimposed on the side of the OH group. The stretching vibrations of amide II or to the presence of atmospheric CO_2 are found at wavelengths 2995 and 2350 cm^{-1}.

16.4.7   CALCULATION OF AVERAGE CRYSTALLINE SIZE

Normal crystalline sizes of ZnO and Fe_2O_3 NPs were evaluated utilizing the Debye–Scherrer recipe (Hall et al. 2000) as follows:

$$D = 0.9\lambda/\beta\cos\theta$$

where λ is the frequency of X-beams (0.1541 nm), β is the full width at half maximum, θ is the diffraction edge, and D is the molecule breadth size.

16.5   INFLUENCED TIIE FUNGUS TREATMENT IN SOIL BY BIODEGRADABILITY OF CZNCPBCS

Aspergillus spp. was seen as powerfully compelling as considered with *Trichoderma* spp. in spite of soil types in corrupting CZNCPBCs in terms of solidified CO_2-carbon movement all through the period of incubation (Mandal et al., 2018). This may be endorsed by the fact that the multiplication rate *of Aspergillus* spp. is higher than that of *Trichoderma* spp. (Alexander 1977) producing more biomass consequently rapid decomposition of bionanocomposites. Decrease in decomposition with the increase in the content of clay might be explained by the fact that higher density crosslinking of clay caused difficulty to microbes in degrading the polymeric structure. Nanoclay present in CZNCPBCs lessened evolution of CO_2-carbon when percentage of the clay content was higher. Chitosan additionally implied amide in its structure resulting in that the C:N ratio is lower and causing quick decomposition of nanomaterials under inoculation of microbial species.

16.6 EFFECT OF NANO-ZNO PARTICLE SUSPENSION ON ROOT AND SHOOT GROWTH

Mahajan et al. (2001) reported that the nano-ZnO response curves on the seedlings of roots and shoots of mung bean and chickpea showed that the growth of roots and shoots increased with the increase the concentration of nano-ZnO.

However, after a certain level of concentration, the root and the shoot growth were decreased. In mung bean seedlings, the best root growth (42.0%, p value 0.0498) and sprouts (97.9%, p value 0.0444) was observed at 20 mg kg^{-1}. At the highest concentration (2000 ppm), there was a delay in the growth of roots (93.3%, p value 1.736×10^{-14}) and shoots (14.8%, p value 2.46×10^{-11}) in seedlings of mung bean observed compared to the control treatment.

Similarly, in sowing chickpeas, a concentration of 1 mg kg^{-1} showed a significantly increase in the root (53.1%, p value 1.125×10^{-7}) and shoots (6.4%, p value 0.026) compared to the witness. Below this level, however, root and shoot growth has been delayed. At 2000 mg kg^{-1}, a significantly decrease in the growth of root (74.2%, p value 1.32×10^{-10}) and shoot (22.5%, p value 0.020) was observed. The minimizing in growth root and shoot at higher concentration can be attributed to the toxic level of the NPs. This shows that the mung bean and chickpea plants react to the added nano-material in a limited concentration range, from which a decrease in growth occurs due to the toxic effect.

16.7 EFFECT ON DIETHYLENETRIAMINEPENTAACETIC ACID (DTPA)-EXTRACTABLE OF ZN AND FE CONTENT BY NANO-ZN CARRIER DURING INCUBATION

Zn content (DTPA-extractable) was on a very basic level contrasted with the treatment. Most noteworthy Zn content was recorded after incubation with the treatment of ZNCPC. There was diminishing of Zn content at the completion of the incubation. ZNCPC went about as slow release Zn plan where soil went about as scattering barrier. Initially, content of Zn was higher in commercial $ZnSO_4 \cdot 7H_2O$ treatment, anyway a tiny bit at a time diminished inferable from participation of released Zn with soil sections (Adhikary et al., 2020a). Upkeep of Zn content in soils was expressed by starting quick reactions in which adsorbtion of Zn on the solid particles trailed by moderate reactions

(Montalvo et al. 2016). Most extraordinary DTPA Fe was found introductory, where the least was seen after the incubation. The incubation completion of the DTPA-extractable Fe content was generally declined in all the concentration treatments with respect to control. Availability of Fe content in soil plan might be limited in light of the fact that DTPA-Zn content was inferable from advancement of insoluble complex franklinite ($ZnFe_2O_4$). Most outrageous Zn^{+2} dissolvability was, in addition, related to the franklinite ($ZnFe_2O_4$) solubility, which was obliged by the Fe^{+3} availability being limited by the dependence on pH solubility of Fe(III) hydroxides (Wisawapipat, et al. 2017).

16.8   EFFECT OF NANO-ZNO PARTICLES SUSPENSION ON BIOMASS

Mahajan et al. (2011) demonstrated that the creation of biomass of root and shoot of mung bean and chickpea seedlings was found as an element of the length of the roots and went for the comparing nano-ZnO treatment. The treatment of ZnO NPs at 20 mg kg^{-1} mung bean seedlings indicated a 40.9% expansion in root mass and a 76.0% increment in shoot mass contrasted with the control. Chickpea plantings indicated a 37.1% expansion in root mass and a 26.6% increment in sprout mass contrasted with the control at 1 ppm treatment. At the most elevated convergence of 2000 ppm, a reduction in root mass of 44.7% and shoot mass of 31.9% was seen in the mung bean seedlings contrasted with the control. In chickpea seedlings, a 32.2% lessening in root mass and a 52.8% diminishing in sprout mass were noticed. Expanding the biomass at a specific focus recommends the ideal portion limit for the development of mung bean and chickpea plants. In any case, a decline in the biomass past this fixation shows the harmful impact of the nano-ZnO particles.

16.9   SOIL DEHYDROGENASE, ALKALINE PHOSPHATASE ACTIVITY (MG TPF HOUR^{-1} G^{-1}SOIL) AS WELL AS OLSEN-P AFFECTED BY DIFFERENT ZNO NANOFERTILIZER SOURCES

The activity DHA enzyme was reflected the dynamic population of microorganism in soil (Nannipieri et al., 2012), and exactly as expected, it was greater in rhizospheric soil as compared with bulk soil. The effectiveness of ZNCPC in increasing DHA activity in the rhizosphere was probably because of positive plant-mediated feedback mechanisms caused by increased plant root activity in response to the greater Zn and P availability. More critical AP and ALP practices in the rhizosphere could be explained by the appearance of root exudates, which animated chemical exercises. Moreover, AP and ALP practices under a *Cyamopsistetra gonoloba* L. rhizosphere with usage of ZnO NPs has also been prompted point by point (Raliya and Tarafdar, 2013b). A factually noteworthy increment in Olsen-extractable-P content under ZNCPC medications in the laboratory incubation studied that ZNCPC debased inside the dirt and discharged citrate into the dirt arrangement with progress of hatching.

During the corruption of ZNCPCs in soil-solubilized local P present, citrate discharged, which thus improved Olsen-extractable P content in the soil (Mandal et al.,2019).

16.10   EFFECT OF ZNCPC ON YIELD ATTRIBUTING CHARACTERISTICS

ZNCPC mode of Zn application significantly increased yield attributing characters (plant height, number of panicles m^{-2} and panicle length) as compared to $ZnSO_4 \cdot 7H_2O$ (see Table 16.1) (Adhikary et al., 2020a; Adhikary et al., 2020b). ZNCPC improved higher DTPA Zn, DTPA-Fe, Olsen-P, as well as microbial properties, which also enhanced the yield. Grain yield and straw yield were recorded maximum under 100% ZNCPC treatment. ZnO NPs also improved the yield of grain and straw. Application of nano-ZnO significantly increased maize yield in solution culture experiment carried out by Adhikari et al. (2015).

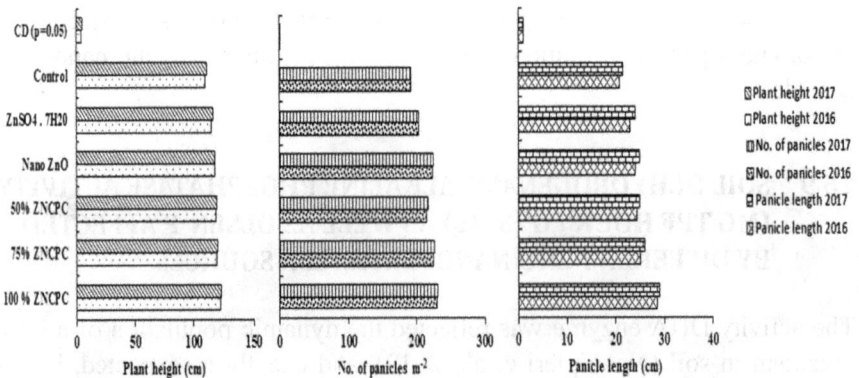

FIGURE 16.3   Effect of various treatments on height of panicle, panicles number m^{-2}, and length of panicle.

Source: Reprinted from Adhikari et al. 2015. © Taylor & Francis.

TABLE 16.1 Effect of Different Treatments on Grain and Straw Yield (Adhikary et al., 2020b)

Treatment	Grain Yield (t ha^{-1})		Straw Yield (t ha^{-1})	
	2016	2017	2016	2017
100 % ZNCPC	6.80	6.82	8.16	8.21
75% ZNCPC	6.54	6.55	7.84	7.95
50% ZNCPC	6.49	6.54	7.31	7.35
Nano-ZnO	6.45	6.47	7.74	7.73
ZnSO$_4$, 7H$_2$0	5.56	5.53	6.02	6.02
Control	4.72	4.71	5.67	5.75
CD ($p = 0.05$)	0.19	0.21	0.24	0.25
CV(%)	5.11	5.12	7.11	6.57
SEM (±)	0.09	0.08	0.11	0.15

16.11 CONCLUSION

Positive effect of ZnO NPs on fungal polysaccharides, phosphatases, and phytase secretion was documented. Microorganisms can be used effectively to produce nano-MgO and FeO. *Pochonia chlamydosporium* has formed both smaller extracellular and intracellular NPs in MgSO$_4$ and, therefore, may be a better choice than *Aspergillus fumigatus*. Biosourced synthesis of Fe NPs may be reproducible, and the resulting NPs have been further proteins, and reducing agents are stabilized by NPs and secreted by the fungus.

ZNCPC at half dose of ZnSO$_4$·7H$_2$O increased DTPA-extractable content of Zn, agronomic characteristics as well as Zn content in rice grain and straw. Application of ZNCPC at the rate of 100% observed highest Zn concentration in grain and apparent zinc recovery. Foliar spray of nano-ZnO at 40 mg kg^{-1} would be a viable option in increasing concentration of Zn in grain. DTPA-Zn, Olsen-P, DHA, and AP and ALP activities in rhizosphere soil increased with the application of ZNCPC.

16.12 FUTURE CHALLENGES

Field experiments should be undertaken to judge the efficiency of nanoclay-coated controlled fertilizer on long run in the field conditions. Studies should be undertaken for standardization of an economic method for mass production of nanoclay composite fertilizers for field application. Research is needed in the field of nanotoxicology for the safety evaluation of NPs through detailed study of the mechanisms and biokinetics causing adverse effects on organisms. Commercialization of such products is the most important future challenge. ZNCPC and ZnO NPs should be assessed in a long-term experiment in different types of soil and harvest.

KEYWORDS

- nano-ZnO
- Fourier transform infrared spectroscopy
- scanning electron microscopy
- transmission electron microscope
- X-ray powder diffraction
- zincated nanoclay polymer composite

REFERENCES

Adhikari, T., Kundu, S., Biswas, A. K., Tarafdar, J. C., and Rao, S. (2015). Characterization of zinc oxide nano particles and their effect on growth of maize (*Zea mays* L.) plant. *J. Plant Nutr.*, *38*, 1505–1515.

Adhikary, S., Mandal, N., Rakshit, R., Das, A., Kumari, N., and Kumar, V. (2020a). Nano zinc carriers influences release kinetics of zinc and iron in a laboratory incubation experiment under Inceptisol and Alfisol. *J. Plant Nutr.*, *43*, 1968–1979. doi:10.1080/01904167.202 0.1766068

Adhikary, S., Mandal, N., Rakshit, R., Das, A., Kumar, V., Kumari, N., and Homa, F. (2020b). Field evaluation of Zincated nanoclay polymer composite (ZNCPC): Impact on DTPA-extractable Zn, sequential Zn fractions and apparent Zn recovery under rice rhizosphere. *Soil and Tillage Research*, *201*, 104607. doi:10.1016/j.still.2020.104607

Alexander, M. (1977). *Introduction to Soil Microbiology*. Malabar, FL, USA: Krieger Pub Co, pp. 151–167.

Bai, H. J., Zhang, Z. M., and Gong, J. (2005). Biological synthesis of semiconductor zinc sulfide nanoparticles by immobilized Rhodobacter sphaeroides. *Biotechnol. Lett.*, *28*, 14, 1135–1139.

Cunningham D. P., and Lundie Jr. L. L. (1993). Precipitation of cadmium by Clostridium thermoaceticum. *Appl. Environ. Microbiol.* *59*, 1, 7–14.

Duhan, J. S., Kumar, R., Kumar, N., Kaur, P., Nehra, K., and Duhan, S. (2017). Nanotechnology: The new perspective in precision agriculture. *Biotechnol. Rep. 15*, 11–23.

Hall, B. D., Zanchet, D., and Ugarte, D. (2000). Estimating nanoparticle size from diffraction measurements. *J. Appl. Crystallogr. 33*, 121–132.

Kaur, P., Duhan, J. S., Thakur, R., and Chaudhury, A. (2018). Comparative pot studies of chitosan and chitosan-metal nanocomposites as nano-agrochemicals against fusarium wilt of chickpea (*Cicerarietinum L.*). *Biocatalysis Agricultural Biotechnol. 14*, 466–471. http://dx.doi.org/10.1016/j.bcab.2018.04

Mahajan, P., Dhoke, S. K., Khanna, A. S., and Tarafdar J. C. (2011). Effect of nano-ZnO on growth of mung bean (*Vignaradiata*) and chickpea (*Cicer arietinum*) seedlings using plant agar method. *Appl. Biological Res. 13*, 54–61.

Mandal, N., Datta, S. C., Manjaiah, K. M., Dwivedi, B. S., Kumar, R., and Aggarwal, P. (2018). Zincated nanoclay polymer composites (ZNCPCs): Synthesis, characterization, biodegradation and controlled release behaviour in soil. *Polymer-Plastics Technol. Eng. 57*, 1760–1770, doi:10.1080/03602559.2017.1422268.

Mandal, N., Datta S. C., Manjaiah, K. M., Dwivedi B. S., Kumar R., and Aggarwal P. (2019). Evaluation of zincated nanoclay polymer composite in releasing Zn and P and effect on soil enzyme activities in a wheat rhizosphere. *Eur. J. Soil Sci. 70*, 1164–1182. https://doi.org/10.1111/ejss.12860

Montalvo, D., Degryse, F., Silva, R. C., Baird, R., and McLaughlin, M. J. (2016). Agronomic effectiveness of zinc sources as micronutrient fertilizer. *Adv. Agron. 139*, 215–267.

Nannipieri, P., Giagnoni, L., Renella, G., Puglisi, E., Ceccanti, B., Masciandaro, G., and Marinari, S. (2012). Soil enzymology: Classical and molecular approaches. *Biol. Fertility Soils 48*, 743–762.

Raliya, R., and Tarafdar, J. C. (2013a). Biosynthesis of gold nanoparticles using Rhizoctonia bataticola TFR-6. *Adv. Sci., Eng. Med. 5*, 1–4.

Raliya, R., and Tarafdar, J. C. (2013b). ZnO nanoparticle biosynthesis and its effect on phosphorous-mobilizing enzyme secretion and gum contents in clusterbean (*Cyamopsistetra gonoloba* L.). *Agricultural Res. 2*, 48–57.

Tarafdar, J. C., Agrawal, A., Raliya, R., Kumar, P., Burman, U., and Kaul, R. K. (2012a). ZnO Nano-particles induced synthesis of polysaccharides and phosphatases by Aspergillus fungi. *J. Adv. Sci., Eng. Med. 4*, 324–328.

Tarafdar, J. C., and Raliya, R. (2012b). Rapid, low-cost, and ecofriendly approach for iron nanoparticle synthesis using *Aspergillus oryzae* TFR9. *Nanotechnology 2013*, 141274.

Wisawapipat, W., Janlaksana, Y., and Christl I. (2017). Zinc solubility in tropical paddy soils: A multi-chemical extraction technique study. *Geoderma 301*, 1–10. doi: 10.1016/j.geoderma.2017.04.002.

Yang, H., Santra, S., and Holloway P. H. (2005). Syntheses and applications of Mn-doped II-VI semiconductor nanocrystals. *J. Nanosci. Nanotechnol. 5*, 9, 1364–1375.

CHAPTER 17

Hydrogel Formulations for Increasing Input Use Efficiency in Agriculture

DHRUBA JYOTI SARKAR[1*] and ANUPAMA SINGH[2]

[1]*ICAR-Central Inland Fisheries Research Institute, Kolkata, West Bengal 700120, India*

[2]*ICAR-Indian Agricultural Research Institute, New Delhi, Delhi 110012, India*

Corresponding author. E-mail: dhruba1813@gmail.com

ABSTRACT

Hydrogels are hydrophilic crosslinked polymeric structures with versatile properties. There are numerous ways to synthesize hydrogel materials, namely physical crosslinking involving freeze and thawing, ionic interaction, etc., and chemical crosslinking involving free radical polymerization, esterification, etherification, etc. Based on the chemical structure, hydrogels can be classified into homopolymeric hydrogels, copolymeric hydrogels, interpenetrating hydrogels, etc.. Due to their polymeric versatility, they have been in use as carrier for various bioactive materials, including drugs, nutraceuticals, and agricultural inputs such as fertilizers, pesticides, etc.. Hydrogels application alone finds its superiority in enhancing water use efficiency in water stress farming. Furthermore, fertilizer- and pesticide-encapsulated hydrogels leads to slow release of the same leading to less environmental pollution and judicious use of agricultural inputs. Despite their well-recognized role in improving water use efficiency and fertilizer application explored in individual research and academic pieces of work, the hydrogels still find difficulty in finding a widespread adoption in tropical agriculture, particularly in India.

17.1 INTRODUCTION

Agriculture continues to be the major consumer of available fresh water resources, with irrigation holding 70% of the total share (FAO, 2017). However, in developing countries such as India, the crop productivity per unit applied water is still less and is a major cause of concern for the resource-stressed farmers, researchers, and policy makers (Sharma et al., 2018). Along with it, issues such as low nutrient use efficiency and ever building pesticide load in the environments need to be addressed properly to gain profitable agriculture, keeping health of natural ecosystem in a sustainable manner (Kumar et al., 2017; Schütz et al., 2018). In this regard, application of innovative water management strategies, including integrated water and nutrient management plan, is of urgent importance. Hydrogels, a special class of hydrophilic polymers with distinct three-dimensional (3-D) matrix properties, have been reported extensively to fulfill the criteria of innovative tools for enhancing water use efficiency (Chu et al., 2013; Ahmed, 2015). They have also been reported as smart carriers of fertilizers, pesticides, and others for their controlled delivery mechanism (Sarkar et al., 2017; Sarkar et al., 2019a). Besides this, hydrogel material was reported to reduce the leaching loss of pesticide in soil column, thus reducing the probability of ground water contamination upon application of pesticides (Sarkar et al., 2012).

Hydrogels are synthetic or biopolymeric networks, which swell extensively in water (Chu et al., 2013; Ahmed, 2015). In physical terms, hydrogel can be defined as a hydrophilic polymeric network, which is sometimes found as colloidal gels in which water is dispersed in the solid polymeric matrix forming a 3D structure (Enas et al., 2013). It can also be defined as a water-swollen crosslinked polymeric network produced by polymerization reaction of one or more monomers. The material has been receiving tremendous attention in the present time due its wide range of application in the field of cosmetics, pharmaceuticals, surgical instruments, wound dressing, etc. (Klouda and Mikos, 2008). However, despite its promising application in the delivery strategies of bioactive substances, its application is very limited in the agriculture and is in very nascent stage. The property of water absorption or water holding of hydrogel comes from the hydrophilic functional groups attached to the polymeric backbone, and the 3D structure of the swollen materials comes from the crosslinks between network chains, which resist their dissolution (Enas et al., 2013). The crosslinking of the hydrophilic polymeric network can be either covalent bonds or physical crosslinks arising from entanglements, association bonds, or strong van der Waals interactions between chains or crystallite bringing together two or more macromolecular chains. Recently, there have been shift from use of synthetic hydrogel to natural hydrogels with high absorption capacity and mechanical strength (Vashist et al., 2014). However, synthetic hydrogels are having long service life with a well-defined structure that can be tailor made with controlled degradability and functionality. Though there are numerous reports on the hydrogels, its synthesis, and application, works lack on the documentation on the types of agriculturally suitable hydrogels and their synthetic protocols. This chapter presents the chemistry of various suitable hydrogel materials suitable for agricultural application. The application of hydrogel materials in enhancing the water productivity in agriculture and their role in the minimizing the fertilizer loss and better nutrient use efficiency is also discussed. Furthermore, this chapter discuses briefly the ecosafety issues of hydrogels application in agricultural with respect to their various chemical compositions.

17.2   CLASSIFICATION OF HYDROGELS

Hydrogels can be classified based on the method of preparation, ionic charge or physical structure and appearance, source, sensitivity, etc. Few of the classifications have been discussed here.

17.2.1   CLASSIFICATION BASED ON THE PHYSICAL STATE

Based on the physical state, hydrogels can be classified into chemical and physical hydrogels. Chemical hydrogels are the covalently crosslinked gel matrices and are permanent in nature (Hennink and Nostrum, 2002). The attainment of the swelling equilibrium in an aqueous medium by this type of hydrogels depends on the polymer–water interaction and the crosslink density of the hydrogel matrix (Rosiak and Yoshii, 1999). On the contrary, when the hydrophilic polymeric chains are held together by various physical forces such as molecular entanglements, ionic bonding, hydrogen bonding, hydrophobic interaction, etc., the physically crosslinked hydrogel network is formed (Hennink and Nostrum, 2002). Since these physical forces are reversible, they can be transformed into a sol system in various physiochemical environments or during application of stress.

17.2.2   CLASSIFICATION BASED ON SOURCE

Based on sources, the hydrogels can be classified into synthetic and natural hydrogels (Kulicke and Nottelmann, 1989). Synthetic hydrogels are prepared from synthetic monomers such as acrylic acid, acrylamide, methacrylic acid, and many others (Ratner and Hoffman, 1976). Synthetic linear chain polymers such as poly(ethylene glycol), poly(2-hydroxypropyl methacrylamide), poly(vinyl alcohol) (PVA), and poly(hydroxyethyl methacrylate) have also been used to synthesize synthetic hydrogels (Mathur et al., 1996).

Natural polymers such as collagen, gelatine, fibrin, silk, agarose, hyaluronic acid, chitosan, dextran, cellulose derivatives, and alginates are being used for natural hydrogel preparation (Gyles et al., 2017). The natural hydrogels are biocompatible and easily biodegrade in nature; however, the main disadvantage of these hydrogels is batch-to-batch variable physical and chemical properties based on the kind of natural polymer used. Furthermore, sometimes, fine structural modification of hydrogels is not possible due to the complex structure of natural polymers.

17.2.3   CLASSIFICATION BASED ON CRYSTALLINE NATURE, ELECTRIC CHARGE, AND PHYSICAL APPEARANCE

Based on the crystalline nature, the hydrogels can be classified into amorphous, semicrystalline, and crystalline hydrogels (Hu and Huang, 2003). Even it was reported that hydrogels are possible with tunable crystalline and amorphous properties. Based on the electric charge, hydrogels can be classified into four groups: nonionic (neutral), ionic (anionic and cationic), amphoteric (ampholytic), and zwitter-ionic hydrogels (Zhang et al., 2013). Based on the physical structural appearance, hydrogels can be classified into films, microsphere, porous, and other 3-D hydrogels.

17.2.4   CLASSIFICATION BASED ON RESPONSIVE PROPERTIES

Hydrogels show responsive behavior to the surrounding environment by changing its swelling properties, shape, etc., depending on its chemical structure. These changes are brought by external stimuli such as pH, temperature, lights, and ionic strength (Kim et al., 2010). Thus, based on the responsive behaviors, the hydrogels can be classified too. Ionic hydrogels display marked changes in the swellability with changes in pH of external swelling medium as compared to nonionic hydrogels (see Figure 17.1) (De et al., 2002). Similarly, thermoresponsive hydrogels show changes in the physical and chemical properties in response to temperature (Klouda and Mikos, 2008). Besides these, many other responsive hydrogels are reported such as hydrogels sensitive to chemical or enzymatic reaction (Aimetti et al., 2009), magnetically responsive hydrogels (Popa et al., 2016), mechanical-stress-responsive hydrogels (Das et al., 2016), light- or radiation-sensitive hydrogels (Zhu et al., 2012), etc..

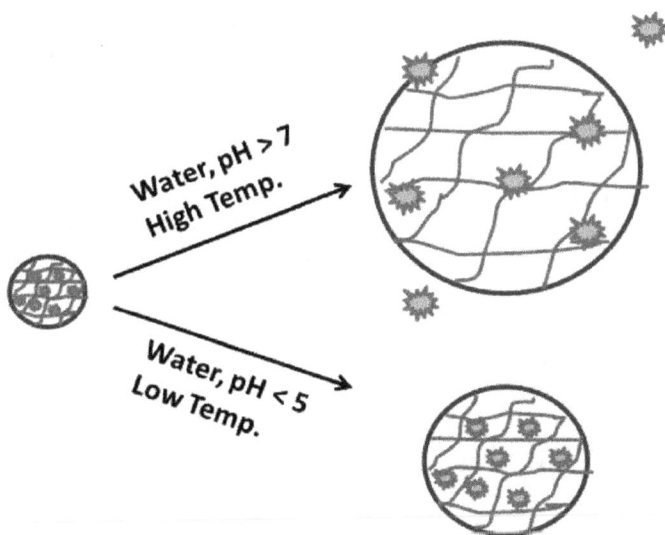

FIGURE 17.1   Schematic of a stimuli responsive hydrogel.

17.2.5   *CLASSIFICATION BASED ON THE METHOD OF PREPARATION*

Based on the synthesis method, the hydrogels can be categorized into homopolymeric hydrogels, copolymeric hydrogels, and interpenetrating hydrogels.

1. *Homopolymeric hydrogels*: This type of hydrogels is prepared with single kind of monomer. The single monomeric species build up the entire polymeric network of the hydrogel; however, homopolymer may have a chemically crosslinked structure based on the nature of the monomer and polymerization technique. Methylene bis acrylamide crosslinked acrylic acid or acrylamide is a typical example of

this type of hydrogel (Katima et al., 1999). Similarly, homopolymeric
hydrogel can be prepared with many other monomers, for example,
methyl methacryalate, vinyl pyrrolidone, etc.

2. *Copolymeric hydrogels*: This type of hydrogel is prepared with two or
    more different types of monomeric species arranged in either random
    or block configuration (Huglin et al., 1986) (see Figure 17.2).

3. *Interpenetrating polymeric hydrogel (IPN)*: IPN is commonly
    defined as a close combination network of two polymers, at least one
    of which is synthesized or crosslinked in the immediate presence of
    the other (Zhang and Peppas, 2000). In general, the hydrogels of this
    type are made of two different crosslinked synthetic and/or natural
    polymers, contained in a network form (see Figure 17.3). In the semi-
    IPN hydrogels, one component is a crosslinked polymer, while the
    other component is a noncrosslinked polymer (Zhang et al., 2009).
    The interlock structure of the crosslinked IPN component provides
    mechanical stability to the structure. The IPNs are synthesized with
    the objective to produce hydrogels with stiffer and tough mechanical
    properties, controllable chemical and physical properties, etc.

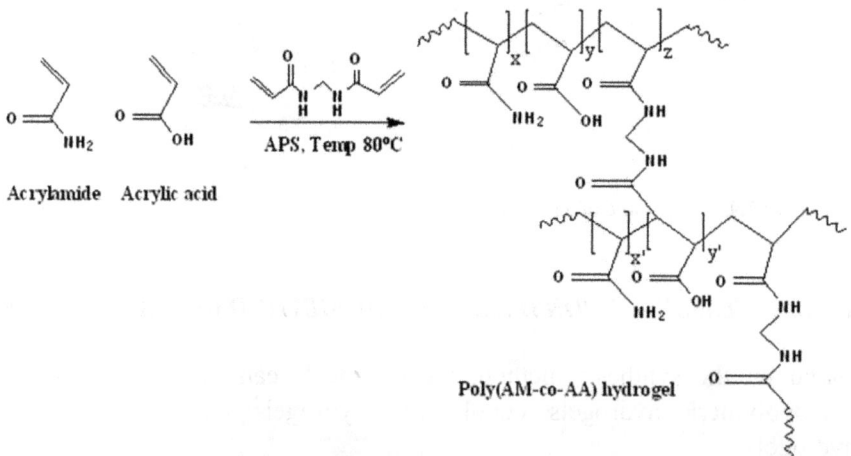

FIGURE 17.2   Copolymeric hydrogels synthesized from acrylamide and acrylic acid.

17.3   METHODS OF HYDROGEL SYNTHESIS

Hydrogels can be prepared either through the physical crosslinking method or
through the chemical crosslinking method, which is initiated in the presence

of a chemical initiator or UV or any other radiation (Peppas et al., 2000). Generally, the chemical crosslinking process involves the reaction of one or more bifunctional molecules, called crosslinkers, with the small molecular weight monomer/s. The process can also link two or more macromolecular polymeric chains through chemical functional groups. Copolymerization along with crosslinking of monomer also leads to the synthesis of hydrogels. The crosslinking reaction can be initiated in the presence of UV light or ionizing radiation (Chapiro, 1962).

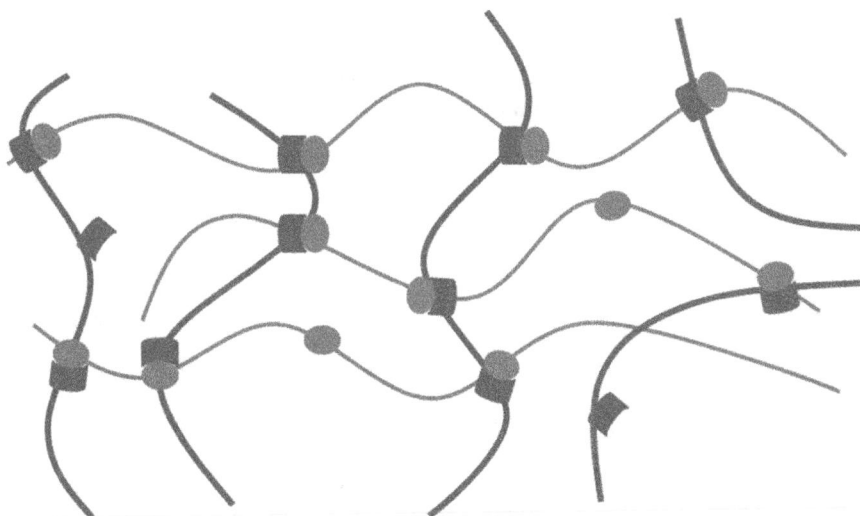

FIGURE 17.3   Schematic of interpenetrating polymeric hydrogel.

17.3.1 PHYSICAL CROSSLINKING

The process involves synthesis of hydrogel without any chemical cross-linking, and the process is generally followed in the food industries. Some of the physical crosslinking methods are the following.

17.3.1.1 HEATING/COOLING A POLYMER SOLUTION

Some polysaccharides, namely, gelatine and carrageenan, form physically crosslinked hydrogels by cooling of their hot solution (Deng et al., 2018). The gel-forming mechanism during cooling involves occurrence of helix formation in the polymeric chain followed by the association of the helices and the formation of the junction zones (Funami et al., 2007). The helices aggregate in the presence of any cations (K^+, Na^+, etc.) to form the stable gels.

17.3.1.2   IONIC INTERACTION

In this approach, physically crosslinked hydrogels are prepared by cross-linking of ionic polymers by the addition of di- or trivalent counter ions, namely. Ca^{2+}, Mg^{2+}, Al^{3+}, etc.. Sodium alginate gels crosslinked by Ca^{2+} ions are typical examples of the process (Mandal et al., 2010). Hydrogels can also be prepared by mixing polyanion with polycation. Two polymers with opposite charges bind together to form a complex that can be soluble or insoluble depending on the concentration and pH of the respective solutions (see Figure 17.4). Polyanionic xanthan coacervates with polycationic chitosan to form hydrogel (Chornet and Severian, 2005).

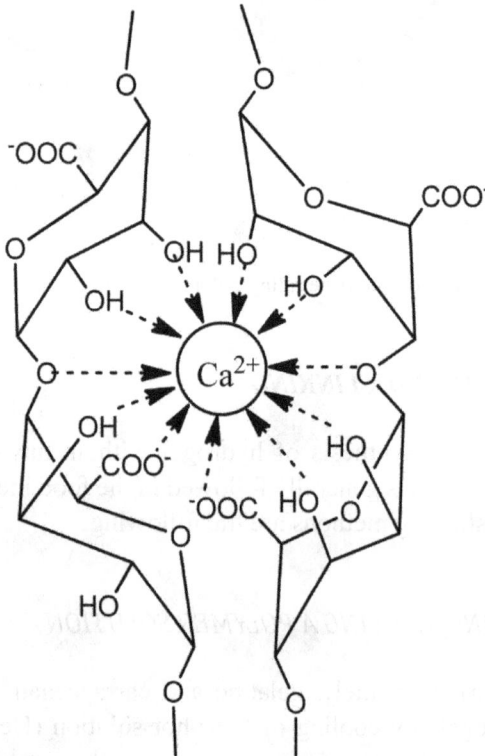

FIGURE 17.4   Ionotropic gelation of sodium alginate with calcium ions.

17.3.1.3   HYDROGEN BONDING

Some cellulosic derivatives, namely, carboxymethylcellulose, form hydro-gels by lowering the pH of the aqueous solution (Takigami et al., 2007). In the low-pH solution, the sodium ions in carboxymethylcellulose are replaced with hydrogen to promote hydrogen bonding, which decreases the polymer solubility and results in the formation of an elastic hydrogel (see Figure 17.5).

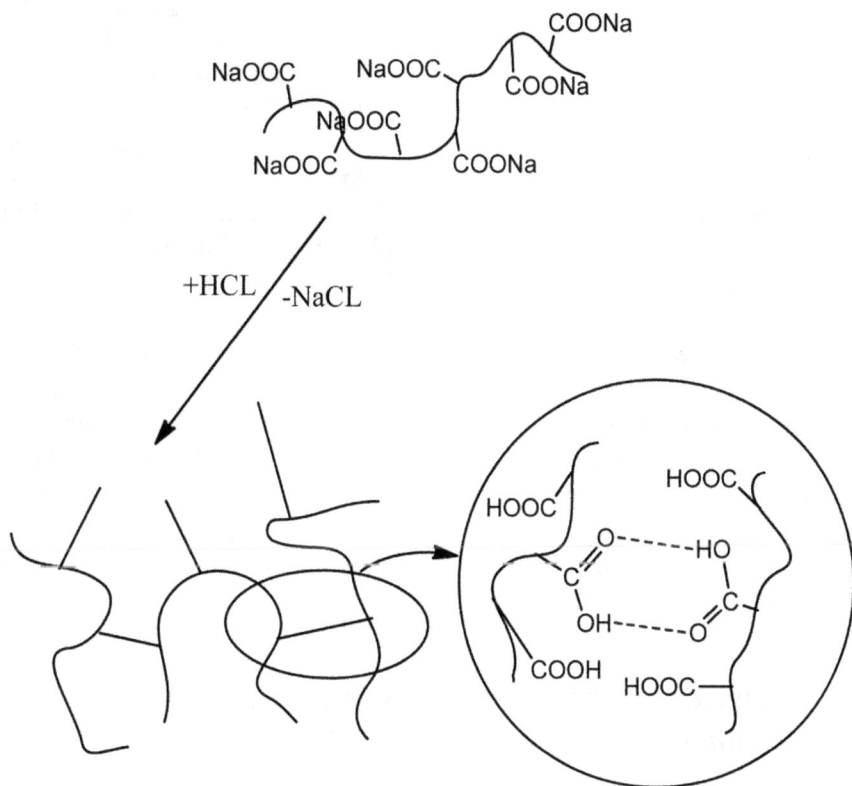

FIGURE 17.5   Hydrogel formed by hydrogen bonding in sodium salt of carboxy methyl cellulose.

17.3.1.4   FREEZE THAWING

This method involves physical crosslinking of a polymer, namely, PVA, xanthane, etc., to transform into hydrogel by freeze-thaw cycles (Shiroodi et al., 2015). Disadvantages associated with these types of hydrogels are dissolution of polymer chain, melting out of crystallites, and production of additional or secondary crystallization over a longer period of time (Hassan and Peppas, 2000).

17.3.2 CHEMICAL CROSSLINKING

This process involves polymerization of monomers or copolymerization of polymeric chains in the presence of a suitable crosslinking agent (Crescenzi et al., 2002). It can also be done while grafting of monomers on the backbone of a polymer. In the free radical polymerization reaction, crosslinking is introduced by the addition of small amount of bifunctional monomers (Pourjavadi et al., 2004). A typical example is synthesis of polyacrylamide hydrogels, where acrylamide monomer along with a small amount of bifunctional crosslinker molecules, methylene-bis-acrylamide, is reacted in the presence of a chemical initiator (persulfate initiator) (Sarkar et al., 2015). The crosslinking of polymers can also be achieved through the reaction of functional groups (such as OH, COOH, and NH_2) with crosslinkers having aldehyde moiety such as glutaraldehyde and adipic acid dihydrazide (Bouhadir et al., 1999).

Crosslinkers such as glutaraldehyde, epichlorohydrin, and N,N-methylene bisacrylamide are being widely used to obtain hydrogels of various synthetic and natural polymers. Carboxymethyl cellulose and acrylic monomer can be crosslinked in the presence of methylene-bis-acrylamide leading to formation of biodegradable semisynthetic hydrogels for fertilizer delivery systems (Sarkar et al., 2015, 2019; Singh et al., 2019).

17.4 SUPERABSORBENT HYDROGELS AS TOOL FOR WATER MANAGEMENT IN SCARCITY ZONES

Use of polymer in agriculture is abundant and found three major applications: the covering of greenhouse, mulch film, and controlled-release fertilizer formulations (Sarkar et al., 2018). Use of hydrogels or superabsorbent polymers (SAPs) in agriculture was first initiated by the United States Department of Agriculture (USDA) in the late 1960s, and commercial use of these materials in agriculture and horticulture was first initiated in the early 1970s in the form of starch/acrylonitrile/acrylamide-based hydrogels (Nnadi, 2012). It was reported that in 2005, the total production of SAPs in Asia, North America, and Europe was 623,000, 490,000, and 370,000 tons

(Zohourian and Kabiri, 2008). The major use of these materials is fixed to public hygiene and biomedical use.

Different precision irrigation tools such as sprinkler, dripper, etc., have been in recent decade for effective water management in the water scarce agricultural system. These tools have been proved as most effective for enhancing crop productivity with minimized use of irrigation water. However, the only disadvantage of these tools is primarily high cost involvement (Rad et al., 2018). Another way out is to alter the water-holding capacity of soils using soil amendment materials (Mangrich et al., 2015). These materials alter the water-holding capacity of the soils, particularly in the rhizosphere zone, by changing the hydraulic conductivity, which is the measure of ability of the soil to transmit water. Various types of soil amendment materials are available at present in the market, among which SAP material or superabsorbent hydrogel is of novel approach (Chandrika et al., 2017; Lejcuś et al., 2018). High water-absorbing and water-holding capacity of hydrogel materials results from the electrostatic repulsion between charges on the polymer chains and osmotic imbalance between the interior and the exterior of the polymer (Zohourian and Kabiri, 2008). Additionally, certain functional groups in the polymeric chain also form hydrogen bonds with water molecules. Use of superabsorbent material as soil amendment was first developed by the USDA by alkaline hydrolysis of starch-graft-polyacrylonitrile in the 1960s (Nnadi, 2012). Subsequently, there have been reports of many SAPs for application in agriculture. The high water-absorbing and water-holding properties of SAP lead to significant decrease in saturated hydraulic conductivity upon mixing them with soil (Narjary et al., 2011). Upon absorbing water, the hydrogel particles swell, leading to reduction of large pores in the soil especially in the sandy soils. It was reported that an increase in both saturated and residual water contents, water-holding capacity, and available water content has been observed due to hydrogel addition in loose textured soils (Dorraji et al., 2010; Singh et al., 2011). Similarly, the water-holding capacity of soil less media was also reported to get enhanced by the addition of hydrogel. Addition of biopolymeric hydrogel and hydrogel clay composite at 0.5% and 0.75% level was reported to significantly enhance the water-holding capacity of sandy loam soil and soilless media (Singh et al., 2011) (see Figure 17.6). Hydrogel clay composite was reported to show greater water-holding capacity due to high water absorbency under load (Singh et al., 2011). The release of absorbed water from the swollen hydrogel matrix in the soil environment is a very important criterion. If water release from the hydrogel is restricted, then the plant will not get sufficient available water, and the purpose of addition of hydrogel will be defeated. It can be seen from Figure 17.7 that hydrogel–clay composite retains more water at a given pressure as compared to hydrogel both at soil and soilless media. However, the moisture retainment effect is more pronounced in soil rather than in soilless media. This implies that hydrogel is better in releasing plant available water in soil than hydrogel clay composite, whereas both the materials perform at par in soilless media.

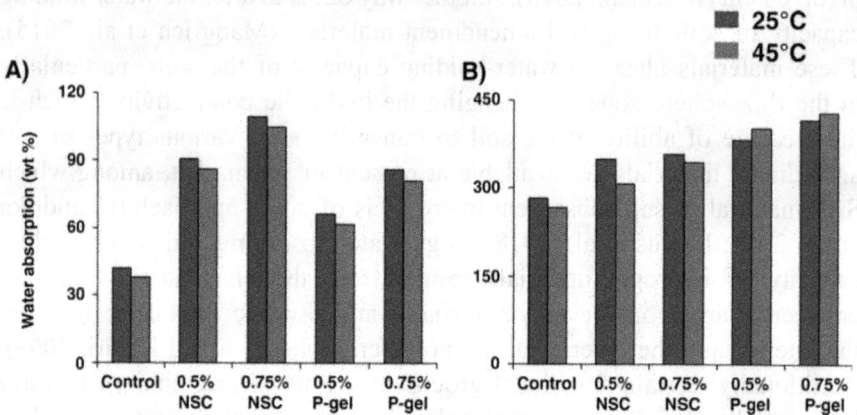

FIGURE 17.6 Water absorption capacity of sandy loam soil (A) and soilless media (B) mixed with different doses of biopolymeric hydrogel (P-gel) and biopolymeric hydrogel clay composite (NSC). Source: Singh et al., 2011. Reprinted by permission of Wiley Online Library (http://onlinelibrary.wiley.com).

It was described that the beneficial effect of hydrogel application to plant depends upon the type of application method being used (Narjary et al., 2011). In nursery preparation, application of hydrogel into growing media or soil was reported to reduce the loss of water through evaporation and leaching (Landis and Haase, 2012). It was further noticed that hydrogel particles could retain the nutrient ions that could have leached out of the root zone (Mikkelsen, 1994). During the root dipping method, hydrogels are able to coat fine roots and, thus, protect them from desiccation (Thomas, 2008). The hydrogel dipping of roots mimics the process of the natural polymeric mucilage production by the healthy roots. Mucilage production by roots was reported to weaken the water potential at the root–soil interface, thus enhancing the water flow between soil and roots leading to the reduction of energy needed for the water uptake by the plant (Carminati and Moradi, 2010). Sowing of seed mixed with hydrogel will lead to retain of moisture around the germinating seed and, thus, improves the establishment of the seedlings in nurseries as well as in the field. It was found that application of hydrogel in the root zone during transplanting operation is much more effective than band or layer placement of hydrogel (Nirmala and Guvvali, 2019). For agricultural application, the ideal characteristics of an hydrogel material found to be high absorption capacity (maximum equilibrium swelling) in saline and hard water conditions, optimized AUL, lowest soluble and residual monomer content, low price, high durability and stability in the swelling environment and during the storage, gradual biodegradability without formation of toxic species, pH neutrality after swelling in water, photostability, and rewetting capability (Singh et al., 2011).

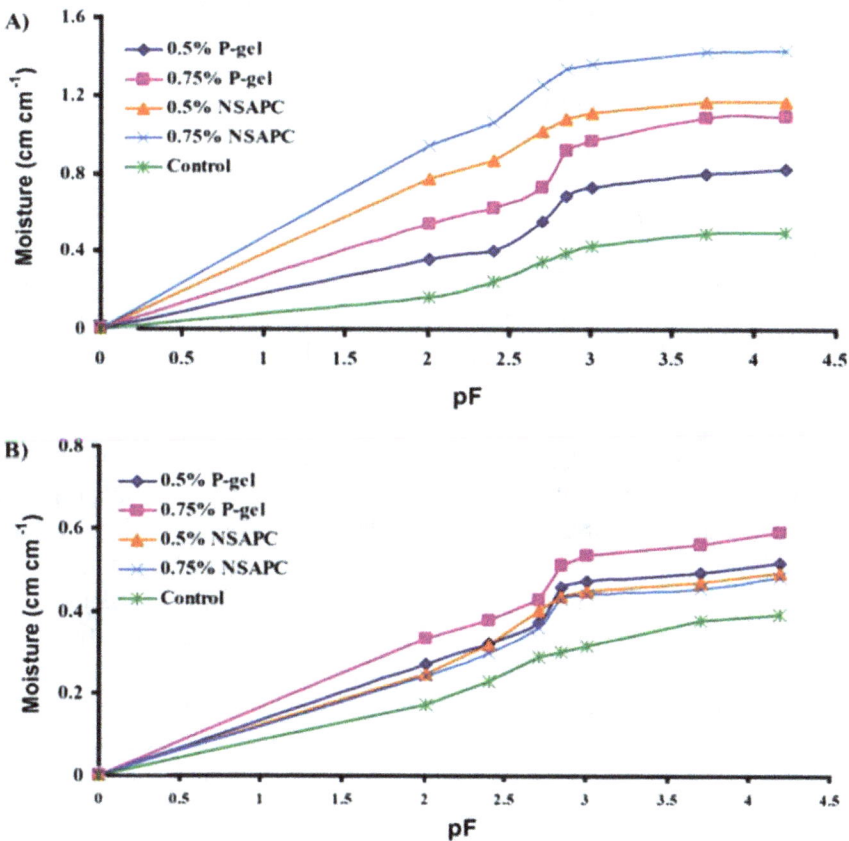

FIGURE 17.7 Moisture release curves sandy loam soil (A) and soilless media (B) mixed with different doses of biopolymeric hydrogel (P-gel) and biopolymeric hydrogel clay composite (NSC). Source: Singh et al., 2011. Reprinted by permission of Wiley Online Library (http://onlinelibrary.wiley.com).

Application of hydrogel for soil amendment has proven to be beneficial in many respects; some of them are listed as follows:

1. Amendment with hydrogel in soil reduces the irrigation frequency particularly in coarse-textured soils. Thus, their application is most suitable for water management practices in arid and semiarid regions (Abedi-Koupai et al., 2008; Dass et al., 2013).
2. Application of hydrogels leads to reduction of irrigation-induced soil erosion and water seepage (Dorraji et al., 2010).
3. Plant survivability increases in hydrogel-amended soil, and the fact was demonstrated in tree seedlings and transplanted trees in arid regions of China, Africa, and Australia (Hüttermann et al., 2009).

17.5 HYDROGEL-BASED FERTILIZER FORMULATION APPROACHES FOR SMART DELIVERY

Fertilizers are recognized as essential agro-inputs, which are required in large quantities (Rao et al., 2006). The large requirement of fertilizers is due to poor nutrient use efficiency by the crop plant and loss of large amount of nutrients due to various environmental processes such as leaching, volatilization, etc. (Singh et al., 2012). For this reason, there has been a great interest on the development of smart delivery tools to release nutrients to plant root zone without losing subsequent amount to the environment (Shaaban et al., 2017). The inefficient fertilizer use has an environmental pollution aspect too. One of the most alarming problems is nutrient leaching and subsequent pollution of ground water. To solve the above problems, development of controlled-release formulation of fertilizers (CRFF) using various polymers as carrier matrix may produce a viable option (Jarosiewicz and Tomaszewska, 2003). CRFFs allow the release of nutrients to the root zone in synchronization with requirement of the plant (Malhi et al. 2010). Thus, it increases the nutrient uptake efficiency by plants. Furthermore, the loss of nutrients to the environment also gets reduced, and it improves the benefit cost ratio of applied fertilizers. Among the various CRFF approaches, hydrogel-based fertilizer formulations are being viewed with lot of interest (Teodorescu et al., 2009). Because of the tunable properties of the hydrogel, the material property can be tailored designed and can express environment-responsive properties. In pharmaceutical science, this property of hydrogel has been extensively studied and being applied for mainly targeted drug delivery (Vaghani et al., 2012). Similarly, this property can be easy induced on the hydrogel-based

CRFF, thus achieving total synchronization of nutrient release behavior in connection to external environment of root zone.

It was reported extensively that to enhance the barrier property in the hydrogel matrix, generally various inorganic filler materials such as clay or nanoclay were introduced (Bounabi et al., 2016). The improvement of barrier properties had led to more slow release of nutrient from the hydrogel matrix. Incorporation of kaoline nanopowder into SAP was reported to decrease the diffusion coefficient of urea release from urea-loaded hydrogel composite (Liang and Liu, 2006). Similarly, incorporation of nanosized clay into superabsorbent polymer (NCPC) matrix decreased the rate of nutrient release from fertilizer-loaded NCPC in distilled water (Sarkar et al., 2014). Furthermore, it was demonstrated that fertilizer-loaded NCPC can reduce the loss of N and P and, hence, maintain better availability of P and N in the soil (Sarkar et al., 2014). Biochar encapsulated with polymeric alginate beads was used as slow release formulation of phosphorus, which showed extended release as compared conventional fertilizer (Domingues et al., 2015). The formulation was showed better minimization of phosphorous loss through leaching leading to less water pollution and eutrophication. Not only major nutrients, hydrogel-based micronutrients (Zn^{2+} and BO_3^{3-}) was reported to demonstrate sustained release of micronutrients from hydrogel matrix over extended periods of time (Sarkar et al., 2016, 2019). However, it was also exhibited that due to incorporation of cationic nutrient species, the water-absorbing capacity of the hydrogel got reduced as compared.

Slow release urea formulation was developed by encapsulating urea with superabsorbent with dual purpose of introducing delivery of hydrogel for moisture preservation (Sarkar et al., 2019). Urea granule was coated with hydrogel through in situ crosslinking of acrylic acid with different concentration of crosslinker, that is, methylenebisacrylamide in the presence of heat (see Figure 17.8) (Sarkar et al., 2019). The hydrogel granules showed both high water absorbency and slow urea release properties. The kinetic parameter of urea release from these coated urea granules was studied and predicted the mechanism using different mathematical models, namely. first-order kinetics, Ritger–Peppas model, Gallagher–Corrigan model, and Sigmoid function, which showed that urea release followed typical sigmoid function with sudden release (see Table 17.1). Similarly, a coated nitrogen fertilizer with slow release and water retention was prepared (Liang et al., 2006). A double-layer polymeric system, inner-layer urea-formaldehyde and outer-layer crosslinked poly(acrylic acid)/organ-attapulgite, was used to coat urea granules for slow release of urea. Similarly, a crosslinked calcium-alginate-(inner layer) and ammonium-acrylate-based superabsorbent (outer layer) was used to coat urea-formaldehyde for extreme slow release of urea along with delivery of superabsorbent for moisture preservation (Guo et al., 2005).

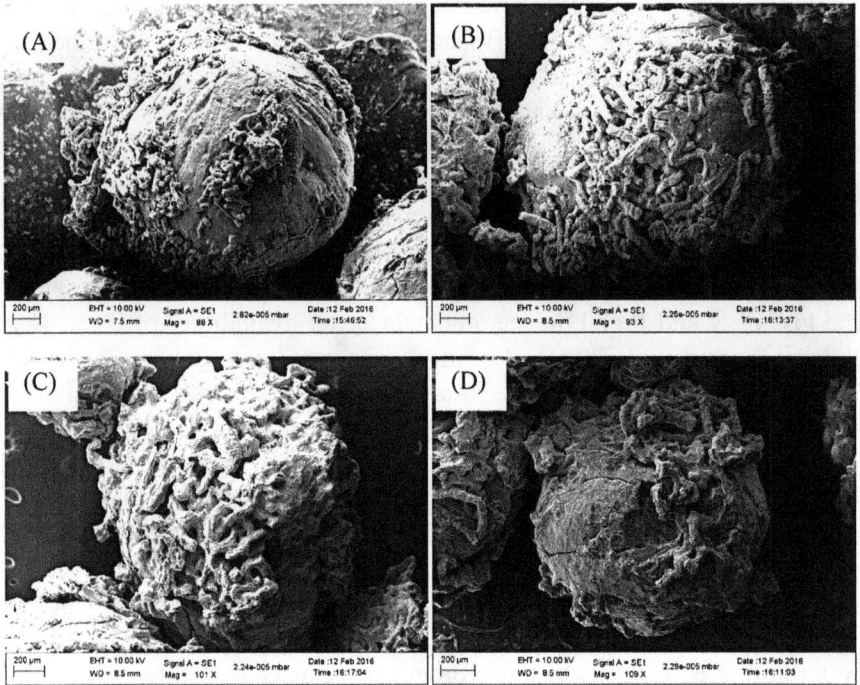

FIGURE 17.8 SEM images of urea granules coated with different concentrations of crosslinking agent (methylenebisacrylamide) 0.002 g/10 g urea (A) 0.004 g/10 g urea, (B) 0.006 g/10 g urea, (C) and 0.008 g/10 g urea, (D) Source: Sarkar et al., 2019. Reprinted by permission of Taylor & Francis Ltd, http://www.tandfonline.com.

Slow release nitrogen fertilizer was also reported to be produced by dissolving urea in acrylic acid followed by crosslinking with N,N'-methylene bisacrylamide (Liu et al., 2006). A novel trilayered controlled-release nitrogen, phosphorous, and potassium fertilizer based on hydrogel was reported that releases 84 ± 18, 63 ± 12, and $36 \pm 15\%$ of the N, P, and K nutrients, respectively, after one-month water immersion (Noppakundilograt et al., 2015). The release phenomena of the N, P, and K nutrients from the hydrogel formulation obeyed a pseudo-Fickian diffusion mechanism. Similarly, zeolite clay was used in the biopolymeric hydrogel composite to introduce more barrier properties for phosphate release (Singh et al., 2019). The scanning electron micrograph of the phosphate-loaded zeolite

TABLE 17.1 Kinetic Parameters of Urea Release From Hydrogel-Coated Urea Granule With Different Concentration of Crosslinker, BCPCU-1 (0.002 g/10 g Urea), BCPCU-2 (0.004 g/10 g Urea), BCPCU-3 (0.006 g/10 g Urea), and BCPCU-4 (0.008 g/10 g Urea) in Neutral Water (pH 7.0)

Compositions	First Order			Ritger–Peppas				Gallagher–Corrigan					Sigmoid function			
	k (min)	n	R^2	$(M_f/M_o)_{max}$	k_f	R^2	f_b	k_1	k_2	t_{2max}	R^2	$(Mt/M_o)_b$	$(Mt/M_o)_{max}$	k_s	m	R^2
Urea	0.60	0.16	0.64	1.00	1.00	0.99	0.99	0.46	1.00	0.99	0.99	0.13	0.87	0.50	1.39	0.99
BCPCU-1	0.33	0.31	0.75	1.04	1.04	0.93	0.92	0.12	4.68	7.65	0.95	0.16	0.83	0.53	21.05	0.99
BCPCU-2	0.24	0.38	0.83	1.04	1.04	0.97	0.92	0.09	4.88	7.59	0.98	0.04	0.95	0.21	2.87	0.99
BCPCU-3	0.22	0.40	0.86	1.02	1.02	0.98	1.02	0.10	2.57	-3.48	0.98	-0.03	1.02	0.13	1.45	0.99
BCPCU-4	0.16	0.48	0.87	1.06	1.06	0.97	1.02	0.08	3.02	-2.10	0.98	0.002	0.99	0.15	2.72	0.99

Source: Sarkar et al., 2019.

biopolymeric hydrogel is presented in Figure 17.9. It can be seen that introduction of zeolite has generated a porous structure in the hydrogel matrix,
and moreover, the pore structure enhances with higher amount of zeolite.
The release of phosphate from the hydrogel matrix shows pH-responsive
behavior; at lower pH (4.0), the phosphate release was less as compared to
neutral pH (7.0) (see Figure 17.10). At neutral pH, the phosphate release
from the hydrogel composite followed the Gallagher–Corrigan equation,
indicating a "brust release" phenomenon at the initial stage in addition to a
simple diffusion mechanism (see Table 17.2).

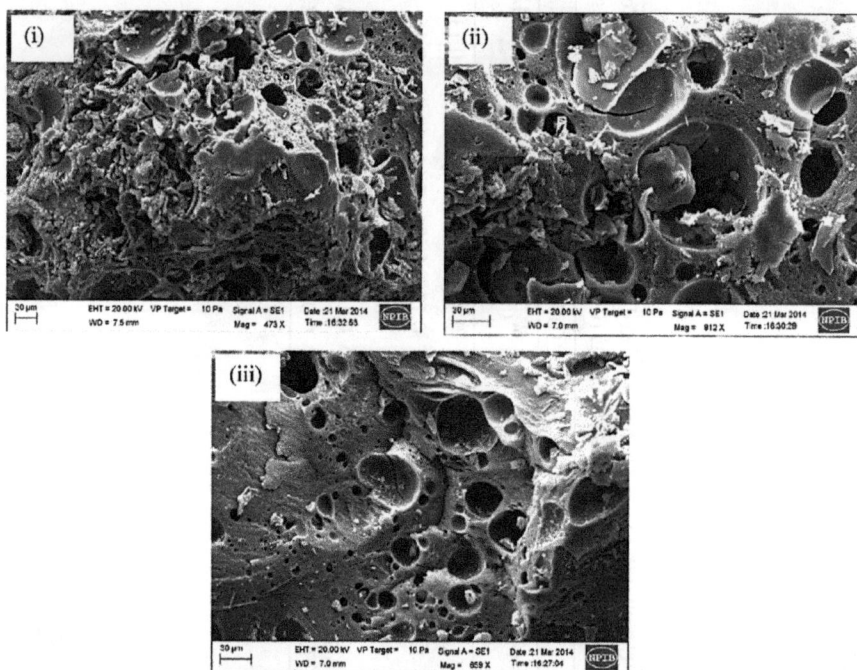

FIGURE 17.9 SEM images of phosphate formulation based on hydrogel/zeolite composite
with different concentrations of zeolite [(i) 25%, (ii) 10%, and (iii) 5%]. Source: Singh et
al., 2019. Reprinted by permission of Wiley Online Library (http://onlinelibrary.wiley.com).

17.6 ECOSAFETY ISSUES OF HYDROGEL APPLICATION IN AGRICULTURE

Use of hydrogel in agriculture is lagged behind owing to the nonavailability
of agriculturally suitable/designed products and inability of conventional

TABLE 17.2 Kinetic Parameters of Phosphate Release From Hydrogel/Zeolite Composite With Different Concentration of Zeolite (ZMHC-1, 25%; ZMHC-2, 10%; and ZMHC-1, 5%) in Neutral Water (pH 7.0)

Composites	First Order			Korsmeyer–Peppas				Gallagher–Corrigan				
	F_{max}	k_1 (day⁻¹)	R^2	K (day⁻ⁿ)	n	R^2	Mechanism	f_B	k_1 (Day⁻¹)	F_{max}	t_{2max} (Day)	k_2 (Day⁻¹) R^2
ZMHC-1	0.49	18.72	0.97	0.19	0.11	0.96	Fickian diffusion	0.75	0.553	0.51	1.93	2.70 0.88
ZMHC-2	0.36	9.18	0.83	0.29	0.09	0.91	Fickian diffusion	0.08	18.22	0.24	4.95	0.49 0.99
ZMHC-3	0.24	6.79	0.86	0.46	0.03	0.96	Fickian diffusion	0.41	23.57	0.46	8.70	8.00 0.99

Source: Singh et al., 2019.

2

1

2

SAPs to perform under harsh agro-climatic condition, particularly in arid and semiarid tropical conditions. Off late, the constraints have been addressed with success, and the use of hydrogels in agriculture is expected to get a big boost in the near future not only as water reservoir but also as controlled-release carrier for agro-inputs. Since the basic ingredients are same as like employed in the synthesis of hydrogels for medical and public hygiene, the safety issue of these new products for agriculture is not likely to be affected adversely. Material safety comparison of the common superabsorbents is presented in Table 17.3 (Parmar and Anupama, 2014).

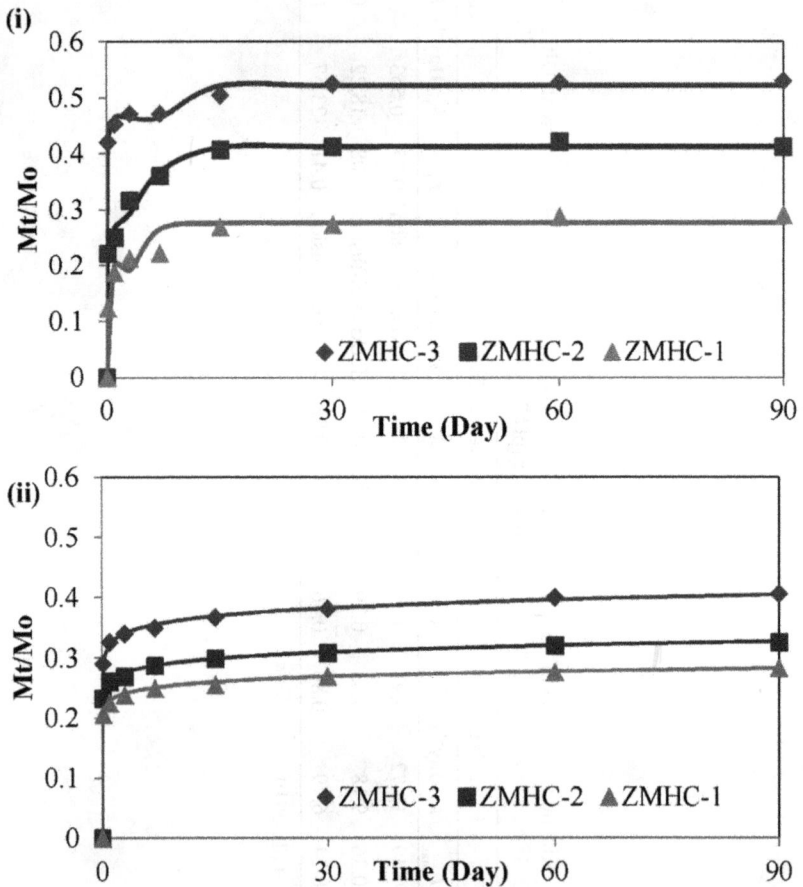

FIGURE 17.10 In vitro release profile of phosphate from hydrogel/zeolite composite with different concentration of zeolite (ZMHC-1, 25%; ZMHC-2, 10%; and ZMHC-1, 5%) in (i) neutral water (pH 7.0) and (ii) acidic water (pH 4.0). Source: Singh et al., 2019. Reprinted by permission of Wiley Online Library (http://onlinelibrary.wiley.com).

TABLE 17.3 Material Safety Information of Some Commercial Major Superabsorbent Polymers

Parameters	Sodium Polyacrylate, crosslinked, 95->99%	Polyacrylamide	Acrylamide and Potassium Acrylate Copolymer, Crosslinked, Anionic	Starch-Grafted Sodium Polyacrylate, >99%
Physical and chemical properties	Appearance: White granular powder, Physical state: solid, Vapor pressure: <10 mmHg, Boiling point: NE, Solubility in water: not soluble, evaporation rate: < 1.0, Order: None, pH: 5.5–6.5 (1% in H_2O), Melting point: > 390 F, Specific gravity: 0.4–0.7 g mL^{-1}	Appearance: granular solid, Color: white, Vapor pressure: NA, solubility in water: not soluble, evaporation rate: < 1.0, Order: None pH: NA, Flash Point: NA	Appearance: clear to yellow crystal, Vapor pressure: <10 mmHg, Solubility (water): insoluble, swells, Oder: None, Evaporation rate: < 1 (Butyl acetate), Flash Point: NA, Boiling point: NE, Melting point: > 390 F, Vapor density: Nil, Bulk density: 0.4–0.7	Appearance: Granular powder, Color: off white, Vapor pressure: NA, Solubility in water: slightly soluble, Evaporation rate: < 1.0, Order: None, Evaporation rate: NA, Melting Point: NA Bulk Density: 25–31 lbs ft^{-3}
Potential Health effect (PHE)	*Eyes:* Dust may cause burning, drying, itching, and other discomfort *Skin:* Exposure to dust may aggravate skin problems due to drying effect	Irritation to eyes and skin	*Inhalation:* may cause irritation of upper respiratory tract and lung if large amount of dust inhaled. *Skin:* prolonged contact causes reddening, and drying of skin or eyes. No toxic effect. *Ingestion:* large amount may cause nausea.	Contact with eyes, skin, or clothing may cause irritation.
Recommended exposure limit	0.05 mg m^{-3} for 8 h	—	—	0.05 mg m^{-3} for 8 h
HMIS rating*	Health 1, Fire 0, Reactivity 0	—	Health 2, Flammability 1 Fire 0, Reactivity 0	

TABLE 17.3   (Continued)

Parameters	Sodium Polyacrylate, crosslinked, 95–>99%	Polyacrylamide	Acrylamide and Potassium Acrylate Copolymer, Crosslinked, Anionic	Starch-Grafted Sodium Polyacrylate, >99%
First Aid	*Eyes:* immediately flush with plenty of water for 15 min. *Skin:* remove polymer dust using soap and water *Ingestion:* no toxic effect by ingestion. Seek medical attention if large quantities ingested *Inhalation:* move to fresh air.	Same	Same	Same
Toxicological information	*Acute toxicity:* LD_{50} (oral) rat: 40 g kg^{-1}, *Chronic toxicity:* chronic inhalation to rat for a lifetime (2 year) to respirable 10-μm-size material produced nonspecific inflammation and chronic lung injury at 0.2 and 0.8 mg m^{-3}. No adverse effect detected at 0.05 mg m^{-3}. *Carcinogenicity:* no information, *Mutagenicity:* no effect	*Acute toxicity:* LD_{50} (oral) rat > 5 g kg^{-1} LD_{50} (dermal) rabbit > 2 g kg^{-1} *Chronic toxicity:* two-year feeding study in rat and one-year study in dogs showed no adverse health effect.	Dry product contains 0.1% or less free residual monomer, i.e., acrylamide which is a suspected human carcinogen (Carcinogen A3: with ACGIH classification). *Chronic toxicity:* lifetime (2 year) inhalation exposure to rats with sodium polyacrylate (<10 μm) produced nonspecific inflammation and chronic lung injury at 0.2 and 0.8 mg m^{-3}. No adverse effect detected at 0.05 mg m^{-3}.	*Chronic toxicity:* Lab rat exposed over lifetime to <10 μm sodium polyacrylate dust at 0.8 and 0.05 mg m^{-3} showed local chronic inflammation leading to tumors in some animals at 0.8 mg m^{-3}. At 0.2 mg m^{-3}, there was evidence of local inflammation of lung tissue but no tumorigenic response.

TABLE 17.3 (*Continued*)

Parameters	Sodium Polyacrylate, crosslinked, 95->99%	Polyacrylamide	Acrylamide and Potassium Acrylate Copolymer, Crosslinked, Anionic	Starch-Grafted Sodium Polyacrylate, >99%
Ecological information	*Ecotoxicity:* Composted polyacrylate absorbents are nontoxic to aquatic or terrestrial organism at predicted exposure levels from current application rates. *Environmental fate:* Relatively inert in aerobic and anaerobic conditions. Immobile in landfills and soil systems (>90% retention) with mobile fraction showing biodegradability. Compatible with incineration of municipal solid waste.	*Ecotoxicity:* ecological injuries neither known nor expected under normal use. Aquatic toxicity unlikely due to low solubility *Persistence/ degradability:* not readily biodegradable (Zahn–Wellens Test); < 10% after 28 days)	—	—

*HMIS: Hazardous Material Identification in System rating; Scale: Minimal 0, Slight 1, Moderate 2, Serious 3, Severe 4, = chronic hazard.

Source: Parmar, 2014.

17.6.1   ECOTOXICITY OF POLYACRYLATE HYDROGELS

Studies showed that polyacrylate-based hydrogels get precipitated in the presence of cations and stay adsorbed on sewage sludge, sediment, or soil (Hamilton et al., 1997). Environmental fate study (lysimeter testing) of polyacrylate was done in soil using C14-labeled sodium polyacrylate (Mw. 4500), which showed that the highest amount (~90%) of the polymer get accumulated in the first 15 mm of the ground and thus cannot percolate into ground water (Chiaudani et al., 1990). However, around 10% of the sodium polyacrylate stay mobile and is difficult to precipitate or adsorbed on to soil. The same study with lower molecular weight sodium polyacrylate (Mw 1000) showed that it was more prone to biodegradation, which is measured by the release of C^{14}-labeled CO_2. Similar result was also found, where sodium polyacrylate oligomer of only less than seven monomeric units (Mw. 500–700) are metabolized by microorganisms (Kawai et al., 1994).

Sodium polyacrylate (Mw. 4500) showed no significant acute toxicity to aquatic test species under United States Environmental Protection Agency (USEPA) toxicity classification guidelines (Freeman and Bender, 1993), and it showed LC_{50} or EC_{50} value of >100 mg L^{-1} in all acute toxicity tests involving bacteria, algae, daphnia, and fish. Sodium polyacrylate (Mw. 4500) showed a chronic no-observed effect concentration (NOEC) of approximately 6 mg L^{-1} in Daphnia and 56 mg L^{-1} in fish. In the worst-case scenario, the sodium polyacrylate concentration at the out fall of paper mill's on-site sewage treatment plants was found to be below 0.04 mg L^{-1} (Pokhrel and Viraraghavan, 2004). Therefore, a considerable safety margin (> 150×) exists between the NOEC of the most sensitive species and polyacrylate concentration found in the natural water.

17.6.2   ECOTOXICITY OF POLYACRYLAMIDE HYDROGELS

Polyacrylamide SAPs are used for water clarification, sludge treatment and sewage sludge thickening, primary settlement, mineral processing, and others. It was reported that while certain amount of polyacrylamide may be lost to environment, it is highly unlikely that polyacrylamide will degrade to acrylamide, which is highly soluble in water (Smith et al., 1996). However, during synthesis of polyacrylamide, certain amount of acrylamide does not polymerize and remain as such residual monomer. It was found that locally produced polyacrylamide has the residual monomer content of 0.1% or less (Barvenik, 1994; Levy and Warrington, 2015); however, for portable water treatment, the residual content of 0.05% is recommended (Guzzo and Guezennec, 2015). The World Health Organization has set up a guidance value of acrylamide in the drinking water as 0.5 µg L^{-1} (Cavalli et al., 2004). The overseas occupational threshold limit value exposure standards for acrylamide range from 0.03 to 0.3 mg m^{-3} (NIOSH, 1988). It was reported that after leaching of the residual acrylamide into the environment, it undergoes through hydrolysis to produce acrylic acid (Shanker et al., 1990).

The USEPA had estimated that acrylamide has a half-life of 6.6 h mostly due to photochemical activities (US EPA, 1994). Due to the high solubility of acrylamide, its volatilization from air is unlikely. The half-life of acrylic acid due to photochemical reactions is also estimated to be 6.6 h (GEMS, 1986). Under aerobic conditions with high acrylamide content (e.g., 25 mg/kg), its half-life was estimated to be 18–45 h, whereas under anaerobic conditions, it resulted in longer half-life (Lande et al., 1979; Brown et al., 1980). In contrast to high acrylamide condition, EU risk assessment report estimated the half-life for the degradation of acrylamide in soil to be 30 days (mostly anaerobic) (European Union, 2002). This report has reviewed a list of studies in acrylamide biodegradation, which specified acrylamide as readily biodegradable.

Acrylic acid is also reported to be readily biodegradable. Under aerobic condition, it rapidly oxidized, and a 28-day study with spiked sandy loam soil with ^{14}C-labeled acrylic acid showed that no residues exists after three days. The half-life was found to be less than one day (Hawkins et al., 1992). Under anaerobic condition, it also degraded by microbes, and it was found to be readily biodegradable with over 75% of theoretical methane production within eight weeks of incubation in the presence of 10% sludge (Shelton and Tiedje, 1984). Based on the ecotoxicity endpoints compared to the calculated environmental concentration, it is likely that acrylic acid would not pose significant adverse impacts on the environment.

17.6.3 ECOTOXICITY OF STARCH-BASED HYDROGELS

Starch is completely biodegradable in all kinds of environment. It mainly consists of D-glucose in two homopolymeric arrangements, namely. amylase, linear chain of α-D(1, 4')-glucan and amylopectin, branched chain of α-D(1,4')-glucan with many α-1,6'-linked branch points (Pareta and Edirisinghe, 2006). Starch shows hydrophilic properties and strong intermolecular hydrogen bonding due to presence of numerous hydroxyl groups at its polymeric chain. There are two secondary hydroxyl group at C2 and C3 and one primary hydroxyl group at C6. The degradation of starch is many caused by hydrolysis through chemoenzymatic reaction and metabolized into carbon dioxide and water (Primarini and Ohta, 2000).

The development of starch-based biodegradable polymers offers less introduction of synthetic or petrochemical product in the environment. They have the potential to be used as carrier for fertilizers for controlled release of the same without any possible harm to soil and soil microorganisms (Chen et al., 2008; Kumbar et al., 2001). A starch-based polymeric film has also been used for mulching of cultivated land, and they can be ploughed into soil and disposed directly. Starch-based polymers are not found to produce any toxic residues on degradation. It was reported that the biodegradation profile of hydrogel, for example, chitosan glycerol-phosphate hydrogel, can be improved by introduction of starch (Sa-Lima et al., 2018). These hydrogels are found to show biocompatibility through *in vitro* cytotoxicity screening.

17.6.4   ECOTOXICITY OF CELLULOSE-BASED HYDROGELS

Cellulose is found abundantly in nature as constituent of plant tissue. Even some bacteria, namely, *Acetobacter xylinum,* are able to synthesize cellulose (Masaoka et al., 1993). Both plant and bacterial cellulose are chemically identical with certain differences in the morphophisycal properties (Czaja et al., 2007). In cellulose, glucose units are held together by 1,4-β-glucosidic linkages imparting high crystallinity causing insolubility in almost every polar and nonpolar solvents. Derivatization of cellulose is done by chemical modification, namely, esterification or etherification of the hydroxyl group to produce cellulosic derivatives, which are useful for industrial application. They are found to be highly ecofriendly and can be easily degraded by several bacteria and fungi. Due to high biocompatibility, cellulose and cellulose derivatives are now being used widely in the synthesis of cellulose-based hydrogels (Miyamoto et al., 1989; Tomšič, et al., 2007). An acute oral toxicity test of bacterial cellulose/acrylamide hydrogels on ICR mice showed that these hydrogels are nontoxic up to 2000 mg/kg upon oral administration (Pandey et al., 2014).

17.7   CONCLUSION

Different kinds of hydrogels based on synthetic monomers such as acrylamide, acrylic acid, etc., and natural polymers such as starch, alginate, derivatized cellulose, etc., are being synthesized depending on their suitability and applicability. The physical and chemical properties of these hydrogel materials can be tailor modified depending on their building blocks and the preparation techniques. Use of natural polymers imparts more biocompatibility and environment friendly properties in the hydrogel product. The synthesis protocol of the hydrogel materials varies tremendously depending on the type of monomer and the nature of desired properties to be included in the materials. Till now, the hydrogel materials available in the market for agricultural application are chemically identical with the hydrogel materials used for the pharmaceutical applications. However, there are some new inventions, which have been solely developed for the application in agriculture with much needed desired properties. A variety of semisynthetic hydrogels have been developed for agricultural applications, which have been proven to efficiently increase the water-holding capacity of the soil under water scarce situation. In agriculture, hydrogel materials can be used for broad range of applications, such as water management in precision farming, slow release of nutrients, delivery of pesticides, waste water management, etc. Various new invention such as use of biopolymers in the hydrogel matrix, use of clay materials to impart more barrier properties, encapsulation of plant nutrients in the hydrogel matrix, crosslinking using various fertilizer, etc., are being tried to reduce the cost of application of hydrogels and to impart more usability for crop production. Though hydrogels are being proven as smart materials for developing controlled-release formulation of different bioactive molecules, including fertilizers and pesticide, their efficacy at field level is still scarce. There is some literature on hydrogel-based fertilizer formulations such as nitrogen, phosphate, micronutrients (boron, zinc, etc.), and some of their efficacy was proven at field level to enhance both water and nutrient use efficiency. Despite these, there is still lack of interest in hydrogel-based formulations, which may be due to high cost associated with them, and moreover, the process of the preparation of fertilizer formulation based on hydrogel is still cumbersome. For these natural polymers, cellulose- and starch-based hydrogels may play a vital role.

KEYWORDS

- **fertilizer**
- **hydrogel**
- **formulation**
- **superadsorbent polymer**

REFERENCES

Abedi-Koupai, J.; Mostafazadeh-Fard, B.; Afyuni, M.; Bagheri, M. R. Effect of treated wastewater on soil chemical and physical properties in an arid region. *Plant Soil Environ.* 2006, *52*(8), 335.

Ahmed, E. M. Hydrogel: Preparation, characterization, and applications: A review. *J. Adv. Res.* 2015, 6(2), 105–121.

Aimetti, A. A.; Machen, A. J.; Anseth, K. S. Poly (ethylene glycol) hydrogels formed by thiol-ene photopolymerization for enzyme-responsive protein delivery. *Biomaterials,* 2009, *30*(30), 6048–6054.

Barvenik, F. W. Polyacrylamide characteristics related to soil applications. *Soil Sci.* 1994, *158*, 235–243.

Bouhadir, K. H.; Hausman, D. S.; Mooney, D. J. Synthesis of cross-linked poly(aldehyde guluronate) hydrogels. *Polymer*, 1999, *40*(12), 3575–3584.

Bounabi, L.; Mokhnachi, N. B.; Haddadine, N.; Ouazib, F.; Barille, R. Development of poly(2-hydroxyethyl methacrylate)/clay composites as drug delivery systems of paracetamol. *J. Drug Deliv. Sci. Tech.*, 2016, *33*, 58–65.

Brown, L.; Rhead, M. M.; Bancroft, K. C. C.; Allen, N. Model studies of the degradation of acrylamide monomer. *Water Res.* 1980, *14*(7), 775–778.

Carminati, A.; Moradi A. How the soil-root interface affects water availability to plants. *Geophys. Res. Abstr.*, 2010, *12*, 10677.

Cavalli, S.; Polesello, S.; Saccani, G. Determination of acrylamide in drinking water by large-volume direct injection and ion-exclusion chromatography–mass spectrometry. *J. Chromatogr. A*, 2004, *1039*(1–2), 155–159.

Chandrika, K. P.; Singh, A.; Sarkar, D. J.; Rathore, A.; Kumar, A. pH-sensitive crosslinked guar gum-based superabsorbent hydrogels: Swelling response in simulated environments and water retention behavior in plant growth media. *J. Appl. Polym. Sci.*, 2014, *131*(22), 41060.

Chapiro, A. *Radiation Chemistry of Polymeric Systems*. New York: Interscience. 1962.

Chen, L.; Xie, Z.; Zhuang, X.; Chen, X.; Jing, X. Controlled release of urea encapsulated by starch-g-poly(L-lactide). *Carbohyd. Polym.* 2008, *72*(2), 342–348.

Chiaudani, G.; Poltronieri, P. *Study on the Environmental Compatibility of Polycarboxylates Used in Detergent Formulations*. Denver, CO: CIPA, 1990.

Chornet, E.; Severian D. Chitosan-xanthan based polyionic hydrogels for stabilization and controlled release of vitamins. Receptors. U.S. Patent 6,964,772. Washington, DC, USA, 2005.

Chu, L.Y.; Xie, R.; Ju, X. J.; Wang, W. *Smart Hydrogel Functional Materials*. Berlin: Springer, 2013.

Crescenzi, V.; Dentini, M.; Bontempo, D.; Masci, G. Hydrogels based on pullulan derivatives crosslinked via a "living" free-radical process. *Macromol. Chem. Phys.* 2002, *203*(10–11), 1285–1291.

Czaja, W. K.; Young, D. J.; Kawecki, M.; Brown, R. M. The future prospects of microbial cellulose in biomedical applications. *Biomacromolecules*, 2007, *8*(1), 1–12.

Das, R. K.; Gocheva, V.; Hammink, R.; Zouani, O. F.; Rowan, A. E. Stress-stiffening-mediated stem-cell commitment switch in soft responsive hydrogels. *Nat. Mater.* 2016, *15*(3), 318–325.

Dass, A.; Singh, A.; Rana, K. S. In-situ moisture conservation and nutrient management practices in fodder-sorghum (Sorghum bicolor). *Ann. Agric. Res.* 2013, 34(3), 254–259.

De, S. K.; Aluru, N. R.; Johnson, B.; Crone, W. C.; Beebe, D. J.; Moore, J. Equilibrium swelling and kinetics of pH-responsive hydrogels: Models, experiments, and simulations. *J. Microelectromech. S.* 2002, 11(5), 544–555.

Deng, Y.; Huang, M.; Sun, D.; Hou, Y.; Li, Y.; Dong, T.; Wang, X.; Zhang, L.; Yang, W. Dual physically cross-linked κ-carrageenan-based double network hydrogels with superior self-healing performance for biomedical application. *ACS Appl. Mater. Inter.* 2018, 10(43), 37544–37554.

Domingues, M. T.; Bueno, C. D. C., Fraceto, L. F.; Loyola-Licea, J. C.; Rosa, A. H. Short-term effect of alginate-biochar microbeads in corn germination. In *Simposio em Tecnologia, Inovação e Sustentabilidade Ambiental*, 2014, p. 142.

Dorraji, S.S.; Golchin, A.; Ahmadi, S. The effects of hydrophilic polymer and soil salinity on corngrowth in Sandy and loamy soils. *Clean - Soil, Air, Water,* 2010, 38(7), 584–591.

European Union, *European Union Risk Assessment Report: Acrylamide*, European Commission–Joint Research Centre Institute for Health and Consumer Protection European Chemicals Bureau, 1st Priority List Volume 24, 2002.

FAO, Water for Sustainable Food and Agriculture: A report produced for the G20 Presidency of Germany, Food and Agriculture Organization of the United Nations, Rome, 2017.

Freeman, M. B.; Bender, T. M. An environmental fate and safety assessment for a low molecular weight polyacrylate detergent additive. *Environ. Technol.* 1993, 14(2), 101–112.

Funami, T.; Hiroe, M.; Noda, S.; Asai, I.; Ikeda, S.; Nishimari, K. Influence of molecular structure imaged with atomic force microscopy on the rheological behavior of carrageenan aqueous systems in the presence or absence of cations. *Food Hydrocoll.* 2007, 21, 617–629.

GEMS, Graphical Exposure Modeling System, FAP, Fate of Atmospheric Pollution. Data base, Office of toxic substances, U.S. Environment Protection Agency, 1986.

Guo, M.; Liu, M.; Zhan, F.; Wu, L. Preparation and properties of a slow-release membrane-encapsulated urea fertilizer with superabsorbent and moisture preservation. *Ind. Eng. Chem. Res.* 2005, 44(12), 4206–4211.

Guzzo, J.; Guezennec, A. G. Degradation and transfer of polyacrylamide based flocculent in sludge and industrial and natural waters. *Environ. Sci. Poll. Res.* 2015, 22, 6387–6389.

Gyles, D. A.; Castro, L. D.; Silva Jr, J. O. C.; Ribeiro-Costa, R. M. A review of the designs and prominent biomedical advances of natural and synthetic hydrogel formulations. *Eur. Polym. J.* 2017, 88, 373–392.

Hamilton, J. D.; Morici, I. J.; Freeman, M. B. Polycarboxylates and Polyacrylate Superabsorbents. / Polycarboxylates. In. *Ecological Assessment Polymers: Strategies for Product Stewardship and Regulatory Programs;* Hamilton, J. D.; Sutcliffe, R., Ed.; Hoboken: Wiley, 1997; pp. 87–102.

Hassan, C. M.; Peppas, N. A. Cellular PVA hydrogels produced by freeze/thawing. *J. Appl. Polym. Sci.* 2000, 76(14), 2075–2079.

Hawkins, D. R.; Kirkpatrick, D.; Aikens, P. J.; Saxton, J. E. The metabolism of acrylic acid in soil under aerobic conditions. Huntingdon, United Kingdom, Huntingdon Research Centre (HRC Confidential Report No. 93A/920625), 1992.

Hennink, W. E.; Nostrum, C. F. V. Novel crosslinking methods to design hydrogels. *Adv. Drug Deliv. Rev.* 2002, 54, 13–36.

Hu, Z.; Huang, G. A new route to crystalline hydrogels, guided by a phase diagram. *Angew. Chem. Int. Ed.* 2003, 42(39), 4799–4802.

Huglin, M. B.; Zakaria, M. B. Swelling properties of copolymeric hydrogels prepared by gamma irradiation. *J. Appl. Polym. Sci.,* 1986, *31*, 457–475.

Hüttermann, A.; Orikiriza L. J. B.; Agaba, H. Application of superabsorbent polymers for improving the ecological chemistry of degraded or polluted lands, *Clean-Soil, Air, Water,* 2009, *37*(7), 517–526.

Jarosiewicz, A.; Tomaszewska, M. Controlled-release NPK fertilizer encapsulated by polymeric membranes. *J. Agri. Food Chem.* 2003, *51*(2), 413–417.

Katime, I.; Novoa, R.; de Apodaca, E. D.; Mendizábal, E.; Puig, J. Theophylline release from poly(acrylic acid-*co*-acrylamide) hydrogels. *Polym. Test.* 1999, *18*(7), 559–566.

Kawai, F. Biodegradation of polyethers and polyacrylate. In: *Studies in Polymer Science.* Amsterdam: Elsevier, 1994; vol. 12, pp. 24–38.

Kim, J.; Yoon, J.; Hayward, R. C. Dynamic display of biomolecular patterns through an elastic creasing instability of stimuli-responsive hydrogels. *Nat. Mater.* 2010, *9*(2), 159–164.

Klouda, L.; Mikos, A.G. Thermoresponsive hydrogels in biomedical applications. *Eur. J. Pharm. Biopharm.* 2008, *68*(1), 34–45.

Kulicke, W. M.; Nottelmann, H. *Structure and Swelling of Some Synthetic, Semisynthetic, and Biopolymer Hydrogels.* In: *Polymers in Aqueous Media,* 1989, ch. 2, pp. 15–44.

Kumar, V.; Singh, A.; Das, T. K.; Sarkar, D. J.; Singh, S. B.; Dhaka, R.; Kumar, A. Release behavior and bioefficacy of imazethapyr formulations based on biopolymeric hydrogels. *J. Environ. Sci. Health, Part B,* 2017, *52*(6), 402–409.

Kumbar, S. G.; Kulkarni, A. R.; Dave, A. M.; Aminabhavi, T. M. Encapsulation efficiency and release kinetics of solid and liquid pesticides through urea formaldehyde crosslinked starch, guar gum, and starch+ guar gum matrices. *J. Appl. Polym. Sci.* 2001, *82*(11), 2863–2866.

Lande, S. S.; Bosch, S. J.; Howard, P. H. Degradation and leaching of acrylamide in soil. *J. Environ. Qual.* 1979, *8*(1), 133–137.

Landis, T. D.; Haase, D. L. Applications of hydrogels in the nursery and during outplanting. In *National Proceedings: Forest and Conservation Nursery Associations-2011.* Haase, D. L.; Pinto, J. R.; Riley, L. E., Eds. Fort Collins, CO: USDA Forest Service, Rocky Mountain Research Station, 2012, pp. 53–58.

Lejcuś, K.; Śpitalniak, M.; Dąbrowska, J. Swelling behaviour of superabsorbent polymers for soil amendment under different loads. *Polymers,* 2018, *10*(3), 271.

Levy, G. J.; Warrington, D. N. Polyacrylamide addition to soils: impacts on soil structure and stability. In *Functional Polymers in Food Science: From Technology to Biology*; Cirillo, G., Spizzirri, U. G., Iemma, F. Eds. Salem, MA: Scrivener, 2015, Ch. 2.

Liang, R.; Liu, M. Preparation and properties of coated nitrogen fertilizer with slow release and water retention. *Ind. Eng. Chem. Res.* 2006, *45*(25), 8610–8616.

Liu, M.; Liang, R.; Zhan, F.; Liu, Z.; Niu, A. Synthesis of a slow-release and superabsorbent nitrogen fertilizer and its properties. *Polym. Adv. Technol.* 2006, *17*(6), 430–438.

Malhi, S. S.; Soon, Y. K.; Grant, C. A.; Lemke, R.; Lupwayi, N. Influence of controlled-release urea on seed yield and N concentration, and N use efficiency of small grain crops grown on Dark Gray Luvisols. *Can. J. Soil Sci.* 2010, *90*(2), 363–372.

Mandal, S.; Basu, S. K.; Sa, B. Ca^{2+} ion cross-linked interpenetrating network matrix tablets of polyacrylamide-grafted-sodium alginate and sodium alginate for sustained release of diltiazem hydrochloride. *Carbohyd. Polym.* 2010, *82*(3), 867–873.

Mangrich, A. S.; Cardoso, E. M. C.; Doumer, M. E.; Romão, L. P. C.; Vidal, M.; Rigol, A.; Novotny, E.H. Improving the water holding capacity of soils of Northeast Brazil by biochar augmentation. In *Water Challenges and Solutions on a Global Scale*. Washington: American Chemical Society, 2015, pp. 339–354.

Masaoka, S.; Ohe, T.; Sakota, N. Production of cellulose from glucose by Acetobacter xylinum. *J. Ferment. Bioeng.* 1993, *75*(1), 18–22.

Mathur, A. M.; Moorjani, S. K.; Scranton, A. B. Methods for synthesis of hydrogel networks: A review. *J. Macromol. Sci. Polymer Rev.* 1996, *36*(2), 405–430.

Mikkelsen R. L. Using hydrophilic polymers to control nutrient release. *Fertilizer Res.* 1994, *38*, 53–59.

Miyamoto, T.; Takahashi, S. I.; Ito, H.; Inagaki, H.; Noishiki, Y. Tissue biocompatibility of cellulose and its derivatives. *J. Biomed. Mater. Res.* 1989, *23*(1), 125–133.

Narjary, B.; Aggarwal, P.; Singh, A.; Chakraborty, D.; Singh, R. Water availability in different soils in relation to hydrogel application. *Geoderma*, 2012, *187–188*, 94–101.

National Institute for Occupational Safety and Health (NIOSH): "NIOSH Recommendations for Occupational Safety and Health Standards 1988" Morbidity and mortality weekly report (MMWR) 37(S-7). Atlanta: U.S. Department of Health and Human Services, 1988.

Nirmala, A.; Guvvali, T. Hydrogel/superabsorbent polymer for water and nutrient management in horticultural crops. *Int. J. Chem. Stud.* 2019, *7*(5), 787–795.

Nnadi, F. N. Super Absorbent Polymer (SAP) and irrigation water conservation. *Irrig. Drain. Syst. Engin.* 2012, *1*, 11–23.

Noppakundilograt, S.; Pheatcharat, N.; Kiatkamjornwong, S. Multilayer-coated NPK compound fertilizer hydrogel with controlled nutrient release and water absorbency. *J. Appl. Polym. Sci.* 2015, *132*(2), 41249.

Pandey, M.; Mohamad, N.; Amin, M. C. I. M. Bacterial cellulose/acrylamide pH-sensitive smart hydrogel: development, characterization, and toxicity studies in ICR mice model. *Mol. Pharm.* 2014, *11*(10), 3596–3608.

Pareta, R.; Edirisinghe, M. J. A novel method for the preparation of starch films and coatings. *Carbohyd. Polym.* 2006, *63*(3), 425–431.

Parmar, B. S.; Anupama Superabsorbent polymers: Material safety of the major chemical groups. *Pesticide Res. J.* 2014, *26*(2), 119–127.

Peppas, N. A.; Huang, Y.; Torres-Lugo, M.; Ward, J. H.; Zhang, J. Physicochemical foundations and structural design of hydrogels in medicine and biology. *Ann. Rev. Biomed. Eng.* 2000, *2*, 9–29.

Pokhrel, D.; Viraraghavan, T. Treatment of pulp and paper mill wastewater—A review. *Sci. Total Environ.* 2004, *333*(1–3), 37–58.

Popa, E. G.; Santo, V. E.; Rodrigues, M. T.; Gomes, M. E. Magnetically-responsive hydrogels for modulation of chondrogenic commitment of human adipose-derived stem cells. *Polymers*, 2016, *8*(2), 28.

Pourjavadi, A.; Harzandi, A. M.; Hosseinzadeh, H. Modified carrageenan 3. Synthesis of a novel polysaccharide-based superabsorbent hydrogel via graft copolymerization of acrylic acid onto kappa-carrageenan in air. *Eur. Polym. J.*, 2004, *40*(7), 1363–1370.

Primarini, D.; Ohta, Y. Some enzyme properties of raw starch digesting amylases from Streptomyces sp. No. 4. *Starch-Stärke*, 2000, *52*(1), 28–32.

Rad, S.; Gan, L.; Chen, X.; You, S.; Huang, L.; Su, S.; Taha, M. R. Sustainable water resources using corner pivot lateral, a novel sprinkler irrigation system layout for small scale farms. *Appl. Sci.* 2018, *8*(12), 2601.

Rao, R. N.; Finck, A.; Blair, G. J.; Tondan, H. L. S. *Plant Nutrition for Food Security: A Guide for Integrated Nutrient Management*. Rome, Italy: Food and Agriculture Organization, 2006.

Ratner, B. D.; Hoffman, A. S. Synthetic hydrogels for biomedical applications. In *Hydrogels for Medical and Related Applications*. Washington, DC: American Chemical Society, 1976, Ch. 1, pp. 1–36.

Rosiak, J. M.; Yoshii, F. Hydrogels and their medical applications. *Nucl. Instrum. Meth. Phys. Res. B* 1999, *151*, 56–64.

Sá-Lima, H.; Caridade, S. G.; Mano, J. F.; Reis, R. L. Stimuli-responsive chitosan-starch injectable hydrogels combined with encapsulated adipose-derived stromal cells for articular cartilage regeneration. *Soft Matter*, 2010, *6*(20), 5184–5195.

Sarkar, D.J.; Barman, M.; Bera, T.; De, M.; Chatterjee, D. Agriculture: Polymers in crop production mulch and fertilizer. In *Encyclopedia of Polymer Applications*, Mishra, M. Ed. Boca Raton, FL: CRC Press, 2018.

Sarkar, D. J.; Bera, T.; Singh, A. Release of urea from cellulosic hydrogel coated urea granule: Modeling effect of crosslink density and pH triggering. *Polym-Plast Technol.* 2019b, *58*(17), 1914–1926.

Sarkar, D. J.; Majumder, S.; Kaushik, P.; Shakil, N. A.; Kumar, J. *Controlled Release of Pesticide Formulations*. In *Applications of Encapsulation and Controlled Release*. Boca Raton, FL: CRC Press, 2019c, pp. 229–247.

Sarkar, D. J.; Singh, A. Base triggered release of insecticide from bentonite reinforced citric acid crosslinked carboxymethyl cellulose hydrogel composites. *Carbohyd. Polym.* 2017, *156*, 303–311.

Sarkar, D. J.; Singh, A. pH-triggered release of boron and thiamethoxam from boric acid crosslinked carboxymethyl cellulose hydrogel based formulations. *Polym-Plast Technol.* 2019a, *58*(1), 83–96.

Sarkar, D. J.; Singh, A.; Gaur, S. R.; Shenoy, A. V. Viscoelastic properties of borax loaded CMC-g-cl-poly(AAm) hydrogel composites and their boron nutrient release behavior. *J. Appl. Polym. Sci.* 2016, *133*(38) 43969.

Sarkar, D.J.; Singh, A.; Mandal, P.; Kumar, A.; Parmar, B. S. Synthesis and characterization of poly(CMC-g-cl-PAam/Zeolite) superabsorbent composites for controlled delivery of zinc micronutrient: swelling and release behavior. *Polym-Plast Technol.* 2015, *54*(4), 357–367.

Sarkar, D. J.; Singh, A.; Singh, N.; Parmar, B. S.; Kumar, A. Effect of addition of superabsorbent hydrogels on sorption and mobility of metribuzin in sandyloam soil. *Pesticide Res. J.* 2012, *24*(2), 138–143.

Sarkar, S.; Datta, S. C. Influence of fertilizer loaded nanoclay superabsorbent polymer composite (NCPC) on dynamics of P and N availability and their uptake by pearl millet (Pennisetum glaucum) in an Inceptisols. *Int. J. Bio-res. Stress Manage.* 2014, *5*(2), 221–227.

ENasScott, G. 'Green' polymers. *Polym. Degrad.* 2000, *68*, 1–7.

Shaaban, A.; Yatim, N. M.; Dimin, M. F.; Yusof, F.; Abd Razak, J. Multiwalled carbon nanotubes enhancing nitrogen uptake and use efficiency of urea fertilizer by paddy. *J. Teknol.*, 2017, *79*(5-2), 254–259.

Shanker, R.; Ramakrishna, C.; Seth, P. K. Microbial degradation of acrylamide monomer. *Arch. Microbiol.* 1990, *154*(2), 192–198.

Sharma, B. R.; Gulati, A.; Mohan, G.; Manchanda, S.; Ray, I.; Amarasinghe, U. Water Productivity Mapping of Major Indian Crops, National Bank for Agriculture and Development, 2018.

Shelton, D. R.; Tiedje, J. M. General method for determining anaerobic biodegradation potential. *Appl. Environ. Microbiol.*, 1984, *47*(4), 850–857.

Shiroodi, S. G.; Rasco, B. A.; Lo, Y. M. Influence of xanthan–curdlan hydrogel complex on freeze-thaw stability and rheological properties of whey protein isolate gel over multiple freeze-thaw cycle. *J. Food Sci.* 2015, *80*(7), E1498–E1505.

Singh, U.; Sanabria, J.; Austin, E. R.; Agyin-Birikorang, S. Nitrogen transformation, ammonia volatilization loss, and nitrate leaching in organically enhanced nitrogen fertilizers relative to urea. *Soil Sci. Soc. Am. J.* 2012, *76*(5), 1842–1854.

Singh, A; Sarkar, D. J.; Mittal, S.; Dhaka, R.; Maiti, P.; Singh, A.; Raghav, T.; Solanki, D.; Ahmed, N.; Singh, S. B. Zeolite reinforced carboxymethyl cellulose-Na+-g0cl-poly(AAm) hydrogel composites with pH responsive phosphate release behavior. *J. Appl. Polym. Sci.* 2019, *136*(15), 47332.

Singh, A.; Sarkar, D. J.; Singh, A. K.; Parsad, R.; Kumar, A.; Parmar, B. S.; Studies on novel nanosuperabsorbent composites: Swelling behavior in different environments and effect on water absorption and retention properties of sandy loam soil and soil-less medium. *J. Appl. Polym. Sci.* 2011, *120*, 1448–1458.

Smith, E. A.; Prues, S. L.; Oehme, F. W. Environmental degradation of polyacrylamides. 1. Effects of artificial environmental conditions: temperature, light, and pH. *Ecotox. Environ. Safe.* 1996, *35*(2), 121–135.

Takigami, M.; Amada, H.; Nagasawa, N.; Yagi, T.; Kasahara, T.; Takigami, S.; Tamada, M. Preparation and properties of CMC gel. *Trans. Mater. Res. Soc. Japan,* 2007, *32*, 713–716.

Teodorescu, M.; Lungu, A.; Stanescu, P. O. Preparation and properties of novel slow-release NPK agrochemical formulations based on poly(acrylic acid) hydrogels and liquid fertilizers. *Ind. Eng. Chem. Res.* 2009, *48*(14), 6527–6534.

Thomas, D. S. Hydrogel applied to the root plug of subtropical eucalypt seedlings halves transplant death following planting. *Forest Ecol. Manage.* 2008, *255*, 1305–1314

Tomšič, B.; Simončič, B.; Orel, B.; Vilčnik, A.; Spreizer, H. Biodegradability of cellulose fabric modified by imidazolidinone. *Carbohyd. Polym.* 2007, *69*(3), 478–488.

US EPA. Chemicals in the environment: Acrylamide. EPA-749-F-94-005. Washington DC, 1994.

Vaghani, S. S.; Patel, M. M.; Satish, C. S. Synthesis and characterization of pH-sensitive hydrogel composed of carboxymethyl chitosan for colon targeted delivery of ornidazole. *Carbohyd. Res.* 2012, *347*(1), 76–82.

Vashist, A.; Vashist, A.; Gupta, Y. K.; Ahmad, S. Recent advances in hydrogel based drug delivery systems for the human body. *J. Mater. Chem. B,* 2014, *2*(2), 147–166.

Zhang, J. T.; Bhat, R.; Jandt, K.D. Temperature-sensitive PVA/PNIPAAm semi-IPN hydrogels with enhanced responsive properties. *Acta Biomater.* 2009, *5*(1), 488–497.

Zhang, L.; Cao, Z.; Bai, T.; Carr, L.; Ella-Menye, J. R.; Irvin, C.; Ratner, B. D.; Jiang, S. Zwitterionic hydrogels implanted in mice resist the foreign-body reaction. *Nat. Biotechnol.* 2013, *31*(6), 553–556.

Zhang, J.; Peppas, N. A. Synthesis and characterization of pH-and temperature-sensitive poly(methacrylic acid)/poly(*N*-isopropylacrylamide) interpenetrating polymeric networks. *Macromolecules,* 2000, *33*(1), 102–107.

Zhu, C. H.; Lu, Y.; Peng, J.; Chen, J. F.; Yu, S. H. Photothermally sensitive poly(N-isopropylacrylamide)/graphene oxide nanocomposite hydrogels as remote light-controlled liquid microvalves. *Adv. Funct. Mater.* 2012, *22*(19), 4017–4022.

Zohourian, M. M.; Kabiri, K. Superabsorbent polymer materials: A review. *Iran. Polym. J.* 2008, *17*(6), 451–477

CHAPTER 18

Nanotechnological Interventions for Increasing Nutrient Use Efficiency in Horticultural Crops

NINTU MANDAL[1*], SATDEV[1], CHANDINI[2], MANOJ KUNDU[3], and HIDAYATULLAH MIR[3]

[1]Department of Soil Science and Agricultural Chemistry, Bihar Agricultural University, Sabour, Bhagalpur, Bihar, India

[2]Department of Agronomy, Bihar Agricultural University, Sabour, Bhagalpur, Bihar, India

[3]Department of Horticulture (Fruit and Fruit Technology), Bihar Agricultural University, Sabour, Bhagalpur, Bihar, India

*Corresponding author. E-mail: nintumandal@gmail.com

ABSTRACT

Studying matter at the atomic or nanometer scale is called nanoscience. Nanotechnology is the design, fabrication, and control structure for specific application. Nanotechnology is a new area having lot of potentials. Horticultural crops face lots of stress, starting from biotic stresses to abiotic stresses either at preharvest condition or postharvest condition. Nanomaterials are currently being applied in horticulture for alleviation of soil salinity, biosensing, and controlled release formulation of nutrients and pesticides. Nanocomposite positively affected various growth, yield, and quality of horticultural crop such as nanofertilizer increased photosynthetic pigments, proximate components (carbohydrates, protein, lipids, fat, ash, etc.), equilibrium water absorbency as well as nanoparticles improved nutrient use efficiency of soil minerals due to its control released properties that plant could easily uptake many available macro- and micronutrients to the soil.

Horticultural wastes (i.e., leaf of fruits) are also a rich source for production of nanomaterials in a greener way. Most of work just started, lots of research work is necessary for standardization of dose of nanoparticles, application mode, and fact of these particles in the environment is yet to be standardized. Nanotoxicology is also a very important aspects that needs due consideration while working with nanomaterials.

18.1   INTRODUCTION

Nanoscience and its related field is known as nanotechnology and this technology has the amazing success potential to revolutionize agri-biased natural resource management with the help of nanotechnology to transform science and engineering in agri-food systems. Level change has started as a new interdisciplinary enterprise (Lal, 2008). The study of the occurrence and manipulation of materials at the atomic and molecular level is called nanoscience. Nanotechnology produces and uses many structures that are controlled at the nanometer scale level. Nanotechnology has immense potential in horticulture field, starting from reduction of preharvest losses to postharvest lossless reduction.

18.2   WHAT ARE NANOPARTICLES?

The definition of the term nanoparticles (NPs) is the subject of debate and is continuing to evolve. Banfield and Zhang (2001) suggested that within an earth science context, nanoparticles are explained based on the size of the particles and it is differ from bulky material based on fundamental properties. Hochella et al. (2008) define the nanopaticles; the size range of the nanocompound will vary different from other materials, but nanoparticles generally size range often is between 1 and 100 nm for earth materials. According to EPA (2005), NPs are, therefore, considered as substances that particles size range is 1–100 nm in more than one dimension.

18.3   PREPARATION OF NANOMATERIALS: TECHNIQUES

Synthesis of nanomaterials generally has two types of fabrication techniques, that is, top–down approaches and bottom–up approaches. First technique is working with macroscopic materials and coming gradually to nanoscale, relatively easy but there may be imperfection in the crystals. Second one, that is, bottom–up approach is working with atomic level particles and gradually buildup of nanoscale materials. It is time taking, but more perfection of structure is obtained. There is summary of these two techniques in Table 18.1.

TABLE 18.1 Synthesis of Nanoparticles by Different Approaches (Shah and Shah, 2013)

Synthesis of Nanomaterials

Top–down approach   Bottom–up approach

Processes	Materials Synthesized	Processes	Materials Synthesized
Arc discharge method	Fullerenes and carbon nanotubes	Homogeneous nucleation	Noble metal particles and gold nanoparticles
Laser ablation	Broad range including	Chemical vapor deposition	Widely used to produce CNTs, fullerenes, and boron nanotubes
Ball milling	Composites and mixtures of elemental powders	Molecular beam epitaxy	Compound semiconductor and thin films
Inert gas condensation	Oxides, alloys, and semiconductors	Sol–gel synthesis	Colloidal and oxide nanoparticles
		Hydrothermal synthesis	Elemental nanopowders and mostly oxides
		Microwave method	Oxide nanoparticles including TiO_2

18.4   SOME NANOSCALE TECHNIQUES: CHARACTERIZATION TECHNIQUES

The development and continual refinement of nanoscale techniques have allowed for much of the revolution in nanoscience and nanotechnology. A wide array of microscopes, for example, scanning electron microscopy (SEM) and epifluorescence microscopy, along with numerous analytical methods ranging from atomic absorption and inductively coupled plasma emission spectrophotometry to extended X-ray absorption fine structure and X-ray absorption near edge structure spectroscopies, are crucial for understanding composition and for proper context and scaling. Here, a few key techniques are highlighted.

18.5   APPLICATION OF NANOTECHNOLOGY IN HORTICULTURE

There is numerous possible application of nanotechnology in horticultural research. Starting from increase of vase life of cut flowers, food packaging, biosensors, control release fertilizers and pesticide formulations, micro-propagation as well as horticultural waste materials can be a good source for nanoparticles synthesis for other applications. Various aspects of application of nanotechnology are represented as hereunder (Figure 18.1).

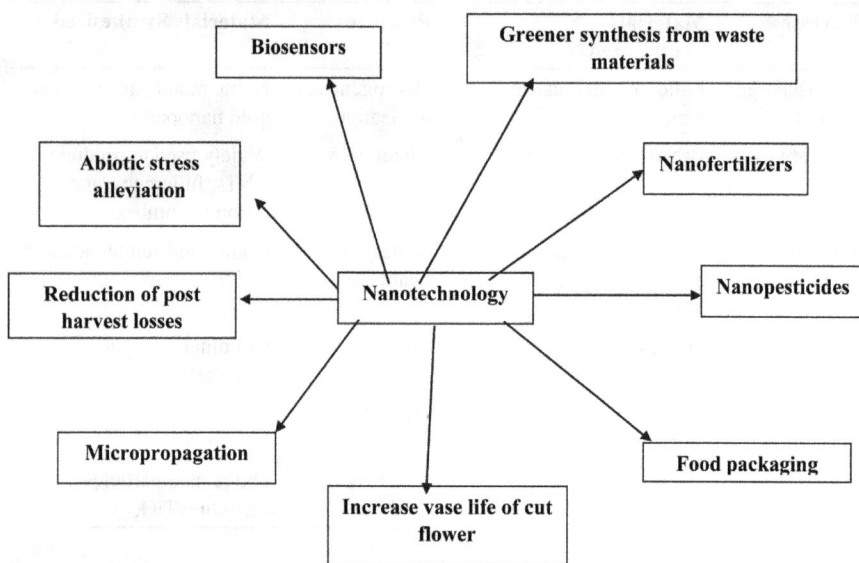

FIGURE 18.1   Prospects of nanotechnology in horticulture.

18.5.1   GREENER SYNTHESIS OF NANOMATERIALS

The development of green production method in nanotechnology is one of the major methods and since then interest in zero-valent iron nanoparticles is growing in which the use of natural products or waste is well done. However, there is still a lack of thorough study and knowledge of this application which allow us to understand the production and application processes. Machado et al. (2013) reported that assess the viability of the utilization of various trees leaves to formation extracts, which are adequate of reducing Fe^{+3} in the aqueous solution to form nanoscale zero valent iron (nZVI). The quality extracts of antioxidant capacity are evaluated from extract. The results were expressed that: (1) the antioxidant capacities are produced higher in dried leaves; (2) most favorable condition (contact time, volume:mass ratio, and temperature) for extraction was identified for every leaves; (3) the green development was main aim, but low cost method and water were used as solvent; (4) based on antioxidant capacity, the extracts were classified in three categories (showed as Fe(II) concentration): 2–10 mmol L^{-1}, 20–40 mmol L^{-1}, and >40 mmol L^{-1} with pomegranate, oak, and green leafy tea produced the richest extracts; and (5) (Figure 18.2) transmission electron microscope provides that nZVIs ($d =$ 10–20 nm) can be extracts produced through the trees leaves (Figure 18.3).

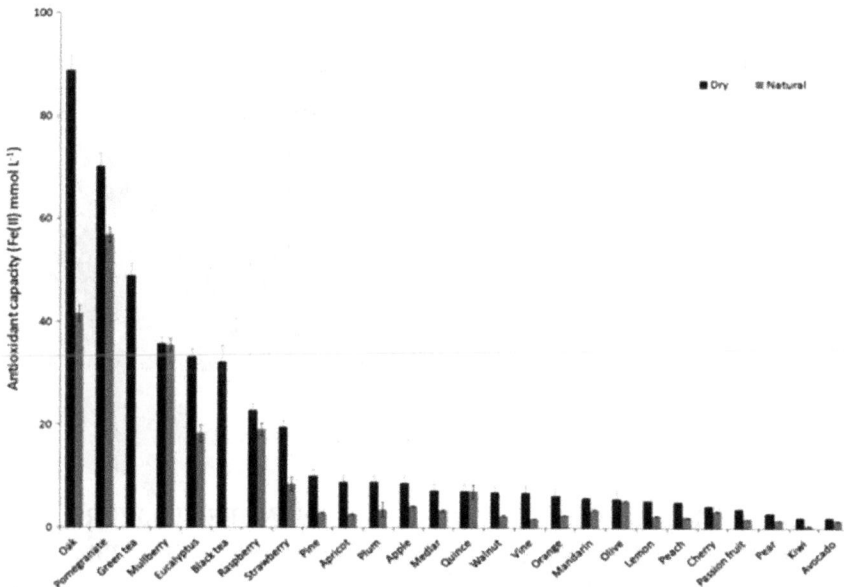

FIGURE 18.2   Antioxidant production capacity of different leaf wastes.

Source: Reprinted with permission from Machado 2013. © Elsevier.

18.5.2   EFFECT OF NANO-BORON OXIDE ON VARIOUS GROWTH STAGES OF SUGAR BEET

Effects of nano-B on the various physicochemical attributes of the sugar beet were evaluated by Pirzad et al. (2019). Foliar application of nano-B oxide (0, 2, 3, and 4 g L^{-1}) at several growth stages (20%, 40%, 60%, 80%, and 100%

of ground cover) under the randomized block design design with three replications was examined at Urmia University, Iran. The application of nano-B
(3 g L^{-1}) recorded significantly highest soil plant analysis development value,
relative water content, and leaf number. The higher B concentration reduced
the number of leaf and low chlorophyll content due to toxicity of boron (Armin
and Asgharipour, 2012). Ground cover of 40% resulted significantly in the
highest root yields, optimum leaf area (86.47 cm^2), and sugar content such
as white sugar and technological sugar with the foliar application (4 g L^{-1}) of
nano-B. Concentration of nano-B increased with various growth stages with
sugar and white sugar were also increased on the contrary impurities (sodium,
potassium, and α-amino N) loss. Increase sugar and white sugar contents at
ground cover of the 40% with the foliar spraying of nano-boron (3 g L^{-1}) were
also reported. Nano-B provided the facilities for the sugar transportation in
plant and due to this it had a major role in the auxin biosynthesis (Dugger,
1973; Ullah et al., 2013). In overall view, the foliar spaying of 3 g L^{-1} of nano-B
gave significantly results for improvement in vegetative growth of sugar beet.

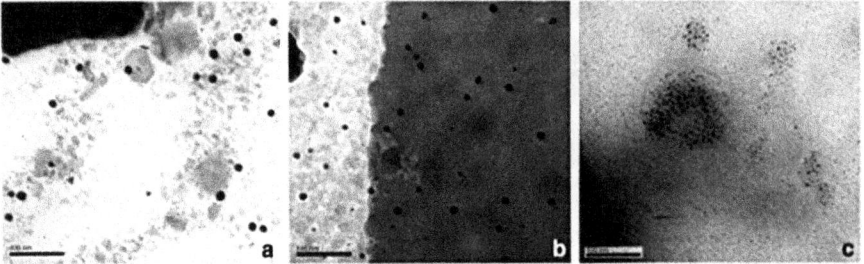

FIGURE 18.3   TEM image of produced ZVIs Fe.

Source: Reprinted with permission from Machado 2013. © Elsevier.

18.5.3   EFFECT OF GREEN SYNTHESIS OF NANOFERTILIZERS ON PHOTOSYNTHETIC PIGMENTS AND PROXIMATE COMPONENTS OF THE CUCURBITA PEPO L.

Two season 2017 and 2018 experiments were conducted on effect of green
synthesized nanomaterials on *Cucurbita pepo* (Shebl et al., 2019). Microwave-
assisted hydrothermal green synthesis of nanomaterial using bulk source of
analytical grade salt such as ferric nitrate [Fe(NO$_3$)$_3$·9H$_2$O], manganese nitrate
[Mn(NO$_3$)$_2$·4H$_2$O], and zinc nitrate [Zn(NO$_3$)$_2$·6H$_2$O].. After synthesized,
NPs characterized through XRD, SEM, and TEM. Results were expressed in
Figure 18.4. The X-ray diffraction was characterized and confirms the purity
and proves the successful synthesized nano-oxides such as Fe$_2$O$_3$, Mn$_3$O$_4$, and

ZnO and its average particle size was 51.6, 56.0, and 36.8 nm, respectively. The TEM and SEM image were analysis of the morphology and textural of nanoparticles. The average crystalline size of synthesized nano-oxide was about 20–60 nm. Moreover, the synthesized zinc oxide nanoparticles (ZnO-NPs) were SEM characterized that result in small particles size and close packing texture as compared with other nano-oxide. The all characterized values indicated that all the synthesized materials were in the nanoscale range. The foliar application of the synthesized MnO-NPs @ 20 ppm on the photosynthetic pigments content of squash plants such as Chla, Chlb, total chlorophyll, and carotenoids was enhanced in the leafs followed by application of FeO- and ZnO-NPs compared with other treatments (Figure 18.5). The Mn plays an important role in photosynthesis process, which sunlight energy turn or change in photosynthetic energy, that is, plant energy which major affected organs of cell is the green plastids in the leaves and chlorophyll production (Schmidt et al., 2016). Proximate components were significantly increased such as organic matter (OM), protein content, and total energy with the foliar application of Mn-NPs, while the application of nano-Fe significantly enhances percentage of ash content for the leaves fiber content.

(a)　　　　　　　　　(b)　　　　　　　　　(c)

FIGURE 18.4 Characterized image of nano-oxide by SEM (a) Fe_2O_3, (b) Mn_3O_4, and (c) ZnO.

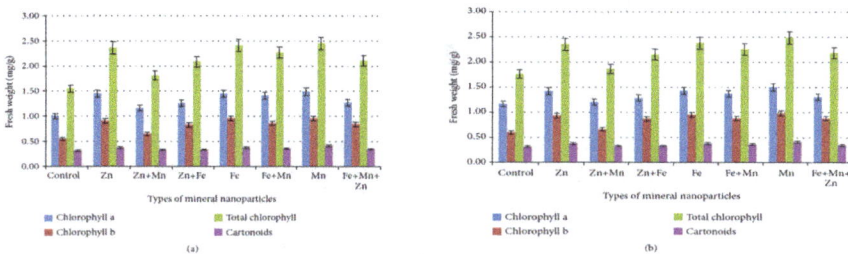

FIGURE 18.5 Effect of nano-FeO, -MnO, and -ZnO on photosynthetic pigments of *Cucurbita pepo* L. leaves during two seasons.

Source: Reprinted from Shebl et al., 2019. https://creativecommons.org/licenses/by/4.0/

18.5.4 EFFECT OF NANOFERTILIZER, FULVIC ACID, AND BORON FERTILIZER ON YIELD AND SUGAR CONTENT YIELD OF SUGAR BEET (BETA VULGARIS L.)

Two researches were conducted in split–plot system with three replications at Abis region, Alexandria, Egypt by Kandil et al. (2020). Monogerm sugar beet cultivar was used for experiments with four foliar treatments such as falvic acid (FvA), nitrogen, phosphorous, potash (NPK)-NPs, and FvA + NPK-NPs. The root yield (22.5 and 22.0 t/fed), shoot yield (6.3 and 6.4 t/fed), and biological yield (28.8 and 28.4 t/fed) (Figure 18.6) of sugar beet significantly increased with the interaction application of FvA, NPs, and NPK with B fertilizer. Similar results were obtained with previous researchers (Dewdar et al., 2018; Abdelsalam et al., 2019) reporting that yield and yield component of many crops increased with increase in the concentration of NPK-NPs. In contrast, El-Hassanin et al. (2016) studied that the growth and yield attributed enhanced many crop species with the application of FvA. The foliar application of FvA + NPK-NPs treatments positively enhances the quality attributes such as sugar content as well as the percentage of sucrose, total soluble solids, and purity of the sugar beet plants during both seasons (Figure 18.7). The significantly lowest mean value was found in the control treatment (water application).

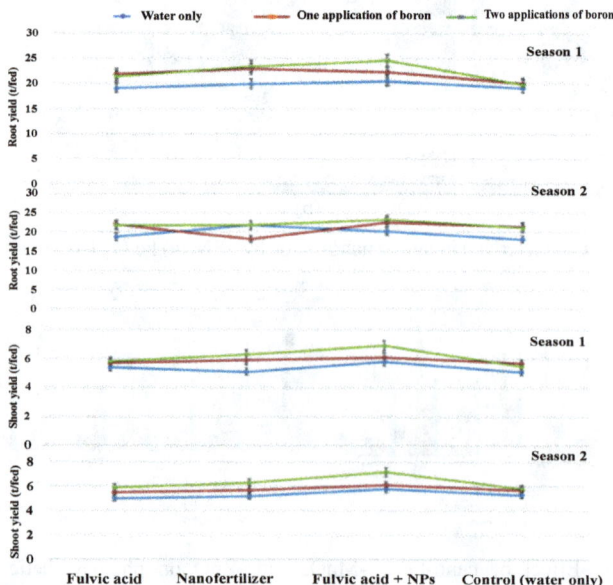

FIGURE 18.6   Effect of interaction applications between FvA + NPs and B fertilizer applications on yield attributes (root and shoot) of sugar beet in two seasons.

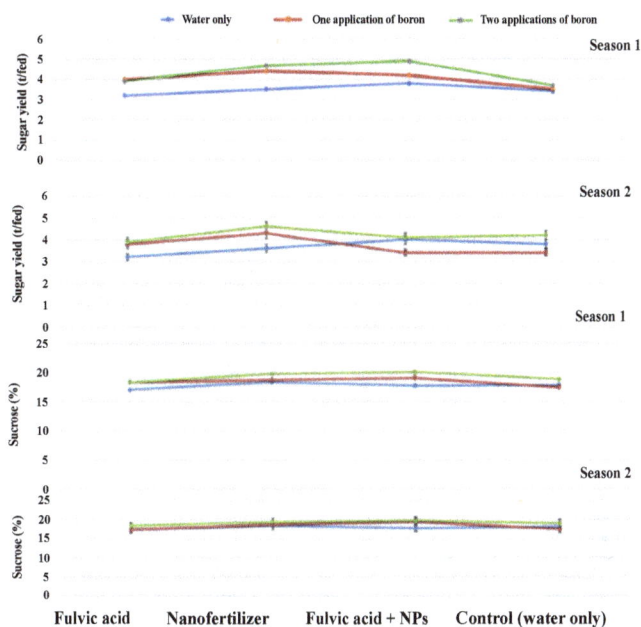

FIGURE 18.7   Effect of interaction applications between FvA + NPs and B fertilizer applications on quality content of sugar beet in two seasons.

18.5.5   IMPACTS OF NANO- AND NONNANOFERTILIZERS ON YIELD, QUALITY, AND NUTRIENT USE EFFICIENCY OF POTATO

According to Azeima et al. (2020), the field experiments were laid out in a design of factorial based on complete randomized block design with eight treatments under three replication and evaluated the yield and yield attributes of potato crop that were influenced by the application of nano-NPK fertilizer and nonnanofertilizer. Results showed that the potato fresh yield (23.71 ton ha^{-1}) and nutrient use efficiency were significantly increased with the foliar application of NPK-nanofertilizers (NFs) (50% half recommended dose) compared to other soil addition fertilizer. The tuber quality of potato was determined for standard parameter such as starch and NO$_3$ content production. Among the application treatments, foliar application of 100% nano-NPK fertilizers, even though paralleled with 50% NPK nanofertilizers and soil added 100% NPK nanofertilizers, recorded more starch yield (81.34%) whereas the foliar application of 25% NPK nanofertilizers recorded lowest starch yield (74.15%). The application of NFs will have the great future potential role in improved crop yield production and quality of potato by removing future threats and reducing the cost fertilizer and pollution hazards of ecosystem (Figure 18.8).

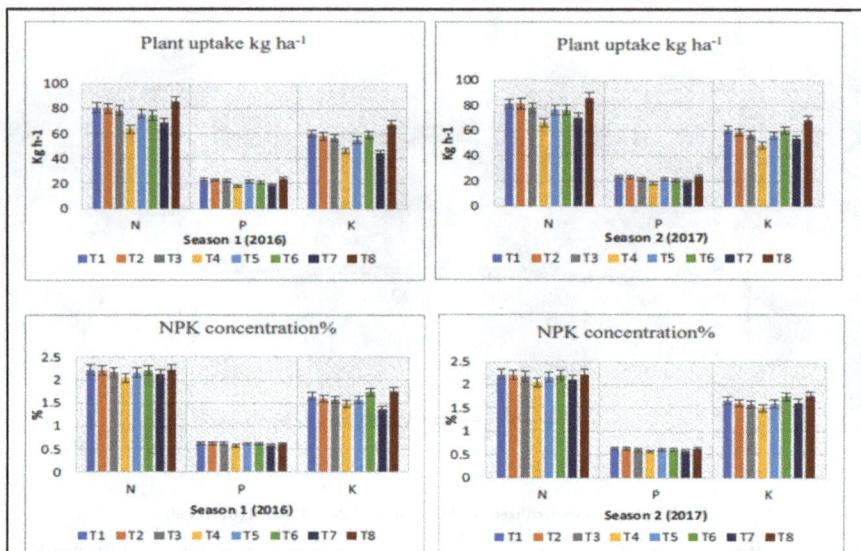

FIGURE 18.8   Potato vegetative uptake and concentrations of NPK (kg ha^{-1}) as affected by NPK nanofertilizers and nonnanofertilizers for both seasons.

18.5.6   EFFECT OF NANOFERTILIZER FOR ITS CONTROL RELEASE PROPERTIES

Chhowalla (2017) studied that today urea leaching is a major problem faced by farmers; thus, major solution for this problem improves the urea availability for plant and minimizes the adverse environmental effects. Nanofertilizer was developed by using urea coated hydroxyapatite NPs (Kottegoda et al., 2017). This production significantly reduced the urea requirements for fertilization, since it can be applied locally.

Perhaps these methods are very impressively, the researcher demonstrates that rice yield is significantly increased even when application of 50% less urea used. Hydroxyapatite NPs are urea coated that are achieved by controlled release of phosphoric acid into a Ca(OH)$_2$ suspension and urea followed by drying faster. The laboratory value for urea released from nanohybrids with a ratio of 1:6 hydroxyapatite to this ratio in which urea released 12 times more slowly as compared to pure urea.

18.5.7 IMPACT OF CARRIER NANO-ZN FORMULATION ON RELEASE KINETIC OF ZN AND FE IN THE SOIL ORDER

Adhikary et al. (2019) reported that the experiments were laboratory incubation with two distinct soils order, namely, *Inceptisol* (*Typic Haplustepts*) and *Alfisol* (*Typic Paleustalfs*) were conducted at BAU, Sabour, Bihar. Zinc acetate (0.1 M) and $Na(OH)_2$ chemical synthesized the ZnO-NPs following the method outlined by Aneesh et al. (2007) and the another synthesized zincated nanoclay polymer composite (ZNCPC) by following standard procedure as outlined by Mandal et al. (2018) by using chemical nanoclay, monomers ($C_3H_4O_2$ and C_3H_5NO), *N,N'*-methylenebisacrylamide (cross-linker), and initiator $[(NH_4)_2S_2O_8]$. Subsequently, synthesized nanocompounds were characterized by TEM images, XRD, and Fourier transform infrared spectroscopy (FTIR). Debye–Scherrer formula used average size of crystalline (Figure 18.9) ZnO-NPs and ZNCPC (Hall et al., 2000). The ZnO-NPs and ZNCPC powder

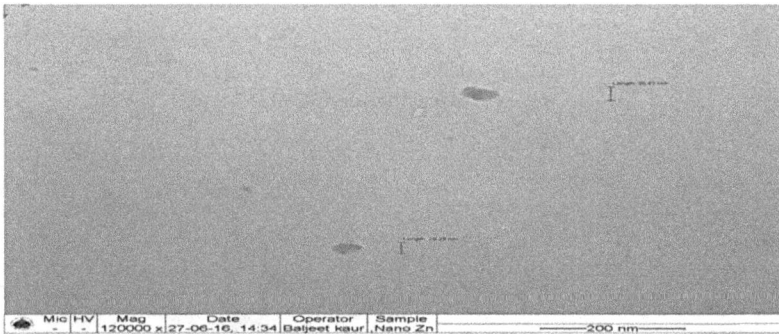

(a)

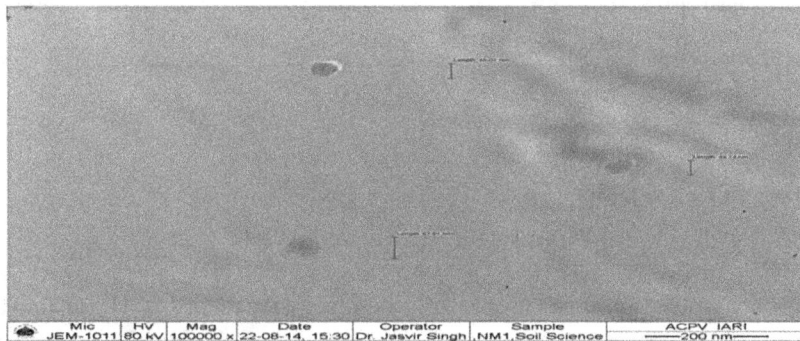

(b)

FIGURE 18.9   TEM image of (a) n-ZnO and (b) nanoclay.

Source: Reprinted from Adhikary et al., 2020. © Taylor & Francis.

diffractions were taken and scanned by X-ray diffractometer (Figure 18.10). FTIR characterized spectral region of 400–4000 cm^{-1} for the characterization of ZnO-NP and ZNCPC (Figure 18.11). Number of the five treatments were applied including four several Zn formulations, namely, (T$_1$) ZNCPC, (T$_2$) n-ZnO, (T$_3$) Zn-citrate, (T$_4$) zinc sulfate (ZnSO$_4$· 7H$_2$O) @ 10 mg Zn kg^{-1} soil, and a control (T$_5$). Result showed that application of nano-Zn formulation was influenced the pH and Zn and Fe of diethylene triamine penta acetate acid (DTPA) extractable of the soil during incubation. Carrier nano-Zn (ZNCPC and Zn-citrate) in the pH of the soil is more potent than conventional ZnSO$_4$·7H$_2$O.

(a)

(b)

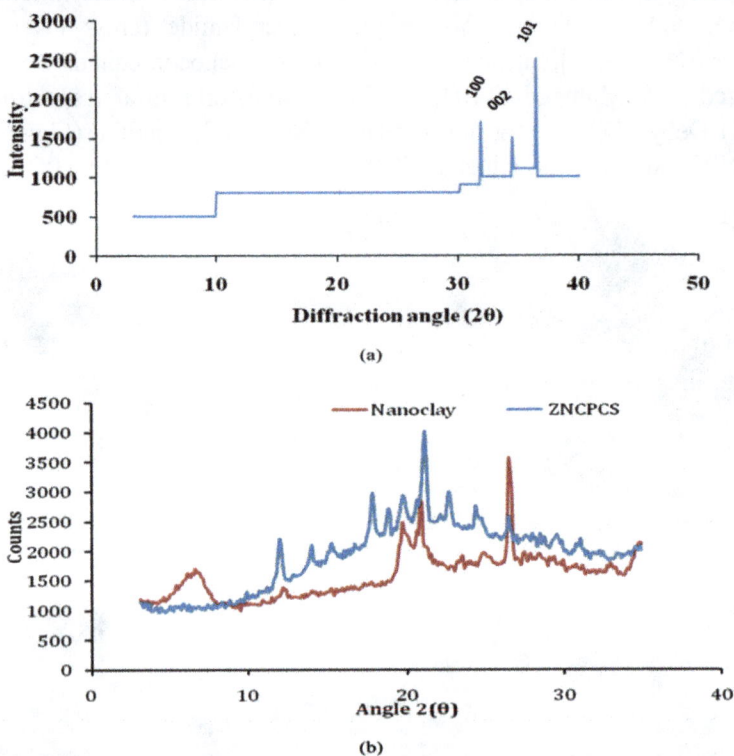

FIGURE 18.10   XRD characterized n-ZnO (a) and nanoclay and ZNCPC (b).

Source: Reprinted from Adhikary et al., 2020. © Taylor & Francis.

As the incubation speed progresses, the successive release of H$^+$ ion into the soil due to the deprotonation of citrate may cause the soil pH value to decrease (Mandal et al., 2015; Roy et al., 2016). The application of ZNCPC significantly higher DTPA extractable zinc content (2.99 mg kg^{-1}) was recorded at 45 days after incubation. ZNCPC doing slow release Zn formulation where diffusion barrier performed by clay. The DTPA extractable Fe content (48.70 mg kg^{-1}) during incubation is significantly maximum was observed at incubation during initial days, while lowest was found at incubation after 45 days (30.18 mg kg^{-1}). Iron content was reduced with DTPA-Zn increase owing to formation of franklinite type of minerals in the *Alfisol* as well as *Inceptisol* (Lindsay, 1979). In overall view, the ZNCPC was observed to be most efficient in comparison to the Zn-NPs carrier and conventional ZnSO$_4$ 7H$_2$O.

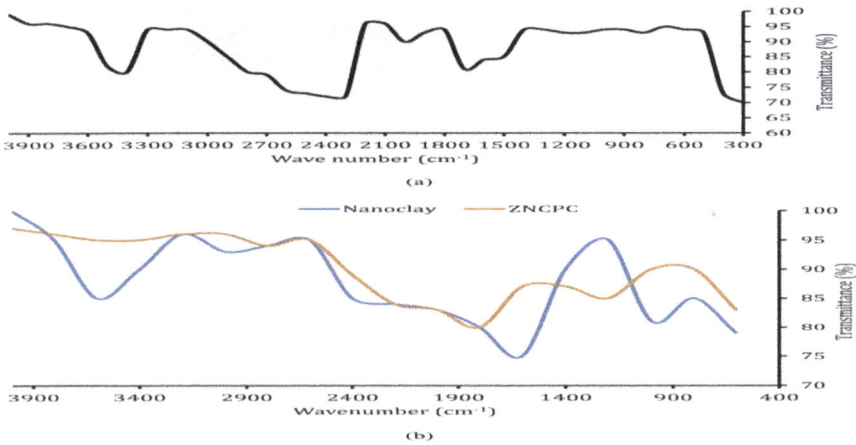

FIGURE 18.11   FTIR spectra at 400–4000 cm^{-1} of n-ZnO (a) and nanoclay and ZNCPC (b).

Source: Reprinted from Adhikary et al., 2020. © Taylor & Francis.

18.5.8   EFFECT OF NOVEL CHITOSAN GRAFTED ZINC CONTAINING NANOCLAY POLYMER BIOCOMPOSITE (CZNCPBC) AND NANOCLAY ON EQUILIBRIUM WATER ABSORBENCY, ZN^{+2} CONTENT, OLSEN-P CONTENT, AND CO$_2$-C EVOLVED

Mandal et al. (2018) experiments were conducted at Indian Agricultural Research Institute, New Delhi. Synthesis of nanoclay (fractionated bentonite) and synthesis of CZNCPBCs take place in laboratory through ultracentrifugation by several analytical grade chemical compounds and subsequently characterization was taken through the TEM image, XRD, FTIR, and SEM image. Results showed that the nanoclay recorded higher specific surface area and cation exchange capacity as comparison to clay. TEM image of the nanoclay and clay confirmed that bentonite fractionated was in nanosize (Figure 18.12). Random powder diffraction method was characterized by XRD (Figure 18.12). FTIR characterized nanoclay and clay as the strong

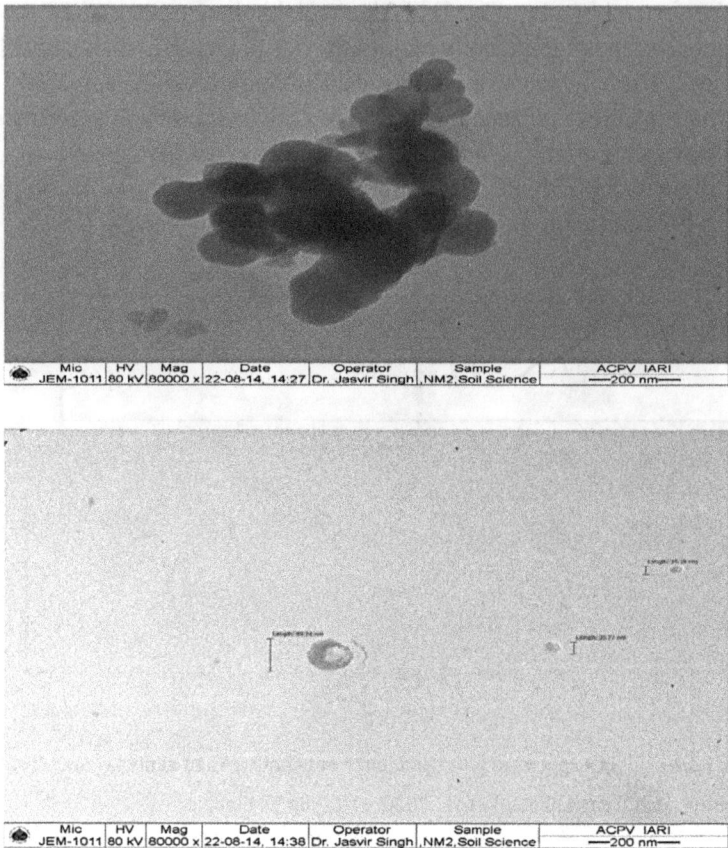

FIGURE 18.12   TEM image of locally bentonite (a) and nanobentonite (b).

Source: Reprinted from Adhikary et al., 2020. © Taylor & Francis.

bound around 3400–3800 cm^{-1} and chitosan peak recorded at 3648 cm^{-1}. Surface morphology of nanoclay and CZNCPBCs was studied by SEM. In this research experiments, seven treatments were taken, namely, T_1-8% clay, T_2-10% clay, T_3-12% clay, T_4-8% nanoclay, T_5-10% nanoclay, T_6-12% nanoclay, and T_7—conventional Zn sources such as $ZnSO_4 \cdot 7H_2O$. The equilibrium water absorbency was significantly affected by type of clay and clay content. Significantly, highest water equilibrium absorbency and water content (gravimetric water content in g mg^{-1} of material) were recorded under the bentonite clay type and clay content of monomer (8%), while lowest equilibrium water absorbency and water content were observed in the case of nanobentonite type of clay which has 12% of monomer of the content of clay (Figure 18.13). Increase the content of clay with decrease in equilibrium water absorbency

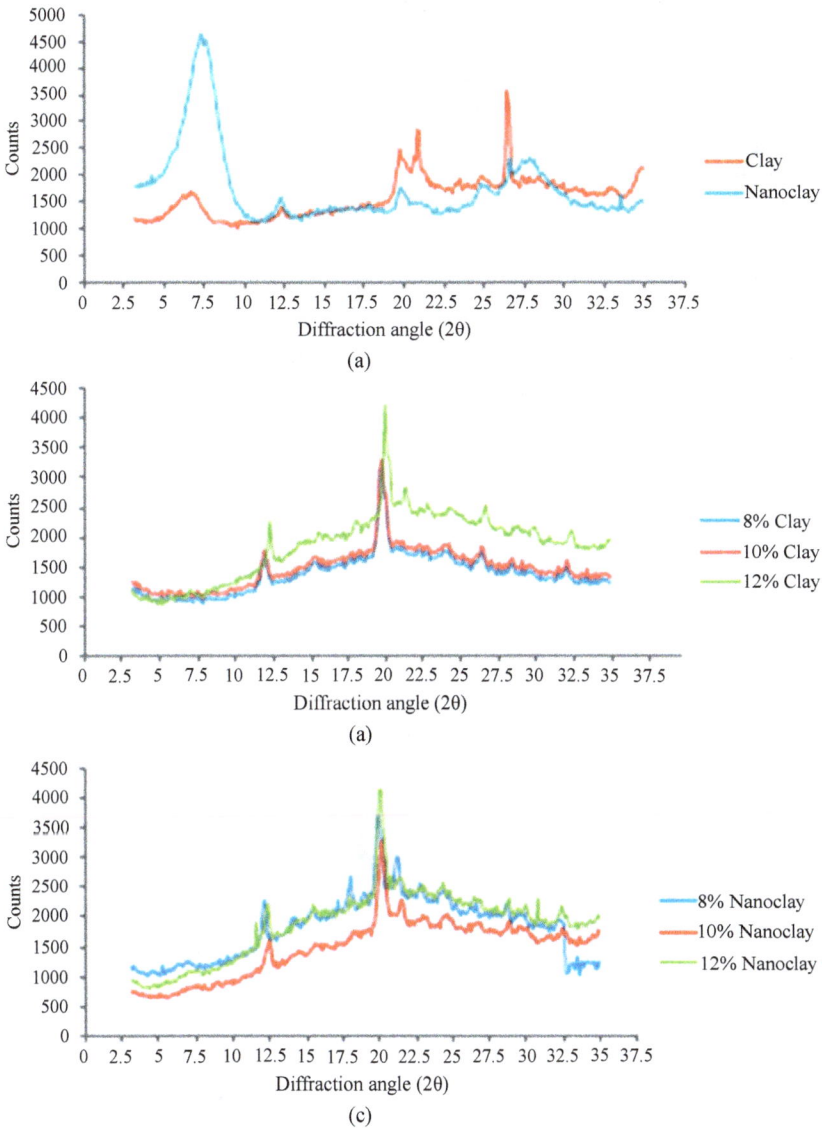

FIGURE 18.13 XRD of (a) locally bentonite and nanobentonite (b) clay and (c) nanoclay containing CZNCPBCs.

Source: Reprinted with permission from Mandal, et al. 2018. © Elsevier.

due to cross-linking density of polymeric matrices (Sarkar et al., 2014; Singh et al., 2011). During incubation period in soil I, application of 8% clay type gave significantly maximum Zn^{+2} content (7.30 mg kg^{-1}), while lowest was recorded under the 8% clay (2.50 mg kg^{-1}) at 60 days after incubation and similar extractable Zn^{+2} content pattern found in the soil II (Figure 18.14). In case of Olsen-phosphorus content (6.02 mg kg^{-1}), it was mentioned that the maximum under the application of 8% clay whereas lowest Olsen-P content (2.50 mg kg^{-1}) was found under the conventional Zn sources (ZnSO$_4$·7H$_2$O) at 60 days after incubation in soil I. Similar result was found in soil II such as the highest Olsen-extractable phosphorus content (7.89 mg kg^{-1}) and minimum Olsen-P content (2.43 mg kg^{-1}) (Figure 18.15). The biodegradability of CZNCPBC was affected by fungus. Inoculation of the fungal significantly increased cumulative CO$_2$-carbon evolution across all treatment types and incubation periods (30–90 days) as compared to conventional Zn sources (ZnSO$_4$·7H$_2$O). The highest CO$_2$-C (817 mg g^{-1} of CZNCPBCs) was found under the 8% clay type at the end periods during time of incubation (90 days).

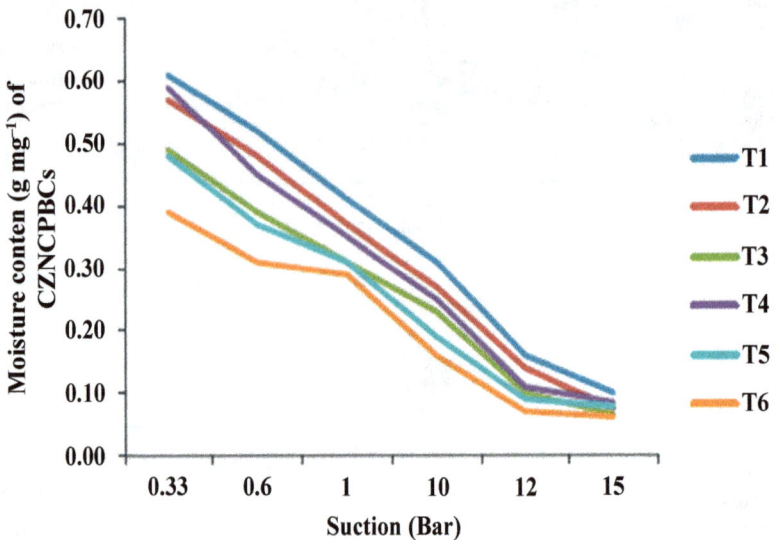

FIGURE 18.14   Moisture characteristics of CZNCPBCs under all treatments.

Source: Reprinted with permission from Mandal, et al. 2018. © Elsevier.

18.5.9   EFFECT OF SYNTHESIS OF MANGANESE ZINC FERRITE AS A NANOFERTILIZER ON YIELD, PROXIMATE COMPONENTS, AND MINERALS CONTENT OF THE SQUASH PLANT (CUCURBITA PEPO L.)

Shebl et al. (2020) revealed that the synthesized manganese zinc ferrite acts as a nanofertilizer by template-free microwave-assisted hydrothermal

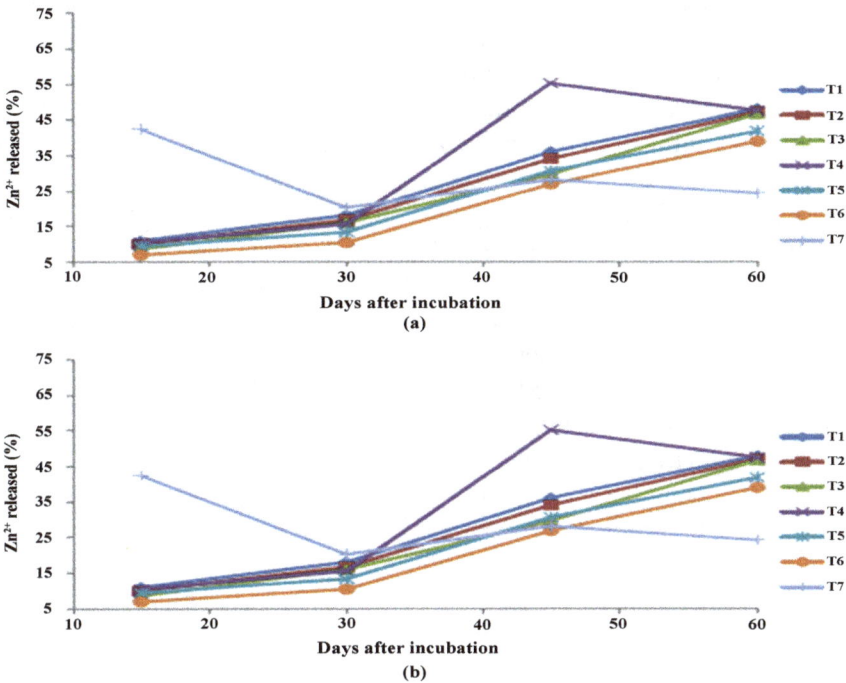

FIGURE 18.15 Percentage of Zn^{2+} released during incubation period in soil I (a) and soil II under the 1–7 treatment.

Source: Reprinted with permission from Mandal, et al. 2018. © Elsevier.

at several temperatures holding 100 °C, 120 °C, 140 °C, 160 °C, and 180 °C and its foliar spraying under various concentrations (0, 10, 20, and 30 mg kg^{-1}) under *Cucurbita pepo* L. crop. Several characterization processes to prove the successful production of the nanofertilizer via XRD, field emission scanning electron microscopes (FE-SEM), high-resolution field emission gun (HRFEG), and HR-TEM as well as the surface area and pore structure were analyzed by using nitrogen gas sorption isotherms. The X-ray diffraction peaks at degree of angle (2θ) values 30°, 35.3°, 42.9°, 53.1°, 56.8°, 62.4° and 73.7°, which were identified as showed that the crystal structure all the samples synthesized manganese zinc ferrite (Figures 18.16 and 18.17). Morphology and textural analysis were analyzed by FE-SEM and HR-TEM (Figures 18.18 and 18.19) and all the prepared ferrite NPs showed a shape as a cubic like and enhance the crystalline and regularity with increased in the holding synthesis temperature. The foliar application of manganese zinc ferrite significantly enhances the attributes of growth and fruit of the *Cucurbita* during two successive seasons, 2017 and 2018.

The interaction of temperature-160 and 10 mg kg^{-1} concentration gave significantly higher yield. The proximate component and minerals content of squash fruit such as protein, fibers, OM, lipids and total energy and carbohydrate and nitrogen, phosphorus, potash and zinc, iron and manganese were significantly influenced by the interaction of temperature and concentration of MnO·6ZnO·4Fe$_2$O$_4$ nanofertilizer. *Cucurbita* plant treated with the T-180 at the concentrations of 30 mg kg^{-1} and 10 mg kg^{-1} gave significantly higher protein and fibers content as well as higher N and K content, while higher OM content, lipids, and total energy obtained with T-160 at 20 mg kg^{-1} but P and Mn content were influenced by treatment with T-160 and 20 and 30 mg kg^{-1}, respectively. The percentage of ash increased with T-100 on 10 ppm, while carbohydrate improved with T-120 and 30 mg kg^{-1}. Zinc content was found highest in the treatment T-100 at 30 mg kg^{-1} whereas the Fe content increased with treatment with T-140 at 30 mg kg^{-1}.

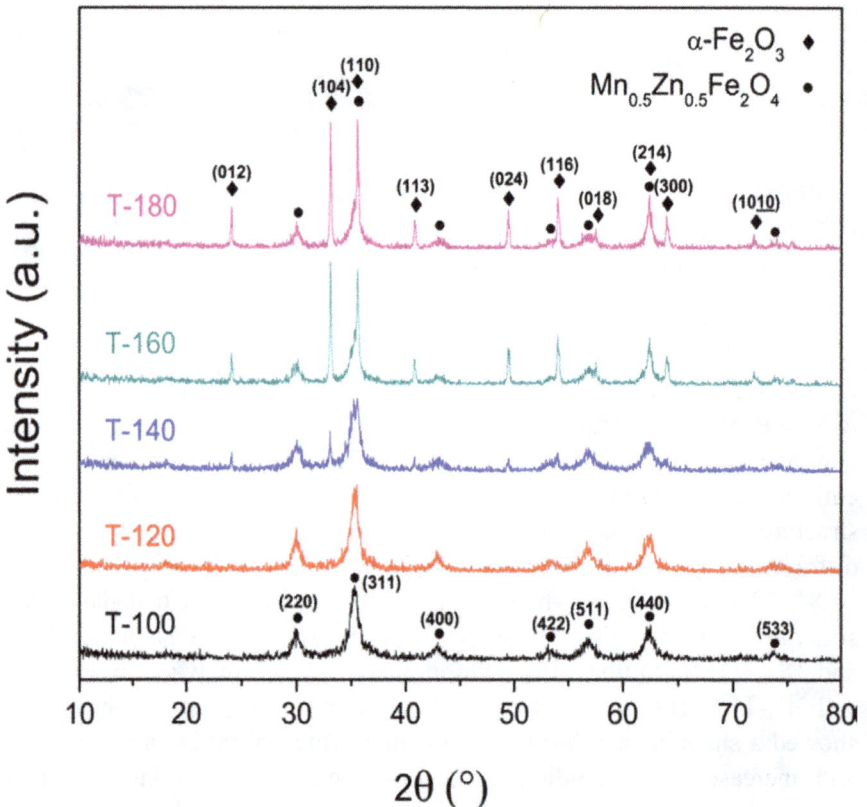

FIGURE 18.16 XRD characterized nanofertilizer samples.

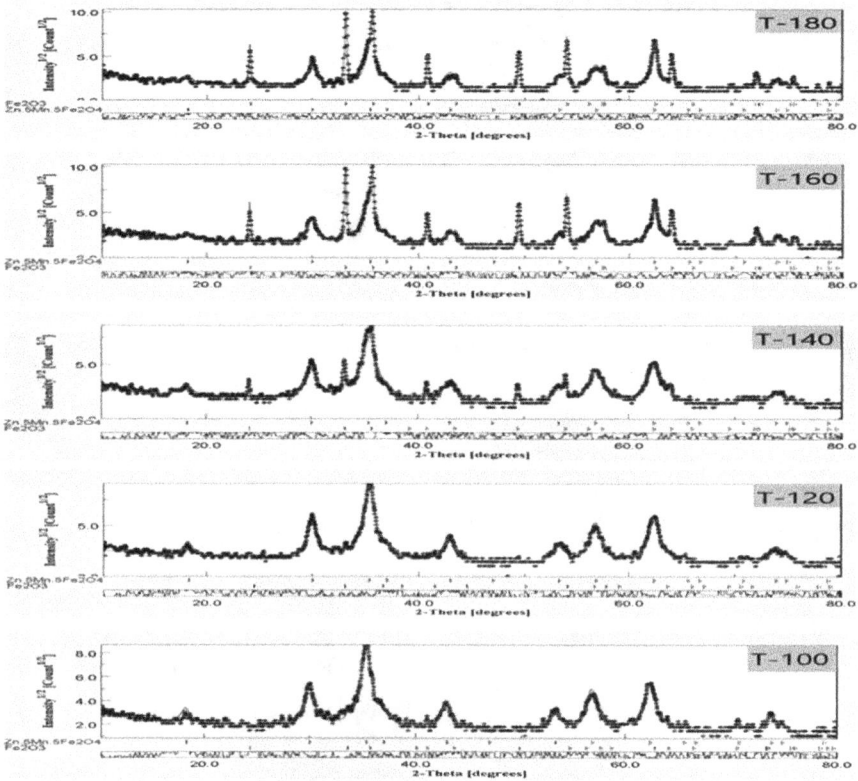

FIGURE 18.17 MAUD refinement of the XRD data of ferrite nanofertilizer samples.

18.5.10 EFFECT OF NANOSILICA AND SILICA ON SALINITY ALLEVIATION

Haghighi and Mohammad (2013) experiments were conducted at hydroponics system in greenhouse. The different concentration of the application of the Si and nano-Si was positively influenced on growth and some key characteristics of gas exchange of tomatoes under the stress condition of salt. The factorial complete randomized design experiment was arranged with six replications at 0, 25, and 50 mM NaCl levels and the concentration of Si and NPs-Si was taken at 0, 1, and 2 mM. The application of NaCl is a major addition to the nutrient solution after the application of silicon and non-Si. After application, nutrient assessed the growth, photosynthetic parameters as well as moisture content in the plants. Results expressed that fresh weight and dry weight (weights) of plant, root volume, and diameter

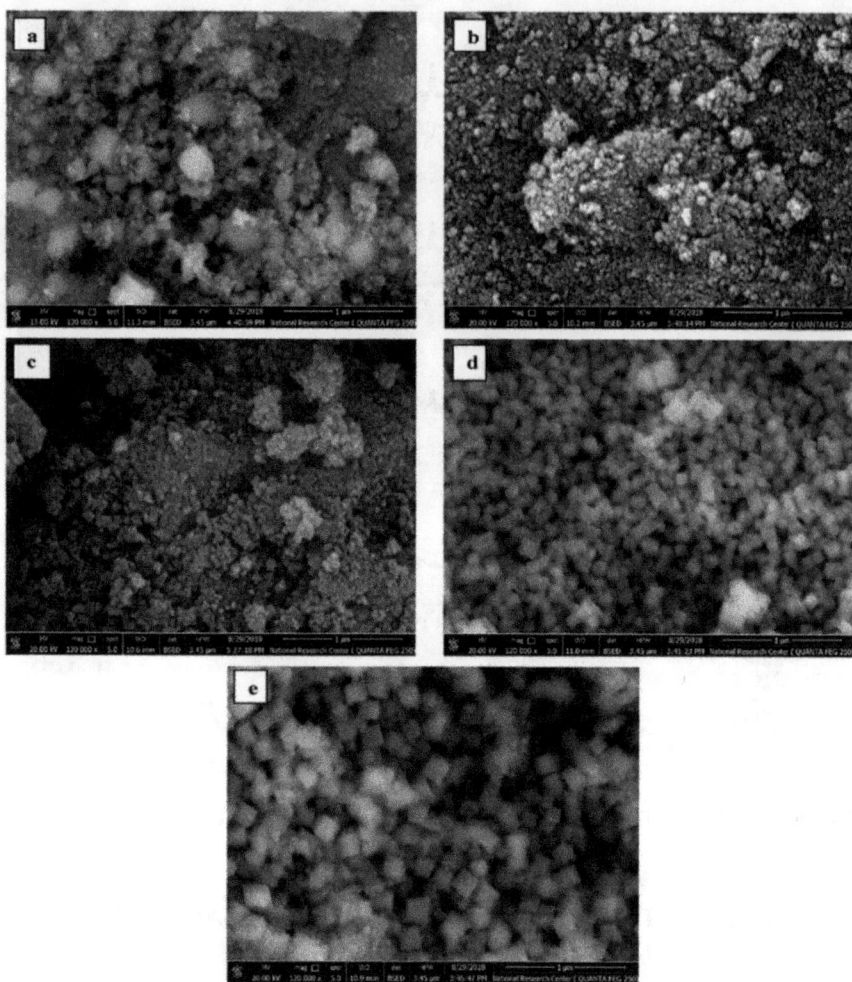

FIGURE 18.18   FE-SEM images of ferrite nanofertilizer samples.

of stem were deleterious effects by salinity, while application of silicon was increased by dry weight and fresh weights of plants, volume of root, concentration of chlorophyll, rate of photosynthesis, and the leaf moisture content due to the silicon that reduce the stress effect of salinity. Conversely, more salinity levels were increased by electrolyte leakage. The substomatal CO_2, rate of photosynthesis (Figure 18.20), mesophyll conductance, and photosynthetic water use efficiency (WUE) (Figure 18.21) were also diminished with increased in the levels of salinity, but the application of Si was decreased the CO_2 of substomatal and conductance of stomatal. The application of nano-Si and non-Si was enhanced the photosynthesis rate, mesophyll conductance, and WUE under the condition of saline stress. Any significantly difference was not observed between application of NPs-Si and Si. Overall view, the application of Si was positively improved tomato plant under salinity stress condition. These experiments studied that Si governs or plays an important role in salinity regulating responses and indicating that protects the activity of photosynthesis of tomato plants against the salinity hazardous.

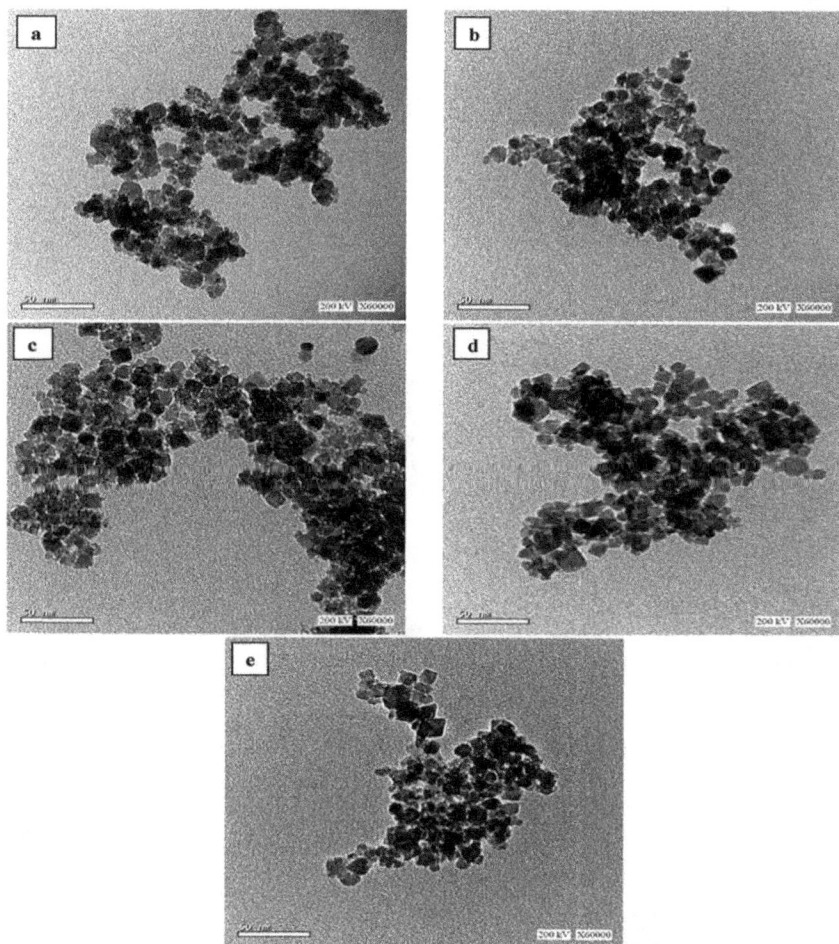

FIGURE 18.19 HR-TEM images of ferrite nanofertilizer samples.

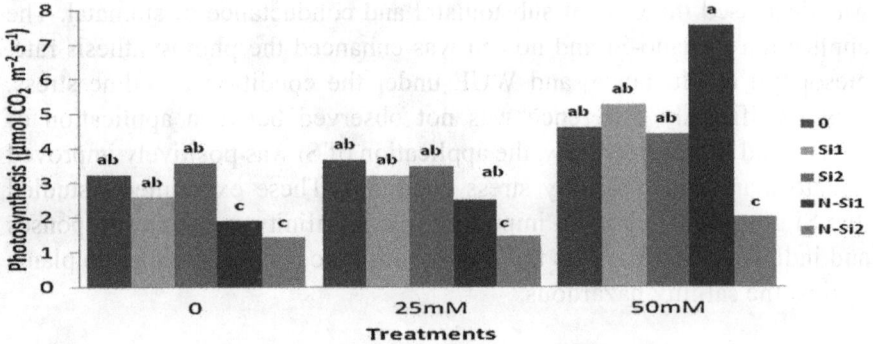

FIGURE 18.20 The effect of salinity stress of NaCl and Si and NPs-Si on photosynthesis rate of tomato plants.

Source: Reprinted with permission from Haghighi and Moh 2013. © Elsevier.

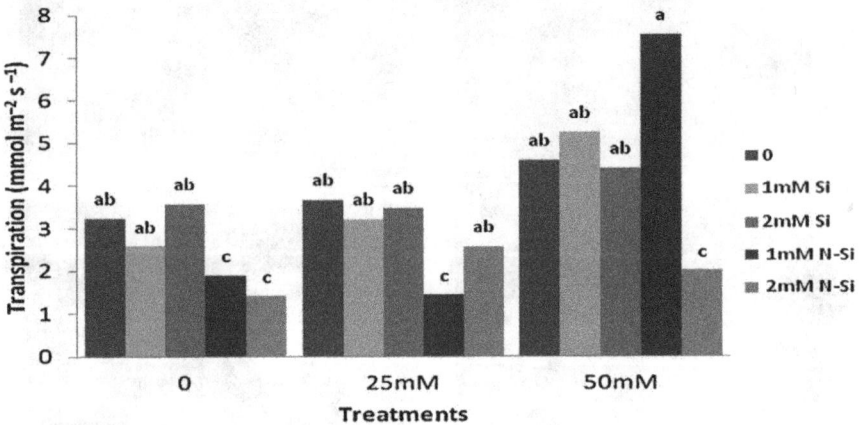

FIGURE 18.21 The effect of salinity stress of NaCl and Si and NPs-Si on the transpiration of tomato plants.

Source: Reprinted with permission from Haghighi and Moh 2013. © Elsevier.

18.5.11 FERTILIZER FROM WASTE PRODUCTS

The developed country facing many environmental challenges such as disposal of rubber tires (Moghaddasi et al., 2013). A new technology introduced to use rubber as a Zn source for plant. According to the previous studies, the particle size of rubber is a major factor for calculation of the nutrient release rate from the waste of the rubber. Rubber waste was used for the preparation of nanoparticles and applying various treatments. The ball mill method is best suitable approach for formation of nanoparticles from waste tire rubber in the presence of Si waste (Figure 18.22). The efficiency of the Zn source of nanoparticles prepared through rubber was determined by nutrient solution culture in cucumber and it is compared with marketable $ZnSO_4 \cdot 7H_2O$. Ground rubber tire, NPs of tire rubber, and ash of rubber were used in hydroponic experiments. Nutrient solution of NPs of tire rubber and ash of rubber had more yields of shoot and fruit of cucumber and also higher accumulation of Zn content in tissues as comparison with $ZnSO_4 \cdot 7H_2O$ grown crop (Figure 18.23). Thus, synthesized NPs from the waste rubber tire and ash of rubber were higher value of Zn containing fertilizer for *Cucumber* grown in the solution of nutrients.

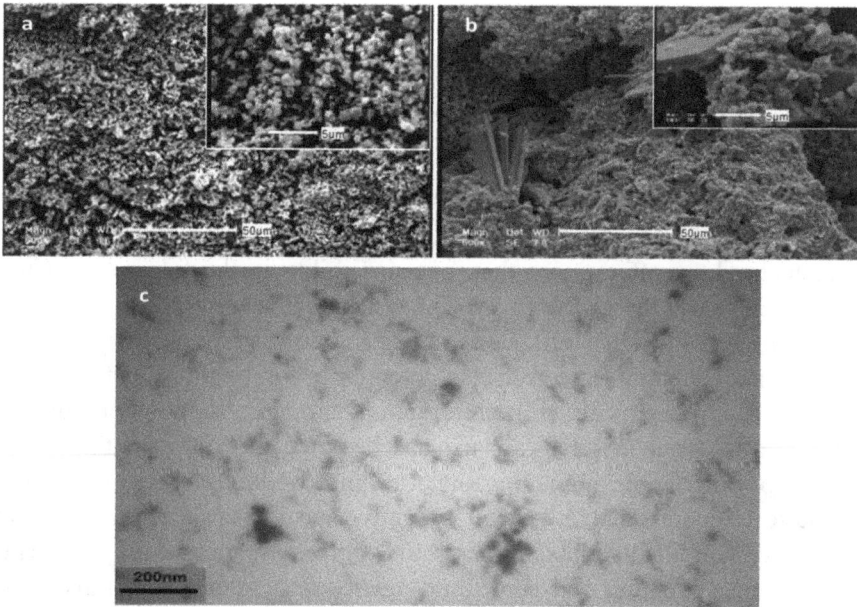

FIGURE 18.22 (a) The SEM image of ash of waste tire rubber before milling, (b) after milling for 5 h with silicon wastes 1:1 and (c) TEM image of ash of waste rubber tire after milling (<50 nm).

Source: Reprinted with permission from Moghaddasi et al., 2013. © Elsevier.

18.5.12 CONTROLLED RELEASE FERTILIZER

Controlled or slow release fertilizers are those which synchronize the nutrient release pattern of fertilizers with the nutrient requirement of plants. Nanoclay composites can be used as a superb coating material for the preparation of these controlled release fertilizers.

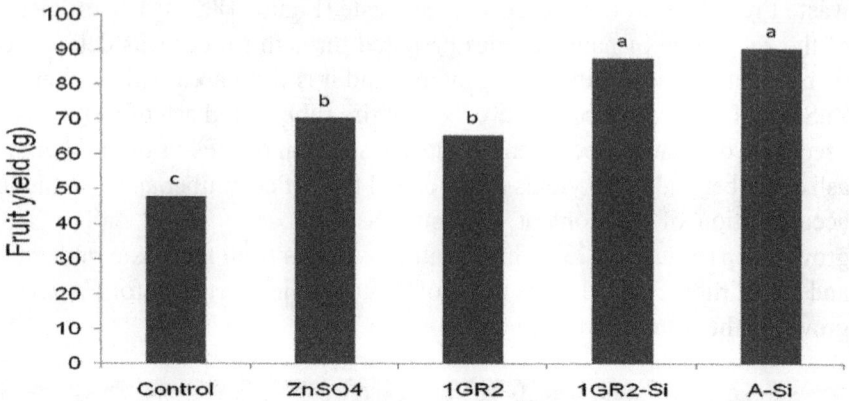

FIGURE 18.23   The different source of Zn in nutrient solution was affected by fruit yield of cucumber.

Source: Reprinted with permission from Moghaddasi et al., 2013. © Elsevier.

18.6   CONCLUSION

Nanotechnology offers more than one possibility in the future but at the same time, it will largely be governed by the simultaneous advances in new, fast, simple, and highly efficient characterization techniques for nanoparticles. Nanomaterials have immense potential in formulations of fertilizers, pesticides, agrochemicals, fruit preservation, and biosensing in horticulture. Horticultural waste materials are also a rich source for synthesis of nanomaterials for other application. Nanomaterial improved nutrient use efficiency of the soil minerals due to its greater reactive specific surface area and slow release system that available minerals are more easily uptake by plant. The production of coating and cementing agents such as nanomaterials were used and its use for slow release nutrient fertilizer, they are proved best to other coating agents. Nanoclay formation by polymer/clay superabsorbent composite and its application in agricultural and horticulture sectors can enhance excellent water retention capacity at lower production cost.

18.7   FUTURE ASPECTS

Field experiments should be undertaken to judge the efficiency of nanoclay-coated controlled fertilizer on the long run in the field conditions. Further studies are to be undertaken for standardization of an economic method for mass production of nanoclay composite fertilizers for field application.

Further research is to be done in the field of nanotoxicology for the safety evaluation of engineered nanoparticles through detailed study of the mechanisms and biokinetics causing adverse effects on organisms. Research on nanotechnology in horticulture is in infancy and lots of efforts on the fabrication, characterization, standardization, biodegradability, and eco-friendly nature of nanoparticles are needed.

KEYWORDS

- equilibrium water absorbency
- proximate components
- nanomaterials, nutrient use efficiency

REFERENCES

Abd El-Azeima, M. M., Sherif, M. A., Hussein, M. S., Tantawy, I. A. A. and Bashandy, S. O. (2020) Impacts of nano- and nonnanofertilizers on potato quality and productivity. *Acta Ecologica Sinica*, 10:388–397.

Abdelsalam, N. R., Fouda, M. M. G., Megeed, A. A., Ajarem, J., Allam, A. A. and Naggar, M. E. (2019) Assessment of silver nanoparticles decorated starch and commercial zinc nanoparticles with respect to their genotoxicity on onion. *International Journal of Biological Macromolecules*, 133:1008–1018.

Adhikary, S., Mandal, N., Rakshit, R., Das, A., Kumari, N. and Kumar, V. (2020) Nano zinc carriers influences release kinetics of zinc and iron in a laboratory incubation experiment under *Inceptisol* and *Alfisol*. *Journal of Plant Nutrition*, 43:1968–1979.

Aneesh, P. M., Vanaja, K. A. and. Jayaraj, M. K. (2007). Synthesis of ZnO nanoparticles by hydrothermal method. *Nanophotonic Materials IV*, 6639. https://doi.org/10.1117/12.730364.

Armin, M. and Asgharipour, M. (2012) Effect of time and concentration of boron foliar application on yield and quality of sugar beet. *American-Eurasian Journal of Agricultural and Environmental Sciences*, 12(4):444–448.

Banfield, J. F. and Zhang, H. (2001) Nanoparticles in the environment. In: "Nanoparticles and the Environment" (Banfield J. F. and Navrotsky, A., Eds.), *Mineralogical Society of America, Washington*, DC, Chapter 1, pp. 1–58.

Chhowalla, M. (2017) Slow release nanofertilizers for bumper crops. *ACS Central Science*, 3:156–157.

Dewdar, M. D. H., Abbas, M. S., El-Hassanin, A. S. and El-Aleem, H. A. A. (2018) Effect of nano micronutrients and nitrogen foliar applications on sugar beet (*Beta vulgaris* L.) of quantity and quality traits in marginal soils in Egypt. *International Journal of Current Microbiology and Applied Sciences*, 7:4490–4498.

Dugger, W. M. (1973) Functional aspects of boron in plants. *Advances in Chemistry Series*, 123:112–129.

EPA. (2005) U.S. Environmental Protection Agency: Nanotechnology White Paper, Science Policy Council, Washington, DC http://www.epa.gov/osa/pdfs/EPA_nanotechnology_white_paper_external_review_draft_12-02-2005.pdf. (downloaded July, 2008).

Haghighi, M, Mohammad, P. (2013) Influence of silicon and nano-silicon on salinity tolerance of cherry tomatoes (*Solanum lycopersicum L.*) at early growth stage. *Scientia Horticulturae*, 161:111–117.

Hall, B. D., Zanchet, D. and Ugarte, D. (2000) Estimating nanoparticle size from diffraction measurements. *Journal of Applied Crystallography*, 33(**6**):1335–1341. doi: 10.1107/S0021889800010888.

Hochella, M. F., Lower, S. K., Maurice, P. A., Penn, R. L., Sahai, N., Sparks, D. L. and Twining, B. S. (2008) Nanominerals, mineral nanoparticles, and earth chemistry. *Science*, 21:1631–1635.

Kandil, E. E., Abdelsalam, N. R., Aziz, A. A. A. E., Ali, H. M. and Siddiqui, M. H. (2020) Efficacy of nanofertilizer, fulvic acid and boron fertilizer on sugar beet (*Beta vulgaris* L.) yield and quality. *Sugar Tech*, 22:782–791.

Kottegoda, N., Sandaruwan, C., Priyadarshana, G., Siriwardhana, S., Rathnayake, U. A., Arachchige, D. M. B., Kumarasinghe, A. R., Dahanayake, D., Karunaratne, V. and Amaratunga, G. A. J. (2017) Urea-hydroxyapatite nanohybrids for slow release of nitrogen. *ACS Nano*, 11:1214–1221.

Lal, R. (2008) Soils and India's food security. *Journal of Indian Society of Soil Science*, 56:129–138.

Machado, S., Pinto, S. L., Grosso, J. P., Nouws, H. P. A., Albergaria, J. T. and Delerue-Matos, C. (2013) Green production of zero-valent iron nanoparticles using tree leaf extracts. *Science of the Total Environment*, 445–446:1–8. doi:10.1016/j.scitotenv.2012.12.033.

Mandal, N., Datta, S. C., Manjaiah, K. M., Dwivedi, B.S., Kumar, R. and Aggarwal, P. (2018) Zincated nanoclay polymer composites (ZNCPCs): synthesis, characterization, biodegradation and controlled release behavior in soil. *Polymer-Plastics Technology and Engineering*, 57(**17**):1760–1770.

Mandal, N., Datta, S. C. and Manjaiah K. M. (2015) Synthesis, characterization and controlled release study of Zn from zincated nanoclay polymer composites (ZNCPCs) in relation to equilibrium water absorbency under Zn deficient *Typic Haplustepts*. *Annals of Plant and Soil Research*, 17:187–195.

Mandal, N., Datta, S.C., Manjaiah, K. M., Dwivedi, B. S., Lata, N., Kumar, R. and Aggarwal, P. (2018) Novel chitosan grafted zinc containing nanoclay polymer biocomposite (CZNCPBC): controlled release formulation (CRF) of Zn^{2+}. *Reactive and Functional Polymers*, 127:55–66.

Moghaddasi, S., Khoshgoftarmanesh, A. H., Karimzadeh, F. and Chaney, R. L. (2013) Preparation of nanoparticles from waste tire rubber and evaluation of their effectiveness as zinc source for cucumber in nutrient solution culture. *Scientia Horticulturae*, 160:398–403.

Pirzad, A., Mamyandi, M. M. and Khalilzadeh, R. (2019) Physiological and morphological responses of sugar beet (*Beta vulgaris L.*) subjected to nano-boron oxide at different growth stages. *Acta Biologica Szegediensis*, 63(**2**):103–111.

Roy, T., Biswas, D. R., Datta, S. C. and Sarkar. A. (2016) Phosphorus release from rock phosphate as influenced by organic acid loaded nanoclay polymer composites in an alfisol. *Proceedings of the National Academy of Sciences, India, Section B: Biological Sciences*, 88:121–132. doi: 10.1007/s40011-016-0739-6.

Sarkar, S., Datta, S. C. and Biswas, D. R. (2014) Synthesis and characterization of nanoclay–polymer composites from soil clay with respect to their water-holding capacities and nutrient-release behavior. *Journal of Applied Polymer Science*, 131:39951. http://dx.doi. org/10. 1002/app.39951.

Schmidt, S. B., Jensen, P. E. and Husted, S. (2016) Manganese deficiency in plants: the impact on photosystem II. *Trends in Plant Science*, 21(7):622–632.

Shah, M. A. and Shah, K. A. (2013) Nanotechnology: The Science of Small, John Wiley, India, Pvt. Ltd.

Shah, V. and Belozerova, I. (2009) Influence of metal nanoparticles on the soil microbial community and germination of lettuce seeds. *Water Air and Soil Pollution*, 197:143–148.

Shebl, A., Hassan, A. A., Salama, D. M., Aziz, M. E. A. E. and Elwahed, M. S. A. A. (2019) Green synthesis of nanofertilizers and their application as a foliar for *Cucurbita pepo* L. *Journal of Nanomaterials*, 2019:1–11.

Shi, S., Wang, W., Liu, L., Wu, S., Wei., Y and Li, W. (2013) Effect of chitosan/nano-silica coating on the physicochemical characteristics of longan fruit under ambient temperature. *Journal of Food Engineering*, 118:125–131.

Singh, A., Sarkar, D. J., Singh, A. K., Prasad, R., Kumar, A. and Parmar, B. S. (2011) Studies on novel nanosuperabsorbent composites: swelling behavior in different environments and effect on water absorption and retention properties of sandy loam soil and soil-less medium. *Journal of Applied Polymer Science*, 120:1448–1458.

Ullah, S., Khan, A. S., Malik, A. U., Afzal, I., Shahid, M. and Razzaq, K. (2013) Foliar application of boron influences the leaf mineral status, vegetative and reproductive growth, yield and fruit quality of 'Kinnow' mandarin. *Journal of Plant Nutrition*, 36(10):1479–1495.

CHAPTER 19

Advances in Micronutrient Fertilizer Production and Efficacy in Plant Nutrition

MANDIRA BARMAN[1*], RAHUL MISHRA[2], and VIVEK TRIVEDI[1]

[1]*Division of Soil Science and Agricultural Chemistry, ICAR-Indian Agricultural Research Institute, New Delhi, India*

[2]*ICAR-Indian Institute of Soil Science, Bhopal, India*

Corresponding author. E-mail: mandira.ssaciari@gmail.com

ABSTRACT

There is widespread deficiency of micronutrients in soil across the globe which is impairing human health. Application of micronutrient fertilizer has become crucial to achieve optimum crop yield. However, the use efficiency of applied micronutrient by crops is abysmally low (<5%) due to very high rate of fixation, loss, and no synchronization of nutrient release with crop demand. Use of water-soluble micronutrient fertilizer further worsens the situation. Various slow release and controlled-release fertilizers have appeared to increase the use efficiency of fertilizer micronutrients. However, a major breakthrough could not be achieved through these fertilizers in enhancing the use efficiency. Therefore, use of newly synthesized micronutrient fertilizers with innovative technologies like nanotechnology is one of the potentially helpful options. The review of available literature indicates that some nanotechnology-enabled micronutrient fertilizers like metallic nanoparticles, nanoclay polymer composites, etc., can augment growth of plants in certain ranges of concentration and may be used as potential micronutrient fertilizer in agriculture to achieve optimum crop yields as well as reduce environmental losses. Significant benefits from the synthesis and implementation of such nanotechnology-enabled micronutrient fertilizers include increased use efficiency, reduced rate of application, and decreased environmental impacts. However, sporadic data based on merely laboratory or greenhouse studies, without proper field trials, are only indicative of the potential of the technology. More evidence, especially field evaluations on large scale covering various agro-climatic zones and a wide range of crops are required to confirm such crop production enhancements and further adoption in crop production technology. Also mechanism of growth enhancement of crops due to the application of such fertilizers needs to be examined thoroughly. In this chapter, the potential of various advanced micronutrient fertilizers in enhancing use efficiency vis-a-vis. phyto availability of micronutrients will be discussed.

19.1   INTRODUCTION

Essential nutrients that are required in small quantities for the growth and development of plants, microorganisms, animals, and human beings are called micronutrients. Iron (Fe), zinc (Zn), copper (Cu), manganese (Mn), nickel (Ni), boron (B), molybdenum (Mo), and chlorine (Cl) are the essential micronutrients for plants and also known as "trace elements." There are two types of micronutrients viz. cationic (Fe, Zn, Cu, Mn, and Ni) and anionic (B, Mo, and Cl). They are classified based on the form in which they are absorbed by the plants. During the past five decades, a widespread micronutrient deficiency was observed in most of the Indian soils. Increasing the demand of crop production due to ever increasing population has encouraged the cultivation of fertilizer responsive improved crop varieties that produce high biomass besides, use of micronutrient-free high-analysis fertilizers. These results in mining of nutrients from soil at a higher rate in one hand, and on the other hand replenishment of mined nutrients by conventional fertilizers are reduced. Therefore, besides, macronutrients, micronutrient deficiencies are also coming up nowadays, which have negative consequences for agronomic crop productivity. Use of nitrogen, phosphorus, and potash for growing of food crops (mainly cereals), over several decades in India has resulted in high crop yields. But this has also contributed to the continuous extraction of micronutrients from soils by crops to a great extent. Hence, the bioavailability of micronutrients particularly of Zn is becoming so inadequate that it is creating problems for Zn-nutrition of humans (Monreal et al., 2016). Since, nutrition of animals and humans is directly associated to nutrition of plants, production of micronutrient-deficient crops/foods will have a direct impact on the health of animal and human, particularly whose primary diet is cereal based with less or very limited access to animal proteins (Cakmak et al., 2010; Li et al., 2016). Assuring sufficient level of micronutrients in food is very important in preventing micronutrient-malnutrition worldwide. Production of nutritious-foods needs ensured supply of all essential macronutrients and micronutrients in soil in balanced proportion. However, balanced fertilization needs proper fertilizer recommendation based on soil test values and availability of efficient micronutrient fertilizers. In this chapter, we will discuss the soil-test methods for micronutrients and use of various modern efficient micronutrient fertilizers for crop production.

Plants absorb micronutrients mainly in the ionic form present in soil solution (except B). However, the amount of micronutrients present in soil solution is far less than their total amount present in soil. The total micronutrient content in soil is distributed into various organic and inorganic fractions of different characteristics. Their availability to plants differs widely. However, a major portion of total micronutrient content in soil is not reactive enough; thus, it is not considered an excellent plant availability indicator. Therefore, as far as plant nutrition is concerned, assessing the presence of a sufficient amount of available micronutrients in soil is vital. Proper delineation of plant-available micronutrient status in soil is very important for its management.

19.2 FUNCTIONS OF MICRONUTRIENTS IN PLANT NUTRITION

Important physiological functions of micronutrients are described below:

- Iron-containing enzymes are cytochrome oxidase, catalase, and peroxidase.
- Manganese is an essential element in photosystem II and participates in photolysis. It is a part of all decarboxylases and dehydrogenases of the TCA cycle.
- Zinc is associated with carbonic anhydrase, dehydrogenases, proteinases, and peptidases. It also plays important role in synthesis of few growth-promoting hormones and reproductive process of several plants which are very important for grain formation.
- Copper-containing enzymes are cytochrome oxidase, polyphenol oxidase, and ascorbic acid oxidase.
- Molybdenum is associated with nitrate reductase and nitrogenase enzymes and is essential in nitrogen metabolism in plants.

- Cobalt is essential for symbiotic N fixation and rhizobial growth and formation of coenzyme cobalamin.
- Nickel is a component of the enzyme urease which presents in a wide range of plants and participates in N metabolism of plants.
- Role of B is also well established in phenol metabolism, maintaining integrity of plasma membrane and cell division, growth of pollen tube, etc. It is associated with water absorption rate and translocation of sugars in plants.

Thus, the micronutrients are essential for various physiological activities inside the plants and are very important for plants and microorganisms growth and development.

19.3 SOIL TESTS FOR AVAILABLE MICRONUTRIENTS

The presence of substantial amounts of micronutrients in soil does not assure sufficient availability to plants, but plants availability is mediated by different chemical, physical, and biological soil factors such as pH, presence of competing ions, organic matter content, geomorphology of soil, etc. To identify the fields deficient/sufficient in micronutrient contents is of utmost importance for their management. Extracting solutions containing chelating agents, mainly dithelyne-triamine-penta-acetic acid (DTPA) and ethylene-diamine-tetra-acetic acid (EDTA) are used to extract micronutrient cations from soil. These chelating agents reduce the free-micronutrient cations activity in soil solution (intensity factor) through formation of a soluble metal–chelate complex during extraction. In response to this depletion of micronutrient ions in solution, more micronutrient ions from solid phases move toward solution to replenish this depletion (capacity factor). Such depletion in activity of metal ions in soil solution is similar to the reduction in activity of micronutrient ions in solution due to plant uptake (Datta et al., 2018). Thus, these chelates are capable of assessing both the intensity and quantity of a particular micronutrient cation in soil. The DTPA (pH 7.3) soil test is most widely used method for extraction of micronutrient cations in soil. However, this method is a good soil test for calcareous and near neutral pH soils. If the soil test is applied to low pH soils, buffering capacity of extracting solution may go beyond (O'Connor, 1988). Thus, for such soils, pH of extracting solution has been modified to 5.3 (Norvell, 1984).

Extraction of soil available B with hot water is commonly used method (Berger and Truog, 1939). However, during colorimetric estimation, interference of color due to organic matter and turbidity due to suspended clay particles create a problem. This method is tedious as well and therefore, is difficult to adopt for routine soil test. Thus, mannitol $CaCl_2$ (Cartwright et al., 1983) for alkaline/calcareous soils and salicylic acid (Datta et al., 1998) for acid soils can be used as replacement of tedious hot water method as these methods extract almost equal amount of B which is available to the plants. For Mo, ammonium oxalate is used as a measure of plant availability for quite a long time. Soils deficient or sufficient in available micronutrient content are generally decided from critical limits/threshold values in soil and plant (Barman et al., 2020). However, the suitability of soil-test method for available nutrients along with their critical limits varies from crop to crop and soil to soil because each method is different with respect to shaking time, soil:solution ratio and mode of action. Important soil-test methods for micronutrients with their critical limits of deficiency are given in Table 19.1.

19.4 USE EFFICIENCY OF MICRONUTRIENT FERTILIZERS

The most commonly used micronutrient fertilizers are sulfate salts of Fe, Zn, Cu, and Mn. After application, these elements react rapidly with various soil constituents by means of oxidation/precipitation or may react with various mineral complexes and clay colloids, making them less available to crops (Marschner, 2012). Therefore, the use efficiency of applied soluble micronutrient fertilizers (micronutrient use efficiency [MUE]) is abysmally low and seldom exceeds 5% (Ryan et al., 2013). Further, applying micronutrient fertilizer in soil in small quantity (<5 kg ha^{-1}) results in uneven distribution (spatial) of the microelements in soil, which contributed to reduction in the use efficiency. However, various chelated micronutrient fertilizers appeared to improve MUE (Liu et al., 2012). These are ring-type compounds, remain in plant-available pool for a longer period as micronutrients inside the chelates exhibit fewer interactions with soil components resulting in less fixation in soils. Chelated micronutrients also show greater mobility in soils compared to inorganic salts. About 90%–95% of applied inorganic form of Zn in soil is rendered unavailable to the plants due to various reactions with different soil components and thus, a chelated form of Zn (e.g., Zn-EDTA) is one of the most important sources which can increase the use efficiency of applied Zn with the concurrent increase in yield of crops.

TABLE 19.1 Important Soil Test Methods for Micronutrients with Critical Limits of Deficiency

Element	Extractant/Soil Test Method	Soil:Solution ratio	Shaking Period (min)	Critical Limit of Deficiency (mg kg-1)
Iron	0.005 M DTPA + 0.1 M TEA* + 0.01 M CaCl$_2$ (pH 7.3)	1:2	120	2.5-5.8
Manganese	0.005 M DTPA + 0.1 M TEA + 0.01 M CaCl$_2$ (pH 7.3)	1:10	120	2-4
Zinc	0.005 M DTPA + 0.1 M TEA + 0.01 M CaCl$_2$ (pH 7.3)	1:2	120	0.5-1.0
Copper	0.005 M DTPA + 0.1 M TEA + 0.01 M CaCl$_2$ (pH 7.3)	1:2	120	0.2-0.5
Nickel	0.005 M DTPA + 0.1 M TEA + 0.01 M CaCl$_2$ (pH 7.3)	1:2	120	0.17
Boron	Boiling hot water	5:10	5	0.5-1
	0.01 M CaCl$_2$ + 0.05 M mannitol	1:2	60	0.25
	0.1 M salicylic acid	1:2	60	0.45
Molybdenum	0.2 M Ammonium oxalate (pH 3.0)	1:10	360	0.05-0.2

Source: Adapted from Datta et al. (2018).

*TEA, Triethanol amine.

Besides soil application, application of foliar sprays of these sulfate, and chelated micronutrient fertilizers are also popularly practiced. However, foliar absorption of micronutrients is more from inorganic sulfate salts than from chelates; but translocation of micronutrients inside plant is better for chelated ones (Rengel et al., 1999). Various factors determine the use efficiency of micronutrient applied through foliar spray. Occurrence and thickness of cuticle waxes on leaf may regulate nutrient penetration by virtue of their hydrophobicity. Number and position of stomata on leaf surface also determine the rate of absorption of micronutrient applied through foliar spray. Various environmental factors such as rain, temperature may affect the micronutrients absorption in foliar applications (Kaiser, 2011). Nutrient uptake through stomata has been extensively studied as a biological process; however, variability associated with the regulating-factors are still unsolved. Application of micronutrient fertilizers in soil is more effective in field crops than spray application. However, in the case of fruit trees, both are considered as effective methods of application.

19.5   MICRONUTRIENT FERTILIZATION OF CROPS

19.5.1   COMMON MICRONUTRIENT FERTILIZERS

To supply micronutrients to the crops both soluble and insoluble sources are used. A list of common micronutrient fertilizers used in India is given in Table 19.2 along with their rates of application. The most commonly used fertilizer for Fe, Zn, Cu, and Mn in India are water-soluble sulfate salts, that is, $FeSO_4·7H_2O$, $ZnSO_4·7H_2O$, $CuSO_4·5H_2O$, and $MnSO_4·H_2O$, respectively. Besides inorganic salts, water-soluble chelated micronutrient fertilizers, like Fe-EDTA and Zn-EDTA, are also being used in India. However, as these chelated fertilizers are costlier than sulfate salts, farmers still prefer to use sulfate salts. Both $Na_2B_4O_7·10H_2O$ and H_3BO_3 are used as fertilizer for B. Molybdenum is usually applied as $(NH_4)_6Mo_7O_{24}$ which is water soluble. Use of micronutrient fortified fertilizers like zincated urea, boronated single-super-phosphate are also being used in limited scale.

19.5.2   ADVANCED MICRONUTRIENT FERTILIZERS

A fairly high rate of nutrient release into the soil is the main problem of conventional micronutrient fertilizers. As most of the micronutrients used

TABLE 19.2 Common Micronutrient Fertilizers Used in India

Nutrient	Fertilizer	Chemical Formula	Minimum Nutrient Content (%)*	General Recommended Dose	
				Soil Application	Foliar Application
Zinc	Zinc sulfate hepta hydrate	$ZnSO_4 \cdot 7H_2O$	21	2.5–10 kg Zn ha^{-1}	0.5% Zn solution; 2–3 sprays
	Zinc sulfate mono hydrate	$ZnSO_4 \cdot H_2O$	33		
	Chelated zinc	Zn-EDTA	12		
	Zincated urea (43% N)	—	2		
Iron	Ferrous sulfate	$FeSO_4 \cdot 7H_2O$	19	5–20 kg Fe ha^{-1}	0.5% ferrous solution; 3–4 sprays
	Chelated iron	Fe-EDTA	12		
Manganese	Manganese sulfate	$MnSO_4 \cdot H_2O$	30.5	10–25 kg Mn ha^{-1}	0.2–0.5% manganese sulfate; 2–3 sprays
Copper	Copper sulfate	$CuSO_4 \cdot 5H_2O$	24	1–5 kg Cu ha^{-1}	0.5% copper sulfate solution; 2–3 sprays
Molybdenum	Ammonium molybdate	$(NH_4)_6Mo_7O_{24} \cdot$	52	0.05–1 kg Mo ha^{-1}	-
Boron	Borax	$Na_2B_4O_7 \cdot 10H_2O$	10.5	2–5 kg B ha^{-1}	0.05–1.0 % B solution; 3–4 sprays
	Boric acid	H_3BO_3	17		
	Di-sodium octa borate tetra hydrate	$Na_2B_8O_{13} \cdot 4H_2O$	20		
	Boronated single super- phosphate (16% P_2O_5 powdered)	—	0.18		

*As per FCO (1985); FAI (1998).

in conventional agriculture are water soluble, nutrients release is very quick from these and thus, plants are unable to take the total amount at the time of release. A huge amount of nutrient penetrates deep into the soil and then either get fixed in soil or leached down to groundwater. Therefore, to reduce the amount discharged to the environment and at the same time increase the use efficiency of applied micronutrient fertilizers, the nutrient release from fertilizers should be slower down and the release should be synchronized with the nutritional requirement of plants to the maximum possible extent. Such fertilizers include slow-release-fertilizers (SRF) and controlled-release-fertilizers (CRF). These fertilizers can release nutrients into soil in a continuous manner at an adequate amount to ensure optimum growth of plants without several applications. Various types of CRF, including hardly soluble fertilizers, encapsulated fertilizers, and nanofertilizers (NF), that is, various advanced micronutrient fertilizers are discussed in the following subsections.

19.5.3 MICRONUTRIENT FERTILIZERS WITH LOW SOLUBILITY

According to Chandra et al. (2009), a micronutrient fertilizer insoluble in water, but may become soluble by the action of plant roots, that is, by means of ion exchange or extracellular organic acid secretions would be an ideal SRF. Chelated micronutrients are highly available to the plants. The oldest example of micronutrient fertilizers with low water solubility are double phosphates that contain nutrients with general formula of M(II) M(I)PO$_4$·XH$_2$O ([M(II)]: Mg, Fe, Mn, Zn, Cu, Co, and M(I): NH$_4$, K, Tl, Rb]. These metal-ammonium phosphates have very low water solubility. Various polyphosphates have a capacity to sequester cationic micro-nutrients in their structure in the form of chelates which can maintain higher micronutrient concentration in soil solution. Arslanoglu (2019) reported that synthesis of particles of struvite with controlled-diameters for Cu(II)-adsorption is possible. Various hydrolyzed fertilizers, like metaphosphates, glass phosphates, etc. have been used as SRFs of micronutrients. Bandyopadhyay et al. (2008) tested them on plants and found a significant yield enhancement at a relatively lower fertilizer dose. Bandyopadhyay et al. (2014) prepared a SRF for micronutrients which is a crystalline-polyphosphate of Zn-Cu-Fe-Mn-Mg-NH$_4$, using hematite, zinc ash, pyrolusite, roasted magnesite, copper sulfate, and phosphoric acid. This had low water solubility, but sufficient solubility in organic acids. This suggests a good accessibility of the nutrients for plants. They tested it in field crops (rice and potato) and observed significant higher yields in synthesized SRF-treated plots over control and conventional micronutrients-treated plots. Uptake of micronutrients by rice and potato was also higher with newly synthesized SRF.

Abat et al. (2015) synthesized cogranulated mono-ammonium phosphate with B which showed slow release property. Release of B from this fertilizer was sufficiently slow and concentration of B around the granule was likely to be safe for most crops. Chandra et al. (2009) synthesized a polymeric phosphate structure containing iron and magnesium with low solubility in water; however, maximum amount of Fe^{3+} (90%–100%) was in plant-available form as these were acid-soluble. In greenhouse pot experiments, this fertilizer increased paddy yield by 46.9% when iron was applied @ 2.01 kg ha^{-1}. These are water-insoluble molecules but are actively dissolved by plant roots, mainly by ion-exchange or chelation. Therefore, results are indicative of a promising slow release iron fertilizer.

19.5.4   ENCAPSULATED MICRONUTRIENT FERTILIZERS

Encapsulation of inorganic fertilizers uses natural or synthetic polymer films that protect nutrients from various rapid reactions with soil components and slower down their release rates into the soil solution. Such encapsulation improves the use efficiency of applied micronutrient fertilizer further as the coating act as a physical barrier that obstruct nutrient transport. Nutrient release from such fertilizers coated with natural or synthetic polymer depends on soil pH, moisture content, temperature, ionic concentration, microbial activity, and demand of crops. Structural properties of polymers used for coating or encapsulation are also important that control the release of micronutrients in soil (Abedi-Koupai et al., 2012). Few examples of such polymers include ethyl cellulose, glycerol mono stearate, ethylene-vinyl-acetate, chitosan, alginate, lignosulfate, starch, pectin, agar, gelatin, palm stearin, etc. (Monreal et al., 2016). However, the coating materials should be less costly, biodegradable, and the products of degradation should not be toxic to the environment. In spite of the fact that potential of these encapsulated micronutrient fertilizers as controlled-release fertilizer has demonstrated in laboratory, but the field evaluation of these polymer-coated micronutrient fertilizers is very limited. Besides encapsulation, various aluminosilicates are being used as carrier of Cu (Huo et al., 2014).

19.6   NANOTECHNOLOGY-ENABLED MICRONUTRIENT FERTILIZERS

Nanotechnology-enabled micronutrient fertilizers or NFs for micronutrients are expected to enhance MUE by decreasing nutrient loss from fertilizers and thus resulting higher plant uptake. One of the components of such fertilizers are nano-sized particles which are very reactive due to the smaller size and greater surface area. As compare to conventional micronutrient fertilizers, NFs for micronutrients are more and more being evaluated as nutrient source for achieving quantitative–qualitative crop improvement. A number of recent studies have compared the effects of NFs vs. conventional micronutrient fertilizers in terms of crop response (Tolaymat et al., 2017, Dimkpa and Bindraban, 2018). Various nanocomposite polymers can bind fertilizer nutrients into their matrix which releases nutrients very slowly to get better nutrient use efficiency more than present controlled-release fertilizers. Nano-biosensor-based fertilizer also releases nutrients in response to chemical signals from soil microbes in a plant rhizosphere. These are DNA based nanobiosensor which, when incorporated into a biopolymer film coating micronutrients fertilizer, will read the electrochemical signals of plant roots. Once the nanobiosensor identifies the signal, it causes the biopolymer to become semipermeable and release the nutrient as per the requirement of plant at the time it is required. However, evaluation of this technology for nutrient delivery to the plants is in its infancy. Hence, physically or biologically synthesized nano-sized chemical fertilizers are most commonly micronutrient NFs used in agriculture. The outcomes of many discrete studies suggest the potential use of NFs for micronutrients as nutrient sources in crops grown in micronutrient-deficient soils.

Recently, a nanoclay polymer composite (NCPC)-based controlled-release Zn fertilizer formulation has been synthesized and evaluated in a greenhouse pot experiment which showed promising results (Mandal et al., 2019). These NCPCs are mainly superabsorbents which are three-dimensionally crosslinked water insoluble hydrophilic polymers capable of absorbing large amounts of aqueous fluids. By diffusion and mass flow the absorbed water along with dissolved salt can move in or out of the composite freely in response to gradient. However, this movement is largely regulated by clay content in NCPC which act as a barrier and causes tortuosity inside the composite and makes them slow release NCPC (Figure 19.1). However, synthesis and evaluation of NF are still at preliminary stages. Few examples of crop responses toward nanotechnology-enabled micronutrient fertilizers are given below:

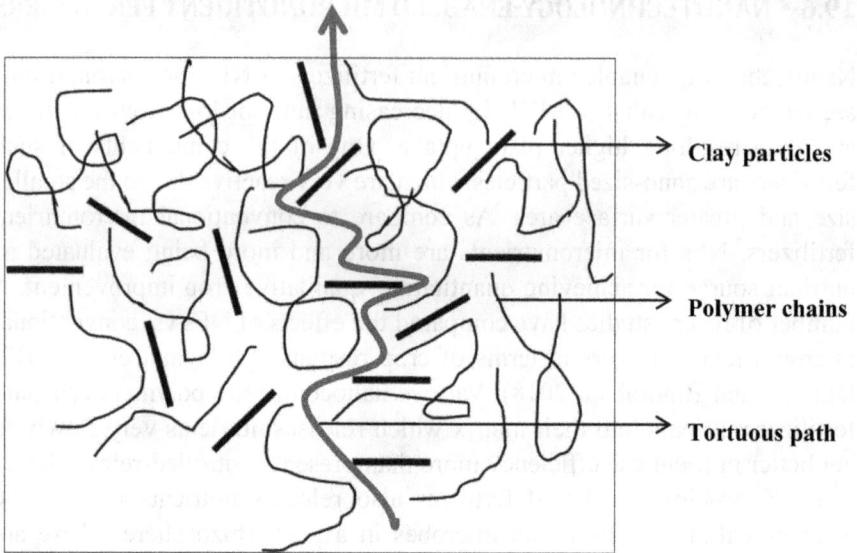

FIGURE 19.1   Matrix of nanoclay polymer composite.

19.6.1   IRON

Rui et al. (2016) evaluated the efficacy of Fe_2O_3-NPs as a micronutrient fertilizer in a greenhouse pot experiment using peanut (*Arachis hypogaea*), as a test crop which is highly sensitive to Fe deficiency. They found that application of Fe_2O_3-NPs-enhanced biomass, root length, plant height, and Fe contents of peanut plants. They also observed that application of Fe_2O_3-NPs improved the growth of plants by mediating the contents of various phytohormones and antioxidant enzyme activity. In a greenhouse experiment, Ghafariyan et al. (2013) observed that Fe-NPs at low concentrations could enhance chlorophyll contents in soybean leaves (subapical) significantly in a solution culture experiment. Thus, soybean could assimilate applied Fe-NPs and decrease Fe deficiency. Delfani et al. (2014) observed that application of Fe-NPs (500 mg L^{-1}; foliar spray) to peas improved the number of pods/plant, 1000-seeds weight, chlorophyll content, and leaf Fe content by 47%, 7%, 10%, and 34%, respectively, over control. Application of Fe-NPs also enhanced the above parameters over application of a conventional Fe salt by 28%, 4%, 12%, and 45%, respectively.

19.6.2   MANGANESE

Dimpka et al. (2018) examined the influence of nano-Mn (Mn_2O_3) on yield of wheat and nutrient uptake, as compared to bulk-Mn and ionic-Mn (6 mg/kg/plant). Both soil and foliar applications were done for nano-Mn. In case of soil application, all Mn types significantly reduced shoot Mn as compared to control. However, nano-Mn contributed to a higher grain Mn-translocation efficiency (22%), as compared to bulk-Mn (21%), salt-Mn (20%), and control (16%). Foliar application of nano-Mn resulted in better grain (12%) and shoot (37%) Mn contents, suggesting a greater response of crop toward foliarly applied nano-Mn. According to Pradhan et al. (2013), metallic Mn-NPs improved mung bean (*Vigna radiata*) growth and increased its photosynthesis over commercially-available $MnSO_4$ salt when the mung bean seedlings were allowed to grow in Hoagland solution containing 0.05 mg Mn L^{-1} for 15 days. An improvement in root and shoot length, fresh biomass, dry biomass, number of rootlets, by 52%, 38%, 38%, 100%, and 71%, respectively, were observed due to application of Mn-NPs over control and 2%, 10%, 8%, 100%, and 28%, respectively, over $MnSO_4$ salt treatment.

19.6.3   ZINC

Influence of Zn-NPs on growth of plants and other organisms are being studied extensively. For example, Mahajan et al. (2011) observed improved growth of crops (mung bean and chickpea) in plant-agar method at low concentrations. A 42% increase in root length, 41% in root biomass, 98% in shoot length, and 76% in biomass of shoot was recorded due to application of ZnO-NPs @ 20 mg L^{-1} over control. In the case of chickpea seedlings, 1 mg Zn L^{-1} caused significant improvement in root length (53%), root biomass (37%), shoot length (6%), and shoot biomass (27%) over control. In a greenhouse experiment, Zhao et al. (2013) observed that ZnO-NPs application @ 400, 800 mg kg^{-1} to soil improved cucumber growth in terms of root and shoot biomass, gluten, starch, and Zn content. Foliar sprays of ZnO-NPs enhanced phytase and phosphatase enzyme activities, resulting in increased P uptake by clusterbean, with no external P fertilizer application (Raliya and Tarafdar, 2013). Similar observations on improvement of cotton plant growth in response to ZnO-NPs application along with P supplements were recorded by Venkatachalam et al. (2016). Tarafdar et al. (2014) tested biologically synthesized Zn-NF on pearl millet and observed significant phenological growth improvements in the plants, improved chlorophyll content, plant dry biomass, total soluble leaf protein, etc. in 42 days old plants grown in field. Lin and Xing (2007) reported that exposure of ZnO-NPs @ 2 mg L^{-1} improved root elongation of crops (radish and rape) over control. However, few reports suggest that exposure of Zn-NPs at high concentrations can cause inhibitory effects or phytotoxicity (Lin and Xing, 2007; Zhao et al., 2014). Exposure of even 10 mg L^{-1} of Zn-NPs was too high to show any positive effects on ryegrass growth, as most of the plants require only 0.05 mg L^{-1} Zn in growing media for optimum growth. Application at higher levels may results in phytotoxicity. (Lin and Xing, 2008).

19.6.4   COPPER

According to Shah and Belozerova (2009), application of Cu-NPs @ 130 and 600 mg kg^{-1} to soil enhanced lettuce-seedling growth by 40% and 91%, respectively. However, increased concentrations (200–1000 mg L^{-1}) showed phytotoxicity to wheat, mung bean, and yellow squash seedlings (Lee et al., 2008; Musante and White, 2012). Nekrasova et al. (2011) reported that waterweed supplied with lower concentrations (0.25 mg Cu L^{-1}) of Cu-NPs improved rate of photosynthesis by 35% over control in a three day incubation study. Stampoulis et al. (2009) reported that a high concentration of (1000 mg Cu L^{-1}) of Cu-NPs in Hoagland solution decreased biomass of zucchini by 90% over control after 14 days. They suggested that Cu concentration in aqueous media for optimal growth of plant is only 0.02 mg L^{-1}, and toxic effect may take place at high concentrations.

19.6.7   MOLYBDENUM

Taran et al. (2014) examined the influence of Mo-NPs on growth of chickpea. The experiment was conducted with four treatments viz. water, Mo-NPs, nitrogen-fixing bacteria, and microbes plus Mo-NPs. Before sowing seeds of chickpea were treated with the four treatments for 1–2 h in a loamy soil. Microbiological assay of rhizosphere soil demonstrated the development of almost important groups of microorganisms by 2–3 times under fourth treatment over control and other treatments. Thus, the fourth treatment proved to be the optimal treatment for chickpea plant root nutrition. Number of roots and mass of nodules per plant were also higher for the fourth treatment over other treatments.

Thus, it can be conferred that NFs for micronutrients have a promising potential as micronutrient fertilizers for improving crop production and decreasing environmental losses.

19.7 CONCLUSIONS

Widespread deficiency of micronutrients not only affects the quantity–quality of food production but also affects human and animal health. Also, use efficiency of applied micronutrient fertilizer hardly exceeds 5%, which results in huge economic loss. Therefore, precision techniques which supply nutrients to the plant with minimum loss are need of the hour. Research effort is being made to enhance MUE, paying special importance on various advanced technologies such as smart controlled-release fertilizers, nanotechnology-enabled micronutrient-fertilizers, etc. However, application of such fertilizers in field scale is at a nascent stage. Various discrete studies show the potential of such advanced fertilizers in enhancing MUE, but practicality of field applications such as dosage, time of application, and other associated cultural practices need further research. The effects of delivering micronutrients through encapsulated and nanofertilizers on yield, grain quality of crops are largely unknown. However, promising possibilities are definitely there for these technologies especially for nanotechnology.

KEYWORDS

- **micronutrient**
- **slow release fertilizers**
- **nanotechnology**
- **nanoclay polymer composites**

REFERENCES

Abat, M.; Degryse, F.; Baird, R.; McLaughlin,M. J. Slow-release boron fertilisers: Co-granulation of boron sources with mono-ammonium phosphate (MAP). Soil Res. 2015, 53, 505.

Abedi-Koupai, J.; Varshosaz, J.; Mesforoosh, M.; Khoshgoftarmanesh, A. H. Controlled release of fertilizer microcapsules using ethylene vinyl acetate polymer to enhance micronutrient deficiency and water use efficiency. J. Plant Nutr. 2012, 35, 1130–1138.

Arslanoglu, H. Adsorption of micronutrient metal ion onto struvite to prepare slow release multi-element fertilizer: Copper(II) doped-struvite. Chemosphere. 2019, 217, 393–401.

Bandyopadhyay, S.; Bhattacharya, I.; Ghosh, K.; Varadachari, C. New slow-releasing molybdenum fertilizer. J. Agric. Food Chem. 2008, 56, 1343–1349.

Bandyopadhyay, S.; Ghosh, K.; Varadachari, C. Multi-micronutrient slow-release fertilizer of zinc, iron, manganese, and copper. Int. J. Chem. Eng. 2014, 2014, 1–7.

Barman, M.; Datta, S. P.; Rattan, R. K.; Meena, M. C. Critical limits of deficiency of nickel in intensively cultivated alluvial soils. J. Soil Sci. Plant Nutr. 2020, 20, 284–292.

Berger, K. C.; Truog, E. Boron determination in soils and plants. Ind. Eng. Chem., Anal. Ed. 1939, 11, 540–544.

Cakmak, I.; Pfeiffer, W. H.; McClafferty, B. Biofortification of durum wheat with zinc and iron. Cer. Chem. 2010, 87, 10–20.

Cartwright, B.; Tiller, B. A.; Zarcinas, B. A.; Spouncer, L. R. The chemical assessment of the boron status of soils. Aust. J. Soil Res., 1983, 21, 321–332.

Chandra, P. K.; Varadachari, C.; Ghosh, K. A new slow releasing iron fertilizer. Chem. Eng. J. 2009, 155, 451–456.

Datta, S. P.; Bhadoria, P. B. S.; Kar, S. Availability of extractable boron in some acid soils, West Bengal, India. Comm. Soil Sci. Plant Anal. 1998, 29, 2285–2306.

Datta, S. P.; Meena, M. C.; Barman, M.; Golui1, D.; Mishra, R.; Shukla, A. K. Soil tests for micronutrients: Current status and future thrust. Indian. J. Fertilisers. 2018, 14, 32–51.

Delfani, M.; Firouzabadi, M. B.; Farrokhi, N.; Makarian, H. Some physiological responses of black-eyed pea to iron and magnesium nanofertilizers. Commun. Soil Sci. Plant Anal. 2014, 45, 530–540.

Dimkpa, C. O.; Bindraban, P. Nano-fertilizers: New product for the industry. J. Agri. Food Chem, 2018, 66, 6462–6473.

Dimkpa, C. O.; Singh, U.; Adisa, I. O.; Bindraban, P. S.; Elmer, W. H.; Gardea-Torresdey, J. L.; White, J. C. Effects of manganese nanoparticle exposure on nutrient acquisition in wheat (*Triticum aestivum* L.). Agronomy. 2018, 8, 158, 1–16.

FAI. The Fertilizer (Control) Order, 1985 and The Essential Commodities Act, 1995, The Fertiliser Association of India, New Delhi, 1998.

Ghafariyan, M. H.; Malakouti, M. J.; Dadpour, M. R.; Stroeve, P.; Mahmoudi, M. Effects of magnetite nanoparticles on soybean chlorophyll. Environ. Sci. Technol. 2013, 47, 10645–10652.

Huo, C.; Ouyang, J.; Yang, H. CuO nanoparticles encapsulated inside Al-MCM-41 mesoporous materials via direct synthetic route. Sci. Rep. 2014, 4,1–9.

Kaiser, D. Are micronutrients needed in high yield environments? University of Minnesota Extension, 2011. (http://www.extension.umn.edu/agriculture/nutrient-management/docs/MVTL-micros-2011.pdf).

Lee, W.; An, Y.; Yoon, H.; Kweon, H. Toxicity and bioavailability of copper nanoparticles to the terrestrial plants mung bean (*Phaseolus radiatus*) and wheat (*Triticum aestivum*): Plant agar test for water-insoluble nanoparticles. Environ. Toxicol. Chem. 27, 2008, 1915–1921.

Lin, D.; Xing, B. Root uptake and phytotoxicity of ZnO nanoparticles. Environ. Sci. Technol. 2008, 42, 5580–5585.

Lin, D.; Xing, B.; Phytotoxicity of nanoparticles: Inhibition of seed germination and root growth. Environ. Pollut. 2007,150, 243–250.

Liu, G.; Hanlon, E.; Li, Y. Understanding and applying chelated fertilizers effectively based on soil pH. Horticultural Sciences Department, Cooperative Extension Service, Institute of Food and Agricultural Sciences, University of Florida, 2012. (http://edis.ifas.ufl.edu/hs1208, accessed November 2014).

Mahajan, P.; Dhoke, S. K.; Khanna, A. S. Effect of nano-ZnO particle suspension on growth of mung (*Vigna radiata*) and gram (*Cicer arietinum*) seedlings using plant agar method. J. Nanotechnol. 2011, 2011, 7.

Mandal, N.; Datta, S. C.; Manjaiah, K. M.; Dwivedi, B. S.; Kumar, R.; Aggarwal, P. Evaluation of zincated nanoclay polymer composite in releasing Zn and P and effect on soil enzyme activities in a wheat rhizosphere. Eur. J. Soil Sci. 2019, 70(6), 1–19.

Marschner, P. Marschner's mineral nutrition of higher plants, 3rd edn. Elsevier, Oxford, 2012.

Monreal, C. M.; DeRosa, M.; Mallubhotla, S. C.; Bindraban, P. S.; Dimkpa, C. Nanotechnologies for increasing the crop use efficiency of fertilizer-micronutrients. Biol. Fert. Soils. 2016, 52, 423–437.

Musante, C.; White, J. C. Toxicity of silver and copper to *Cucurbita pepo*: Differential effects of nano and bulk-size particles. Environ. Toxicol. 2012, 27, 510–517.

Nekrasova, G. F.; Ushakova, O. S.; Ermakov, A. E.; Uimin, M. A.; Byzov, I. V. Effects of copper (II) ions and copper oxide nanoparticles on Elodea densa Planch. Russ. J. Ecol. 2011, 42, 458–463.

Norvell, W. A. Comparison of chelating agents for metals in diverse soil materials. Soil Sci. Soc. AM. J. 1984, 48, 1285–1292.

O'Connor, G. A. Use and misuse of the DTPA soil test. J. Environ. Qual. 1988, 17, 715–718.

Pradhan, S.; Patra, P.; Das, S.; Chandra, S.; Mitra, S.; Dey, K. K; Akbar, S.; Palit, P.; Goswami, A. Photochemical modulation of biosafe manganese nanoparticles on *Vignaradiate*: A detailed molecular, biochemical, and biophysical study. Environ. Sci. Technol. 2013, 47, 13122–13131.

Raliya, R.; Tarafdar, J. C. ZnO nanoparticle biosynthesis and its effecton phosphorous-mobilizing enzyme secretion and gum contents in clusterbean (*Cyamopsis tetragonoloba* L.). Agric. Res. 2013, 2, 48–57.

Rengel, Z.; Batten, G. D.; Crowley, D. E. Agronomic approaches for improving the micronutrient density in edible portions of field crops. Field Crop Res. 1999, 60, 27–40.

Rui, M.; Ma, C.; Hao, Y.; Guo, J.; Rui, Y.; Tang, X.; Zhao, Q.; Fan, X.; Zhang, Z.; Hou, T.; Zhu, S. Iron oxide nanoparticles as a potential iron fertilizer for peanut (*Arachis hypogaea*) Frontiers. Plant. Sci. 2016, 7, 1–10.

Ryan, J.; Rashid, A.; Torrent, J.; Yau, S. K.; Ibrikci, H.; Sommer, R.; Erenoglu, E. B.; Sparks, D. L. Micronutrient constraints to crop production in the Middle East-West Asia region: Significance, research and management. Adv. Agron. 2013, 122, 1–84.

Shah, V.; Belozerova, I. Influence of metal nanoparticles on the soil microbial community and germination of lettuce seeds. Water Air Soil Pollut. 2009, 197, 143–148.

Stampoulis, D.; Sinha, S. K.; White, J. C. Assay-dependent phytotoxicity of nanoparticles to plants. Environ. Sci. Technol. 2009, 43, 9473–9479.

Tarafdar, J. C.; Raliya, R.; Mahawar, H.; Rathore, I. Development of zinc nanofertilizer to enhance crop production in pearl millet (*Pennisetum americanum*). Agric. Res. 2014, 3, 257–262.

Taran, N. Y.; Gonchar, O. M.; Lopatko, K. G.; Batsmanova, L. M.; Patyka, M. V.; Volkogon, M. V. The effect of colloidal solution of molybdenum nanoparticles on the microbial composition in rhizosphere of *Cicer arietinum* L. Nanoscale Res. Lett. 2014, 9, 289.

Tolaymat, A.; Genaidy, A.; Abdelraheem, W.; Dionysiou, D.; Andersen,C. The effects of metallic engineered nanoparticles upon plant systems: An analytic examination of scientific evidence. Sci. Total Environ. 2017, 579, 93–106.

Venkatachalam, P.; Priyanka, N.; Manikandan, K.; Ganeshbabu, I.; Indiraarulselvi, P.; Geetha, N.; Muralikrishna, K.; Bhattacharya, R. C.; Tiwari, M.; Sharma, N.; Sahi, S. V. Enhanced plant growth promoting role of phycomolecules coated zinc oxide nanoparticles with P supplementation in cotton (*Gossypium hirsutum* L.). Plant, Physiol. Biochem. 2016. doi: 10.1016/j.plaphy.2016.09.004.

Zhao, L.; Peralta-Videa, J. R.; Rico, C. M.; Hernandez-Viezcas, J. A.; Sun, Y.; Niu, G.; Servin, A.; Nunez, J. E.; Duarte-Gardea, M.; Gardea-Torresdey, J. L. CeO_2 and ZnO nanoparticles change the nutritional qualities of cucumber (*Cucumis sativus*). J. Agric. Food Chem. 2014, 62, 2752–2759.

Zhao, L.; Sun, Y.; Hernandez-Viezcas, J. A.; Servin, A. D.; Hong, J.; Niu, G. Influence of CeO_2 and ZnO nanoparticles on cucumber physiological markers and bioaccumulation of Ce and Zn: a life cycle study. J. Agric. Food Chem. 2013, 61, 11945–11951.

PART VII
Remediation of Metals and Pesticides Contaminated Soils

CHAPTER 20

Heavy Metal Pollution in Soil and Remediation Strategies

JAJATI MANDAL[1*], DEBASIS GOLUI[2], PRASENJIT RAY[3], and PRADIP BHATTACHARYYA[4]

[1]*Department of Soil Science and Agricultural Chemistry, Bihar Agricultural University, Sabour, Bhagalpur, India*

[2]*Division of Soil Science and Agricultural Chemistry, ICAR-Indian Agricultural Research Institute, New Delhi, India*

[3]*ICAR-National Bureau of Soil Survey and Land Use Planning, Regional Centre, Jorhat, India*

[4]*Agricultural and Ecological Research Unit, Indian Statistical Institute, Giridih, Jharkhand, India*

Corresponding author. E-mail: jajati.bckv@gmail.com

ABSTRACT

Heavy metal pollution due to the occurrence of As, Pb, Cd, and Cr in the soil system has created, creating, and will create severe health implications to the human population all over the world. To restrain the human population from the ominous outcome, a comprehensive knowledge about the contaminants is highly needed. In this chapter, a modest initiative has been undertaken to address issues comprising of, sources of heavy metals and metalloids in the soil, their behavior in the soil system, approaches for quantification of metal hazard in soil along with the remediation strategies. The authors have tried to collate up-to-date information which will help the readers to exalt their academic acumen.

20.1   INTRODUCTION

The build-up of heavy metals in cultivated field is mainly from the sources like industrial effluents, sewage, sludge, contaminated ground water, and river water. It is of rising apprehension as it induces detrimental effects on soil biota which in turn is a potential risk to human and animal health (Ray et al., 2017; Golui et al., 2019, 2020). Transfer of metals to the palatable portion of crops grown in contaminated soils often renders the food crops unfit for human and animal consumption. Excessive intake of metals and metalloids due to ingestion of food stuffs grown in contaminated soil may result into different physiological and metabolic disorders in human and animal. All the micronutrient cations viz. zinc (Zn), copper (Cu), manganese (Mn), iron (Fe), and nickel (Ni), which are indispensable for plant growth termed as metal (atomic number >20 and specific gravity >5.0). Based on the concentrations, they exhibit both deficiency and toxicity in the plants/organisms. Lead (Pb), cadmium (Cd), chromium (Cr), mercury (Hg), selenium (Se), arsenic (As), and fluorine (F) are other metal, metalloid, and nonmetal of concern, which can cause toxicity to the plant/organisms, when present at an elevated level (Datta et al., 2017). Therefore, a combined effort of assessing the severity of contamination, time-to-time monitoring along with proper remediation, is the need of the hour to judge the suitability of land for agricultural activities as well as to prevent food chain contamination. This chapter has engrossed on the sources of metal and metalloid pollution in soil along with possible remediation measures.

20.2   HEAVY METAL SOURCES IN SOIL

Various geogenic processes are primarily accountable for natural occurrence of heavy metals in trace amounts in soils (Kabata-Pendias and Pendias, 2001; Pierzynski et al., 2000). The numerous pathways of metals and metalloids in the soil is represented schematically in Figure 20.1.

20.2.1   GEOGENIC SOURCES

The process of weathering of rocks (igneous and sedimentary), minerals, and parent materials and also to some extent coal are the prime cause of heavy metals and metalloids in nature. Arsenic is dominated in sulfide-bearing minerals especially in iron ores. Arsenopyrites, realgar, and orpiment are some of the minerals which serve as a source of arsenic. The formation of sedimentary rocks mainly contributes toward arsenic enrichment of the Fe–OH and sulfides. It is due to the precipitation of the hydroxides and sulfides of iron. Therefore, iron deposits and iron ores of sedimentary origin are rich in arsenic and may attain a value greater than 20 mg kg^{-1}. Arsenic contamination of the groundwater is mainly from geological sources, that is, through reductive dissolution of arsenopyrite minerals and has created havoc in countries like India, Bangladesh, China, and Mexico. Cadmium (Cd) content fluctuates with the categories of parent materials. The soils originated from the basaltic rocks contain more cadmium compared to that originated from the granites. Soils free from pollution generally contain 0.01 to 30 mg kg^{-1} of total Cd with a mean range value of 0.06 to 0.5 mg kg^{-1} (Page et al., 1981; Alloway, 1995). Lead (Pb) content in the lithosphere is estimated at 16 mg kg^{-1}, and the total Pb content of agricultural soils lies between 2 and 200 mg kg^{-1}. Soils with levels in excess of this are restricted to a comparatively few areas of India where depositions of Pb minerals are found. Most soils contain trace amount of nickel (Ni), usually less than 100 mg kg^{-1}, and pretty close at which Ni toxicity occurs. The soils originated from ultra-basic igneous rocks and predominantly serpentine, however, contain 20–40 times this amount and nickel toxicity can be observed in plants. In the earth's crust, the mean chromium (Cr) content is 125 mg kg^{-1} with a range of 80–200 mg kg^{-1} (National Academy of Sciences, NAS, 1974).

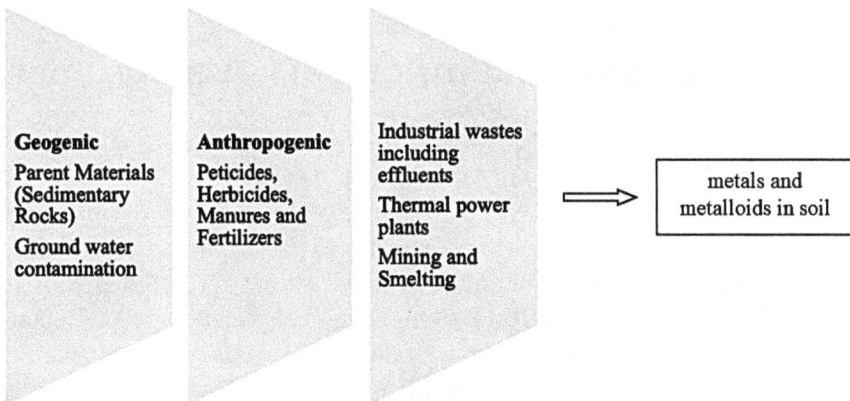

Geogenic
Parent Materials (Sedimentary Rocks)
Ground water contamination

Anthropogenic
Peticides, Herbicides, Manures and Fertilizers

Industrial wastes including effluents
Thermal power plants
Mining and Smelting

⟹ metals and metalloids in soil

FIGURE 20.1 Pathways of metals and metalloids in the soil.

20.2.2 ANTHROPOGENIC SOURCES

Industrial practices involve processing of products, their fabrication, and also the disposal of various wastes and effluents, which are the prime cause of metal pollution of soil. As for example, agricultural land inundated with irrigation by the industrial run-offs originating from zinc-smelter plants of the Hindustan Zinc Limited, Debari, Udaipur, Rajasthan for the last 50 years showed lofty levels of Zn, Cd, and Pb in soil (Ray et al., 2017; Golui et al., 2019; Mishra et al., 2019). Copper (Cu) being a constituent of various formulations of fungicides, growth promoter such as $CuOCl_2$ and "Bordeaux" mixture results in build-up of copper in agricultural soils. Use of Pb-based petrol contributed toward atmospheric contamination in many countries as devoid of any restriction on the usage of lead enriched fuel (Sumner, 2000). Application of municipal solid waste is the key source of metal (Zn and Pb) input in agricultural land located in an around Class I cities of India (Golui et al., 2019; Bhattacharyya et al., 2006, 2008a; Sahariah et al., 2015). Application of phosphorus fertilizer also performs as a cause of metal input, especially Cd, in agricultural land (Kirpichtchikova et al., 2006). Chromium emission from steel making processes is even now the prime contribution to Cr pollution in the environment. Various chemicals like sodium chromate and Na-dichromate have various industrial usages, like manufacture of various pigments, synthesis of organics, tanning of leather products, and preservatives for wood. All these contaminants find its path into the natural ecosystem in various ways like improper disposal of industrial wastes in soils, etc.

20.3 APPROACHES FOR QUANTIFICATION OF METAL POLLUTION IN SOIL

20.3.1 BASED ON TOTAL METAL CONTENT IN SOIL

The use of metal concentration (total) in soil as extracted with tri acid mixture or aqua-regia is used as a sole index of metal hazard for polluted soils in most of the countries (Table 20.1). Comparison between amounts of metals in test soil with the permissible limits of the target metals will determine the suitability of agricultural land for cultivation in relation to human health. Soil can be considered as safe, when amount of metal and metalloids in soil is below permissible limit. Recently, use of total metal concentration in soil as an index of pollution has not been recognized as a good indicator in the view that such limit does not take into consideration impact of important soil physicochemical properties on solubility of metals (Golui et al., 2014, 2017, 2020; Meena et al., 2016; Bhattacharyya et al., 2008a, 2008b, 2011; Bhattacharyya and Reddy, 2012). Following too stern regulatory measures may lead to under-utilization of natural sources like land and also anthropogenic sources for pointless remediation. In case of too lenient regulations, it will lead toward deplorable risk for community health and ecosystem from soil toxins.

TABLE 20.1 Values of maximum allowable limits (MAL) for total heavy metals in soils (mg kg^{-1}) as used in different countries

Element	Austria	Canada	Poland	Japan	Great Britain	Germany
Cd	5	8	3	—	3	2
Co	50	25	50	50	—	—
Cr	100	75	100	—	50	200
Cu	100	100	100	125	100	50
Ni	100	100	100	100	50	100
Pb	100	200	100	400	100	500
Zn	300	400	300	250	300	300

Source: Datta *et al.*, 2017

20.3.2 BASED ON INDICES OF METAL HAZARD

Over the years, various indices were formulated and evaluated for assessing level of metal pollution soil. For example, geo-accumulation index (I_{geo}) (Zhou et al., 2015; Mondal et al., 2017), pollution index (PI) and integrated pollution index (Rodriguez-Seijo et al., 2016; Goswami et al., 2014), pollution load index (Ali et al., 2016; Mondal et al., 2020), contamination factor, ecological risk factor (Ray et al., 2017), etc. are used to categorize polluted soil in different suitability classes. These indices, to some extent aid in the assessment of pollution through comparison of metal content in the spiked and untreated soils. In the recent times, another multivariate approach has evolved that is the concept of soil quality for evaluating the veracity of soil pollution (Golui et al., 2019). However, quantification of soil quality index with respect of metal pollution in the soils is at infancy and virtually nonexistent. Use of such soil quality indexes effectively as an educational tool for students, as a pathway of planning for the farmers, social workers, policymakers, and also for the funding agencies.

20.3.3   BASED ON METAL CONTENT IN CROP

Over the years, different regulatory organization prescribed safe limit of metal in edible portions of crop (food material) (Datta et al., 2017; Karak et al., 2011) (Table 20.2). Suitability of food materials grown in metal polluted soil for human consumption can be evaluated based on such limits. However, such generalized limits are not very effective as food habit of human varies greatly from region to region. Therefore, such limits should be linked with food habit of human in a specific region.

TABLE 20.2   Permissible limit of metal content in food materials

Elements	Food products	Country	Maximum limit (mg kg^{-1})
Cadmium	Unpolished rice	Japan	1
Cadmium	Polished rice	China	0.4
Cadmium	Rice	—	0.5
Cadmium	Rice	EU	0.2
Lead	Rice and other cereals	EU	0.2
Mercury	Rice	Taiwan	0.05
Arsenic	Rice		1
Arsenic	Food	UK and Australia	1

Source: Compiled from various sources

20.3.3.1   BASED ON HAZARD QUOTIENT

Contamination through the food chain is considered as one of the foremost paths for admission of metals into the human system apart from drinking water, ingestion of soil, inhalation of dust, or fume and smoke. Risk assessment of metal polluted soil should constitute all of these pathways of metal entry in human body. Risk can be expressed in terms of hazard quotient (HQ) following the USEPA protocol (IRIS, 2020). According to USEPA, HQ is defined as the proportion of the average daily dose (ADD; milligram intake per kilogram body weight per day) of a metal to a reference dose (RfD, milligram intake per kilogram body weight per day).

$$HQ = ADD/RfD$$

If HQ >1.00, then the ADD of a specific metal outstrips the RfD; there is a probable threat allied with particular metal (Table 20.3). Hazard to human well-being (HQ) for entry of metals through intake of food materials like green vegetables (e.g., spinach), wheat, and rice cultivated on contaminated soil can be calculated using HQ formula as mentioned above (Datta and

Young, 2005; Golui et al., 2014; 2017; Meena et al., 2016; Mandal et al., 2019a; Mondal et al., 2017). However, risk assessment for metal contaminated soil using HQ for food material is not complete. The human health is also vulnerable from the risks associated due to intake of contaminated groundwater, microorganisms present in soil and water, polluted surface water, sometimes direct intake of soil by human which have not been pondered here. Therefore, permissible limit of HQ has to be rationalized and fixed below 1.0, considering the proportion of target food material in total daily diet (Datta et al., 2017; Sahariah et al., 2015).

TABLE 20.3 Safe limit of intake of pollutant elements for humans

Element	RfD (mg/kg/day) for intake
Arsenic	0.0021
Cadmium	0.001
Chromium (VI)	0.003
Manganese	0.14
Methyl mercury	0.0001
Nickel	0.02
Zinc	0.3
Copper	0.5
Lead	0.0035

Source: Datta et al., 2017

20.3.3.2 BASED ON EXTRACTABLE METALS

Over the years, researchers engaged in metal research concluded that total metals and metalloids content in soil is not a worthy indicator for evaluating the level of metal pollution. To overcome this problem, an integrated approach has been developed since late nineties for risk appraisal of metal polluted soils with the ultimate aim of prescribing safe limit of extractable metal with regard to significant soil properties and anthropological wellbeing (Datta and Young, 2005). Precise prediction of transference of metals and metalloids from the soil to plants is very important step in routine risk appraisal of metal polluted soils. Simple methodologies like solubility and free ion activity model to forecast metal uptake by plants grown in polluted soils have been used successfully by several researchers from time to time (Datta and Young, 2005; Golui et al., 2014, 2017; Meena et al., 2016; Mandal et al., 2019b). All these studies combined plant uptake data and population dietary information with the final goal of assessing risk (HQ) for metal ingestion by humans through intake of food materials. Indeed, a protocol was developed to prescribe safe limits of extractable/bioavailable metals and metalloids in soil in relation to soil pH and oxidizable organic carbon content (Meena et al., 2016; Golui et al., 2017) (Table 20.4).

TABLE 20.4   Safe limit of extractable metal and metalloid in relation to soil properties and human health hazard

Element	Extractant	pH	Organic Carbon (%)	Critical value of HQ	Safe limit (mg kg-1)	Crop
Cadmium	0.005 M DTPA	6.0	0.25	0.5	0.15	Rice
Cadmium	0.005 M DTPA	8.0	0.50	0.5	6.90	Rice
Cadmium	0.005 M DTPA	6.0	0.25	0.5	0.16	Wheat
Cadmium	0.005 M DTPA	8.0	0.50	0.5	1.31	Wheat
Arsenic	0.5 M NaHCO3	7.5	0.50	0.5	0.43	Rice
Arsenic	0.5 M NaHCO3	8.5	0.75	0.5	0.54	Rice

Source: Meena *et al.*, 2016; Golui *et al.*, 2017

20.4   SOLID-SOLUTION EQUILIBRIA OF METAL IN CONTAMINATED SOILS

The metals are present both in the soil solution as well as on the clay and surfaces and also bound to the organic matter present in the soil. Therefore, a chemical equilibrium persists between the soil solution and the surface adsorbed metals and this governs the bioavailability of the metals in the soils. Various chemical properties such as the chemical bond (covalent, electrovalent, vander Waal's forces), presence of competing ions, and chelation govern the solubility of the metal ions. Hence, a meticulous information about the solid phase of the metal ions is required to be explored to identify the release conditions and pattern. To ensure proper risk assessment due to pollution, the bioavailable or mobile fractions of the metals should be considered rather than the total metal content which may result in poor correlations (Datta and Young, 2005; Bhattacharyya et al., 2008c; Tripathy et al., 2014). Ray and Datta (2016) and Golui et al. (2020) have endeavored to illustrate the availability of metals considering it's activity in the soil solution, that is, the "intensity factor." The ionic activity of an element present in the soil is necessary to develop a relationship of the element concerned with the minerals present in the soils. In a recent study by Ray and Datta (2016), solid mineral phases controlling solubility of zinc and cadmium in zinc smelter polluted soil have been identified on the basis of free-ion activities of the elements in soil solution. It was reported that willemite (Zn_2SiO_4) and octavite ($CdCO_3$) governed the solubility of zinc and cadmium, respectively, in smelter contaminated soil. Such studies are useful in contriving the potential remediation strategies based on sound chemical background.

20.5 IMPACT OF METAL POLLUTION ON HUMAN HEALTH

In Table 20.5, the human health hazards caused due to the metal pollution in the soils and subsequent transfer to food chain. Nonstop ingestion of arsenic contaminated water and food stuffs leading to chronic arsenic toxicity like pigmentation and keratosis. In advanced cases, it may lead to arsenicosis, combined with weakness, respiratory ailments, fibrosis of liver, diseases of the heart, gangrene, and skin cancer along with cancer of lungs and bladder (Guha Mazumder, 2003; Golui et al., 2017). The association between the incidence of itai-itai disease and level of Cd effluence within the rife area around parts of the Jinzu River was recounted from Japan in 1912. Presence of Pb in digestive tract and respiratory tract may disturb and trouble many of the human tissues and systems. Nickel and copper are tumor fostering elements, whose carcinogenesis influence has appealed worldwide apprehensions. Persons who are in intimate interaction with the powdered nickel are more likely to writhe from lung cancer, and the content of nickel in the environment is optimistically associated with nasal cancer (Chen, 2011).

TABLE 20.5 Effect of metal toxicity on human health

Element	Impact on human health due to toxicity
Arsenic	Skin cancer, hyperkeratosis, hyperpigmentation, black foot, rashes, cancer of internal organs
Cadmium	Renal tubular dysfunction, proteinuria, glucosuria, aminoaciduria, itai-itai disease
Chromium	Renal dysfunction, lung cancer
Fluoride	Calcification of the ligaments, osteosclerosis, endochondral ossification, thickening of the flat bones, osteomalaciaand osteoporosis
Copper	Wilson's disease and cirrosis, haemolysis, hepatic necrosis, renal demage and salivary gland swelling
Lead	Encephalopathy (damage to brain), failure in reproduction, metabolic disorder, neurophysical deficit in children, affects the haematologic and renal system
Mercury	Neurological defects, depression, irritability, confusion, tremor, visual and auditory defects
Selenium	Persistent, adverse clinical signs developed with as high as 50% morbidity
Zinc	Interferes with reproduction, impair the growth of embryo

Source: Golui *et al.*, 2019

20.6   REMEDIATION APPROACHES FOR METAL POLLUTED SOILS

The remediation of the soils contaminated with metals and metalloids, although challenging, can be achieved following physiochemical and biological approaches. The physical approach involves removal of soil from contaminated site, decontamination of soil, and return of clean soil residue to the site (in some cases). In the chemical approach, heavy metals in contaminated soil are mostly immobilized or fixed through alteration of soil chemistry and facilitating the formation of insoluble chemical species using amendments or additives. The biological approach involves use of plants or microbes to trim down the level of contamination in soils. Mostly, the physical approach of remediation is based on the principle of ex-situ techniques. However, chemical and biological approaches of remediation are mostly based on the principles of in-situ techniques. Ex-situ techniques involve exclusion of contaminated soils for treatment on- or offsite, and in-situ techniques involve remediation without excavation of polluted soils. Usually, remediation of metal polluted soils encompasses physical removal (decontamination) or immobilization of metals, rather degradation of metals.

20.6.1   PHYSICAL APPROACH

Most commonly used techniques under physical approach of soil remediation are as follows:

(1) *Excavation and land-filling*: It involves bulk digging out of contaminated soil and burial (land-filling) at a hazardous waste site (Huang and Cunningham, 1996); (2) *Soil washing:* This is an ex-situ technique of transferring metals from soil matrix to washing fluids (mostly acids, e.g., HCl) followed by precipitation of metals as metal salts and revert of unpolluted soil residue to the site. Although chemical extractants (washing fluids) are used in this method, it does not detoxify or considerably revise the contaminants (USEPA, 1997); (3) *Soil replacement*: It is the process of using unpolluted soil to swap or partially replenish the polluted soil with the objective of diluting metal intensity (Qian and Liu, 2000); and (4) *Thermal treatment*: It is the technique of heating contaminated soils using steam, microwave, infrared radiation to make the pollutant (e.g., Hg, As) unstable and collecting unstable metals using void negative pressure or

carrier gas for the abstraction of metals and metalloids from the soil (Li et al., 2010).

In common, the physical approach of metal remediation (ex-situ technique) is very expensive. Normally, remediation expenditure range from US$60–300/m³ of soil in case of land-filling or low-temperature management, and US$200–700/m³ in case of materials requiring unique landfill machines/instruments or high-temperature thermal treatments. Soil exclusion and substitution with unpolluted soil is obviously more expensive, costing between $8 and $24 million per hectare per meter of soil depth removed as reported by (Cunningham et al., 1995; Glass, 2000). The physical approach (e.g., soil washing) is disruptive to soil and ecosystem, and also accompanied with the production of secondary waste. The chances of ground water contamination due to trickling of hazardous metals and metalloids (in case of land-filling) and secondary pollution (in case of soil replacement) are also very high in this approach. Moreover, the ex-situ techniques as followed in physical approach are pertinent to moderately undersized areas. These techniques are not suitable for hefty areas, such as mining or smelting sites.

20.6.2   CHEMICAL APPROACH

Most commonly used technique under chemical approach of soil remediation is in-situ immobilization or steadying of metals and metalloids using amendments. Other techniques include vitrification and electrokinetic remediation.

Immobilization (stabilization) of metals: The technique involves addition of reagents or amendments to the polluted soil to yield more chemically durable components. Immobilization of metals and metalloids is achieved mainly through chemical adsorption, precipitation of the metals, and complexation/chelation reactions which cause in the reallocation of contaminants from the soil solution phase to solid phase, and thereby reducing their bioavailability (Basta and McGowen, 2004). Several studies across the globe have reported successful in-situ control of metals in contaminated soils using inorganic and organic amendments. Based on these studies, the potential amendments to arrest or immobilize the metals and metalloids in soil are presented in Table 20.6.

TABLE 20.6   Potential amendments for immobilization of metal(loid)s in soil

Amendments	Metal(loid)s	Observations	References
Phosphate Compounds:			
KH2PO4	Cd	Augmenting the immobilization process and reducing plant availability	Bolan *et al.*, 2003a
Phosphates	As	Boosting phosphate availability decreased As uptake.	Wang *et al.*, 2002
H3PO4,Ca(H2PO4), phosphate rock	Pb, Zn, Cu	Increment in residual fraction of Pb, Zn and Cu, and diminishing translocation from root to shoot of plant.	Cao *et al.*, 2003
Apatites, phosphate rock, triple superphosphate, diammonium phosphate	Cd, Pb, Zn	Decrease uptake by plant.	Chen *et al.*, 2007
Liming Materials:			
Ca(OH)2	Cd	Altered to immobile fractions	Bolan *et al.*, 2003b
Lime	Cd	Not efficient in immobilizing the contaminant	Li *et al.*, 1996
Lime	Pb, Cd	Cd uptake was reduced but scanty influence Pb uptake in raddish.	Han and Lee, 1996
Lime	As	Augmenting As immobilization by the formation of Ca–As precipitates	Moon *et al.*, 2004
Lime	Zn, Pb, Cd	Reduced solubility, plant uptake and bioavailability of all the elements	Ray (2016)
CaCO3+Farmyard manure, CaCO3	Zn, Cu, Ni	Immobilized Zn, Cu, Ni in soil, reduced Zn and Ni content in lettuce.	Paulose *et al.* (2007)
Organic Matter:			
Biosolid	Cd	Reduce the Cd bioavailability	Bolan *et al.*, 2003c
Biosolid, manure	Cu	Reduce the Cu availability	Bolan *et al.*, 2003d
Biosolid	Zn	Reduce the Zn availabilty	Shuman *et al.*, 2000
Organic matter	Ni, Cd, Zn	Reduce the Ni uptake but not Zn and Cd in rice	Kashem and Singh, 2001
Compost	Cd, Cu, Pb, Zn	Zn and Cd leaching decreased, but Cu and Pb leaching increased.	Ruttens *et al.*, 2006
Vermicompost	As	Reduction of As accumulation in sesame.	Sinha *et al.*, 2011
Sugarcane Bagasse	As	Reduction of As accumulation in wheat and maize	Mandal *et al.*, 2019
Paddy husk	As	Reduction of As accumulation in wheat	Raj *et al.*, 2020
Humus, compost	Cd	Reduce Cd uptake in rice	Ok *et al.*, 2011

TABLE 20.6 *(Continued)*

Amendments	Metal(loid)s	Observations	References
Coal fly ash, peat	Cu, Pb	Reduced metal(loid) leaching in the field lysimeters.	Kumpiene *et al.* (2007)
Coal fly ash	Zn, Pb, Cd	Reduced solubility, plant uptake and bioavailability of all the elements.	Ray (2016)
Metal Oxides			
Goethite, iron grit, iron (II) and (III) sulphate	As, Cu, Zn	Reduced plant shoot As content	Hartley and Lepp, 2008
Hydrous Mn Oxide	Cd, Zn, Pb	Reduce mobility and uptake in rye grass	Mench *et al.*, 1994
Mn oxide	Pb	Reduce availability of Pb to plants	Hettiarachchi *et al.*, 2000

Source: Golui *et al.*, 2019

The effectiveness of different amendments, namely, phosphate compounds, liming materials, organic ameliorant, and metal oxides in immobilization of metals and metalloids has been attributed to sound chemical reactions in soil by various authors. Phosphate compounds immobilize metals and metalloids in soils due to direct adsorption/substitution of heavy metals by phosphorus compounds, phosphorus anion-induced metal adsorption, and precipitation of metals and metalloids with solution phosphorus as metal and metalloid-phosphates (Bolan et al., 2003a; Basta and McGowen, 2004; Chen et al., 2007). Precipitation as metal-P has been well documented as one of the key means for the immobilization of metals, such as lead and zinc, in soils. It is reported that Pb-hydroxypyromorphite mineral due to its low solubility controls Pb concentration in soil amended with phosphate materials (Santillian-Medrano and Jurinak, 1975; Ma et al., 1993). The addition of phosphorus in zinc-contaminated sediments has been reported to form precipitation of hopeite, a Zn phosphate mineral (Williams et al., 2011; Ndiba et al., 2008). However, ineffectiveness of phosphate materials in reducing availability of Cd in soil is attributed to the fact that solubility of cadmium phosphate in general is excessive to manipulate the solution concentration of Cd (Bolan and Duraisamy, 2003). There are also experimental confirmations advocating incompetence of phosphatic amendments in lowering availability of metals, including Pb, Zn, and Cd in polluted soils (Creger and Peryea, 1994; Paulose et al., 2007; Datta et al., 2007; Ray, 2016).

Liming, a technique to ameliorate the acid soils, its usefulness to check the movement of metals in contaminated soil as reported by various authors is indicated in Table 20.6. Liming causes the rise of soil pH which results in precipitation of the metals as their hydroxides and carbonates and hence reduces its bioavailability (Datta et al., 2007; Paulose et al., 2007; Ok et al., 2011a,b,c). The rise in soil pH causes the rise in pH-dependent surface negative charge which in turn surge the adsorption of metals (McBride et al., 1997; Bolan and Duraisamy, 2003; Bolan et al., 2014). Upsurge in soil pH due to liming triggers the formation of metal and metalloids hydroxides which have more empathy for the adsorption sites than the sole metal cations (Naidu et al., 1994).

Although, contradictory reports are available on the effectiveness of organics or bio-solids in immobilizing heavy metals and metalloids in soil (Bolan et al., 2014), arresting of metals and metalloids by organics is achieved mainly through adsorption of the metals and also through chelation and complexation. Many authors have suggested that alkaline-stabilized bio-solids with low total or bioavailable metal(loid) content can be useful in remediation purpose. Similarly, usefulness of metal oxides in immobilization studies has been attributed to the amphoteric nature of oxides which make them appropriate for adsorption of metals and metalloids in soil.

In chemical immobilization (stabilization) technique, soil ecological conditions such as pH, redox potential (Eh), cation exchange capacity, and soil organic matter content can affect the substantial release of immobilized heavy metals and metalloids in soil solution. Therefore, censoring of long-term stability of the immobilized heavy metal is very important. In general, this technique is affordable, easy to implement and can be followed for immediate effects.

Vitrification: It involves high-temperature (1400–2000 °C) treatment of the contaminated soil to transform into glasslike solids. Heavy metals are encapsulated in to glasslike solid materials. Vitrified substances with particular traits may be attained by using sand, clay, and/or native soil. In-situ vitrification is favored over ex-situ technique due to lower energy requirement and cost (USEPA, 1992). In case of in-situ vitrification, high temperature is generated through high voltage electricity by inserting graphite electrodes in contaminated soil at particular spacing. This is a commercially proven technology by USEPA for disposal of heavy metal wastes. The in-situ vitrification technique is suitable for small area and it is an expensive technique. The estimated cost of vitrification is $330–425 per ton of soil treated (FRTR, 2012). In case of soils having high organic matter content along with volatile contaminants this technique of vitrification is not applicable.

Electrokinetic remediation: This technique of remediation is centered on the principles of electrochemistry, which involves application of direct current (DC) electricity via electrodes injected in the contaminated soil for facilitating the migration of cations to the cathode and negatively charged ions to the anode under an recognized electric field. Under a DC electric field, the metal ions migrate which is primarily via electroosmosis, electro-migration (movement of ions to the counterpart electrode), electrophoresis and diffusion due to gradient force (Page and Page, 2002). Heavy metals loaded at the polarized electrodes are eradicated by electroplating, (co-) precipitation, solution pumping, or ion exchange resin complexation (FRTR, 2012). As this technology requires a high DC voltage for maintaining a strong electric field (~150 V/m) for electroosmosis and low ion migration speed, solicitation of this technology is narrow (Zhou et al., 2006; Peng et al., 2013). This technique of soil remediation is still at developing stage. Although, the effectiveness of this technique has been demonstrated in few studies (Zhou et al., 2006; FRTR, 2012; Hansen et al., 2016), large-scale application is meager.

20.6.3 BIOLOGICAL APPROACH

Biological approach for remediation of metal and metalloid polluted soils includes phytoremediation and bioremediation.

20.6.3.1 PHYTOREMEDIATION

Phytoremediation is an in-situ remediation, involving employment of plants to extract metals from contaminated soil or stabilize them in soil. In general, reduction in metal(loid) contamination in soil is achieved through phytoextraction (removal technology) and phytostabilization (containment technology). Phytoextraction denotes the uptake of metals and metalloids by plant roots into the areal parts of plants. Phytostabilization refers to the immobilization of metals and metalloids in soil by plant roots.

In phytoextraction, certain plants, also known as hyperaccumulators, absorb unusually high concentration of metals from polluted soil without suffering phytotoxic damage. Names of some hyperaccumulators are presented in Table 20.7. Technical challenges associated with hyperaccumulation of metal(loid)s for soil remediation include low metal(loid) removal rates and disposal of the metal accumulating plants. Land filling and incineration are the options to dispose-off metal(loid)s after harvesting of hyperaccumulators. Remediation of soil by this technique may require up to 20 years subject to the extent of metal contamination and the efficacy of hyperaccumulator to confiscate metals from soil (Blaylock and Huang, 2000). To accelerate phytoextraction, plant genetic manipulation may be an option (Khan et al.,

TABLE 20.7 List of potential plant species for phytoremediation

Elements	Plant Species	Maximum concentration (mg kg⁻¹)	Reference
Arsenic	Pteris vittata	22,630	Ma et al., 2001
	Pteris cretica	3030	Zhao et al., 2002
	Pteris umbrosa	7600	Zhao et al., 2002
Cadmium	Thlaspi caerulescens	2130	Reeves and Brooks, 1983
Chromium	Dicoma niccolifera	30,000	Peterson, 1975
	Leptospermum scoparium	20,000	Lyon et al., 1969
	Sutera fodina	48,000	Peterson, 1975
Copper	Haumaniastrum katangese	9222	Brooks et al., 1987
	Ipomea alpina	12,300	Cunningham and Ow, 1996
	Pandiaka metallorum	6270	Brooks et al., 1987
Nickel	Alyssum (48 taxa, all from Sect. Odontarrhena)	29,400	Reeves et al., 1996
	Geissois pruinosa	34,000	Jaffre and Schmid, 1974
	Pimelea leptospermoides	60,170	Reeves et al., 1996
Lead	Agrostis tenuis	13,490	Williams et al., 1977
	Minuartia verna	20,000	Jhonston and Proctor, 1977
	Thlaspi rotundifolium	8200	Cunningham and Ow, 1996
Zinc	Arabidopsis halleri	30,000	Zhao et al., 2000
	Thlaspi caerulescens	51,600	Brown et al., 1994
	Thlaspi calaminare	39,600	Reeves and Brooks, 1983

2000). However, developing genetically engineered hyperaccumulators for faster removal of metals and metalloids is a protracted method.

In phytostabilization, immobilization of metal(loid)s in soil occurs due to rhizospheric reduction, soil stabilization, root adsorption, and exudate complexation/precipitation (Liu et al., 2018). For being potentially suitable for phytostabilization, plants must have high root biomass. *Agrostis tenuis*, *Festuca rubra*, *Gentiana pennelliana*, *Hyparrhenia hirta*, *Zygophyllum fabago*, and *Vossia cuspidate* are some important plant species for phytostabilizing soils contaminated by Pb, Zn, Cr, and Cu (Yoon et al., 2006; Galal et al., 2017; Radziemska et al., 2017).

In general, the phytoremediation technique is based on the assumption of long-term applicability. The cost involvement in this technique is usually less than any other conventional method or technique of soil remediation. Therefore, this technique can have large scale applications, provided highly efficient plant species are involved.

20.6.3.2 BIOREMEDIATION

This technique involves the use of microorganisms for remediation purpose. The mechanisms by which microorganisms detoxify metal(loid)s include intracellular accumulation, extracellular chemical precipitation, oxidation-reduction reaction, sorption on microbial cell surface and volatilization (Garbisu and Alkorta, 2003; Yao et al., 2012). The method is generally useful for detoxifying organic toxins in soil. However, based on numerous studies enlisted the potential microbes, including bacteria, fungi, algae which have the potential for remediation of metals and metalloids contaminated soils. Although the potentiality of these microorganisms for metals removal has mostly been tested in aqueous medium, nevertheless these microorganisms may serve as potential agents for metal(loid) removal from soil. Further, these studies indicate that biosorption is the major mechanism of bioremediation by microorganisms. For example, *Bacillus circulans* and *Bacillus megaterium* have been reported to be potential bacterial strains for biosorption and bioaccumulation of hexavalent (VI) chromium (Srinath et al., 2002). Similarly, biosorption of Pb, Cd, Cu, and Ni by the fungus *Aspergillus niger* has been reported by Kapoor et al. (1999). Besides, volatilization of inorganic mercury (Hg) from soil microcosm using transgenic bacterium *Bacillus cereus* BW-03($_p$PW-05) has been reported by (Dash and Das, 2015). Such study indicates the potentiality of genetically concocted microorganisms for the bioremediation of problematic soils. Reports on the applicability of bioremediation technique for metal(loid) removal from soil using microorganisms is scantly and this technique is still at infancy.

20.7 APPLICABILITY OF REMEDIATION TECHNIQUES

The applicability and assortment of any remediation technique is depended on the nature and source of contamination, contamination level, cost-effectiveness, long-term permanence, general acceptance, remediation goal, time requirement, and ease in implementation. All these criteria are important for selection of remediation technique based on site-specific issues. In developing countries like India with limited funds allocated for environmental restitution, cost-effective and ecologically viable alternatives are required for restoration of metal(loid) contaminated soils in order to diminish associated health risk and enhance food and nutritional security. The ex-situ remediation technologies, due to involvement of high cost, are usually not preferred. Chemical immobilization and phytoremediation are the encouraging tools for remediation of soil polluted by metal(loid)s in developing countries. Chemical immobilization and phytoremediation can be followed for highly contaminated and low to moderately contaminated soils, respectively. In excessively high contaminated soils, for example, mining and smelting sites, chemical immobilization technique can be practiced to minimize the toxicity due to high-accumulation of metal(loid)s in soil for establishing plants, followed by phytoremediation to restore the contaminated site gradually in long-run. For successful implementation of any remediation technology, proper planning in view of the merits and demerits of the proposed techniques is of immense importance. For example, in order to make phytoremediation a successful remediation strategy, proper arrangement should be made to dispose-off the plants after harvesting based on the availability of resources and funds. Besides, health risks associated with any remediation technology must be assessed for its successful implementation.

20.8   SUMMARY AND CONCLUSIONS

Pollution of soil due to accumulation of metals is an important environmental issue. Buildup of dreaded metals in soil, particularly due to human activities causes serious environmental degradation. Soil pollution by metals and metalloids also poses serious menace to human health through food sequence contamination. Therefore, remediation and restoration of metal(loid) contaminated soil is one of the high priority areas of research as far as human health and environment are concerned. Numerous techniques have been established and proposed to diminish contamination of soil by the heavy metals. These techniques involve physical, chemical, and biological activities for remediation of heavy metal contamination in soil. Such techniques possess differential applicability with varying benefits and drawbacks. Therefore, selection of appropriate remediation techniques depends on various factors namely, extent of contamination, performance, and cost-competitiveness. Among the various available strategies for remediation of metal(loid) contaminated soils, chemical immobilization and phytoremediation are the promising technologies with high applicability in developing countries.

In order to develop more efficient systems for remediation of metals and metalloid contaminated soils, following researchable issues need to be addressed:

1. In view of scarce information on long-term constancy of immobilized metals and metalloids in amended soil, more studies are required in this direction. Detailed studies on the formation of solid phases and their solubility in amended soils may help address the issue.
2. In chemical immobilization studies, effectiveness of amendments is mostly assessed by increased or decreased availability of metal(loid)s in contaminated soil. It is seldom studied if the accomplished reduction in available heavy metals is really harmless from human well-being point of view. Remediation studies linking health risk assessment are the priority areas of research.
3. Further studies are required for greater understanding of the process of metal absorption by hyperaccumulators and genes responsible for absorption in order to develop more efficient plant cultivars through genetic engineering for phytoremediation.
4. More studies are required in order to reconnoiter the possibility of large-scale application of the state-of-the-art electro kinetic remediation technology for removal of metals from contaminated soils.
5. In view of limited available information on the applicability of bioremediation technique for metal(loid) removal from soil using microorganisms, more research is desired to screen the prospective microorganisms for remediation of metal contaminated soils.

KEYWORDS

- **hazard quotient**
- **metals**
- **pollution**
- **remediation**

REFERENCES

Ali, M.M.; Ali, M.L.; Islam, M.S.; Rahman, M.Z. 2016. Preliminary assessment of heavy metals in water and sediment of Karnaphuli river, Bangladesh. *Environ. Nanotechnol. Monit. Manage.* 2016, 5, 27–35.

Alloway, B.J. Heavy Metals in Soils, 2nd edition.; Blackie Academic and Professional: London, UK, 1995.

Basta, N.T.; McGowen, S.L. Evaluation of chemical immobilization treatments for reducing heavy metal transport in a smelter-contaminated soil. *Environ. Pollut.* 2004, 127, 73–82.

Bhattacharyya, P.; Chakrabdorty, A.; Chakrabarti, K.; Tripathy, S.; Powell, M.A. Copper and zinc uptake by rice and accumulation in soil amended with municipal solid waste compost. *Environ. Geol.* 2006, 49, 1064–1070.

Bhattacharyya, P.; Chakrabdorty, A.; Chakrabarti, K.; Tripathy, S.; Powell, M.A. Fractionation and bioavailability of lead in municipal solid waste compost and uptake by rice straw and grain under submerged condition. *Geosci. J.* 2008a, 12, 41–45.

Bhattacharyya, P.; Reddy, K.J. Effect of flue gas treatment on the solubility and fractionation of different metals in fly ash of Powder River basin coal. *Water Air Soil Pollut* 2012, 223, 4169–4181.

Bhattacharyya, P.; Reddy, K.J.; Attili, V. Solubility and fractionation of different metals in fly ash of Powder River basin coal. *Water Air Soil Pollut.* 2011, 220, 327–337.

Bhattacharyya, P.; Tripathy, S.; Kim, K.; Kim, S.H. Arsenic fractions and enzyme activities in arsenic-contaminated soils by groundwater irrigation in West Bengal. *Ecotox. Environ. Safe.* 2008b, 71, 149–156.

Bhattacharyya, P.; Tripathy, S.; Chakrabarti, K.; Chakraborty, A.; Banik, P. Fractionation and bioavailability of metals and their impacts on microbial properties in sewage irrigated soil. *Chemosphere.* 2008c, 72, 543–550.

Blaylock, M.J.; Huang, J.W. Phytoextraction of metals. In Phytoremediation of toxic metals: using plants to clean-up the environment; Raskin, I., Ensley, B.D.; John Wiley & Sons Inc., New York, 2000; pp. 53–70.

Bolan, N.; Kunhikrishnan, A.; Thangarajan, R.; Kumpiene, J.; Park, J.; Makino, T.; Kirkham, M.B.; Scheckel, K. Remediation of heavy metal(loid)s contaminated soils—to mobilize or to immobilize? *J. Hazard. Mater.* 2014, 266, 141–166.

Bolan, N.S.; Adriano, D.C.; Duraisamy, A., Mani, P. Immobilization and phytoavailability of cadmium in variable charge soils. III. Effect of biosolid compost addition. *Plant Soil.* 2003a, 256, 23–241.

Bolan, N.S.; D.C. Adriano, D.C.; Mani, P.; Duraisamy, A.; Arulmozhiselvan, S. Immobilization and phytoavailability of cadmium in variable charge soils: I. Effect of phosphate addition. *Plant Soil.* 2003b, 250, 83–94.

Bolan, N.S.; Adriano, D.C.; Mani, P.; Duraisamy, A.; Arulmozhiselvan, S. Immobilization and phytoavailability of cadmium in variable charge soils: II. Effect of lime addition. *Plant Soil.* 2003c, 250, 187–198.

Bolan, N.S.; Duraisamy, V.P. Role of inorganic and organic soil amendments on immobilization and phytoavailability of heavy metals: A review involving specific case studies. *Aust. J. Soil Res.* 2003d, 41, 533–555.

Cao, X.; Ma, L.Q.; Chen, M.; Singh, S.P.; Harris, W.G. Phosphate-induced metal immobilization in a contaminated site. *Environ. Pollut.* 2003, 122, 19–28.

Chen, Y.F. Review of the research on heavy metal contamination of China's city soil and its treatment method. *Chi. Pop. Res. Environ.* 2011, 21, 536–539.

Chen, S.; Xu, M.; Ma, Y.; Yang, J. Evaluation of different phosphate amendments on availability of metals in contaminated soil. *Ecotoxicol. Environ. Saf.* 2007, 67, 278–285.

Creger, T.L.; Peryea, F.J. Phosphate fertilizer enhances arsenic uptake by apricot liners grown in lead-arsenate-enriched soil. *Hort. Sci.* 1994, 29(2), 88–92.

Cunningham, S.D.; Berti W.R.; Huang J.W. Phytoremediation of contaminated soils. *Trends Biotechnol.* 1995, 13, 393–397.

Cunningham, S.D.; Ow, D.W. Promises and prospects of phytoremediation. *Plant. Physiol.* 1996, 110, 715–719.

Dash, H.R.; Das, S. Bioremediation of inorganic mercury through volatilization and biosorption by transgenic Bacillus cereus BW-03 (pPW-05). *Int. Biodeterior. Biodegrad.* 2015, 103, 179–185.

Datta, S.P., Young, S.D. Predicting metal uptake and risk to human food chain from leafy vegetables grown on soils amended by long-term application of sewage sludge. *Water Air Soil Pollut.* 2005, 163, 119–136.

Datta, S.P.; Golui, D.; Sanyal, S.K. Assessing potential threats of soil pollutant elements in relation to food-chain contamination with suggested remedial measures. In Souvenir of 82nd Annual Convention and National Seminar of Indian Society of Soil Science, Kolkata, December 11–14, 2017, 137–150.

Datta, S.P.; Rattan, R.K.; Chandra, S. Influence of different amendments on the availability of cadmium to crops in the sewage-irrigated soil. *J. Indian Soc. Soil Sci.* 2007, 55, 86–89.

FRTR, 2012. Remediation Technologies Screening Matrix and Reference Guide, Version 4.0. Federal Remediation Technologies Roundtable, Washington, DC.

Galal, T.M.; Gharib, F.A.; Ghazi, S.M.; Mansour, K.H. Phytostabilization of heavy metals by the emergent macrophyte *Vossia cuspidada* (Roxb.) Griff.: a phytoremediation approach. *Int. J. Phytoremediation.* 2017, 19, 992–999.

Garbisu, C.; Alkorta, I. Basic concepts on heavy metal soil bioremediation. *Eur. J. Miner. Process. Environ. Prot.* 2003, 3, 58 66.

Glass, D.J. Economic potential of phytoremediation. In Phytoremediation of Toxic Metals Using Plants to Clean Up the Environment. Raskin, I., Ensley, B.D. Ed.; Wiley: New York, 2000; pp. 15–31.

Golui, D.; Datta, S.P.; Dwivedi, B.S.; Meena, M.C.; Trivedi, V.K. Prediction of free metal ion activity in contaminated soils using WHAM VII, baker soil test and solubility model. *Chemosphere*, 2020, 243, 125408.

Golui, D.; Datta, S.P.; Dwivedi, B.S.; Meena, M.C.; Varghese, E.; Sanyal, S.K.; Ray, P.; Shukla, A.K.; Trivedi, V.K. 2019. Assessing soil degradation in relation to metal pollution-A multivariate approach. *Soil Sediment Contam.* 2019, 28(7), 630–649.

Golui, D.; Datta, S.P.; Rattan, R.K.; Dwivedi, B.S.; Meena, M.C. Predicting bioavailability of metals from sludge amended soils. *Environ. Monit. Assess.* 2014, 186, 8541–8553.

Golui, D.; Guha Mazumder, D.N.; Sanyal, S.K.; Datta, S.P.; Ray, P.; Patra, P.K.; Sarkar, S.; Bhattacharya, K. Safe limit of arsenic in soil in relation to dietary exposure ofarsenicosis patients from Malda district, West Bengal—a case study. *Ecotoxicol. Environ. Saf.* 2017, 144, 227–235.

Goswami, L.; Raul, P.; Sahariah, B.; Bhattacharyya, P.; Bhattacharya, S.S. Characterization and risk evaluation of tea industry coal ash for environmental suitability. *Clean- Soil, Air, Water.* 2014, 42, 1470–1476.

Guha Mazumder, D.N. Chronic arsenic toxicity: clinical features, epidemiology and treatment: experience in West Bengal. *J. Environ. Sci. Health.* 2003, 38, 141–163.

Han, D.H.; Lee, J.H. Effect of liming on uptake of lead and cadmium by Raphanus sativa. *Arch. Environ. Contam. Toxicol.* 1996, 31, 488–493.

Hansen, H. K.; Ottosen, L. M.; Ribeiro, A.B. Electrokinetic soil remediation: an overview. In Electrokinetics Across Disciplines and Continents. Ribeiro, A., Mateus, E., Couto, N. (Ed.; Springer, Cham, Switzerland, 2016; pp. 3–18.

Hartley, W.; Lepp, N.W. Remediation of arsenic contaminated soils by iron oxide application evaluated in terms of plant productivity, arsenic and phytotoxic metal uptake. *Sci. Total Environ.* 2008, 390, 35–44.

Hettiarachchi, G.M.; Pierzynski, G.M.; Ransom, M.D. In situ stabilization of soil lead using phosphorus and manganese oxide. *Environ. Sci. Technol.*, 2000, 34, 4614–4619.

Huang, J. W.; Cunningham, J. D. Lead phytoextraction: species variation in lead uptake and translocation. *New. Phytol.* 1996, 134, 75–84.

IRIS, 2020. Integrated Risk Information System Database. US Environmental Protection Agency.

Kabata-Pendias, A.; Pendias, H. Trace Metals in Soils and Plants, CRC Press, Boca Raton, FL, USA, 2nd edition, 2001.

Kapoor, A.; Viraraghavan, T.; Cullimore, D.R. Removal of heavy metals using fungus *Aspergillus niger. Bioresour. Tech.* 1999, 70(1), 95–104.

Karak, T.; Abollino, O.; Bhattacharyya, P.; Das, K.K.; Paul, R.K. Fractionation and speciation of arsenic in three tea gardens soil profiles and distribution of As in different parts of tea plant (*Camellia sinensis* L.). *Chemosphere.* 2011, 85, 948–960.

Kashem, M.A., Singh, B.R. Metal availability in contaminated soil: II. Uptake of Cd, Ni and Zn in rice plants grown under flooded culture with organic matter addition. *Nutr. Cycl. Agroecosyst.* 2001, 61, 257–266.

Khan, A.G.; Kuek, C.; Chaudhry, TM.; Khoo, CS.; Hayes, W.J. Role of plants, mycorrhizae and phytochelators in heavy metal contaminated land remediation. *Chemosphere.* 2000, 41, 197–207.

Kirpichtchikova, T.A.; Manceau, A.; Spadini, L.; Panfili, F.; Marcus, M.A.; Jacquet, T. Speciation and solubility of heavy metals in contaminated soil using X-ray microfluorescence, EXAFS spectroscopy, chemical extraction, and thermodynamic modelling. *Geochim. Cosmochim. Acta.* 2006, 70(9), 2163–2190.

Kumpiene, J.; Lagerkvist, A.; Maurice, C. Stabilization of Pb- and Cu-contaminated soil using coal fly ash and peat. *Environ. Pollut.* 2007, 145, 365–372.

Li, Y.M.; Chaney, R.L.; Schneiter, A.A.; Johnson, B.L. Effect of limestone applications on cadmium content of sunflower (*Helianthus annuus* L.) leaves and kernels. *Plant Soil.* 1996, 180, 297–302.

Li, J.; Zhang, G.N.; Li, Y. Review on the remediation technologies of POPs. *Hebei Environment Sci.* 2010, 65(8), 1295–1299.

Liu, L.; Li, W.; Song, W.; Guo, M. Remediation techniques for heavy metal-contaminated soils: Principles and applicability. *Sci. Total Environ.* 2018, 633, 206–219.

Ma, Q.Y.; Traina, S.J.; Logan, T.J. In situ lead immobilization by apatite. *Environ. Sci. Technol.* 1993, 27, 1803–1810.

Mandal, J.; Golui, D.; Datta, S.P. Assessing equilibria of organo-arsenic complexes and predicting uptake of arsenic by wheat grain from organic matter amended soils. *Chemosphere.* 2019a, 229, 419–426.

Mandal, J.; Golui, D.; Raj, A.; Ganguly, P. Risk Assessment of arsenic in wheat and maize grown in organic matter amended soils of Indo-Gangetic Plain of Bihar, India. *Soil Sediment Contam.* 2019b, 28(8), 757–772.

McBride, M.B.; Suave, S.; Hendershot, W. Solubility control of Cu, Zn, Cd, and Pb in contaminated soils. *Eur. J. Soil Sci.* 1997, 48, 337–346.

Meena, R.; Datta, S.P.; Golui, D.; Dwivedi, B.S.; Meena, M.C. Long-term impact of sewage irrigation on soil properties and assessing risk in relation to transfer of metals to human food chain. *Environ. Sci. Pollut.* Res. 2016, 23, 14269–14283.

Mench, M.; Vangronsveld, J.; Didier, V.; Clijsters, H. Evaluation of metal mobility, plant availability and immobilization by chemical agents in a limed-silty soil. *Environ. Pollut.* 1994, 86, 279–286.

Mishra, R.; Datta, S.P.; Annapurna, K.; Meena, M.C.; Dwivedi, B.S.; Golui, D.; Bandyopadhyay, K.K. Enhancing the effectiveness of zinc, cadmium and lead phytoextraction in polluted soils by using amendments and microorganisms. *Environ. Sci. Pollut. Res.* 2019, 26, 17224–17235.

Mondal, A.; Das, S.; Sah, R.K.; Bhattacharyya, P.; Bhattacharya, S.S. Environmental footprints of brick kiln bottom ashes: geostatistical approach for assessment of metal toxicity. *Sci. Total Environ.* 2017, 609, 215–224.

Mondal, A.; Goswami, L.; Hussain, N.; Barman, S.; Kalita, E.; Bhattacharyya, P.; Bhattacharya, S.S. Detoxification and eco-friendly recycling of brick kiln coal ash using *Eisenia fetida*: a clean approach through vermitechnology. *Chemosphere.* 2020, 244, 125470.

Moon, D.H.; Dermatas, D.; Menounou, N. Arsenic immobilization by calcium–arsenic precipitates in lime treated soils. *Sci. Total. Environ.* 2004, 330, 171–185.

Naidu, R.; Bolan, N.S.; Kookana, R.S.; Tiller, K.G. Ionic-strength and pH effects on the sorption of cadmium and the surface charge of soils. *Eur. J. Soil Sci.* 1994, 45, 419–429.

National Academy of Sciences. Chromium, Report of the subcommittee on chromium. Committee on Biologic Effects of Atmospheric Pollutants, National Academy of Sciences, Washington, D.C., 1974; p. 250.

Ndiba, P.; Axe, L.; Boonfueng, T. Heavy metal immobilization through phosphate and thermal treatment of dredged sediments. *Environ. Sci. Technol.* 2008, 42, 920–926.

Ok, Y.S.; Kim, S.C.; Kim, D.K.; Skousen, J.G.; Lee, J.S.; Cheong, Y.W.; Kim, S.J.; Yang, J.E. Ameliorants to immobilize Cd in rice paddy soils contaminated by abandoned metal mines in Korea. Environ. *Geochem. Health.* 2011a, 33, 23–30.

Ok, Y.S.; Lee, S.S.; Jeon, W.T.; Oh, S.E.; Usman, A.R.A.; Moon, D.H. Application of eggshell waste for the immobilization of cadmium and lead in a contaminated soil. *Environ. Geochem. Health.* 2011b, 33, 31–39.

Ok, Y.S.; Lim, J.E.; Moon, D.H. Stabilization of Pb and Cd contaminated soils and soil quality improvements using waste oyster shells. *Environ. Geochem. Health.* 2011c, 33, 83–91.

Page, A.L.; Bingham F.T.; Chang, A.C. Effect of Heavy Metal Pollution on Plants. Applied Science Publication: London, 1981; Vol. 1, pp. 77–109.

Page, M.M.; Page, C.L. Electroremediation of contaminated soils. *J. Environ. Eng.* 2002, 128, 208–219.

Paulose, B.; Datta, S.P.; Rattan, R.K.; Chhonkar, P.K. Effect of amendments on the extractability, retention and plant uptake of metals on a sewage-irrigated soil. *Environ. Pollut. Res.* 2007, 146, 19–24.

Peng, C.; Almeira, J.; Abou-Shady, A. Enhancement of ion migration in porous media by the use of varying electric fields. Sep. *Purif. Technol.* 2013, 118, 591–597.

Pierzynski, G.M.; Sims, J.T.; Vance, G.F. Soils and Environmental Quality, 2nd ed.; CRC Press: London, UK, 2000.

Qian, S.Q.; Liu. Z. An overview of development in the soil-remediation technologies. *Chem. Industr. Eng. Proc.* 2000, 4, 10–12.

Raj, A.; Mandal, J.; Kumari, P.B. Organic amendment can reduce arsenic uptake in wheat. *J Pharmacogn Phytochem.* 2020, 9(2), 1355–1360.

Radziemska, M.; Vaverkova, M.D.; Baryla, A. Phytostabilization—management strategy for stabilizing trace elements in contaminated soils. *Int. J. Environ. Res. Public Health.* 2017, 14, 958.

Ray, P. Dynamics of Heavy Metals and Their Immobilization in Zinc Smelter- Effluent Contaminated Soils. Ph.D Dissertation, Indian Agricultural Research Institute, New Delhi, India, 2016.

Ray, P.; Datta, S.P. Solid phase speciation of Zn and Cd in zinc smelter effluent irrigated soils. *Chem. Speciat. Bioavailab.* 2016, 29(1), 6–14.

Ray, P.; Datta, S.P.; Dwivedi. B.S. Long-term irrigation with zinc smelter effluent affects important soil properties and heavy metal content in food crops and soil in Rajasthan, India. *Soil Sci. Plant Nutr.* 2017, 63, 628–37.

Rodriguez-Seijo, A.; Alfaya, M.C.; Andrade, M.L.; Vega. F.A. Copper, chromium, nickel, lead and zinc levels and pollution degree in firing range soils. *Land Degrad. Dev.* 2016, 27, 1721–1730.

Ruttens, A.; Colpaert, J.; Mench,M.; Boisson, J.; Carleer, R.; Vangronsveld, J. Phytostabilization of a metal contaminated sandy soil. II. Influence of compost and/or inorganic metal immobilizing soil amendments on metal leaching. *Environ. Pollut.* 2006, 144, 533–539.

Sahariah, B.; Goswami, L.; Imran, U.; Farooqui; Raul, P.; Bhattacharyya, P.; Bhattacharya, S.S. Solubility, hydrogeochemical impact, and health assessment of heavy metals in municipal wastes of two differently populated cities. *J. Geochem. Explor.* 2015, 157, 100–109.

Santillian-Medrano, J.; Jurinak, J.J. The chemistry of lead and cadmium in soil: Solid phase formation. *Soil Sci. Soc. Am. Proc.* 1975, 39, 851–856.

Shuman, L.M.; Dudka, S., Das, K. Zinc forms and plant availability in a compost amended soil. *Water Air Soil Pollut.* 2000, 128, 1–11.

Singh, R.P.; Agrawal, M. Effects of sewage sludge amendment on heavy metal accumulation and consequent responses of Beta vulgaris plants. *Chemosphere.* 2007, 67, 2229–2240.

Srinath, T.; Verma, T.; Ramteke, P.W.; Garg, S.K. Chromium(VI) biosorption and bioaccumulation by chromate resistant bacteria. *Chemosphere.* 2002, 48(4), 427–435.

Sumner, M.E. Beneficial use of effluents, wastes, and biosolids. *Commun. Soil Sci. Plant Anal.* 2000, 31(11–14), 1701–1715.

USEPA. Recent developments for in situ treatment of metal contaminated soils. Tech. Rep. EPA-542-R-97-004, United States Environmental Protection Agency, Office of Research and Development, Washington, DC, USA, 1997.

USEPA. Vitrification technologies for treatment of Hazardous and radioactive waste handbook; Tech. Rep. EPA/625/R-92/002, United States Environmental Protection Agency, Office of Research and Development, Washington, DC, USA, 1992.

Williams, A.G.B.; Scheckel, K.G.; McDermott, G.; Gratson, D.; Neptune, D.; Ryan, J.A. Speciation and bioavailability of zinc in amended sediments. *Chem. Speciation Bioavailability.* 2011, 23(3), 143–154.

Wang, J.; Zhao, F.J.; Meharg, A.A.; Raab, A.; Feildmann, J.; McGrath, S.P. Mechanisms of arsenic hyperaccumulation in *Pteris vitata*. Uptake kinetics, interactions with phosphate and arsenic speciation. *Plant Physiol.* 2002, 130, 1552–1561.

Yao, Z.; Li, J.; Xie, H.; Yu, C. Review on remediation technologies of soil contaminated by heavy metals. *Procedia Environment. Sci.* 2012, 16, 722–729.

Yoon, J.; Cao, X.; Zhou, Q.; Ma, L.Q. Accumulation of Pb, Cu, and Zn in native plants growing on a contaminated Florida site. *Sci. Total Environ.* 2006, 368, 456–464.

Zhou, D.; Cang, L.; Alshawabkeh, A.; Wang, Y.; Hao, X. Pilot-scale electrokinetic treatment of a Cu contaminated red soil. *Chemosphere.* 2006, 63, 964–971.

Zhou, M.; Liao, B.; Shu, W.; Yang, B.; Lan. C. Pollution assessment and potential sources of heavy metals in agricultural soils around four Pb/Zn mines of Shaoguan city, China. *Soil Sediment. Contam.* 2015, 24, 76–89.

CHAPTER 21

Advancement in Pesticide Chemistry for Sustainable Agriculture

PRITAM GANGULY[1*] and PRITHUSAYAK MONDAL[2]

[1]*Department of Soil Science & Agricultural Chemistry, Bihar Agricultural University, Sabour, Bhagalpur, India*

[2]*Regional Research Station (Terai Zone), Uttar Banga Krishi Viswavidyalaya, Pundibari, Cooch Behar 736165, West Bengal, India*

Corresponding author. E-mail: pritam0410@gmail.com

ABSTRACT

The rapid and promising performances of synthetic pesticides attracted farmers all over the world for their immediate application in successful pest management. However, higher pest pressure and subsequent misuse of synthetic chemicals have resulted in numerous negative environmental impacts such as pesticide residues on environmental components causing environmental pollution, pest resurgence following resistance to continuous applications. The major portion of the applied pesticides reportedly remains unused and moves to unwanted segments resulting in an adverse effect on human health and ecosystems. So, there is a growing trend of semisynthetic and natural pesticides due to their low environmental loads. Newer and innovative strategies such as the use of controlled-release formulations, botanicals, and chemicals from the biological origin are already showing promising results. Biopesticides, unlike the synthetics, are earth friendly and usually have slow actions and long-lasting effects on the target with little or no mammalian toxicity, and therefore, the chance of residue build-up is less. Smart formulations developed from nanotechnological approaches are also gaining importance. There is a need for public awareness of these rapidly growing technologies to guide them on the choice of agricultural pesticides and their impacts on the environment. This chapter deliberates an insight about the role of pesticides in crop protection, and their development and their application methods, the comparison of new technologies with conventional agrarian methods, and their impact on the target pests, food, and environments. Additionally, critical information on the challenges encumbered and the future research scopes in this emerging area are also highlighted.

21.1 INTRODUCTION

Production of food and its safe storage for the long-term use are the indispensable processes of human civilization. It is well known that until a country achieves self-sufficiency in food grain production, it is not at all considered independent. Therefore, agriculture has a pivotal role in the country's economic system. To ensure food security to a huge population, we must enforce a strong defense machinery to combat against the huge pest population. Here, pesticides are the most destructive weapon that farming community can rely upon. But, being a toxic compound, it has some negative impact on the environment. The traditional molecules have been perceived as persistent organic pollutants which can disturb the ecological balance. As agriculture is still heavily dependent on nature, the use of these poisonous chemicals can destroy soil fertility by killing beneficial microbes, causing an imbalance in nutrient flow. Thus, it is easily foreseen that crop productivity can be hampered if pesticides are being used indiscriminately; especially those are highly persistent in nature.

Sustainable agriculture demands use of farm inputs that can conserve the natural resources. It cannot allow any ingredient that may cause degradation of existing biotic and abiotic components of the farming processes. Therefore, research on pesticide chemistry and its application have taken a sharp turn to make them fit for use in sustainable farming. In this chapter, we will discuss the importance of pesticide use and the important research advancement that has been taken place globally in past years.

21.2 BENEFITS OF PESTICIDE USE

Food plants around the world are damaged by around 30,000 weeds, 100,000 diseases (caused by bacteria, fungi, viruses, and other pathogens) and more than 10,000 insects and nematodes (Dhaliwal et al., 2010). To check these pest populations, pesticide application is inevitable (Table 21.1).

TABLE 21.1   Pesticide-wise Consumption of Indigenous Pesticides During 2014–2015 to 2018–2019

Pesticide Group	Unit: M.T. (Tech. Grade)				
	2014–2015	2015–2016	2016–2017	2017–2018	2018–2019
Insecticide	6740.02	7121.26	11542.65	3945.29	9478.05
Fungicide	6322.64	7400.14	8789.28	9078.84	8962.21
Herbicide	4475.85	3543.28	4075.28	2969.62	3998.45
Rodenticide	270.73	274.55	409.39	185.92	291.16
Plant growth regulator	49.32	42.14	110.19	33.03	85.34
Biopesticide	2941.94	2999.94	4385.79	3472.92	3195.94
Others	–	37.33	59.45	1.16	2.52
Total	20800.50	21418.64	29372.03	19686.78	26013.67

"–" = Data not found; data provided up to 30th November, 2019.

Source: States/UTs Zonal conference on inputs of Plant Protection for Kharif and Rabi Seasons (DPPQS, 2020).

Pesticide application has numerous benefits such as improved yield, minimization of losses before and after the harvest against pests and diseases, quality and quantity enhancement of foods and reduction in the degree and frequency of vector-borne and other diseases in humans and animals. These benefits can be categorized into two types—primary and secondary (Aktar et al., 2009). The primary benefits are the direct consequences resulting from pesticide use those are immediate and easily comprehensible. For example, the destruction of brinjal fruit and shoot borer leads to primary benefits of greater yield and superior quality of brinjal whereas, the secondary benefits are basically indirect benefits which are achieved from the primary benefits. These may look a little fuzzy at first but their effects sustain for an extended period. However, in such cases, it becomes harder to establish a cause–effect relationship. Still, these can undoubtedly be a positive side of pesticide usage. For example, higher crop yield may aid in higher revenue generation that can effectively be utilized for the child education or healthcare facilities.

21.2.1   PRODUCTION AND PRODUCTIVITY ENHANCEMENT

There is a significant contribution of pesticide on agriculture which is one of the vital pillars of the Indian economy. Apart from agriculture, forestry, public health, and other domestic sectors are also benefitted from pesticide application. Food grain productions have been increased in many folds although the area of arable land almost remained the same. The use of high-yielding variety seeds, modern irrigation technologies, and new generation agrochemicals have pivotal role in it. Similarly, productivity has also been increased owing to certain factors like use of balanced fertigation, improved varieties and sophisticated machinery and implements. Pesticides are indispensable defense tools in minimizing losses from diseases, weeds, and insect pests that may contribute to postharvest losses also. This spectacular spike in economics due to yield increase was also reported by various researchers in the 20th century (Warren, 1998; Webster et al., 1999). Moreover, many pesticides in the environment undergo photochemical and other transformations and ultimately generate nontoxic metabolites those are considered relatively safe for mammals including humans and the Earth (Kole et al., 1999).

21.2.2   CROP PROTECTION

In the medium-land ecosystem, even under puddled conditions during the critical stage rice guaranteed an efficient and inexpensive weed control practice to check reduction in rice yield due to weeds ranging from 28% to 48%, based on comparisons with control (weedy) plots (Behera and Singh, 1999). Weeds may reduce the yield of dryland crops (Behera and Singh, 1999) by 37%–79%. Heavy infestation of weeds, particularly in the early stage of crop growth and development, ultimately accounts for around 40% yield reduction. Herbicides provide both labor and financial benefits.

21.2.3   VECTOR DISEASE CONTROL

Many insects act as vectors of different diseases. In such cases, the only remedy from vector-borne diseases is killing the vectors. Insecticides have already been practised to control animal vectors causing deadly diseases like malaria which was estimated to take a toll of 500 lives each day (Ross, 2005). Malaria is one of the principal causes of mortality in the developing world and a considerable public health issue in India (Bhatia, 2004). Disease prevention and management strategies are equally vital for livestock also.

21.2.4 QUALITY OF FOOD

In First-World countries, it has been perceived that a regular diet containing fresh fruit and vegetables counteracts the potential risks from a low amount of pesticide residues in crops. Growing evidences suggest that eating fresh fruit and vegetables regularly reduces the risk of many chronic diseases including cancers, diabetes, and cardiovascular diseases. Lewis et al. (2005) reviewed the nutritional potentials of apples and blueberries and concluded that regular intake of them in the US diet may act as protectants against cancer and heart disease due to higher antioxidant contents. Lewis attributed doubling in wild blueberry production and subsequently increases in consumption chiefly to herbicide use for improved weed management.

21.3 ADVANCES IN PESTICIDE SYNTHESIS

21.3.1 ADVANCES IN SYNTHESIS PROCESS

Development of pesticides with newer mode of action is very essential for the sustainable production. Researchers, across the world, continuously put their efforts to find new chemistry-based molecules in pest management, so that

1. New compounds would be more ecofriendly, producing less or no detrimental effects on environment.
2. Chances of pests getting resistant against the compounds become less.
3. Products should be more economic.
4. Products should be efficient in tackling new or emergent pests.

The main driving force behind the discovery of new compounds is to fight against resistance (Ujváry, 2006). Till date, 262 weed species in 92 crops have been found resistant against 167 herbicides in 70 countries (Heap, 2020). In the case of insects, more than 500 species have been found resistant to different insecticides (Pesticide Environmental Stewardship, 2020). More than 450 cases of fungicide resistance reported against plant pathogens (FRAC, 2018), attacking various crops. So, this leads to the continuous search for new chemistry-based molecules which can effectively control the pest population. Moreover, the role of certain regulatory bodies across the world is very crucial in this aspect. They will not allow a new molecule for getting registered unless and until the compound could outperform the efficacy of the existing compounds. Therefore, the challenge is tough for the researchers to bring in new chemistry-based pesticides which would be more efficient, cost-effective, and ecofriendly.

Agrochemical industries across the world have been engaged to find new chemistry molecules and spend handsome money for that. Screening of new compounds is a herculean task where different routes are being examined. There may be the screening of natural products; designing vital enzyme inhibition in cells; structural modification in existing molecules; 3D computer modeling or even by serendipity (Ujváry, 2006).

Traditional methods involving synthesis and screening at the mass level were time taking and costly too. There were several approaches like optimizing structure–activity relationship (SAR), conventional synthesis, analogue screening which led to the discovery of molecules like DDT, organophosphates, 2,4-D, thirum, etc. (Das, 2016). But, since new techniques like combinatorial chemistry, chemical genetics, high-throughput screening (HTS), chemoinformatics, and gene expression profiling have come up which allow the researchers to screen enormous molecules in a very short period with less investment. This has been something revolutionary for agrochemical industries.

HTS has been successfully introduced for searching novel pesticides and to maintain the flow of compounds to their marketplace (Ridley et al., 1998). Both in-vivo and complementary in-vitro HTS approaches have been established in pesticides research (Drews et al., 2012). The modern instruments can allow more than 100 plates in a day for the single point enzymatic assay with a run time of 15 min (Das and Mukherjee, 2014). In-vivo HTS permits screening a huge number of compounds for assessing biopotency in fully automated mode. If something found encouraging that is allowed for the next phase of testing and isolation for further development. With this technique, several ecofriendly fungicides have been discovered like strobilurins, anilinopyrimidines, phenylpyrroles, and phenoxyquinolines (Hansch and Leo, 1995).

Applying techniques of combinatorial chemistry to develop a library of more number of compounds in less time have been found quite encouraging to the scientists. It trims the long list of molecules on the basis of certain physicochemical parameters like hydrogen bonding, molecular mass, electronic configuration, log P, etc., fully supported by computer designing and automation. Like HTS, this technique is also rapid, robust, and time saving.

Pesticides, such as insecticides and postemergent weedicides, have been scroutinized which are safer to the environment and specific in action (Duart et al., 2001).

Molecular modeling at three dimensional level has been made possible by the modern computerized designing. Molecules of interest along with target site of action (enzymes/receptors) can be structurally obtained by using X-ray crystallography technique. Pesticides with more ecofriendly can be designed and their binding sites can also be investigated.

Another important development in this field of research is chemical genetics. It is basically a study of interaction between small molecules and biological systems which cause perturbation to see possible outcome. This technique is found useful in pesticide discovery which helps to find novel proteins coupled with in-vivo chemical ligands (Walsh, 2007). It has several advantages over conventional screening.

21.3.2 EVOLUTION OF SYNTHETIC COMPOUNDS

The use of agrochemicals started before 1940. These compounds, known as first-generation agrochemicals, primarily consisted of heavy metals like arsenic (As), mercury (Hg), and lead (Pb). Due to their extreme toxicity and inefficiency, these pesticides had to be banned shortly. The second-generation pesticides were developed via synthetic routes after the failure of earlier pesticides. The first pesticide of this group is DDT which was synthesized by Zeidler in 1874, but its extensive use started since 1939 onward when its insecticidal property was discovered by Paul Muller that ultimately earned him the prestigious Nobel Prize. The most common members of second-generation pesticides are chlorinated hydrocarbons, organophosphates, and carbamates. The third-generation pesticides are basically the modified forms of the insect hormones that show a great specificity to the pest or insect targeted. The juvenile hormone analogs such as methoprene, hydroprene, and fenoxycarb do not kill insects directly but prevent the development of their larvae. Gradually, they decrease the population of the insect as the adult insects will never be formed.

Novel group of insecticides is chosen considering their selectivity to insects and ecosafety profile (Casida and Durkin, 2013). Other groups of pesticides are also selected similarly. Some of the new generation pesticides along with their structures are mentioned in Table 21.2.

TABLE 21.2   Few New Generation Chemicals as Pesticides

Name of the Chemical	Chemical Family	Primary Function	Structure
Imidacloprid	Neonicotinoid	Insecticide	
Fipronil	Phenyl pyrazole	Insecticide	
Flubendamide	Diamide	Insecticide	
Novaluron	Benzoyl urea	Insecticide	
Spiromesifen	Tetronic acid derivative	Insecticide	
Metazosulfuron	Sulfonylurea	Herbicide	
Diclofop-methyl	Aryloxyphe-noxy-propionate ("FOPs")	Herbicide	

TABLE 21.2   *(Continued)*

Name of the Chemical	Chemical Family	Primary Function	Structure
Alloxydim-sodium	Cyclohexane-dione ("DIMs")	Herbicide	
Pinoxaden	Phenylpyrazoline ("DEN")	Herbicide	
Tiadinil	Thiadiazole-carboxamide	Fungicide	
pyraziflumid	Pyrazine-carboxamides	Fungicide	
Kresoxim-methyl	Strobilurin derivative	Fungicide	

21.3.3   BOTANICALS, BIOPESTICIDES, AND BIO-RATIONAL PESTICIDES—THE SAFER ALTERNATIVE

The synthetic routes of pesticide development play havoc with the environment and a serious concern for today's agriculture (Patra et al., 2020), so the researchers are striving hard toward sustainable agricultural practices for a better tomorrow. The United States Environmental Protection Agency (US EPA) recommends green approaches like using biorational pesticides that are inherently different from conventional ones. The term biorational means any materials or substances having a natural origin (or man-made materials resembling those of natural origin) that exhibit adverse or lethal effects on specific target pest(s), for example, insects, weeds, plant diseases (including nematodes), and vertebrate pests. They are nontoxic to humans and other nontarget plants and animals and may possess a unique mode of action and most importantly they maintain a safe ecological profile.

Biorational pesticides are classified into two categories: (1) biochemicals (enzymes, hormones, pheromones, and natural agents, such as insect and plant growth regulators) and (2) microbials (bacteria, fungi, nematodes, viruses, and protozoans). The US EPA in the 1990s started emphasizing upon a class of products known as biopesticides. According to them, biopesticides can be categorized into three different types:

1. Microbial pesticides, consisting of pathogenic microorganisms (e.g., fungi, bacteria, virus, etc.);
2. Biochemicals, consisting of natural substances having potential to destroy pests via nontoxic mechanisms (e.g., insect pheromones);
3. Plant incorporated protectants, consisting of genetically modified or transgenic plants (e.g., Bt cotton).

Distinguishing characteristics between biorational/biopesticides and conventional pesticides include: target specificity, very low to almost nil toxicity to nontarget species, generally low use rate, fast degradation in the environment, compatible with integrated pest management programs and contraction in dependence on conventional agrochemicals.

21.4   ADVANCES IN FORMULATION TECHNOLOGIES

A considerable attention has been paid in the formulation aspect of agrochemicals also. Pesticide formulations, which are conventionally used, basically found as dust, emulsion, or wettable powder (EP) form. The majority of these formulations, especially liquids, are containing organic solvents which are inflammable and toxic in nature. These are also called traditional or classical formulation (Hazra et al., 2017). It is having no reason to believe that these end-products are safe and suitable for sustainable agriculture. These old formulations may deteriorate soil fertility by disturbing the ecological balance and may cause environmental pollution. Thus, scientists have been searching for some safer ecofriendly formulations, which are broadly coming out as water-based (Knowles, 2008). Moreover, pesticide release system in target area has been customized or more specifically, slow- or controlled-release formulations are continuously being developed.

21.4.1 LIQUID FORMULATION—NEW DEVELOPMENTS

Among the new liquid formulations, suspension concentrates (SC) has become widely popular for added safety and less cost involved. These are basically water-based formulations where high amount of toxicants are present (Hazra et al., 2017). Examples of this category are kresoxim methyl 50% SC, bifenthrin 10% SC, etc. Absence of hazardous organic solvents makes these formulations more environment friendly. But, there are certain limitations like stability issues in storage which may cause aggregation of inherent particles. Another popular water-based formulation is the oil in water emulsion (EW; e.g., tricontanol 0.1 EW, butachlor 50 EW). The prerequisite for this composition is that the active compound must be having low solubility in water to resist crystallization (Tadros, 1995). To overcome this problem, and also to make out combination of active ingredients, a new type of formulation, called suspoemulsion (SE), has been developed. Here, certain surfactant and thickeners are added to arrest flocculation and separation of phases, respectively (Tadros, 1995). Examples of SE are Carbendazim 25% and Flusilazole 12.5% SE, Propiconazole 10.7% and Tricyclazole 34.2% SE, etc. Thereafter, research has been progressed further and comparatively safer compositions are developed such as microemulsion (ME; e.g., emamectin benzoate 5% ME and azadirachtin 1% ME). This is thermodynamically stable formulation consisting of two immiscible liquids (Hiromoto, 2007), having fine size (<50 nm) of droplets. In this case, concentration of active ingredient is generally less than that of surfactant. Consequently, the end-product shows increased efficacy with a lower negative impact on environment.

Latest development in this field has been witnessed as oil dispersion (OD) formulation. After adding water, OD formulation becomes emulsion or SE. The oil phase is consisted of various oils like vegetable oil or its esters, mineral oils, etc. The adjuvants are required for ODs like oil-compatible emulsifiers and dispersing agent, must be chosen with special care for maintaining the stability of the formulation.

Another important development are flowables (F) which are concentrated suspensions (40–70% w/w) of micronized insoluble active ingredient (a.i.) or pesticide in water (aqueous flowables, denoted by AF) or other liquids (nonAF). Most flowables are prepared by first impregnating them onto a dry carrier, such as clay. Then, the a.i. with or without any carrier is crushed into fine powders. After that, those powders are suspended in a very less amount of liquid (and perhaps other inert ingredients) which ultimately leads to formation of a thick liquid suspension. Before spraying on target areas, AFs are diluted with water in a spray tank to get minimum effective pesticide concentration. To facilitate easy transfer to spray tank, there must be of low viscosity and good fluidity. Therefore, an effective dispersing agent along with an efficient wetting agent is sometimes required to ensure adequate dispersion of a.i. in the water (Castro, 1998). Good suspension stability is a challenge for this type of formulation, and care should be taken to stop sediment formation which ultimately makes the formulation inefficient for application (Hazra, 2015). It has been found that a blend of smectite clay (aka bentonite) and xanthan gum works synergistically to assure excellent suspension stability for a longer period at low cost and at lost viscosity. Basically, flowables combine many of the characteristics of liquid emulsifiable concentrate (EC) and dry WP.

21.4.2 NEW SEED TREATMENT FORMULATIONS

Seed treatment formulations are used to prevent the entry and simultaneous growth of pests by forming a coat around the seeds. This layer of coating is prepared by using active ingredient, wetting agent, dispersant, film former, antifreeze, defoamer, pH regulator, water, and other auxiliaries (Dayer et al., 2007). After dilution, it can be applied directly on the surface of seeds to create a protective coating with firm strength, thickness, and permeability. Generally, pesticide application is not required for upto 45 days of sowing to protect the plants from the pests that have been targeted by theseed coating agent (S) itself. As the dose rate for seed coating is nearly 1/50 of the field application rates, the new formulation saves the toxicant load on the crop (Hazra, 2015).

Seed coating agent can be prepared into many forms like water flowable S (FS), suspended emulsion S (SES), water-emulsion S (EWS), microcapsule S (CS), water-dispersible granule type S (WGS), dry flowable S (DFS), etc., (El-Mohamedy et al., 2008). Among these, FS is a special type of SC, widely used across the world. The ingredients of this formulation are crushed in a manner so that the particle size comes less than 4 µm.

21.4.3 NEW DRY PRODUCT TECHNOLOGY

To get safer and more efficient formulation than WP and SC, researchers have developed water dispersible granules (WDG/WG) or may often be known as DF (Kim et al., 2003). The advantages of using this technology are convenient packaging as well as handling, nondustiness, free flowing, and moreover, quick dispersal into water before spray (Marcroft et al., 2008). Although being a granular formulation, DF behaves like a liquid formulation in terms of handling. Extrusion granulation is mostly favored by the manufacturer because of its safety, versatility, and cost-effectiveness followed by fluid bed spray technique. One of the most important criteria for being good granule is the quick dispersion of the product into water without any major difficulty. It should be dispersed completely into water within 2 min in varied degrees of temperature and water hardness. Examples of this formulation are Mancozeb 75 WG, Captan 83 WG, Endosulfan 50 WG, Cypermethrin 40 WG, Deltamethrin 25 WG, Thiamethoxam 25 WG, and so on.

Dispersion concentrate (DC) is the new kind of formulation where toxicant is dissolved in a polar solvent which is water miscible. Then dispersing/emulsifying agent is added in water to form stable formulation with the finer particle size. Some active ingredients are not fit for getting formulated as SC, EC, soluble liquid, and ME formulations due to their physiochemical or biological incompatibilities. DC formulation can act as alternatives to these formulations. Here also, selection of effective dispersing agent is important for proper dilution. It should be finely dispersed in water and remain stable for minimum 24 h to prevent blocking in spray equipment and reduced biopotency.

21.4.4 CONTROLLED RELEASE FORMULATIONS

Microencapsulation technique (polymer membrane) has been gaining popularity in recent times (Beestman, 2003). Microencapsulation process is based on the principle called interfacial polymerization. The toxicant can be released in a controlled manner by adjusting the membrane thickness, droplet size and the extent of cross-linking or polymer porosity. Therefore, the pesticide release rate is basically a diffusion controlled process. Researchers are continuously working in this field to improve the process for having more safe products (Fernández, 2007). The common examples of this capsule suspension (CS) formulation are Lambda Cyhalothrin 25 CS, Lambda Cyhalothrin 10 CS, etc.

Another significant development is the discovery of combined (mixed) ZW formulation for the user and environment-friendly agrochemical application. The combination has been made by using lambda cyhalothrin CS and chlorpyrifos-concentrated emulsion (Hazra et al., 2013). This is a unique combination where lambda cyhalothrin causes quick knockdown of target insects and chlorpyrifos is responsible for slow but systemic long-term pest management (Takeshita et al., 2001). Therefore, this formulation has broad spectrum use in different crops to control a large number of insect pests. As this is polymer membrane based microencapsulated product, applicator can use two insecticides simultaneously in a sole application.

21.4.5 NANO-FORMULATIONS

The objectives of using nanotechnological tools to develop pesticide formulation are to increase the efficacy and decrease environmental load of toxicants (Mondal and Kumar, 2016). The particle size of nanoemulsion is less than 200 nm, thus the end product becomes inherently transparent/translucent and thermodynamically stable (Nair et al., 2010). Here, the surfactant concentration is less in comparison to ME formulation which makes it more ecofriendly and cost effective formulation (Kuzma et al., 2006). This nanoemulsion is produced by following low-energy-based emulsification method which results into small sized nanoparticles of extended shelf life (Zabkiewicz, 2000; O' Sullivan et al., 2010; Sarwar, 2014).

In nanodispersions, solid nano-sized particles of pesticide molecules are dispersed in fluids. Like nanoemulsions, nanodispersions are also stabilized for uniform distribution in water using surfactant molecules localized at the particle surface with a typical arrangement of the polar (hydrophilic) portions extending toward water molecules and the nonpolar (hydrophobic) portions approaching toward pesticide molecules (Acosta, 2009; Muller and Junghanns, 2006; Muller et al., 1995).

While nanodispersions with long-term stability may be prepared with particles as large as 200 nm in diameter, smaller particles (diameter less than 50 nm) are preferred, as they yield the greatest improvement in pesticide solubility. In contrast to nanoemulsions, only kinetically stable nanodispersions can be prepared (Muller and Junghanns, 2006).

Nanoencapsulation is defined as the process of encapsulating substances with various coating materials at the nanoscale range. Nanocapsule consists of two parts, one is primary material, known as active or core, containing solid, liquid, or gaseous nanoparticles, and another secondary material, known as matrix or shell, encasing the core materials (Augustin et al., 2009). The core contains the active ingredients (e.g., drugs, pesticides, vitamins, etc.), while the shell sets apart and safeguards the core from the neighboring environment. This protection can be temporary or permanent. In case of former, the core is generally released by diffusion or in response to a trigger, such as shear, pH, or enzyme action, thus enabling their controlled release and prompt delivery to a targeted site (Desai and Park, 2005; Jyoti et al., 2010).

A novel method of entrapping active ingredients in polymer matrix via wet bottom-up approach has been developed that can deliver desired active ingredient(s) at a controlled rate (Gopal et al., 2011; Mondal et al., 2017).

21.5 FUTURE PROSPECTS

Research and development institutions have been constantly engaged in seeking new strategies toward the development of novel molecules which could be target specific, easily biodegradable with minimal mammalian toxicity. The prime theme for this development is to give adequate protection to the plants along with proper safety to the environment, which is imperative for sustainable agriculture. Numerous researches have been carried out and are still going on to develop safer molecules which could undergo smooth degradation in presence of light, microbes, or other chemicals leaving trace amounts of residues in the environment. There may be certain conditions like high partition co-efficient (Ganguly et al., 2017), high leaching potential, presence of toxic metabolites (Barik et al., 2018), etc., which can lead to bioaccumulation of compounds in humans. It is highly essential to examine the toxic behavior of new compounds into the body of human before doing any recommendation (Patra et al., 2018). The toxic residues of pesticides must be found below critical limits in soil and harvest samples, so that these may be concluded as safe compounds (Bhattacharyya et al., 2011). This is very much essential to maintain the soil fertility and to support growth of beneficial microbes. Restoration of natural resources has also been taken care. But, the search must be going on to find new solutions as the problem also continues to modify itself to pose new challenges. Future efforts need to be made more precisely as there would be greater population pressure which in turn will increase the demand for food grains. It may insist the indiscriminate use of pesticides to protect the crops and consequently increases the yield.

The biotechnological intervention will be the key factor in upcoming days to develop pesticide products in order to achieve sustainable production. Crop–pest interaction and effects of microclimate play determining role in this aspect. The use of biorational pesticides must be encouraged. More accurate target delivery system needs to be in place to reduce the residue loads in the environment. Highly effective, safe, low-dose chemicals should be developed to meet out the demand. Nanotechnology may play an important role in reducing the dosage of the toxicants to a considerable amount. But, presently, the number of nanopesticide, available in the market, is very less.

The major agrochemical companies are always in search for potent formulations that could be accepted worldwide (Mulqueen, 2003). So, this is a big challenge for the researchers who, besides doing basic research, have to develop the formulations that can be applied globally. Although lots of new chemistry molecules as well as formulation aids are coming up, but their on-field testing, cost—benefit ratio, etc., need to be worked out carefully.

KEYWORDS

- **pest management**
- **new generation**
- **formulation technology**
- **bio-efficacy**
- **sustainability**

REFERENCES

Acosta, E. Bioavailability of nanoparticles in nutrient and nutraceutical delivery. *Curr. Opin. Colloid Interface Sci.* 2009, 14, 3–15.

Aktar, M.W.; Sengupta, D.; Chowdhury, A. Impact of pesticides use in agriculture: their benefits and hazards. *Interdisc. Toxicol.* 2009. 2(1), 1–12.

Augustin, M.A.; Hemar, Y. Nano- and micro-structured assemblies for encapsulation of food ingredients. *Chem. Soc. Rev.* 2009, **38**, 902–912.

Barik, S.R.; Ganguly, P.; Patra, S.; Dutta, S.K.; Goon, A.; Bhattacharyya, A. Persistence behavior of metamifop and its metabolite in rice ecosystem. *Chemosphere*, 2018, 193, 875–882.

Beestman, G.B. Controlled release in crop protection: past experience and future potential. In Chemistry of Crop Protection, Progress and Prospects in Science and Regulation; Voss, G.; Ramos, G., Eds.; Wiley-VCH Verlag GmbH & Co: Weinheim. 2003; pp. 272–279.

Behera, B.; Singh, S.G. Studies on weed management in monsoon season crop of tomato. *Indian J. Weed Sci.* 1999, 31(1–2), 67–70.

Bhatia, M.R.; Fox-Rushby, J.; Mills, M. Cost-effectiveness of malaria control interventions when malaria mortality is low: insecticide-treated nets versus in-house residual spraying in India. *Soil Sci. Med.* 2004, 59, 525.

Bhattacharyya, A.; Ganguly, P.; Barik, S.R.; Kundu, C. Studies on the persistence of diclosulam in soybean crop. In Proceedings of 23rd Asian-Pacific Weed Science Society Conference, The Sebel Cairns, Queensland, Australia, September 26–29, 2011; McFadyen, R.; Chandrasena, N.; Adkins, S.; Hashem, A.; Walker, S.; Lemerle, D.; Weston, L.; Lloyd, S. Eds.; Queensland, 2011.

Casida, J.E.; Durkin, K.A. Neuroactive insecticides: targets, selectivity, resistance and secondary effects. *Annu. Rev. Entomol.* 2013, 58, 99–117.

Castro, B.A.; Riley, T.J.; Leonard, B.R. Evaluation of Gaucho® seed treatment and soil insecticides for management of the red imported fire ant on seedling grain sorghum during 1994–1996: Agricultural Center, Louisiana Agricultural Experiment Station Research Report: 1998, 101, 1–4.

Das, S.K. Screening of bioactive compounds for development of new pesticides: A mini review. *Univers. J. Agric. Res.* 2016, 4(1), 15–20.

Das, S.K.; Mukherjee, I. Influence of microbial community on degradation of flubendiamide in two Indian soils. *Environ. Monit. Assess.* 2014, 186(5), 3213–3219.

Dayer, A.; Burrows, M.; Johnston, B.; Tharp, C. Small grain seed treatment guide, MSU Extension Publications Mont Guide MT199608AG. 2007.

Desai, K.G.H.; Park, H.J. Recent developments in microencapsulation of food ingredients. *Drying Technol.* 2005, **23**, 1361–1394.

Dhaliwal, G.S.; Jindal, V.; Dhawan, A.K. Insect pest problems and crop losses: changing trends. *Indian J. Ecol.* 2010, 37(1), 1–7.

DPPQS (2020). Directorate of Plant Protection, Quarantine & Storage, Govt. of India Statistical Database. http://ppqs.gov.in/sites/default/files/pesticidewise_consumptionindig_0.xls (accessed 20 April 2020).

Drews, M.; Tietjen, K.; Sparks, T.C. High-throughput screening in agrochemical research. In Modern Methods in Crop Protection Research; Jeschke, P.; Krämer, W.; Schirmer, U.; Witschel, M., Eds.; John Wiley & Sons: New Jersey, 2012; pp. 3–20.

Duart, M.J.; Garcia-Domenech, R.; Anton-Fos, G.M.; Galvez, J. Optimization of a mathematical topological pattern for the prediction of antihistaminic activity. *J. Comput. Aided Mol. Des.* 2001, 15(6), 561–572.

El-Mohamedy, R.S.R.; Abd El-Baky, M.M.H. Evaluation of different types of seed treatment on control of root rot disease, improvement growth and yield quality of pea plant in Nobaria province. *Res. J. Agric. Biol. Sci.* 2008, 4, 611–621.

Exploring QSAR: Fundamentals and applications in chemistry and biology; Hansch, C., Leo, A., Eds.; ACS Professional Reference Book. American Chemical Society, 1995.

Fernández-Pérez, M. Controlled release systems to prevent the agro-environmental pollution derived from pesticide use. *J. Environ. Sci. Health.* 2007, 42(7), 857–862.

FRAC, 2018. List of plant pathogenic organisms resistant to disease control agents, https://www.frac.info/docs/default-source/publications/list-of-resistant-plant-pathogens/list-of-resistant-plant-pathogenic-organisms_may-2018.pdf?sfvrsn=a2454b9a_2 (accessed 12 April, 2020).

Ganguly, P.; Barik, S.R.; Patra, S.; Roy, S.; Bhattacharyya, A. Persistence of chlorfluazuron in cabbage under different agro-climatic conditions of India and its risk assessment. *Environ. Toxicol. Chem.* 2017, 36(11), 3028–3033.

Gopal, M.; Roy, S.C.; Roy, I.; Pradhan, S.; Srivastava, C.; Gogoi, R.; Kumar, R.; Goswami, A. Nanoencapsulated Hexaconazole: A Novel Fungicide and the Process for Making the Same, India. Patent 2051/DEL/2011. (07 July 2011).

Hazra, D.K. Recent advancement in pesticide formulations for user and environment friendly pest management. *Int. J. Res. Rev.* 2015, 2, 35–40.

Hazra, D.K.; Karmakar, R.; Poi, R.; Bhattacharya, S.; Mondal, S. Recent advances in pesticide formulations for eco-friendly and sustainable vegetable pest management: A review. *Arch. Agri. Environ. Sci.* 2017, 2(3), 232–237.

Hazra, D.K.; Pant, M.; Raza, S.K.; Patanjali, P.K. Formulation technology: key parameters for food safety with respect to agrochemicals use in crop protection. *J. Plant Protect. Sci.* 2013, 5(2), 1–19.

Heap, I. The International Herbicide-Resistant Weed Database. http://www.weedscience.org/Home.aspx. (accessed 04 March, 2020).

Hiromoto, B. Pesticide microemulsions and dispersant/penetrant formulations. U.S. Patent US 7297351. (20 November, 2007).

Jyothi, N.V.N.; Prasanna, P.M.; Sakarkar, S.N.; Prabha, K.S.; Ramaiah, P.S.; Srawan, G.Y. Microencapsulation techniques, factors influencing encapsulation efficiency. *J. Microencapsul.* 2010, 27, 187–197.

Kim, D.S.; Koo, S.J.; Lee, J.N.; Hwang, K.H.; Kim, T.Y.; Kang, K.G.; Hwang, K.S.; Joe, G.H.; Cho, J.H. Flucetosulfuron: a new sulfonylurea herbicide. Proceeding of International Congress, Crop Science & Technology, BCPC, Farnham, Surrey, UK, 2003, pp. 87–92.

Knowles, A. Recent developments of safer formulations of agrochemicals. *Environmentalist.* 2008, 28(1), 35–44.

Kole, R.K.; Banerjee, H.; Bhattacharyya, A.; Chowdhury, A.; AdityaChaudhury, N. Photo transformation of some pesticides. *J. Indian Chem. Soc.* 1999, 76, 595–600.

Kuzma, J.; VerHage, P. Nanotechnology in Agriculture and Food Production: Anticipated Applications. Washington, DC; The Project on Emerging Nanotechnologies; 2006. Available from: http://www.nanotechproject.org/process/assets/files/2706/94_pen4_agfood.pdf (accessed 26 October 2019).

Lewis, N.M.; Jamie, R. Blueberries in the American diet. *Nutr. Today.* 2005, 40(2), 92.

Marcroft, S.J.; Potter, T.D. The fungicide fluquinconazole applied as a seed dressing to canola reduces *Leptosphaeria maculans* (blackleg) severity in south-eastern Australia. *Australasian Plant Path.* 2008, 37, 396–401.

Mondal, P.; Kumar, R. Preparation of azomethine based nano-chemicals and antibacterial activity against nitrifying bacteria. *Pestic. Res. J.* 2016, 28(2), 194–200.

Mondal, P.; Kumar, R.; Gogoi, R. Azomethine based nano-chemicals: Development, in vitro and in vivo fungicidal evaluation against *Sclerotium rolfsii*, *Rhizoctonia bataticola* and *Rhizoctonia solani*. *Bioorg. Chem.* 2017, 70, 153–162.

Muller, R.H.; Junghanns, J.U. Drug nanocrystals/nanosuspensions for the delivery of poorly soluble drugs. In Nanoparticles as Drug Carriers. Torchilin, V.P., Ed.; Imperial College Press: London, 2006; p. 307–328.

Muller, R.H.; Peters, K.; Becker, R.; Kruss, B. Nanosuspensions for the IV administration of poorly soluble drugs: stability during sterilization and long-term storage. *P. Control. Release Soc.* 1995, 22, 574–575.

Mulqueen, P. Recent advances in agrochemical formulation. *Adv. Colloid Interface Sci.* 2003, 106, 83–107.

Nair, R.; Varghese, S.H.; Nair, B.G.; Maekawa, T.; Yoshida, Y.; Kumar, D.S. Nanoparticulate material delivery to plants. *Plant Sci.* 2010, 179, 154–163.

O' Sullivan, C.M.; Tuck, C.R.; Butler Ellis, M.C.; Miller, P.C.H.; Bateman, R. An alternative surfactant to nonyl phenol ethoxylates for spray application research. *Asp. Appl. Biol.* 2010, 99, 311–316.

Patra, S.; Ganguly, P.; Barik, S.R.; Goon, A.; Mandal, J.; Samanta, A.; Bhattacharyya, A. Persistence behaviour and safety risk evaluation of pyridalyl in tomato and cabbage. *Food Chem.* 2020, 309, 125711.

Patra, S.; Ganguly, P.; Barik, S.R.; Samanta, A. Dissipation kinetics and risk assessment of chlorfenapyr on tomato and cabbage. *Environ. Monit. Assess.* 2018, 190(2), 71.

Pesticide Environmental Stewardship. Introduction to Insecticide Resistance, https://pesticidestewardship.org/resistance/insecticide-resistance/ (accessed 12 April, 2020).

Ridley, S.M.; Elliott, A.C.; Yeung, M.; Youle, D. High-throughput screening as a tool for agrochemical discovery: automated synthesis, compound input, assay design and process management. *Pestic. Sci.* 1998, 54(4), 327–337.

Ross G. Risks and benefits of DDT. *Lancet.* 2005, 366(9499), 1771.

Sarwar, M. Understanding the importance and scope of agricultural education to the society. *Int. J. Innov. Res. Educ. Sci.* 2014, 1(2), 145–148.

Tadros, T.F. Surfactants in Agrochemicals; Marcel Dekker: New York, 1995.

Takeshita, T.; Noritake, K. Development and promotion of labor-saving application technology for paddy herbicides in Japan. *Weed Biol. Manag.* 2001, 1(1), 61–70.

Ujváry, I. Research and development in pesticide chemistry current status and a glimpse at the future. In Debrecen University, Proceedings of 4th International Plant Protection Symposium at Debrecen University (11th Trans-Tisza Plant Protection Forum), Debrecen, Hungary, 18–19 October, 2006; Kövics, G.J., Ed.; Debrecen, 2006, p 27–38.

Walsh, T.A. The emerging field of chemical genetics: potential applications for pesticide discovery. *Pest Manag. Sci.* 2007, 63(12), 1165–1171.

Warren, G.F. Spectacular increases in crop yields in the United States in the twentieth century. *Weed Tech.* 1998, 12, 752.

Webster, J.P.G.; Bowles, R.G.; Williams, N.T. Estimating the economic benefits of alternative pesticide usage scenarios: wheat production in the United Kingdom. Crop Prod. 1999, 18, 83.

Zabkiewicz, J.A. Adjuvants and herbicidal efficacy—present status and future prospects. *Weed Res.* 2000, 40, 139–149.

CHAPTER 22

Resource Recovery from Biowaste for Agriculture through Composting and Microbial Technology

P. C. MOHARANA[1*], AVIJIT GHOSH[2], M. D. MEENA[3], NINTU MANDAL[4], and D. R. BISWAS[5]

[1]ICAR-National Bureau of Soil Survey and Land Use Planning, Regional Centre, Udaipur 313001, India

[2]ICAR-Indian Grassland and Fodder Research Institute, Jhansi 284003, Uttar Pradesh, India

[3]ICAR-Directorate of Rapeseed-Mustard Research, Bharatpur 321303, Rajasthan, India

[4]Nanosynthesis and Nanoformulations Laboratory, Bihar Agricultural University, Sabour 813 210, Bihar

[5]Division of Soil Science and Agricultural Chemistry, ICAR-Indian Agricultural Research Institute, New Delhi 110012, India

*Corresponding author. E-mail: pravashiari@gmail.com

ABSTRACT

As waste management is urgent for environmental and economic sustainability the prime focus of this chapter is to signify the current prospects of organic waste management in India and their potential in agriculture. This chapter further emphasizes sustainable and eco-friendly management of organic resources mainly in the form of crop residues, animal waste, and municipal solid waste.

22.1 INTRODUCTION

Resource depletion is among the greatest environmental problems, we face today. Efficient resource recovery becomes relevant aspects of environmental management systems that could assist in addressing these global challenges. In this context, the proper management of waste and efficient nutrients recovery from organic waste can help to reduce the dependence on costly chemical fertilizers (Cofie et al., 2016). In this context, compost is a cheap alternative option (Goyal et al., 2005; Biswas et al., 2009; Moharana and Biswas, 2016). The conservation of organic matter to preserve the soil is a major task. Organic coal typically has a diversive function in soil, for example, to control the supply of plants with nutrients, to buffer, to filter, to regenerate, and maintain the soil safe. The use of high-profile fertilizers in conventional, surface and secondary carbon levels declined (Rasool et al. 2007). In 2017–2018, fertilizer nutrients ($N+P_2O_5+K_2O$) are currently used in India at about 26.59 Mt, and nutrient suppressions from the soil are projected at around 33–35 Mt in plants (FAI, 2019). This ensures that roughly 8–10 Mt of plant nutrients are already available. India would also potentially take approximately 40–45 tonnes by 2025. In addition, other types of nutrients, such as plant waste such as crop residues, green manure, agro-industrial wastes, sewage sludge, biofertilizers and so on, must be bridged. The use of organic manures in the management of the nutrient transition of a soil plant system is therefore necessary to enhance the nutrient status in the soil and to conserve the soil's organic matter. The introduction of nonchemical replenishment of the annual carbon emissions of organic materials in agriculture is indeed very necessary (Goyal et al., 2005; Meena et al., 2019). Efficient management and processing of usable farm waste into organic manures has been reported to be a possible source of vital nutrients for plant production. Compost has been a major component in incorporating plant nutrient applications and is common today with different organic sources of plant nutrients. In India, the usage of biodegradable organic sources like manure, compost, seed residue, and municipal solid waste (MSW) is rising exponentially, taking into account global demands and increasing industrial fertilizer prices. This chapter aims at providing information on the management and recovery of nutrients from organic wastes through composting, taking into account the discussed facts. In turn, the analysis should determine contextual limitations and potential study goals.

22.2   SCENARIO OF BIO-WASTE GENERATION IN INDIA

The success of recycling of biowastes depends upon development and integration of various activities at farm in a way that availability of organic resources for recycling of nutrients is not a constraint. However utilization of off farm resources like industrial effluents, agro industry wastes, and wastes from mineral processing are very much restricted. If these resources are used appropriately and scientifically it can open new vistas in the utilization of the untapped source of nutrients. A variety of solid wastes are available for recycling in agriculture. The major sources of plant nutrients are:

1. *Agricultural residues*: crop residues, tree wastes, and aquatic wastes
2. *Livestock wastes*: cattle shed wastes, piggery and poultry wastes, etc
3. Municipal solid wastes (MSW)
4. *Agro industry wastes*: Press mud, coir pith from coconut industry, rice husk, paper and pulp waste, oil cakes, sawdust, groundnut husk, distillery waste, fruit and vegetable industry waste, etc.

22.3   CROP RESIDUE

The overwhelming majority of property, producing a huge amount of crop residues (noneconomic plant elements) after harvest, is cultivated in India. In the United States, this tremendous volume of residues of crops has an economic benefit. On-farm and off-farm crops generate between 500 and 550 million tons of residues from a total of 110 tons of wheat, 122 tons of rice, 71 tons of maize, 26 tons of millets, 141 tons of sugarcane, 8 tons of fiber crops (jute mesta, cotton) and 28 Mt of pulses (Pathak et al, 2010), respectively. The majority of crop residues were produced by grain, fibers, oils, pulses and sugar cane, with an estimated production of 352 Mt, 66 Mt, 29 Mt, 13 Mt, and 12 Mt, respectively (Table 22.1). A total of 70% of cultivated residues followed by fibrous crops (13%) were produced in grain crops such as rice, wheat , maize and millets. At the moment, the burning of three quarters of the crop residues is mainly dispersed in northern India in order for the left-over straw and stubble following harvests to be removed. As a consequence of the combustion of renewable waste and minerals large quantities are being wasted. The burning of crop residues often results in waste that is harmful to public safety and creates world warming greenhouse gasses consisting of toxins and greenhouse gas pollutants such as CO, CO_2, and CH_4. In recent years, Pathak et al. (2010) have calculated the country's intake of approximately 93 Mt of crop residues. Consequently, proper management of crop residues is important.

TABLE 22.1 Total crop Residue Generation (Tonnes) in Different States of India During 2014–2015

State/UT	Rice	Wheat	Coarse Cereal	Pulse	Oilseed	Sugarcane	Cotton	Jute and Mesta	Total
Andhra Pradesh + Telangana	13.5	0.0	8.1	1.8	1.3	5.2	2.2	0.02	32.1
Assam	5.7	0.1	0.1	0.2	0.2	0.4	0	0.27	6.9
Bihar	7.5	5	3.8	0.7	0.1	5.5	0	0.5	23.1
Chhattisgarh	7	0.2	0.5	1	0.2	0	0	0	8.9
Gujarat	1.9	4	2.6	0.9	4.3	5.5	3.6	0	22.9
Haryana	4.7	14.7	1.4	0.1	0.8	3	0.8	0	25.4
Himachal Pradesh	0.1	0.9	1.4	0.1	0	0	0	0	2.5
Jammu & Kashmir	0.5	0.4	0.8	0	0	0	0	0	1.8
Jharkhand	3.9	0.4	0.8	0.9	0.2	0.2	0	0	6.4
Karnataka	4.3	0.3	11.4	2.3	1.1	16.4	0.7	0	36.6
Kerala	0.7	0	0	0	0	0.1	0	0	0.7
Madhya Pradesh	4.2	17.6	5.1	7.3	8.4	1.8	0.6	0	45
Maharashtra	3.4	1.5	7.6	2.7	3.1	32.1	2.3	0	52.7
Orissa	9.7	0	0.4	0.7	0.2	0.3	0.1	0.02	11.4
Punjab	13	19.6	0.9	0.1	0.1	2.8	0.5	0	36.9
Rajasthan	0.4	12.2	12.9	3	5.8	0.2	0.5	0	35.1
Tamil Nadu	6.8	0	5.1	1	1	9.6	0.3	0	23.8
Uttar Pradesh	14.3	31.3	6.1	2.2	0.9	54.4	0	0	109.2
Uttarakhand	0.7	0.8	0.5	0.1	0	2.4	0	0	4.6

TABLE 22.1 (*Continued*)

State/ UT	Rice	Wheat	Coarse Cereal	Pulse	Oilseed	Sugarcane	Cotton	Jute and Mesta	Total
West Bengal	17.2	1.2	11	0.3	1	0.8	0	3	24.6
Others	3	0.1	0.8	1.3	0.2	0.4	0	0.03	5.8
All-India	122.6	110.3	71.3	26.7	28.9	141.1	11.6	3.85	516.3
Harvest index	0.4–0.6	0.43–0.54	0.28–0.38	0.28–0.40	0.35–0.50	0.269	0.294	0.3	-
Dry Matter Fraction	0.86	0.88	0.83	0.8	0.8	0.88	0.8	0.8	-

Source: Adapted from Devi et al. (2017), data provided by Ministry of Statistics and Program Implementation, 2013–2014

22.4   ANIMAL WASTE

India is the nation dominated by agriculture and the animal life in the Indian farming system has a very significant role. In addition to providing milk and beef, they provide garden compost, fur, poultry, etc. India's overall livestock population has reduced significantly between 2007 and 2012 (Anonymous 2012, Table 22.2), from 529.7 m to 512.1 million. The most recent world-wide figures on livestock estimate the overall amount of livestock to be 1.43 million, 1.87 million sheep and goats, 0.98 bn and 19.6 bn chickens (Robinson et al. 2014). A further emphasis on the use of lands for crop production instead of growing pasture in order to encourage livestock and grazing populations can be attributed to the reduction of livestock in India. In addition, growing urbanization has reduced the land accessible for livestock farming. While the majority of cattle are spread across several small farms, a large number of manure treatment facilities in central processing plants are increasingly important.

TABLE 22.2   Estimated Annual Production of Dung/Excreta by Animals and Poultry

Species	Production of Dung/Excreta (Mt)		
	2003	**2007**	**2012**
Cattle	203.7	218.9	209.9
Buffaloes	132.2	142.2	146.7
Yaks	27.85	0.083	0.077
Mithuns	0.380	0.292	0.330
Sheeps and goats	0.410	31.78	30.01
Horses/ponies	0.510	0.310	0.317
Donkeys and mules	3.380	0.284	0.254
Camels	0.070	0.419	0.324
Pigs	0.310	2.783	2.574
Total (animals)	368.8	402.7	389.3
Poultry	0.690	0.916	1.029
Grand total (Animal +poultry)	369.4	447.0	470.8

[a]*Excreta/dung by animals and poultry estimated for 2007 and 2012 from the base value of 2003.*

Source: Anonymous, 2008. Basic Animal Husbandry Statistics. Ministry of Agriculture, Government of India, New Delhi, 19th Livestock census-2012 All India Report (2012).

22.5   MUNICIPAL SOLID WASTE

The ever increasing population, urbanization, and industrial expansion in Indian cities have created waste managing glitches. India produces almost 70 Mt of MSW per year, expected to increase by around 165 million tonnes, by 2031 and could reach some 436 Mt by 2050 if the pace continues (the planning report of the commission, 2014).). Local pressure in India plays a major role in increasing MSW amounts. India generates more than 55 million tons of MSW manufactured in India annually; an annual growth of about 5% is projected. The results are reported. Annual waste is estimated to be 1.33% per capita per year. Table 22.3 presents a comprehensive MSW scenario generated by the various cities of India. Typically speaking, solid waste may be characterized as waste not borne by water; that is zero additional usage. It covers pollution from manufacturing, forestry, community and farmland (Table 22.4). The MSW is made up of a broad variety of organic matter, ash and dust, paper and plastic, glass, and metals (Sharholy et al., 2007). The MSW potential is produced in large cities of India, as shown in Table 22.3. This contains biological matter (51%), recyclable materials (17.5%) and other neutral products (31%) (Annepu, 2012), for example. Open MSW pollution has a significant detrimental impact on the climate and human safety in all industries (Rathi 2006; Sharholy et al., 2006). MSW is usually eliminated without pretreatments or environmental restrictions in low-lying environments. It is important to sustainably handle MSW as one of Indian megacities' principal environmental problems.

TABLE 22.3   Production Potential of Municipal Solid Waste in Major Cities of India

Name of City	Population (as per 2001)	Waste generation Rate (kg/capita/day)	Compostable (%)	Recyclables (%)	C/N ratio
Kavaratti	10,119	0.30	46.01	27.20	18.04
Shillong	132,867	0.34	62.54	17.27	28.86
Shimla	142,555	0.27	43.02	36.64	23.76
Agartala	18,998	0.40	58.57	13.68	30.02
Gandhinagar	195,985	0.22	34.30	13.20	36.05
Dhanbad	199,258	0.39	46.93	16.16	18.22
Pondicherry	220,865	0.59	46.96	24.29	36.86
Imphal	221,492	0.19	60.00	18.51	22.34
Aizwal	228,280	0.25	54.24	22.97	27.45
Jammu	369,959	0.58	51.51	21.08	26.79

TABLE 22.3   *(Continued)*

Name of City	Population (as per 2001)	Waste generation Rate (kg/capita/day)	Compostable (%)	Recyclables (%)	C/N ratio
Dehradun	424,674	0.31	51.37	19.58	25.90
Asansol	475,439	0.44	50.33	14.21	14.08
Kochi	595,575	0.67	57.34	19.36	18.22
Raipur	605,747	0.30	51.40	16.31	223.50
Bhubaneswar	648,032	0.36	49.81	12.69	22.57
Tiruvanantapuram	744,983	0.23	72.96	14.36	35.19
Chandigarh	808,515	0.40	57.18	10.91	22.52
Guwahati	809,895	0.20	53.69	23.28	17.71
Ranchi	847,093	0.25	51.49	9.86	22.23
Vijaywada	851,282	0.44	59.43	17.40	33.90
Srinagar	898,440	0.48	61.77	17.76	22.46
Madurai	928,868	0.30	55.32	17.25	32.69
Coimbatore	930,882	0.57	50.06	15.52	45.83
Jabalpur	932,484	0.23	58.07	16.61	28.22
Amritsar	966,862	0.45	65.02	13.94	30.69
Rajkot	967,476	0.21	41.50	11.20	52.56
Allahabad	975,393	0.52	35.49	19.22	19.00
Vishakhapatnam	982,904	0.59	45.96	24.20	41.70
Faridabad	1,055,938	0.42	42.06	23.31	18.58
Meerut	1,068,772	0.46	54.54	10.96	19.24
Nashik	1,077,236	0.19	39.52	25.11	37.20
Varanasi	1,091,918	0.39	45.18	17.23	19.40
Jamshedpur	1,104,713	0.31	43.36	15.69	19.69
Agra	1,275,135	0.51	46.38	15.79	21.56
Vadodara	1,306,227	0.27	47.43	14.50	40.34
Patna	1,366,444	0.37	51.96	12.57	18.62
Ludhiana	1,398,467	0.53	49.80	19.32	52.17
Mumbai	1,437,354	0.40	52.44	22.33	21.58
Indore	1,474,968	0.38	48.97	12.57	29.30
Nagpur	2,052,066	0.25	47.41	15.53	26.37
Lucknow	2,185,927	0.22	47.41	15.53	21.41
Jaipur	2,322,575	0.39	45.50	12.10	43.29
Surat	2,433,835	0.41	56.87	11.21	42.16
Pune	2,538,473	0.46	62.44	16.66	35.54
Kanpur	2,551,337	0.43	47.52	11.93	27.64

TABLE 22.3 *(Continued)*

Name of City	Population (as per 2001)	Waste generation Rate (kg/capita/day)	Compostable (%)	Recyclables (%)	C/N ratio
Ahmedabad	3,520,085	0.37	40.81	11.65	29.64
Hyderabad	3,843,585	0.57	54.20	21.60	25.90
Bangalore	4,301,326	0.39	51.84	22.43	35.12
Chennai	4,343,645	0.62	41.34	16.34	29.25
Kolkata	4,572,876	0.58	50.56	11.48	31.81
Delhi	10,306,452	0.57	54.42	15.52	34.87
Greater Mumbai	1,1978,450	0.45	62.44	16.66	39.04

Source: Adapted from Central Pollution Control Board, 2005

TABLE 22.4 Type and Composition of Municipal Solid Waste Generation

Source	Typical Waste Generators	Type of Solid Waste
Residential	Household activities	Food waste, paper, cardboard, plastics, wood, glass, metals, electronic items etc.
Industrial	Manufacturing units, power plants, process industries etc.	Housekeeping wastes, hazardous wastes, ashes, special wastes etc.
Commercial & Institutional	Hotels, restaurants, markets, office buildings, schools, hospitals, prisons etc.	Bio-medical waste, Food waste, glass, metals, plastic, paper, special wastes etc.
Construction and Demolition	New construction sites, demolition of existing structures, road repair etc.	Wood, steel, concrete, dust, etc.
Municipal services	Street cleaning, landscaping, parks and other recreational areas, water and wastewater treatment plants	Tree trimmings, general wastes, sludge, etc.
Agriculture	Crops, orchards, vineyards, dairies, farm etc.	Agricultural wastes, hazardous wastes such as pesticides
Mining	Open-cast mining, underground mining	Mainly inert materials such as ash

22.6 HORTICULTURAL AND FRUIT WASTE

The estimated annual generation of by-products from the horticultural and plantation sectors is estimated to be 263.4 Mt, out of which 134 Mt is considered to be available for recycling (Figure 22.1). The by-products and wastes arising from food processing units are available in substantial quantities.

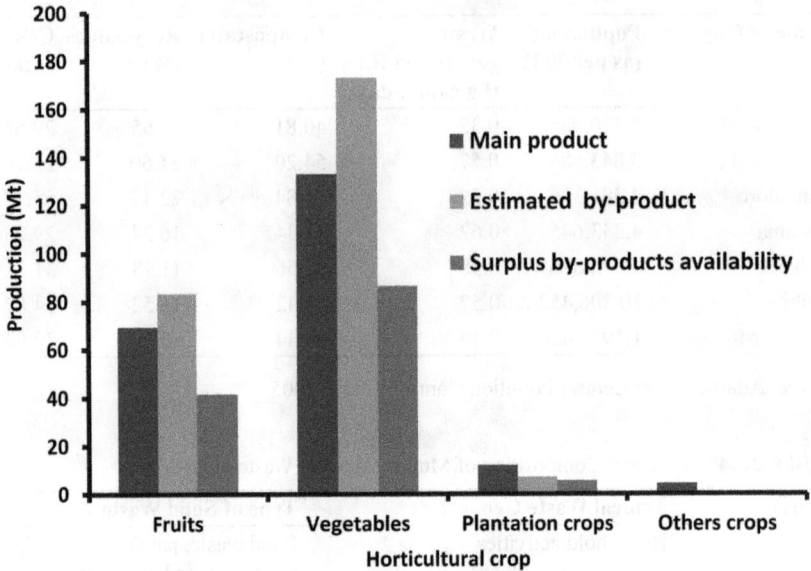

FIGURE 22.1 *Estimated production of crop residues from the horticultural and plantation sectors*

Source: Compiled from Anonymous, 2009. Agricultural Statistics at a Glance. Directorate of Economics and Statistics, Ministry of Agriculture, Government of India, New Delhi. http://www.dacnet.nic.in/eandds.

22.7   COMPOSTING OF BIOWASTES

Composting has emerged as a valuable route for the disposal of agricultural wastes which may be defined as the biological process that uses naturally occurring microorganisms to convert biodegradable organic matter into humus-like product known as compost (Raj and Antil, 2011; Moharana and Biswas, 2016).

22.8   PARAMETERS AFFECTING THE COMPOSTING PROCESS AND QUALITY

The parameters include a drop in temperature, degree of self-heating capacity, oxygen consumption, cation-exchange capacity, organic matter, nutrient contents, and C:N ratio (Tiquia et al., 2002).The parameters playing key role in the composting process are described next.

22.8.1   NATURE OF THE SUBSTRATE

If the conditions for biodegradation are appropriate, all types of organic residues conducive to the microorganism's enzymatic activities can be converted into compost. The existence of the substrate is the most important control factor for any composting method because it is the primary source of food for microorganists in the compost. Many substrates contain primarily of water-insoluble polymers. These polymers are hydrolyzed into monomers by extracellular enzymes produced by the microbes and then dissolved into water and into the microbe cell where further degradation occurs. The compost maturity depends on the substrate's nature. Once the base has its plant origins, carbonaceous substances such as cellulose, hemicellulose, and lignin are the main components (Moharana and Biswas, 2016). A smaller degree of the products nitrogen (proteins) exists. Protein, cellulose, and hemicellulose components are readily decomposed. While cellulosic substrates constitute good composting raw material, lignin being a complex aromatic polymer, is substantially resistant to microbial attack.

22.8.2   C/N RATIO

Considerably microorganisms consume 30 parts of C per unit of N (Biswas et al., 2009); therefore, the optimum C/N ratio is in the range 25–35 for composting.

22.8.3   MOISTURE

Moisture content of 60%–70% is considered ideal for start of the decomposition.

22.8.4   OXYGEN AND TEMPERATURE

For aerobic system, 50–60 °C (70 °C in some cases) is optimum (Benito et al., 2003).

22.8.5   pH

Nakasaki et al. (1993) found the pH range of 7–8 to be optimum for microbial growth, in some cases 6.5–7.5 is good (Bharadwaj, 1995).

22.8.6 MICROORGANISMS

The presence of fungi, bacteria, and invertebrates like earthworms are important.

22.9 EVALUATION OF COMPOST MATURITY AND QUALITY

Compost guidelines valid for a specific location are provided in Table 22.5.

22.9.1 TEMPERATURE, ODOR AND COLOR

Four important features are: (1) temperature rise and fall; (2) change in odor and color; (3) change in texture; and (4) destruction of volatile solids (i.e., organic matter)

22.9.2 WATER SOLUBLE CARBON

Garcia et al. (1992) recorded a decline in water soluble carbon (WSC) over the composting period and the values of 91 days of the gross organic material were 0.41–1.19 trillion. WSC's measurement criteria cannot be restricted by the raw material used for composting, as that is defined by compost (Goyal et al., 2005; Huang et al., 2006; Moharana and Biswas, 2016). Such scientists also estimated that the average amount beyond the maturity of compost is 0.5%. WSC < 1% and WSC < 1.7% were suggested by Hue and Liu (1995) as well as Bernal et al. (1998).

22.9.3 NITRIFICATION

A nitrification index that equals the ratio of $NH_4 + -N$ to $NO_3–N$ may also describe compost maturity. The $NO_3–N$ rise is incremental over a prolonged span of time and it is thus impossible to establish the stage at which the rise starts. Owing to its conversion to $NO_3–N$ or its volatilization as NH_3, the concentration of $NH_4 + -N$ declined because of the strong pH during composting. The final NO_3 amount, though, is based on the source material used in the composting process.

TABLE 22.5 Maturity Indices Used to Assess Maturity of Organic Waste Compost

Parameter	Compost	Findings	References
Nitrification index (NI)	Food waste	NI < 0.5, fully mature 0.5 < NI < 3, mature NI > 3, immature	Zhang and Sun (2016)
	RP-enriched compost	0.64 to 0.93, mature	Moharana and Biswas (2016)
Germination index (GI)	Agro-industrial wastes	Sensitive indicator for maturation and phytotoxicity	Raj and Antil (2011)
	Wastewater produced by the debittering process of green olives	GI < 25, very phytotoxic 26 < GI < 65, phytotoxic 66 < GI < 100, nonphytotoxic	Aggelis et al. (2002)
	RP-enriched compost	GI > 70%, nonphytototoxic	Moharana and Biswas (2016)
Dissolved organic matter and electron transfer capacity (ETC)	Food waste	Decomposition degree is associated with dissolved organic matter. ETC correlated with germination index	Yuan et al. (2012)
WSC/Org-N ratio	Yard waste compost	WSC/Org-N ratio < 0.70, mature	Hue and Liu (1995)
	Animal manures	WSC/Org-N < 0.55, mature	Bernal et al. (2009),
	RP-enriched compost	WSC/Org-N, 0.08–0.20, mature	Moharana and Biswas (2016)
Particle size		Optimum size for mature compost: 0.25–2.0 mm	Zhang and Sun (2016)
Polymerization degree	RP-enriched compost	Formation of simple sugars. Reduction of nonhumic substances	Moharana and Biswas (2016)
Enzymatic assay	RP-enriched compost	Correlated with the germination index	Moharana and Biswas (2016)
Phytotoxicity index	Food waste	Values below100% indicate immaturity or any toxicity degree Values above 100% indicate maturity and no toxicity	Young et al. (2016)

22.9.4   SOLUBLE ORGANIC C/ORGANIC N

A variety of study groups suggested the WSC/OrgN as a sophistication index (Bernal et al., 1998; Raj and Antil, 2011; Moharana and Biswas, 2016). Although the C/N ratio was applied for compost maturity as an indicator, the WSC/OrgN ratio appears to better represent the stability of compost reported by different authors (Pascual et al., 1997; Raj and Antil, 2011; Moharana and Biswas, 2016).

22.9.5   MICROBIAL RATIO INDICES

Changes in microbial activities during composting have been measured to assess the degradation of complex molecules by microbial consortia which are more efficient than by microbes in isolation.

22.9.6   DEHYDROGENASE ACTIVITY

It is a measure of total biological activity in compost. Dehydrogenase activity decreases with composting time of different organic feedstocks (Benito et al., 2003; Pelaez et al., 2004; Moharana and Biswas, 2016). The decrease in dehydrogenase activity to low values toward end of composting indicates maturity (Tiquia, 2005).

22.9.7   HUMIFICATION

A ratio of humic acid and fulvic acid of 1.9 has been proposed as a maturity index of MSW compost (Jimenez and Garcia, 1992).

22.9.8   GERMINATION INDEX (GI)

If GI varies from 26 to 65, substrate is phytotoxic; if GI varies from 66 to 100, substrate is nonphytotoxic, stable and can be used in agricultural purpose; and if GI > 101, substrate is phytonutrient or phyto-stimulant and can be used in agricultural purposes as fertilizer.

22.10   NUTRIENT POTENTIAL OF BIOWASTES

Three major sources of organic waste are commonly used in agriculture, namely, crop residue, animal waste, and MSW in India. In this section, the nutrient potential of three major organic wastes is discussed.

22.10.1   NUTRIENT POTENTIAL OF CROP RESIDUE

Food grain crop residue accumulates 30%–35% of added N–P and 70%–80% of K (Table 22.6). Ten key crops (rice, wheat, Sorghum, barley, finger millets, sugar cane, potato tubers, and pluses) from India are expected to produce approximately 312.5 million Mt of crop residues of approximately 6.46 million tons of NPK nutrients. The supply of crop residues in India in 2025 was estimated to be 496 Mt, providing the trappable nutrients of 3.39 Mt over the subsequent periods.

TABLE 22.6   Generation and Nutrient Potential of Different Crop Residues in India During 2012

Crops	Total Residue (Mt)	Total NPK (Mt)	Surplus NPK (Mt)
Rice	210.48	4.567	1.522
Wheat	140.27	2.553	0.851
Sorghum	15.84	0.331	0.110
Millet	34.96	0.612	0.201
Maize	100.17	2.053	0.684
Bengal gram	11.48	0.305	0.102
Pigeon pea	12.08	0.308	0.103
Lentil	2.26	0.044	0.014
Groundnut	9.40	0.301	0.100
Rap seed	16.06	0.318	0.106
Soybean	14.67	0.519	0.173
Sunflower	1.62	0.042	0.014
Cotton	17.46	0.295	0.098
Sugarcane	102.36	1.904	0.635
Potato	34.91	0.625	0.313
Total	724.02	14.78	5.030

Sources: Manna, M. C.; Rahman, M. M.; Naidu, R.; Sahu, A.; Bhattacharjya, S.; Wanjari, R. H.; Patra, A. K.; Chaudhari, S. K.; Majumdar, K.; Khanna, S. S. Bio-waste management in subtropical soils of India: Future challenges and opportunities in Agriculture. *Adv. Agron.* **2018**, *152*, 87–148.

22.10.2   NUTRIENT POTENTIAL OF ANIMAL WASTE

Nutrient potential of different animals and poultry birds waste in India is presented in Table 22.7. According to Potter et al. (2010), manure provided around 152.6 Mt of N and P on a global scale in 2007, whereas mineral fertilizer use was reported to be 180.1 Mt (N, P, K) for 2011 (FAO, 2012).

22.10.3   NUTRIENT POTENTIAL OF MSW

The MSW comprises of a significant proportion of 30%–57% including 0.56%–0.71% N, 0.52%–0.82% P, and 0.52%–0.83% K. The application on farmland is planned as a twofold target of environmental purification and soil replenishment of their depleting fertility (Figures 22.2 and 22.3).

22.11   MARKETING OF COMPOSTS

The financial feasibility of a program designed to nutrient recovery from organic wastes is a function of the availability, reliability, and location of markets. Compost has a variety of potential applications due to its beneficial characteristics and can be used by several market segments like agriculture (small and large scale);

- landscaping;
- gardening (residential, community);
- nurseries;
- top dressing (e.g., golf courses, parks, median strips);
- land reclamation or rehabilitation (landfills, surface mines, and others); and
- erosion control.

22.12   CONSTRAINTS ON WASTE RECYCLING IN AGRICULTURAL LAND

Recycling of organic waste as fertilizer propounds number of advantages during management of waste, as it reduces the dependency on inorganic fertilizers and subsequently reduces the treatment and disposal of waste

TABLE 22.7 Generation and Nutrient Potential (NPK, million tons) of Different Animals and Poultry Birds Waste in India during the Year 2012

Species	Population During 2012	Production of Dung/Excreta in Mt	TOC	TN	TP	TK	NPK
Cattle	190.9	209.9	72.87	0.945	0.321	0.630	1.896
Buffaloes	108.7	146.7	50.93	0.660	0.225	0.440	1.325
Yaks[a]	0.077	0.077	–	–	–	–	–
Mithuns[a]	0.298	0.330	–	–	–	–	–
Sheeps and goats	200.2	30.01	16.16	0.975	0.225	0.750	1.950
Horses/ponies	0.625	0.317	0.131	0.002	0.001	0.002	0.005
Donkeys and mules[a]	0.515	0.254	–	–	–	–	–
Camels[a]	0.4	0.324	–	–	–	–	–
Pigs	10.29	2.574	1.349	0.084	0.023	0.048	0.155
Total (animal)	512.0	389.3	141.4	2.666	0.795	1.870	5.331
Poultry	729.2	1.029	0.331	0.014	0.007	0.007	0.028
Total (animal + poultry)	1241.2	470.8	141.8	2.680	0.802	1.877	5.359

[a] Data not available for NPK estimation

Sources: Manna, M. C.; Rahman, M. M.; Naidu, R.; Sahu, A.; Bhattacharjya, S.; Wanjari, R. H.; Patra, A. K.; Chaudhari, S. K.; Majumdar, K.; Khanna, S. S. Bio-waste management in subtropical soils of India: Future challenges and opportunities in Agriculture. *Adv. Agron.* **2018**, *152*, 87–148.

(Hargreaves et al., 2008; Lopes et al., 2011). The application of MSW provides soil nutrients that enhance soil organic matter, soil structure improvement and increased nutrient uptake by plants (Singh et al., 2012). But, they also represent significant constraints in application.

- extremes of C:N ratio in farm litters;
- poor nutrient concentration;
- required in large quantity;
- labor intensive;
- threat for pest, sickness and weed spreading;
- scarcity of livestock fodder and firewood in rural areas;
- liquid portion is hazardous and voluminous; and
- contamination with heavy metals in industrial wastes.

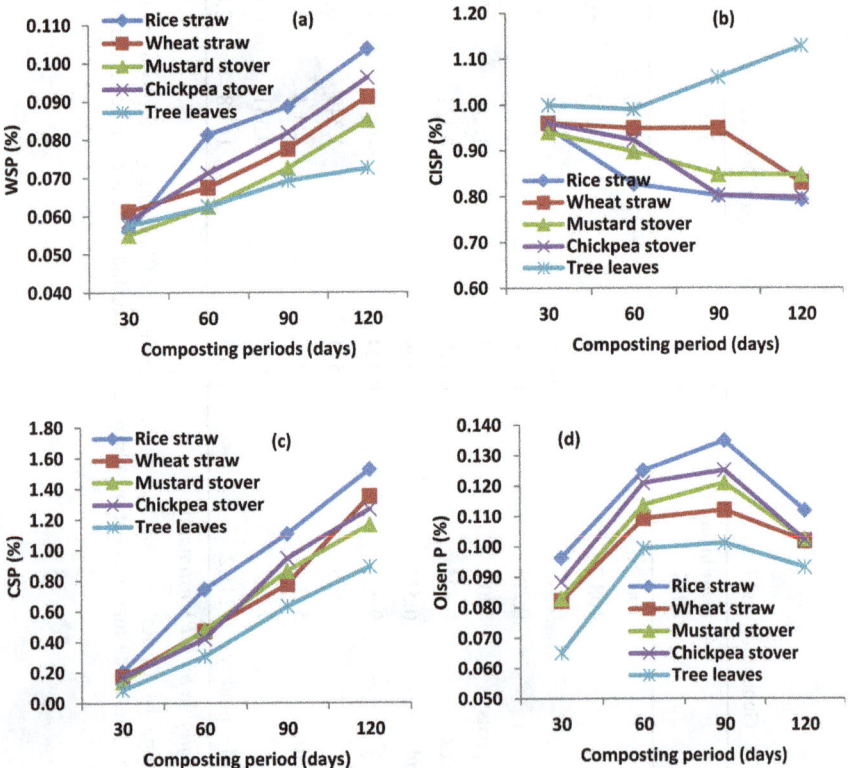

FIGURE 22.2 Changes in P fractions as influenced by crop residues and RP during preparation of various composts (Sources: Adapted from Moharana and Biswas (2016)).

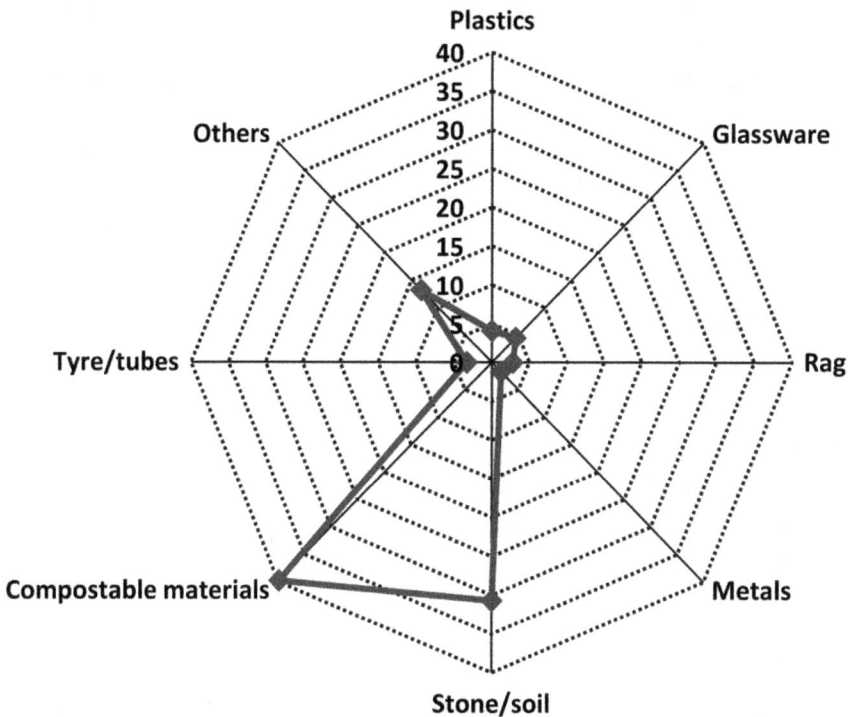

FIGURE 22.3 Composition (%) of unsegregated MSW collected from different cities (Sources: Adapted from Meena et al., 2019)

22.13 CONCLUSION

This chapter explores and represents the plausible approach of bioconversion of biowastes to utilize their ample nutrient contents for upgradation of soil quality and also the productivity and quality of crops. The major dilemmas concerning crop residues burning are soil health, soil biodiversity, crop productivity, and these wastes' heavy contribution to global warming and climate change. Most animal waste is being used as fuel cakes which are a very good source of plant nutrients and soil organic matter. Two of the major problems being encountered are the insufficient collection and inappropriate final disposal of MSW. The unscientific or unsystematic management of all wastes has made the situation worse and led to several environmental and health-related problems simply increasing and not being solved. Their nutrient potential to agriculture as well as threat to environment also varies depending upon source and nature of materials from which waste has been generated. The recycling of crop residues and organic wastes through composting methods is the key technology for disposal and production of organic manures and minimization of environmental pollution. There is a challenge to change the public perspective about waste. Holistic management of these can help achieve the goal of their sustainable use in society with very ultimate waste.

KEYWORDS

- **biowaste**
- **composting**
- **soil conditioner**
- **waste management**

REFERENCES

Aggelis, G.; Ehaliotis, C.; Nerud, F.; Stoychiev, I.; Luberatos, G.; Zervakis, G. Evaluation of white-rot fungi for detoxification and decoloration of effluents from the green olives debittering process. *Appl. Microbiol. Biotechnol.* **2002**, *59*, 353–360.

Annepu, R. K. Sustainable Solid Waste Management in India. MSc Thesis, Earth Engineering Center, Columbia University, New York, **2012**.

Anonymous. 19thLivestock Census-2012 All India Report. Ministry of Agriculture Department of Animal Husbandry, Dairying and Fisheries Krishi Bhawan, New Delhi, **2012**; pp 1–30.

Aparna, C.; Saritha, P.; Himabindu, V.; Anjaneyulu, Y. Techniques for the evaluation of maturity for composts. *Waste Manage.* **2008**, *28*, 1773–1784.

Benito, M.; Masaguer, A.; Moliner, A.; Arrigo, N.; Palma, R. M. Chemical and microbiological parameters for the characterisation of the stability and maturity of pruning waste compost. *Biol. Fertil. Soils* **2003**, *37*, 184–189.

Bernal, M. P.; Lopez-Real, J. M.; Scott, K. M. Application of natural zeolites for the reduction of ammonia emissions during the composting of organic wastes in a composting simulator. *Bioresour. Technol.* **1993**, *43*, 35–39.

Bernal, M. P.; Parades, C.; Monedero, S. A.; Cegarra, J. Maturity and stability parameters of composts prepared with a wide range of organic wastes, *Bioresour. Technol.*, **1998**, *63*, 91–99.

Bharadwaj, K. K. R. Improvements in microbial compost technology: A special reference to microbiology of composting. In *Wealth from Waste*; Khanna, S., and Mohan, K, Eds.; New Delhi: Tata Energy Research Institute, 1995; pp. 115–135.

Biswas, D. R.; Narayanasamy, G.; Datta, S. C.; Geeta, S.; Mamata, B.; Maiti, D.; Mishra, A.; Basak, B. B. Changes in nutrient status during preparation of enriched organomineral fertilizers using rice straw, low-grade rock phosphate, waste mica, and phosphate solubilizing microorganism. *Commun. Soil Sci. Plant Anal.* **2009**, *40*, 2285–2307.

Cofie, O.; Nikiema, J.; Impraim, R.; Adamtey, N.; Paul, J.; Koné, D. *Co-composting of Solid Waste and Fecal Sludge for Nutrient and Organic Matter Recovery.* Colombo, Sri Lanka: International Water Management Institute (IWMI). CGIAR Research Program on Water, Land and Ecosystems (WLE), (Resource Recovery and Reuse Series 3) **2016**; p 47.

Devi, S.; Gupta, C.; Jat, S. L., Parmar, M. S. Crop residue recycling for economic and environmental sustainability: The case of India. *Open Agriculture* **2017**, 2, 486–494.

Dobermann, A.; White, P. F. Strategies for nutrient management in irrigated and rainfed lowland rice systems. *Nutr. Cycl. Agroecosyst.* **1999**, *53*, 1–18.

Epstein, E. *The Science of Composting.* Lancaster, PA: Technomic Publishing, 1997.

FAI. *Fertiliser Statistics (2018–2019).* New Delhi: The Fertiliser Association of India; 2019.

FAO. *Current World Fertilizer Trends and Outlook 2016.* United Nations, Rome: Food and Agriculture Organization, 2012.

Gaind, S.; Nain, L.; Patel, V. B. Quality evaluation of co-composted wheat straw, poultry droppings and oil seed cakes. *Biodegradation* **2009**, *20*, 307–317.

Garcia, C.; Hernandez, T.; Costa, F.; Ayusho, M. Evaluation of maturity of municipal waste compost using simple chemical parameters. *Commun. Soil Sci. Plant Anal.* **1992**, *23*, 1501–1512.

Goyal, S.; Dhull, S. K.; Kapoor, K. K. Chemical and biological changes during composting of different organic wastes and assessment of compost maturity. *Bioresour. Technol.* **2005**, *96*, 1584–1591.

Hargreaves, J. C.; Adl, M. S.; Warman, P. R. A review of the use of composted municipal solid waste in agriculture. *Agric. Ecosyst. Environ.* **2008**, *123*, 1–14.

Hellmann, B.; Zelles, L.; Palojarvi, A.; Bai, Q. Emission of climate-relevant trace gases and succession of microbial communities during open-windrow composting. *Appl. Environ. Microbiol.*, **1997**, *63*, 1011–1018.

Hsu, J. H.; Lo, S. L. Recycling of separated pig manure: Characterization of maturity and chemical fractionation of elements during composting. *Water Sci. Technol.* **1999**, *40*, 121–127.

Huang, G. F., Wong, J. W. C., Wu, Q. T., Nagar, B. B. Effect of C/N on composting of pig manure with sawdust. *Waste Manag.* **2004**, *24*, 805–813.

Huang, G. F.; Wu, Q. T.; Wong, J. W. C.; Nagar, B. B. Transformation of organic matter during co-composting of pig manure with sawdust. *Bioresour. Technol.* **2006**, *97*, 1834–1842

Hue, N.; Liu, J. Predicting compost stability. *Compost Sci. Util.* **1995**, *3*, 8–15.

Iglesias-Jimenez, E.; Perez Garcia, G. Determination of maturity indices for city refuse composts. *Agric. Ecosyst. Environ.* **1992**, *38*, 331–343.

Ipek, U.; Obek, E.; Akca, L.; Arslan, E. I.; Hasar, H.; Dogru, M.; Baykara, O. Determination of degradation of radioactivity and its kinetics in aerobic composting, *Bioresour. Technol.* **2002**, *84*, 283–286.

Jimenez, I.; Garcia, P. Determination of maturity indices for city refuse composts *Agric. Ecosyst. Environ.* **1992**, *38*, 331–343.

Lopes, C.; Herva, M.; Franco-Uría, A.; Roca, E. Inventory of heavy metal content in organic waste applied as fertilizer in agriculture: evaluating the risk of transfer into the food chain. *Environ. Sci. Pollut. Res.* **2011**, 18(6), 918–939.

Marche, T.; Schnitzer, M.; Dinel, H.; Pare, T.; Champagne, P.; Schulten, H. R.; Facey, G. Chemical changes during composting of a paper mill sludge hardwood sawdust mixture. *Geoderma* **2003**, *116*, 345–356.

Meena, M.D.; Yadav, R. K.; Narjary, B.; Yadav, G.; Jat, H.S.; Sheoran, P.; Meena, M.K.; Antil, R. S.; Meena, B.L.; Singh, H.V., Meena, V. S.; Rai, P.K.; Ghosh, A.; Moharana, P. C. Municipal solid waste (MSW): Strategies to improve salt affected soil sustainability: A review. *Waste Manage.* **2019**, *84*, 38–53.

Moharana, P. C.; Biswas, D. R. Assessment of maturity indices of rock phosphate enriched composts using variable crop residues. *Bioresour. Technol.* **2016**, 222, 1–13.

Nakasaki, K.; Yaguchi, H.; Sasaki, M.; Kubota, H. Effects of pH control on composting of garbage. *Waste Manag. Res.* **1993**, *11*, 117–125.

Pascual, J. A.; Ayuso, M.; Garcia, C.; Herna´ndez, T. Characterization of urban wastes according to fertility and phytotoxicity parameters. *Waste Manag. Res.* **1997**, *15*, 103–112.

Pathak H.; Bhatia A.; Jain N.; Aggarwal P. K. Greenhouse gas emission and mitigation in Indian agriculture–A review. In *ING Bulletins on Regional Assessment of Reactive Nitrogen*, Bulletin No. 19; Bijay-Singh (Ed.), New Delhi: SCON-ING, 2010, p. 34.

Pelaez, C.; Mejia, A.; Planas, A. Development of a solid phase kinetic assay for determination of enzyme activities during composting. *Process Biochem.* **2004**, *39*, 971–975.

Planning Commission Report. Reports of the task force on waste to energy (Vol-I) (in the context of integrated MSW management). 2014; https://mnre.gov.in/img/documents/uploads/7c1d4e00bb994cffbdd32d12cae627ce.pdf (accessed June 13, 2020).

Potter, P.; Ramankutty, N.; Bennet, E.; Donner, S. D. Spatial patterns of global fertilizer application and manure production. *Earth Interactions*, **2010**, Vol. 14, Paper No. 2.

Raj, D.; Antil, R. S. Evaluation of maturity and stability parameters of composts prepared from agro-industrial wastes. *Bioresour. Technol.* **2011**, *102*, 2868–2873.

Rasool, R.; Kukal, S. S.; Hira, G. S. Soil physical fertility and crop performance as affected by long term application of FYM and inorganic fertilizers in rice–wheat system. *Soil Tillage Res.* **2007**, *96*, 64–72.

Rathi, S. Alternative approaches for better municipal solid waste management in Mumbai, India. *J. Waste Manage.* **2006**, *26* (10), 1192–1200.

Robinson, T. P.; Wint, G. R. W.; Concheda, G.; Van Boeckel, T. P.; Ercoli, V.; Palamara, E. Mapping the global distribution of livestock. *PLoS One*, **2014**, 9 (5): e96084, Available at http://journals.plos.org/plosone/article?id=10.1371/journal.pone.0096084.

Said-Pullicino, D.; Erriquens, F. G.; Gigliotti, G. Changes in the chemical characteristics of water-extractable organic matter during composting and their influence on compost stability and maturity. *Bioresour. Technol.* **2007**, *98*, 1822–1831.

Saritha, M., Arora, A., Singh, S., Nain, L. *Streptomyces griseorubens* mediated delignification of paddy straw for improved enzymatic saccharification yields. *Bioresour. Technol.* **2013**, 135,12–17.

Sharholy, M.; Ahmad, K.; Mahmood, G.; Trivedi, R. C. Development of prediction models for municipal solid waste generation for Delhi city. In *Proceedings of National Conference of Advanced in Mechanical Engineering (AIME-2006)*, New Delhi, India: Jamia Millia Islamia, **2006**; pp 1176–1186.

Sharholy, M.; Ahmad, K.; Vaishya, R. C.; Gupta, R. D. Municipal solid waste characteristics and management in Allahabad, India. *J. Waste Manage.* **2007**, *27*, 490–496.

Sharma, V. K.; Canditelli, M.; Fortuna, F.; Cornacchia, G. Processing of urban and agro-industrial residues by aerobic composting: Review. *Energy Convers. Manag.* **1997**, *38*, 453–478.

Singh, R. P.; Singh, P.; Ibrahim, M. H.; Hashim, R. Land application of sewage sludge: physicochemical and microbial response. *Reviews of Environmental Contamination and Toxicology*. Springer, New York, 2012; pp. 41–61.

Singh, Y.; Sidhu, H. S. Management of cereal crop residues for sustainable rice-wheat production system in the Indo-Gangetic Plains of India. *Proc. Indian Natl. Sci. Acad.* **2014**, *80*, 95–114.

Smars, S.; Gustafsson, L.; Beck-Friis, B.; Johnsson, H. Improvement of the composting time for household waste during an initial low pH phase by mesophilic temperature control. *Bioresour. Technol.* **2002**, 84, 237–241.

Tiquia, S. M. Microbiological parameters as indicators of compost maturity. *J. Appl. Microbiol.* **2005**, *99*, 816–828.

Tiquia, S. M.; Richard, T. L.; Honeyman, M. S. Carbon, nutrient, and mass loss during composting. *Nutr. Cycl. Agroecosyst.* **2002**, *62*, 15–24.

Wang, P.; Changa, C. M.; Watson, M. E.; Dick, W. A.; Chen, Y.; Hoitink, H. A. J. Maturity indices for composted dairy and pig manures. *Soil Biol. Biochem.* **2004**, *36*, 767–776.

Young, B.; Rizzo, P.; Riera, N.; Torre, V.; López, V.; Molina, C.; Fernández, F.; Crespo, D.; Barrena, R.; Komilis, D.; Sánchez, A. Development of phytotoxicity indexes and their correlation with ecotoxicological, stability and physicochemical parameters during passive composting of poultry manure. *Waste Manage.* **2016**, *54*, 101–109.

Yuan, Y.; Tao, Y.; Zhou, S.; Yuan, T.; Lu, Q.; He, J. Electron transfer capacity as a rapid and simple maturity index for compost. *Bioresour. Technol.* **2012**, *116*, 428–434.

Zhang, L.; Sun, X. Improving green waste composting by addition of sugarcane bagasse and exhausted grape marc. *Bioresour. Technol.* **2016**, 218, 335–343.

Zorpas, A. A.; Arapoglou, D.; Panagiotis, K. Waste paper and clinoptilolite as a bulking material with dewatered anaerobically stabilized primary sewage sludge (DASPSS) for compost production. *Waste Manage.* **2003**, *23*, 27–35.

Zucconi, F.; Pera, A.; Forte, M.; de Bertoldi, M. Evaluating toxicity of immature compost. *Biocycle* **1981**, *22*, 54–57.

Index